Lecture Notes in Computer Science   12013

More information about this series at http://www.springer.com/series/7407

Roberto Moreno-Díaz · Franz Pichler ·
Alexis Quesada-Arencibia (Eds.)

Computer Aided
Systems Theory –
EUROCAST 2019

17th International Conference
Las Palmas de Gran Canaria, Spain, February 17–22, 2019
Revised Selected Papers, Part I

 Springer

Editors
Roberto Moreno-Díaz
University of Las Palmas de Gran Canaria
Las Palmas de Gran Canaria, Spain

Franz Pichler
Johannes Kepler University Linz
Linz, Austria

Alexis Quesada-Arencibia
University of Las Palmas de Gran Canaria
Las Palmas de Gran Canaria, Spain

ISSN 0302-9743   ISSN 1611-3349   (electronic)
Lecture Notes in Computer Science
ISBN 978-3-030-45092-2   ISBN 978-3-030-45093-9   (eBook)
https://doi.org/10.1007/978-3-030-45093-9

LNCS Sublibrary: SL1 – Theoretical Computer Science and General Issues

This Springer imprint is published by the registered company Springer Nature Switzerland AG
The registered company address is: Gewerbestrasse 11, 6330 Cham, Switzerland

Preface

The concept of CAST as a computer-aided systems theory was introduced by Franz Pichler in the late 1980s to refer to computer theoretical and practical development as tools for solving problems in system science. It was thought of as the third component (the other two being CAD and CAM) required to complete the path from computer and systems sciences to practical developments in science and engineering.

Franz Pichler, of the University of Linz, organized the first CAST workshop in April 1988, which demonstrated the acceptance of the concepts by the scientific and technical community. Next, Roberto Moreno-Díaz, of the University of Las Palmas de Gran Canaria, joined Franz Pichler, motivated and encouraged by Werner Schimanovich, of the University of Vienna (present honorary chair of Eurocast), and they organized the first international meeting on CAST (Las Palmas February 1989), under the name EUROCAST 1989. The event again proved to be a very successful gathering of systems theorists, computer scientists, and engineers from most European countries, North America, and Japan.

It was agreed that the EUROCAST international conference would be organized every two years, alternating between Las Palmas de Gran Canaria and a continental European location. Since 2001 the conference has been held exclusively in Las Palmas. Thus, successive EUROCAST meetings took place in Krems (1991), Las Palmas (1993), Innsbruck (1995), Las Palmas (1997), and Vienna (1999), before being held exclusively in Las Palmas in 2001, 2003, 2005, 2007, 2009, 2011, 2013, 2015, 2017 and 2019, in addition to an extra-European CAST conference in Ottawa in 1994. Selected papers from these meetings were published as Springer *Lecture Notes in Computer Science* volumes 410, 585, 763, 1030, 1333, 1798, 2178, 2809, 3643, 4739, 5717, 6927, 6928, 8111, 8112, 9520, 10671, and 10672, respectively, and in several special issues of *Cybernetics and Systems: An International Journal*. EUROCAST and CAST meetings are definitely consolidated, as shown by the number and quality of the contributions over the years.

With EUROCAST 2019 we celebrated our 30th anniversary. It took place at the Elder Museum of Science and Technology of Las Palmas, during February 17–22, and it continued with the approach tested at previous conferences as an international computer-related conference with a truely interdisciplinary character. As in the previous conferences, the participants profiles were extended to include fields that are in the frontier of science and engineering of computers, information and communication technologies, and the fields of social and human sciences. The best paradigm is the Web, with its associate systems engineering, CAD-CAST tools, and professional application products (Apps) for services in the social, public, and private domains.

There were specialized workshops, which, on this occasion, were devoted to the following topics:

1. Systems Theory and Applications, chaired by Pichler (Linz) and Moreno-Díaz (Las Palmas)
2. Pioneers and Landmarks in the Development of Information and Communication Technologies, chaired by Pichler (Linz) and Seising (Munich)
3. Stochastic Models and Applications to Natural, Social and Technical Systems, chaired by Nobile and Di Crescenzo (Salerno)
4. Theory and Applications of Metaheuristic Algorithms, chaired by Affenzeller, Wagner (Hagenberg), and Raidl (Vienna)
5. Model-Based System Design, Verification and Simulation, chaired by Nikodem (Wroclaw), Ceska (Brno), and Ito (Utsunomiya)
6. Applications of Signal Processing Technology, chaired by Huemer, Zagar, Lunglmayr, and Haselmayr (Linz)
7. Artificial Intelligence and Data Mining for Intelligent Transportation Systems and Smart Mobility, chaired by Sanchez-Medina (Las Palmas), del Ser (Bilbao), Vlahogianni (Athens), García (Madrid), Olaverri-Monreal (Linz), and Acosta (La Laguna)
8. Computer Vision and Machine Learning for Image Analysis and Applications, chaired by Penedo (A Coruña), Rádeva (Barcelona), and Ortega-Hortas (A Coruña)
9. Computer and Systems Based Methods and Electronic Technologies in Medicine, chaired by Rozenblit (Tucson), Maynar (Las Palmas), and Klempous (Wroclaw)
10. Advances in Biomedical Signal and Image Processing, chaired by Fridli (Budapest), Huemer, Kovacs, and Böck (Linz)
11. Systems Concepts and Methods in Touristic Flows, chaired by Palma-Méndez (Murcia), Rodriguez, and Moreno-Díaz Jr. (Las Palmas)
12. Systems in Industrial Robotics, Automation and IoT, chaired by Jacob (Kempten), Stetter (Munich), and Markl (Vienna)

In this conference, as in previous ones, most of the credit for the success is due to the workshop chairs. They and the sessions chairs, with the counseling of the International Advisory Committee, selected from 172 presented papers. After oral presentations and subsequent corrections, 123 revised papers are included in this volume.

The event and this volume were possible thanks to the efforts of the workshop chairs in the diffusion and promotion of the conference, as well as in the selection and organization of all the material. The editors would like to express their thanks to all the contributors, many of whom have already been Eurocast participants for years, and particularly to the considerable interaction of young and senior researchers, as well as to the invited speakers: Prof. Paul Cull, from Oregon State University, USA; Prof. Christoph Stiller, from Karlsruhe Institute of Technology (KIT), Germany; and Prof. Bruno Buchberger, from the Research Institut for Symbolic Computation (RISC), Johannes Kepler University Linz, Austria. We would also like to thank the Director of the Elder Museum of Science and Technology, D. José Gilberto Moreno, and the museum staff. Special thanks are due to the staff of Springer in Heidelberg for their valuable support.

November 2019

Roberto Moreno-Díaz
Franz Pichler
Alexis Quesada-Arencibia

Organization

EUROCAST 2019 was organized by the Universidad de Las Palmas de Gran Canaria, Spain, Johannes Kepler University Linz, Austria, and Museo Elder de la Ciencia y la Tecnología, Spain.

Conference Chair

Roberto Moreno-Díaz   Universidad de Las Palmas de Gran Canaria, Spain

Program Chair

Franz Pichler   Johannes Kepler University Linz, Austria

Honorary Chair

Werner Schimanovich   Austrian Society for Automation and Robotics, Austria

Organizing Committee Chair

Alexis Quesada-Arencibia   Universidad de Las Palmas de Gran Canaria, Spain

Plenary Lectures

Tales of Computer and Systems Theory

Paul Cull

Computer Science, Kelley Engineering Center,
Oregon State University Corvallis, OR 97331, USA
pc@cs.orst.edu

Abstract. Were computers invented so that Norbert Wiener would not have to compute ballistic tables? Could a theoretical mapping between analysis and algebra allow scientists to turn animals inside out? What do Euclid and Lewis Carroll have to do with computer aided systems theory? These are some of the improbable questions we will explore in anecdotal history of computers and systems theory. This is not formal history, rather tales that are passed between colleagues and from teachers to students.

Among the questions we would like to answer these standout: Is EVERYTHING possible? Might the IMPOSSIBLE still be PRACTICAL? Does theory drive practice or does practice drive theory?

In the end we may be left with more questions than answers.

Short CV: Paul Cull is a long time contributor to EUROCAST. He has been attending since the 1990s. His background is in mathematical biology and computer science. He did his graduate studies with Nicolas Rashevsky's group at the University of Chicago. His PhD thesis, under the direction of Luigi Ricciardi, was on the use of linear algebra for the analysis of neural nets. In 1970, he joined the faculty of Oregon State University as one of the founding members of the Computer Science Department. After many years of teaching and research, he is now Professor Emeritus.

Promises and Challenges of Automated Vehicles

Christoph Stiller

Karlsruhe Institute of Technology, KIT, Germany
stiller@kit.edu

Abstract. This talk discusses the state of the art and a potential evolution of self-driving cars. It will outline approaches and challenges for achieving full autonomy for self-driving cars and elaborate the potential of cooperativity for automated driving. The talk will look at homologation issues and how these are approached by different stakeholders.

We will look at many examples, including the DARPA and GCDC Challenges.

The talk will draw on lessons learned and challenges to be overcome from the perspective of the German Research Priority Program "Cooperative Interactive Automobiles."

Short CV: Christoph Stiller studied Electrical Engineering in Aachen, Germany, and Trondheim, Norway, and received both his diploma and PhD (Dr.-Ing.) from Aachen University of Technology in 1988 and 1994, respectively. He worked with INRS-Telecommunications in Montreal, Canada, for a post-doctoral year in 1994/1995 and with Robert Bosch GmbH, Germany, from 1995–2001. In 2001, he became Chaired Professor and Director of the Institute for Measurement and Control Systems at Karlsruhe Institute of Technology (KIT), Germany.

Dr. Stiller serves as Senior Editor for the *IEEE Transactions on Intelligent Vehicles* (2015-present) and as Associate Editor for the *IEEE Intelligent Transportation Systems Magazine* (2012-present). He served as Editor in Chief of the *IEEE Intelligent Transportation Systems Magazine* (2009–2011). His automated driving team Annie-WAY were finalist in the Darpa Urban Challenge 2007 as well as first and second winner of the Grand Cooperative Driving Challenge in 2011 and 2016, respectively. He has served in several positions for the IEEE Intelligent Transportation Systems Society including being its President during 2012–2013.

Automated Mathematical Invention: Would Gröbner Need a PhD Student Today?

Bruno Buchberger

Research Institute for Symbolic Computation (RISC),
Johannes Kepler University, Linz, Austria
www.brunobuchberger.com

Abstract. Wolfgang Gröbner (1899–1980) was my PhD advisor back in 1964 at the University of Innsbruck, Austria. The problem he posed to me had been formulated, in a slightly different form, by Paul Gordan in 1899 and was still open in 1964. Roughly, the problem asks for an algorithmic canonical simplifier for the congruence relations with regard to multivariate polynomial ideals. I solved the problem in 1965 in my PhD thesis by introducing what I later called the theory and method of "Gröbner bases". The theory and method found numerous applications both inside mathematics and in basically all areas of science and technology in which non-linear polynomial systems play a role (e.g. robotics, cryptography, computer-aided design, software verification, systems theory, etc.)

In this talk I will, first, give an easy and practical introduction to the theory and method of Gröbner bases for those with no or only little background in this area. My main emphasis, however, will be on my recent research on automating mathematical invention. For this, I will take the theory of Gröbner bases as my main example. I will show how, by recent progress in automated reasoning and, in particular, my method of "Lazy Thinking" for the automated invention and proof of mathematical theorems and algorithms, my theory and method of Gröbner bases today could be "invented" completely automatically. In other words, cum grano salis, Professor Gröbner today would not need a PhD student for solving his problem any more.

From this, I will draw some conclusions on the future of mathematics.

Short CV: Bruno Buchberger is Professor Emeritus of Computer Mathematics at RISC (Research Institute for Symbolic Computation), Johannes Kepler University (JKU) in Linz, Austria.

Founding editor (1985–2000) of the *Journal of Symbolic Computation*. Founding chairman (1987–2000) of RISC. Founder and Director (1989–2013) of the JKU Softwarepark Hagenberg, the first Softwarepark as such, a world leading concept.

Author of the theory of Gröbner bases, established in his PhD thesis 1965 and expanded upon later in his publications. Since then, the theory has been a subject of over 20 textbooks, over 3,000 publications, and of a larger number of citations.

His current main research interest is on automated mathematical theory exploration (the "Theorema Project").

Contents – Part I

Pioneers and Landmarks in the Development of Information and Communication Technologies

Stochastic Models and Applications to Natural, Social and Technical Systems

Theory and Applications of Metaheuristic Algorithms

Model-Based System Design, Verification and Simulation

Contents – Part II

Artificial Intelligence and Data Mining for Intelligent Transportation Systems and Smart Mobility

Computer Vision, Machine Learning for Image Analysis and Applications

Computer and Systems Based Methods and Electronic Technologies in Medicine

Advances in Biomedical Signal and Image Processing

Systems Concepts and Methods in Touristic Flows

Systems in Industrial Robotics, Automation and IoT

Systems Theory and Applications

Design and Implementation of an Autopilot for an UAV

Johannes von Eichel-Streiber, Christoph Weber, and Jens Altenburg[✉]

University of Applied Sciences, Bingen am Rhein, Germany
eichel.streiber@itb-institut.de,
{c.weber,j.altenburg}@th-bingen.de

Abstract. Based on a theoretical model, a flight control system (autopilot) is designed for an UAV. The mathematical basics were analysed and transfered into a closed software system supported by Matlab/Simulink and by practical flight tests. Therefore, essential part of the project are the necessary hardware design of the control electronics and the connected sensor subsystems. The most important sensors will be introduced. The hardware based on the ARM Cortex M4 architecture. For safety reasons each hardware node has its own CPU. In order to maximum flexibility in the application all control electronics are designed as expandable systems of hardware modules. The Futaba SBUS.2 is used for communication between sensors, actuators and autopilot. The flight tests of the entire system are carried out in different flight configurations, i.e. several airframes were tested for optimizing and adapting for different operational scenarios.

Keywords: UAV · Autopilot · Flight simulation · Cascade controller · Futaba SBUS.2 · Real time operating systems · 3D-printing · Aerodynamic optimisation · MATLAB · Simulink · SolidWorks · Embedded systems

1 Preface

At the beginning of this project the initial question was which autopilot is suitable for an UAV. After comparing the market the decision was to develop an own autopilot, because no default autopilot completely fitted to the requirements. The advantage is a high flexibility and an expandable system.

The development of the airframe is closely locked to the development of the autopilot's hard- and software. Several, partly contradictory, requirements have to be brought together to a technically acceptable compromise. Based on experiences and experiments with 3D printing technology, some parts of the airframe were optimized in aerodynamics, others were designed in special lightweight construction and, last but not least, the relevant relationships between electronic and mechanical optimisation were also addressed. The CAD tool SolidWorks gives the possibility for aerodynamic optimisation for that components.

During the design of the autopilot, the implementation of the control loops into the time slices of the RTOS was a major challenge. These time slices allow a huge advantage in terms of flexibility of the software design. A professional flight simulator for

R. Moreno-Díaz et al. (Eds.): EUROCAST 2019, LNCS 12013, pp. 3–10, 2020.
https://doi.org/10.1007/978-3-030-45093-9_1

model aircrafts and the definition of a mathematical model of the controller in MATLAB served to solve this job. The developed autopilot is able to stabilize all three axis (roll, pitch and heading) of the aircraft. Future scenarios will require a fully autonomous flight. For this purpose, a flight planner is part of further software developments. The main task of this flight planner is to define the flight direction, airspeed, altitude and climbrate of the UAV. An addional external program generates the neccessary waypoints (GPS based) for the entire flying route of the flight planner. The autopilot controls the specified airspeed and altitute between the different path sections during the mission. Due to the properties of required controller input data, it proved unavoidable to develop a number of special sensors to measure the flight attitude of the UAV. Important properties of the sensors (IMU, Pitot tube and GPS) and their connection with the control bus (Futaba S-BUS.2) are explained, i.e. real-time processing, signal pre-processing as well as signal conditioning. The control by the autopilot has to be done in most of the weather conditions. The entire system is understood as a complete unit of mechanics, electronics and software components. With this in mind, the system is well prepared for future expansion, such as radar, wide-range telemetry and artificial intelligence, with other words the aircraft is design to be upward scalable (Fig. 1).

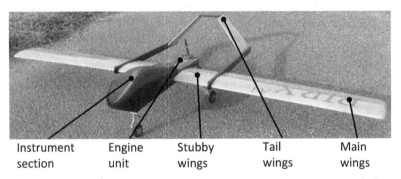

| Instrument | Engine | Stubby | Tail | Main |
| section | unit | wings | wings | wings |

Fig. 1. Prototype of the UAV with 3D printed parts and discreption of the main components of the UAV. 3D Printed Parts are Instrument section, Stubby wings and a part of the tail wings.

2 Optimizing of the Airframe

In [5] the basic design of an airframe was shown on the basis of the exclusive use of 3D printed parts. During the flight test the tails, wings and parts of the airframe were replaced by other manufacturing processes for cost and weight efficiency. The series production is made with special tools and a multi-component foam. The servo drives are embedded in the tails and wings for a aerodynamic surface.

A special feature of the airframe is the division into mechanical modules. The engine unit, stubby wings and tail unit are one fixed unit. Before the start, the wings as well as the instrument section are mounted to this fixed unit. The instrument section includes the most electronic parts and is used for the payload as well. The basic low cost instrument section is only 100 mm high. This is too small for special payload sensors like a camera.

Because of this, the instrument section was analyzed and aerodynamic redesigned in a master thesis. The result with the flow simulation is shown in the following figure (Fig. 2):

Fig. 2. Flow simulation of the instrument section with SolidWorks. The lines are the velocity of the air from, high velocity (dark) to slow velocity (bright) [4].

The lines in pictures are the velocity of the air. The lines on the top are dark. The velocity is high and this means the pressure is low. The lower pressure effects an uplift of the instrument section. So, the instrument section has a characteristic of a small wing. The aerodynamic redesign isn't limited to the instrument section. The stubby wings got a new function and needed a redesign, too. Inside the stubby wings sensors are located. The PCBs of the GPS sensor and the pitot sensor are inside the stubby wings. Furthermore, the parts of the tail wings are made with 3D printed parts.

3 Electronic System of the UAV

Figure 3 shows the complete hardware system. In the concept the idea was to develop a small part of electronic parts. The first experiments showed that it is not possible to buy the parts with our requirements.

The most important communication bus between the electronic parts is the Futaba S-BUS.2. It is used to exchange the commands between the CPUs, servo drives and sensors. The bus is bidirectional and packet-oriented. Furthermore, it is used for power supply with "High-Voltage" which is 7, 4 V. In addition, the Futaba-System is used to have telemetry between the UAV and the remote control, to control the UAV manually.

The board CPU is the main controller where the autopilot is implemented. The IMU is directly connected to the board CPU. An UART interface is connected to the payload CPU. This CPU is used to control the payload like a camera. To exchange important data between these two CPUs the UART is used. In addition, a telemetry CPU is used to control the long range telemetry. The GPS and pitot tube sensor has an extra PCB and is explained in the next chapter.

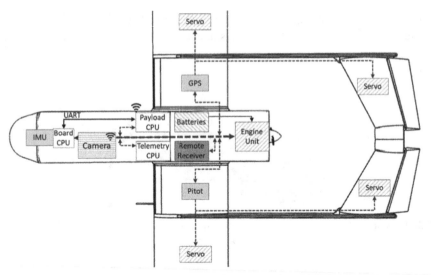

Fig. 3. Block diagram of the electronic concept of the UAV. The main bus system (Futaba S-Bus 2, dashed lines) supports all components in hard real time with commands and power.

3.1 Sensors

IMU
The IMU sensor (Inertial Measurement Unit) is a combination of gyroscope, acceleration sensor and geomagnetic sensor. With these three measurements it is called 9 DOF (Degrees Of Freedom), ever measurement in all three axes directions. The sensor is used to to determine the aircraft position for the controller. Because of that, it is the most important sensor. It is connected to the BoardCPU via UART. The output of the IMU sensor includes three different parameters (euler angle, acceleration and gravitational) in all three directions.

Pitot-Sensor
The Pitot sensor is used to measure the speed of the UAV in the air. This air velocity can differ from the ground velocity because of disturbances like wind. A minimum air velocity is necessary because otherwise the UAV falls down and crashes. For this reason, the Pitot sensor is the second most important sensor for the autopilot. The pitot sensor is connected to the BoardCPU via S-Bus 2. Therefore, the sensor has an extra PCB with microcontroller which is placed in the stubby wing. Due to the placement in the stubby wing, a simple connection with the pitot tube is possible. The pitot tube is the air inlet to measure the air velocity. The inlet must be outside of any aircraft turbulences. Therefore, the stubby wing is an ideal location.

GPS
The GPS (Global Positioning System) is the third sensor required by the autopilot. The GPS data is used to fly to the mission waypoints. For this, the autopilot compares the current GPS data with the waypoints. Like the Pitot sensor, the GPS is placed in the

stubby wing. For this reason, it has also a microcontroller and is connected via S-Bus 2. To reduce the traffic on the bus, a minimum data set is transmitted. This data set includes date, time, longitude and latitude, altitude, course and ground speed.

4 Model of the Autopilot

The implemented sensor units permanently supply the board computer with information about flight attitude, angular velocity, airspeed and groundspeed, as well as air pressure, humidity, temperature and the current GPS position. These information provide the basis for a process identification. The process identification is necessary to optimally setup a controller-algorithm and its parameters.

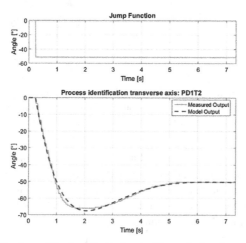

Fig. 4. Process identification of the longitudinal axis. First subplot shows the stimulation of the process. The second subplot visualizes the measured output and the process modulation [1].

Figure 4 subplot 1 shows a jump function which stimulates the process. The deflection is 50° and a right curve is initiated. Subplot 2 shows the process behavior. A PD_1T_2 transfer function for the transverse axis has been shown by a post-modeling. With the knowledge about the process behavior a basis for the controller synthesis is created. The behaviour of the transverse and vertical axis was carried out analogously. The coupling of the individual axes to each other is also really important. For example, if the aiframe is rolling around the longitudinal axis, a moment around the transverse axis is also generated. In order to react against these effects, studies were also carried out in these areas and integrated into the controller algorithmen.

To be able to control the airframe a cascade control system was implemented. A navigation module and an autopilot are representing the form of an outer control loop. The navigation module receives a pre-planned flight route via a SD card. This flight route includes all waypoints with altitude, longitude and latitude. The navigation module is able to calculate a flight course from the individual flight data. This course is needed to reach the next GPS point. The board computer also calculates the distance to the next destination and the desired air and groundspeed by the autopilot. The set values are calculated by the transfer functions. In principle, for cascade controls, the internal controllers should be designed quickly and the outer control loops take over functions for achieving steady-state accuracy. The internal control loop is realized by a basic controller. This basic controller ultimately ensures a stable angular position during the flight. In case of a rotation speed which is too fast, the own movement is damped by an attenuator [3]. The implemented cascade control is visualized in Figure 5.

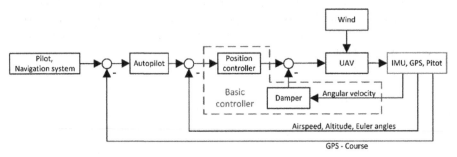

Fig. 5. Cascade controll of the airframe. The outer loop discribes the navigation unit and the pilot. The basic controller, which stabilizes the axis during the flight, represents the inner loop of the cascade controll.

A PID controller structure is selected as the transfer function for the roll and pitch controllers. The PID controller has the advantage of controlling nonlinear systems and can be easily integrated into an embedded system. To implement the controller in the C compiler, the controller must be described mathematically in the frequency domain first. The general notation of a PID controller applies:

$$G_{PID}(s) = \frac{u(s)}{e(s)} = K_P + \frac{K_I}{s} + K_D \cdot s$$

For implementation in the C compiler, the transfer function of the PID controller must now be converted into a difference equation. To do this, the transfer function must first be transformed into the z area by using the shift rule [2].

$$G_{PID}(z) = \frac{u(z)}{e(z)} = K_P + \frac{K_I}{z - 1} + K_D \cdot \frac{z - 1}{z}$$

This is followed by the retransformation into the time domain:

$$u_k = u_{k-1} + e_k \cdot (K_P + K_I + K_D) + e_{k-1} \cdot (-K_P - 2 \cdot K_D) + e_{k-2} \cdot K_D$$

This equation can be implemented in the microcontroller. Here u_k represents the current controller output variable and u_{k-1} the preceding output variable.

5 Flight Test

The implementation of the software as well as the revision of the airframe and the entire hardware of the on-board system took place almost simultaneously. One of the biggest challenges was to comply with the strict limitations of the payload to be transported. The test vehicle has a telemetry interface to the ground. With the help of a specially developed ground station, all important system parameters can be set. These are controller parameters, the determination of the flight mode and the storage of desired data. In addition, all sensors can be checked via the ground station and their parameters can be shown on a display in almost real time. For the pilot this is an enormous help in checking the flight system. Before each start, a safety list must be worked through in which all important parameters must be checked. Important sensor and control parameters are recorded in real time on an SD card during the flight and can then be evaluated carefully. By recording the data, the control behavior could be further optimized until an autonomous flight is possible. Figure 6 shows the control behavior of the roll and pitch controller. The upper plot shows a right-hand curve and then a left-hand curve. The control algorithm reacts to the control difference and initiates a corresponding turn. The lower subplot shows a desired flight up and flight down.

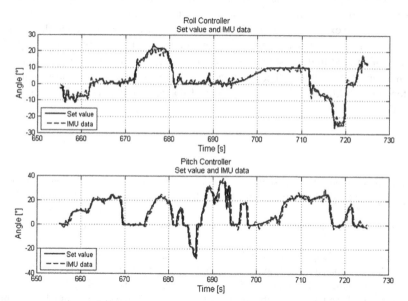

Fig. 6. Reaction of the roll and pitch controller. The first subplot shows the controll behaviour of the roll and the second subplot of the pitch controller.

A major advantage of the commissioning was the implemented cascade control, which allows external control loops to be switched on and off individually. At the start only the internal stabilisation control is activated. As soon as the aircraft arrives at a defined altitude, additional control systems are switched on one after the other. This includes the altitude, airspeed, groundspeed and navigation controller systems. After all systems are switched on, the flight process is controlled by the navigation module and the autopilot, which keeps the airframe on the desired flight path.

6 Summary and Outlook

The difficulties of hardware development and software implementation for the autopilot of a UAV were initially underestimated. In the course of the development, further development problems became apparent. The combination of the aeromechanical properties with the requirements of the autopilot are considerable. Ultimately, solutions could be found for all requirements.

A powerful model exists to describe the control characteristics of flight attitude stabilization. As a result of practical testing, this model is sufficiently validated and adaptable to different flight scenarios. Due to the mechanical construction and the gliding ability of the aircraft, a larger operating range could be proven compared to multicopters. At a cruising speed of 45 … 110 km/h, ranges far beyond the optical range of vision are possible. The desired flight routes are planned in a map software and then transferred to the flight system by using a SD card. In the case of long-range navigation, additional sensors, for example anti-collision radar and wide-range telemetry, are required. The hardware electronic system in the present version 3 and the new 3D printed parts are prepared for these extensions. In the future, the aircraft will be equipped with additional sensors to cover an even wider range of applications. A radar for scanning the ground is already integrated in the system. It is used to detect more accurate information about the condition of the ground. The knowledge about the condition of the ground is to be extended by additional sensors. These are LiDAR systems and cameras for object recognition.

References

1. von Eichel-Streiber, J.: Entwicklung und Inbetriebnahme einer eingebetteten Mehrachs-Regelung zur Flugstabilisierung eines UAV. Master thesis, University of Applied Sciences Bingen (2016)
2. Föllinger, O.: Regelungstechnik: Einführung in die Methoden und ihre Anwendung. ISBN 3-7785-1972-7
3. Buchholz, J.: Regelungstechnik und Flugregler, 2. Ergänzte Auflage
4. Gross, A.: Leichtbaukonstruktion einer Flugdrohne mittels 3D-Drucktechnik unter Beachtung der Strömungsgesetze. Master thesis, University of Applied Sciences Bingen (2018)
5. Altenburg, J., Hilgert, C., von Eichel-Streiber, J.: PIRX^{3D} – pilotless reconfigurable experimental UAV. In: Moreno-Díaz, R., Pichler, F., Quesada-Arencibia, A. (eds.) EUROCAST 2017. LNCS, vol. 10671, pp. 183–190. Springer, Cham (2018). https://doi.org/10.1007/978-3-319-74718-7_22. ISBN 978-3-319-74727-0

ROS-Based Architecture for Multiple Unmanned Vehicles (UXVs) Formation

Raúl Sosa San Frutos[1](✉), Abdulla Al Kaff[2](✉), Ahmed Hussein[3](✉),
Ángel Madridano[2](✉), David Martín[2](✉), and Arturo de la Escalera[2](✉)

[1] Akka Technologies, Paris, France
rsosasanfrutos@gmail.com
[2] Universidad Carlos III de Madrid, Leganés, Madrid, Spain
{akaff,amadrida,dmgomez,escalera}@ing.uc3m.es
[3] IAV GmbH, Berlin, Germany
ahmed.hussein@ieee.org

Abstract. A formation is recognized in different animals as a result of the cooperative behavior among its members, where each member maintains a specific distance and orientation with respect to the others in movement. In the robotics world, a group of robots is defined as Multi-Robot System (MRS), which are used in many applications, such as search and rescue or reconnaissance and surveillance. In these applications, MRS require robots to self-organize to solve complex tasks with better overall performance, in terms of maximizing spatial/temporal coverage or minimizing mission completion time. Consequently, this work presents a robust, heterogeneous and scalable architecture based on ROS for the formation of multiple robots, supported by a dynamic leader and virtual grid algorithms. The algorithm runs on all robots, so they select the leader based on an objective cost function. The leading robot then obtains the formation positions of the other robots and assigns them to the desired position. The assignment of positions is optimized and the proposed architecture has been validated using a simulation environment, which can be easily exported to real robots.

1 Introduction

The field of robotics has focused much of its research on groups of robots who cooperate and work together to complete a mission, i.e. Multi-Robot Systems (MRS). MRS are characterised by having agents who possess collective behaviour, based on communication and iteration between each other and the environment. This type of behaviour provides a set of advantages over a single robot, including the ability to generate and modify formation autonomously and coordinated to better adapt to the environment in real time [1].

From the point of view of adaptation to various situations, robotics has introduced different approaches to meet the necessary requirements. On the one hand,

This work was supported by the Comunidad de Madrid Government through the Industrial Doctorates Grants (GRANT IND2017/TIC-7834).

R. Moreno-Díaz et al. (Eds.): EUROCAST 2019, LNCS 12013, pp. 11–19, 2020.
https://doi.org/10.1007/978-3-030-45093-9_2

in [2], a hybrid approach is developed for the formation of MRS that blends the well-established automata theory, virtual grid and dynamic leader. The leader-follower system in robot formation was also presented in [3] and [4]. On the other hand, in [5], authors presented an approach for robot formation controller, which is scalable, with constraints in communications and sensor ranges.

The main objective of this work is to propose an ROS architecture that allows the formation of a heterogeneous and scalable MRS. This structure will generate a solid and robust solution for the problem of configuration and movement of an MRS in formation. The architecture is defined as Centralized-Distributed since the MRS formation calculations can be executed in a single robot, the leader, as well as in the rest if the first one falls. In each robot two simultaneous algorithms are executed, formation manager and robot, similar in all robots.

This paper is structured as follows: Sects. 2 and 3, describe in detail how the proposed ROS architecture is built. Followed by Sect. 4, which describes the experimental work executed to validate the architecture and Sect. 5 shows the results obtained in the mentioned experiments. Finally, Sect. 6 gives an overall conclusion as well as suggesting further directions to develop.

2   Formation Manager

The formation manager regulates the MRS as a group. It selects a leader in every instance of time, which will be the one in charge of executing all the formation calculations, this is called "Dynamic Leader". The leader allocates the positions for each of the "following" robots at each instant in time. This allocation varies depending on the state of the system, finally, this allocation is optimized.

2.1   Leader Selection

The leader selection is calculated at every instance by every robot with the known leadership rules. The resulting leader robot will be the one executing the calculations in the formation manager, as mentioned above.

$$Objective function = \frac{w_1}{\sum_{i=1}^{N_{robots}} DA_i} + \frac{w_2}{D_{goal}} + \frac{w_3}{CP} + ... \tag{1}$$

This selection computes an objective function seen in Eq. 1 with a set of leadership rules which include parameters; such as distance to the goal (D_{goal}), proximity with the neighboring robots (DA_i) or computational power (CP). The leader will be the robot with best score in the objective function according to the leadership rules, and the weighting of each rule (w_i). However, different leader selection parameters may be added depending on the system requirements. The selector calculates the target function for each robot and, if two robots have the same result, the first robot set as leader will remain in that condition. The non-leader robots wait for the leader to assign them a position.

2.2   Formation Status

The next step in the formation manager is illustrated in Fig. 1 is the status check, which will determine the state of the system.

The system would change states form left to right as seen in Fig. 1, the system is considered *scattered* when there is no intention of forming. Once the system is initialized and receives formation commands, it changes to *forming* state, where the robots are allocated around the leader in the given formation, and the leader holds its position. Once all the leaders have reached their desired position within a certain range r, *check_arrival*, (see Eq. 2) the system is formed, and hence it changes to *moving* state, where the leader moves towards the global goal, and the rest of the positions are calculated with respect to the leader, and hence the movement of the leader produces a movement of the whole system. Once, the leader reaches the global goal, and all the robots are in their desired position, the system changes to *hold* state, where it holds its position until a new global goal or formation command are introduced.

$$|A_{cp} - F_p| > r \tag{2}$$

Where A_{cp} is robots current position, $F_p is$ is the position in formation, and r the acceptance range.

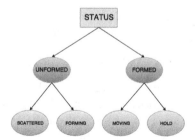

Fig. 1. A hierarchy of the formation status

2.3   Formation Assignment

The formation assignment is divided into two steps: the calculation of the position to form the given formation and the optimization of the positions assigned to the robots.

1. **Positions Calculation:** In this subsection, given a specific formation type, location for robots are calculated in order to acquire the requested formation. In this paper, the formations studied are: line, square and arrow.
   The algorithms for the calculation of the formation positions take the position of the leaders and the type of formation as inputs and generate the positions in the formation. These algorithms also consider as input the diameter of the

robots in order to calculate the distance between robots in the formation and the global goal for the MRS to be formed towards it.

The **line formation** can be horizontal or vertical. As the formation is elaborated around the leader, the original position of the leader is taken as the first position of the formation. The other positions are calculated by adding the distance between robots (sep) which is calculated as three times the robot diameter (A_d) of the largest robot in the system. This sep is added (or subtracted if the global goal is *behind* the leader) in the x-axis if the line type is horizontal and in the y-axis if the line type is vertical.

The **square formation** algorithm also results in a list of positions for the formation. It also varies with respect to the global goal, in both the x and y axis. The leader's position is used as reference to generate the formation.

The distance between corners is assumed to be three times the diameter of the robot if the number of centres is 0, these being the number of robots between corners. The number of centers n_{Center} is calculated with the Eq. 3 for those situations in which you work with more than 4 robots. In those cases, some agents will not be located in the corners and, therefore, it will be necessary to calculate a new distance between corners (Eq. 4) to be able to place robots in the central positions [2].

$$Num.Centers = ceil(\frac{n_A - n_{Corner}}{n_{Lines}}) \qquad (3)$$

Where for the square formation, number of lines n_{Lines} and number of corners n_{Corner} equal 4 since the formation is a square.

$$d_{CornerNew} = (d_{CornerOld} \times n_{Center}) + A_d \qquad (4)$$

Where, n_A is the number of robots, and A_d is the robot diameter.

In the square formation algorithm, knowing the distance between corners, it is added or subtracted to the x-axis based on the location of the global for one robot, to the y-axis for another, and to both axes for the robot in the opposite corner of the square. Then, knowing the number of centers, from Eq. 3, the intermediate locations between corners are obtained in the same way adding (or subtracting) the separation between centers, which is the distance between corners divided by $n_{Centre} + 1$.

The **arrow formation** works in a similar manner, it gets the same inputs and also results in the list of positions to create the formation. This formation can be divided in rows, and each row has N_A number of robots in it, this number increases by one in each row, having 1 robot in row 1, 2 in row 2 and so on. The sequence representing the total number of robot in the system depending in the row number is seen in Eq. 5 below:

$$N_A = \frac{r(r+1)}{2} \qquad (5)$$

The algorithm uses this to calculate the positions where robots are allocated. It checks if the robot number is the last of the row with the method just mentioned, if so it adds the distance between robots in x and y axes using Pythagoras theorem. The algorithm allocates the leader's position to the first member, then it goes through all the robots in each individual row, for the first one it adds the separation in the x and y axes to the robot in the edge of the previous row, and the following, with the same y, it subtracts the separation in the x axes. It repeats this for every row in the formation.

2. **Assignment Optimization:** The assignment optimization is used to obtain the optimal position for each robot in the formation, this is done using an optimization algorithm, Simulated Annealing (SA)[6]. Knowing the current global position of each robot and the positions in the formation, each robot is assigned a position, minimizing the total distance covered by the system. This is the cost of the assignment, or the path length, and it is calculated with the Eq. 6. It calculates the total distance covered by the system from the current location to the assigned positions. It executes a sum of the euclidean distances for every robot, from its current position to the position in the formation it has been assigned.

$$Pathlength = \sum_{i=0}^{N_A} \sqrt{(A_i x_c - A_i x_g)^2 + (A_i y_c - A_i y_g)^2} \qquad (6)$$

3   Robot Manager

The robot manager regulates the robots movement. It executes a path planning to find the shortest clear path between it's current position and the robot goal location, which is the position each robot occupies in the formation, given by the leader and updates the current position. The main input is the robot goal position, which represents the position it occupies in the formation. In the current case, a clear obstacle-free environment is assumed, this means the path planning is simplified to a point follower, and the robot moves in a straight line towards the local goal given by the formation manager. Once the robot reaches the goal position within a certain margin, the robot does not move until the goal is modified. Algorithm 1 explains the steps that the robot manager take, from the moment it receives a goal until it moves towards this position, ending by publishing the updated position to feed the formation manager as mentioned above. It is assumed that the speed of movement is always constant and equal to 1 m/s, which allows to know in a simple way which is the distance covered. In addition, knowing the angle between the target position and the current position, it is possible to calculate with simple trigonometry the direction of movement in both axes (Algorithm 1).

4   Experimental Work

The experimental work is executed to validate the proposed ROS architecture.

Algorithm 1: Robot manager

Input: Goal position G, speed v
Output: New robots current position nA_{cp}
1 **Define:** robots current position A_{cp}, time t, distance covered $dist$, angle $angle$
2 **begin**
3 | $t \leftarrow$ initialize
4 | **if** $Check\_arrival$ **then**
5 | | exit
6 | $t \leftarrow$ get Δt
7 | $dist = v \times \Delta t$
8 | $angle = arctan(G_y - A_{cp}y, G_x - A_{cp}x)$
9 | $d_x = dist \times cos(angle)$
10 | $d_y = dist \times sin(angle)$
11 | $nA_{cp}x = A_{cp}x + d_x$
12 | $nA_{cp}y = A_{cp}y + d_y$

4.1 Setup and Scenarios

The simulation experiments are conducted in a ROS Kinetic environment having individual nodes for each robot, with its respective topics of current position and goal position. The visualization is done in RViz.

The different scenarios proposed to validate the scalability of the architecture include the three different formations: arrow, square and line, with a different number of robots, 12, 8 and 4.

4.2 Evaluation Metrics (EM)

There are two main evaluation metrics taken: FQ (Formation Quality), which is a temporal EM and OFD (Optimal Formation Divergence) which is a spatial EM. The EM selected are Collective behavior-based to validate the swarming behavior of the MRS ROS architecture.

FQ (Formation Quality): It identifies how well the robots in the system collaborate together to build the formation and how long does it take them to reach the final goal in formation. It is a time based evaluation metric calculated with the Eq. 7 which is the average time taken for all the robots to reach the final state compared to the *OptimalBuildingTime* Eq. 8.

$$FQ = \sum_{i=1}^{n} \frac{T_{F_i}}{n} \tag{7}$$

$$OptimalBuildingTime = \sum_{i=1}^{n} \frac{V}{D_{theoretic_n}} \tag{8}$$

Where n is the number of robots in the MRS, T_F is the total time taken to reach the final position, V is the velocity, constant throughout the experiment and $D_{theoretic}$ is the theoretical distance that the robot will cover.

OFD (Optimal Formation Divergence): Compares the difference between desired formation positions against the positions reached by the robots, which will be within the acceptance range mentioned in Sect. 2.2.

$$OFD = \sum_{i=1}^{n} \frac{OAD_i}{n} \tag{9}$$

$$OAD = |P_{D_s} - P_{A_s}| \tag{10}$$

P_{D_s} is the desired position for each robot and P_{A_s} is the final position achieved.

5   Results and Discussion

In this section, the results obtained are shown to validate the working ROS architecture proposed here.

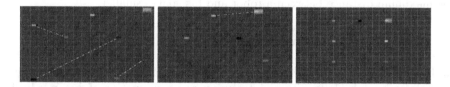

Fig. 2. RViz representation of the steps taken by the MRS (Color figure online)

The Fig. 2 shows an RViz representation of the robots in an empty environment. Three steps of the MRS movement described in the Sect. 2 are shown. The first step shows the initial state of the MRS where the leader (green) is in his target position and the other robots move towards the desired position with respect to the leader. The second step shows how the formation has been achieved and now the goal for the leader is the overall goal, so the leader's movement towards his goal is shown, with the MRS moving into formation. The final step shows the MRS forming a square.

Table 1 displays the average difference in m between the desired position and the final position of the robots at the various scenarios. From this table it can be observed that the OFD always is just below the acceptance range of 0.1 m, which proofs this acceptance range could have been reduced for further accuracy.

Table 2 displays the time taken by the formation to achieve the final position compared to the OBT. The key data observed in this table is the difference between FQ and OBT. The differences occurs due to the variation in forming times between robots, hence some robots having to hold their formed position

Table 1. Optimal formation divergence

Units (m)	Formation type		
Robots No.	Arrow	Square	Line
4	0.0866	0.0866	0.0764
8	0.0837	0.0886	0.0880
12	0.0933	0.0844	0.0859

Table 2. Formation quality

Robots No.	Formation	FQ (s)	OBT (s)	Diff. (s)
4	**Arrow**	22.6001	15.3544	7.2457
	Square	13.2668	11.7950	1.4718
	Line	16.9334	13.5563	3.3772
8	**Arrow**	28.8168	17.0987	11.7181
	Square	20.0584	14.2496	5.8088
	Line	18.6168	13.6526	4.9641
12	**Arrow**	19.2695	15.2289	4.0406
	Square	20.5890	12.8026	7.7864
	Line	24.0834	14.0972	9.9862

until other robots reach their position in the formation. It can be seen how in scenarios with fewer robots, the differences reduces, as a result of being fewer robots to form. However, an anomaly is observed with the arrow formation type. It can be appreciated that the difference increases with fewer robots, this is because after the formation is accomplished there is a change in leader which adds to the forming time. The increase in time for 8 compared to 12 is because the maximum distance between a robot and his goal is smaller.

6   Conclusion and Future Work

In this paper, an architecture is proposed for MRS formation in ROS ready for its application in real robots. The architecture uses a combination of dynamic leadership, and a leader-following virtual grid. It supports three different formation patterns, square, arrow and line. As well as being as heterogeneous and robust, being able to form with a large number and type of different robots in the same system. This is demonstrated in the experimental work, where the architecture is tested in 9 scenarios with results that verify the proposed objectives.

These work includes a series of aspects not dealt with in it and that constitute future lines of research, such as: developing the capacity to form in motion, in order to reduce the difference between the FQ and the OBT of the MRS; extending the architecture by introducing a way of planning ways to use the architecture in environments with greater complexity, and presence of obstacles; increasing the types of formation for specific missions. In addition to exporting this architecture to real robots.

References

1. Yan, Z., Jouandeau, N., Cherif, A.A.: A survey and analysis of multi-robot coordination. Int. J. Adv. Rob. Syst. **10**(12), 399 (2013)
2. Wagdy, A., Khamis, A.: Adaptive group formation in multirobot systems. Adv. Artif. Intell. **2013**, 10 (2013)

3. Karimoddini, A., Lin, H., Chen, B.M., Lee, T.H.: Hybrid three-dimensional formation control for unmanned helicopters. Automatica **49**(2), 424–433 (2013)
4. Jasim, W., Gu, D.: Robust team formation control for quadrotors. IEEE Trans. Control Syst. Technol. **26**(4), 1516–1523 (2017)
5. Yu, H., Shi, P., Lim, C.-C.: Robot formation control in stealth mode with scalable team size. Int. J. Control **89**(11), 2155–2168 (2016)
6. Kirkpatrick, S., Gelatt, C.D., Vecchi, M.P.: Optimization by simulated annealing. Science **220**(4598), 671–680 (1983)

Knowledge Discovery: From Uncertainty to Ambiguity and Back

Margaret Miró-Julià$^{(\boxtimes)}$ ⓘ, Monica J. Ruiz-Miró, and Irene García Mosquera ⓘ

Departament de Ciències Matemàtiques i Informàtica, Universitat de les Illes Balears,
07122 Palma de Mallorca, Spain
{margaret.miro,monica.ruiz,irene.garcia}@uib.es

Abstract. Knowledge Discovery in Databases is concerned with the development of methods and techniques for making sense of data. Its aim is to model the shapes of distributions and to discover patterns. During the knowledge acquisition process choices are made. Uncertainty and ambiguity hinder the process and "poor" choices cannot be avoided. Uncertainty corresponds to situations in which the choices are unclear and/or their consequences difficult to measure. Ambiguity arises from the lack of context, there is not sufficient information to assure the success of the choice, thus causing confusion. And decision making is hampered by perceptions of uncertainty and ambiguity.

Medical diagnosis/prognosis is a complex decision making process. In this paper we present a comparative study of model ambiguity on breast cancer predictions. Automatic classification of breast cancer on mammograms using two models, logistic regression and artificial neural networks are considered. The models were trained and tested to separate malignant and benign tumors for two different scenarios. Ambiguity in the prediction was studied for the different models.

The results show that a measure of uncertainty is practical to explain observable phenomena, such as medical data. Since model ambiguity can rarely be avoided, ordering alternative models by their degree of ambiguity is crucial in medical decision making processes. Furthermore, the levels of uncertainty and ambiguity are relevant in the knowledge representation process and open up new possibilities for richer Data Mining tasks.

Keywords: Decision making · Uncertainty · Ambiguity

1 Introduction to Data Modeling

Artificial Intelligence and Data Mining techniques can be properly combined to analyze complexity in specific domain problems and provide understanding of the phenomena. In real-world applications of Artificial Intelligence, considerable human effort is used in designing the representation of any domain-specific problem in such a way that generalised algorithms can be used to deliver decision-support solutions to human decision makers. On the other hand, Data Mining

© Springer Nature Switzerland AG 2020
R. Moreno-Díaz et al. (Eds.): EUROCAST 2019, LNCS 12013, pp. 20–27, 2020.
https://doi.org/10.1007/978-3-030-45093-9_3

emerges in response to technological advances and considers the treatment of large amounts of data. Data Mining is the discovery of interesting, unexpected or valuable structures in large data sets. Its aim is to model the shapes of distributions and to discover patterns.

Model building is a high level "global" descriptive summary of data sets, which in modern statistics include: regression models, cluster decomposition and Bayesian networks. Models describe the overall shape of the data [1]. Pattern discovery is a "local" structure in a possibly vast search space. Patterns describe data with an anomalously high density compared with that expected in a baseline model. Patterns are usually embedded in a mass of irrelevant data [2]. Most apparent patterns in data are due to chance and data distortion, and many more are well known or uninteresting.

The driving force behind Data Mining is two-folded: scientific and commercial. Problems and intellectual challenges arise from both contexts; tools and software development are especially driven by commercial considerations. In this sense, the growth of Data Mining researchers has been motivated by the company's need to find the knowledge that is hidden in their data, since this knowledge allows companies to compete against other companies.

The need of efficient methods to search knowledge in data has favoured the development of a lot of Data Mining algorithms and Data Mining tools. When modeling a data set, different situations can be considered. From a theoretical point of view, data is used to build models that causally explain observed data and predict future data. The models hope to predict the change, usually averaged over the population, in the outcome variable due to a change in the causal variable. Whereas, from an algorithmic perspective, data is used to build statistical models which hopefully will allow making predictions about the properties of unobserved individuals (or missing attributes of observed individuals) who are assumed to be from the same population that generated the data.

When building models that causally explain observed data and predict future data, both communities come face to face with the concepts of risk (quantifiable uncertainty) and ambiguity (unquantifiable uncertainty). Certainty, and thus uncertainty, is a phenomenon of the mind; humans are uncertain about the world around them. From a traditional Artificial Intelligence perspective, we must conceive a way to computationally model this uncertainty, and how humans decide immersed in it. This has naturally turned to probability theory in both deductive and inductive Artificial Intelligence models of human reasoning with uncertainty. In the deductive frame, probabilistic networks of inference of various kinds have become a dominant knowledge representation [3], and in the inductive frame, statistical learning has become the dominant foundation for learning algorithms [4]. Aware of it or not, most Data Mining problems are solved under conditions of uncertainty and ambiguity. In fact, no differentiation between uncertainty and ambiguity is made, and the degree of uncertainty or ambiguity is considered exogenous to the problem-solving process. There is a growing need to be able to represent the uncertainties and ambiguities in Data Mining.

In this paper we propose that knowledge discovery under ambiguity involves fundamentally different tasks than knowledge discovery under uncertainty. Consequently, different representation structures are appropriate and different techniques are needed. Furthermore, the levels of uncertainty and ambiguity are determined in the knowledge representation process. These patterns open up new possibilities for richer Data Mining tasks. Probabilistic knowledge representation methods are used to study correlation structures to determine levels of uncertainty and ambiguity.

2   Knowledge Discovery and Big Data

The starting point of any Knowledge Discovery process is the database or raw data. A large collection of unanalyzed facts from which conclusions may be drawn. In the Big Data era, massive amounts of data, both structured and unstructured, are collected across social media sites, mobile communications, business environments and institutions. But it's not the amount of data that's important. It's what organizations do with the data that matters. Big Data analytics offer insights that lead to better decisions and strategic business moves [5]. There is a shift from logical, causality-based reasoning to the acknowledgment of correlation links between events. These insights generated in companies, universities and institutions provide for an adaptation to a new culture of decision making and an evolution of the scientific method, both of which are starting to be built and will provide opportunities for future research [6].

Not too long ago, it was difficult to collect data. Now it's no longer an issue Big Data becomes a truly valuable commodity only when the data (qualitative and quantitative variables) is of high quality. Knowledge Discovery in Databases (Data Mining) offers support to modern organizations on how to make decision from massively increased information so that they can better understand their markets, customers, suppliers, operations and internal business processes. It also studies how to make decisions from massively increased information.

So much data is collected that the responsible thing to do is to preserve data integrity in such a way that it facilitates reliable data-driven decisions can that steer the company in the right direction. Data Integrity refers to the trustworthiness of data, its overall completeness, accuracy, consistency, timeliness, validity and uniqueness. Data of high quality that also expected to be attributable, legible, contemporaneous, original and accurate.

3   Uncertainty vs. Ambiguity

What's the difference between uncertainty and ambiguity? Uncertainty refers to epistemic situations involving imperfect or unknown information (data and/or knowledge). Synonyms are doubt, hesitancy, not known precisely, not clearly determined, unstable, likely to change. Ambiguity is a type of meaning in which a phrase, statement or resolution is not explicitly defined, making several interpretations plausible. Context plays an important role in resolving ambiguity.

Synonyms are doubtfulness or uncertainty of meaning or intention, more than one meaning and it is not clear which meaning is meant.

Ambiguity and uncertainty are contributory factors in many areas of our life, ... When we think we are going at one speed and our instruments tell us something different ..., which is correct? Do you follow you instinct or the instruments? The point here is: "ambiguity leads to uncertainty." Uncertainty and confusion are the reactions we have to things like ambiguity. An ambiguity is anything that can be interpreted in more than one way. Uncertainty is the reaction, doubt, confusion, ... We use the term uncertainty for situations where we know what we're measuring (i.e. the facts) but are in doubt about its numerical or categorical values. We use the word ambiguity to refer to situations in which we are in doubt about what the facts are or which facts are relevant.

4   Uncertainty, Ambiguity and Decision Making

The distinction between uncertainty and ambiguity is a subtle and important one for decision making. There are a couple of problems with uncertainty. The first one is its effect on our cognitions, our ability to think, to solve problems and to act. And secondly,and this might sound strange, you have to know you are uncertain to be able to do anything about it. We don't realize that there is an uncertainty (or ambiguity), ... and we carry on thinking things are correct and certain ... until things go badly wrong. Due to the lack of recognition of uncertainties (and ambiguities) patterns where decisions fail emerge. These patterns created a form of paralysis and decision makers don't know how to respond.

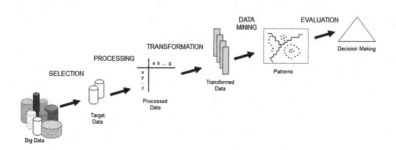

Fig. 1. Decision making process

A decision making process involves several steps, see Fig. 1. The starting point of the process is the data, massive amounts of different types of data. Next, target data is determined by selection and subsampling of the raw data. Data is processed and transformed into useful features so that new models can be found by means of Data Mining techniques. The mined patterns are interpreted, evaluated and converted into useful decision making.

In decision making doubt can manifest itself in a variety of ways: one could have ... doubts about the data itself (data accuracy); doubts about what data is needed (data relevance); doubts about the available processes and transformations; doubts about the possible models; doubts about the decision criteria; ... or even doubts on one's own preferences for these options.

How would you classify the different steps in the decision making process, as uncertain or ambiguous? How can one make well-informed decisions in such conditions?

Uncertainty in the Data. Due to measurement inaccuracy, sampling discrepancy, outdated data sources, or other errors. This is especially true for location-based services and sensor monitoring. These various sources of uncertainty have to be considered in order to produce accurate results.

Ambiguity in Data Selection and in Available Processes and Transformations. Feature selection is used in order to reduce dimensionality, eliminate irrelevant or redundant data and sharpen accuracy. Defining whether a feature is "relevant" is complicated and ambiguous. Traditional approaches have limited capability.

Ambiguity About the Model. Model formulation assumes the existence of a "true" model in a pre-specified known form. In practice, model ambiguity has received little attention. Furthermore, we are modeling uncertain data with traditional techniques.

Ambiguity About the Decision Criteria. Rational decision making approaches assume understanding of the decision problem as well as the facts and assumptions around it. These conditions are clearly violated in the case of ambiguous decision problems.

... And ambiguity leads us back to uncertainty. When everything goes the way they are supposed to ... choices are easy. Unfortunately most of the decisions we face each day address situations that don't go exactly as planned. Decision making is more often hampered by perceptions of uncertainty and ambiguity than by the lack of data. The decision making process is based on precise choices, but in everyday life, "poor" choices are made. Uncertainty corresponds to situations in which the choices are unclear and/or their consequences difficult to measure. Ambiguity arises from the perceived lack of explicit context. Situations are never ambiguous, but our perceptions of situations can be and open to more than one interpretation. If we improve the process in which we take decisions, we reduce our perceptions of ambiguity.

5   Comparative Study of Model Ambiguity

A major class of problems in medical science involves the diagnosis of disease, based upon various tests performed upon the patient. The evaluation of data taken from patients and complex decision making are the most important factors in diagnosis. Breast cancer is a fatal disease causing high mortality in women. Constant efforts are being made for creating more efficient techniques for its early and accurate diagnosis.

Mammographic results are often used to make an early diagnosis of breast cancer. An effective diagnostic procedure is dependent upon high levels of consistency between physicians' ratings of screening and/or automatic methods (supervised and unsupervised) used as computer-aided systems.

The classification of mammographies using a binary categorical scale (diseased or not diseased) is highly susceptible to interobserver variability and human errors, resulting in a suitable problem for examining how uncertainty and ambiguity might play an influential role on the consistency of predictions.

5.1   The Data

A publicly available database https://data.world/uci/mammographic-mass [7] was used to examine accuracy and disagreement between two machine/statistical learning methods. The database contains 961 mammograms with 516 benign cases and 445 malignant cases. The attributes reported in the database are: Patient's age in years: $[18, 45) = 1$, $[45, 57) = 2$, $[57, 66) = 3$ and $[66, 96] = 4$. Mass shape: round $= 1$, oval $= 2$, lobular $= 3$, irregular $= 4$. Margin: circumscribed $= 1$, micro lobulated $= 2$, obscured $= 3$, ill-defined $= 4$ and spiculated $= 5$. Density: Mass density, high $= 1$, iso $= 2$, low $= 3$ and fat-containing $= 4$. Malignant (biopsy result): benign $= 0$, malignant $= 1$.

The database is split in training set (70%) and test set (30%). The models were trained on 580 cases (298 benign and 282 malignant cases) and tested on 250 cases (121 benign and 129 malignant cases). En order to compare results, two different scenarios $S1$ and $S2$ with same proportions are considered.

5.2   The Models

Two models considered here: artificial neural networks and logistic regression.

Artificial Neural Networks (ANN). ANNs are computer algorithms whose structure and function are based on models of the structure and learning behavior of biological neural networks. These algorithms are typically employed to classify a set of patterns into one of several classes. The classification rules are not written into the algorithm, but are learned by the network from examples. A variety of medical tasks has been successfully performed using such networks. In particular, breast cancer analysis and decision making in mammography.

The ANN for malignancy prediction was implemented as multilayer feed-forward architecture with one and two hidden layers. Input feature values are the mammographic variables described above. The network was trained using a backpropagation training algorithm. Different network architectures for both scenarios have been considered.

Logistic Regression (LR). Logistic regression is a statistical learning technique for binary classification problems. LR involves a linear discriminant to predict the probability that the given input point belongs to a certain binary class (Malignant 1 or Malignant 0). Specifically, we fit the model with the logit linkage function, binomial family and the parameters model were estimated iteratively with least squares. We compute the optimal probability cutoff score in the training dataset that minimizes the misclassification error for the above model. Next, whenever the estimated probability is greater or equal to 0.6 in the test dataset, then the corresponding observation is classified as Malignant.

5.3 The Results

A measure of uncertainty is necessary to explain observable phenomena, such as medical data. Since model ambiguity can rarely be avoided, ordering alternative models by their degree of ambiguity is crucial in medical decision making processes. Both ambiguity and uncertainty affect medical decisions, but in different and sometimes opposite ways.

Table 1 summarizes the results obtained for Sensitivity, Specificity, Similarity, and Distance to sensitivity = 1 and specificity = 1 by model and scenario.

Table 1. Comparative results by model against biopsy results

Model	Sensitivity	Specificity	Similarity	Distance
LR	0.8347	**0.845**	0.84	**0.226**
ANN S1M3	**0.8843**	0.7519	0.816	0.2737
ANN S1M4	0.876	0.7597	0.816	0.2704
ANN S1M5	0.876	0.7752	0.824	0.256
ANN S2M3	0.8595	0.7985	**0.828**	0.254
ANN S2M4	0.8265	0.7907	0.808	0.272

In order to compare the ambiguity related to the model selection FP (false positive) and FN (false negative) are used. Table 2 shows these results. For example, there are 18 observed ambiguities between LR and ANN S1M3.

Table 2. Ambiguity results between models

FN\FP	Malignant	LR	ANN S1M3	ANN S1M4	ANN S1M5
Malignant	0	20	32	31	29
LR	**20**	0	18	17	14
ANN S1M3	14	**0**	0	5	0
ANN S1M4	15	1	7	0	5
ANN S1M5	15	0	4	7	0

6   Conclusions and Future Work

Most real-life decisions are filled with uncertainty and/or ambiguity. The failure to recognize uncertainty or ambiguity, whatever its source, may lead to erroneous and misleading interpretations. Conventional scientific studies attempt to reduce uncertainty. However, in recent studies that try to accommodate it explicitly, ambiguity is often ignored. The key point is to figure out the kind of uncertainty and/or ambiguity one is dealing with and to choose an approach that works for it. Improving the decision making process requires dealing with uncertainty and ambiguity. In dealing with uncertainty, we need to understand the fundamentals of probability to make the better choice. To reduce ambiguity, we need to understand the quality of the data at hand, the nature of the problems we're facing, and which methodologies will produce the best results. The challenge behind decision making has less to do with the quantity of data and models at our disposal than our capacity to improve the understanding and representation of the mental processes behind decision making.

It is crucial that the importance of dealing with uncertainty or ambiguity be recognized widely so that the extra effort that this requires is not seen as an unnecessary burden but as something that may usefully enhance research. Data uncertainty and model ambiguity are fundamentally intertwined and must be studied together.

Further study, focused on methods that accommodate uncertainty and ambiguity and provide uncertainty-aware systems, as well as, methods that extract useful knowledge from ever-expanding datasets, is required. Bayesian Belief Networks capture structural and numeric probabilistic relationships among random variables and conditional probability tables. Bayesian networks are valid models for representing and reasoning about uncertainty in complex applications. As future work, a methodology for ambiguity measurement based on Bayesian Networks will be considered.

References

1. Hand, D.J.: Principles of data mining. Drug Saf. **30**(7), 621–622 (2007)
2. Bishop, C.M.: Pattern Recognition and Machine Learning (Information Science and Statistics). Springer, Berlin (2006)
3. Pearl, J.: Probabilistic Reasoning in Intelligent Systems: Networks of Plausible Inference. Morgan Kaufmann Publishers Inc., San Francisco (1988)
4. Hastie, T., Tibshirani, R., Friedman, J.: The Elements of Statistical Learning: Data Mining, Inference and Prediction, 2nd edn. Springer, Heidelberg (2009)
5. Ward, J.S., Barker, A.: Undefined by data: a survey of big data definitions. CoRR, 1309.5821 (2013)
6. Anderson, C.: The end of theory: the data deluge makes the scientific method obsolete. Wired (2008). http://www.wired.com/science/discoveries/magazine/16-07/pb_theory. Accessed 12 May 2019
7. Elter, M., et al.: The prediction of breast cancer biopsy outcomes using two CAD approaches that both emphasize an intelligible decision process. Med. Phys. **34**(11), 4164–4172 (2007)

From Linear Systems to Fuzzy Systems to Perception-Based Systems

Rudolf Seising$^{(\boxtimes)}$ ⓘ

Deutsches Museum, Munich, Germany
r.seising@deutsches-museum.de

Abstract. This is another insight into the genesis of the theory of fuzzy sets and systems in the life of its founder Lotfi A. Zadeh. We present some historical roots of the theory of fuzzy systems and their further development to Computing with Words (CWW) and the Computational Theory of Perceptions (CTP).

Keywords: Lotfi A. Zadeh · System theory · Fuzzy systems · Computing with words · Perception-based systems · Computational Theory of Perceptions

1 Introduction

In 1954, the electrical engineer Lotfi A. Zadeh wrote a paper entitled "SystemTheory." Zadeh was then assistant professor at the Columbia University in New York that was published in the *Columbia Engineering Quarterly* [1]. This paper explained the brand-new system theory to the students of Electrical Engineering in New York's Columbia University: "If you have never heard of system theory, you need not feel like an ignoramus. It is not one of the well-established branches of science. In fact, it has not yet been officially recognized as a scientific discipline. It does not appear on programs of meetings of scientific societies nor in indices to scientific publications. It does not have well-defined boundaries, nor does it have settled objectives" [1, p. 34]. Here Zadeh characterized systems as black boxes with inputs and outputs. He asserted that, if these inputs and outputs are describable as time-dependent functions, then the dynamic of the system can be studied mathematically, and, as such, the input–output relationship of the system states that the output is just a function of the input. Between the heading and the text of his paper, there is a black box (Fig. 1).

As examples of a system, he listed

- a set of particles exerting attraction on one another
- a group of humans forming a society or an organization
- a complex of interrelated industries
- an electrical network
- a large-scale electronic computer.

Zadeh emphasized that all scientific disciplines are concerned with systems but that the new branch, called system theory, considers systems as mathematical constructs

© Springer Nature Switzerland AG 2020
R. Moreno-Díaz et al. (Eds.): EUROCAST 2019, LNCS 12013, pp. 28–35, 2020.
https://doi.org/10.1007/978-3-030-45093-9_4

System Theory

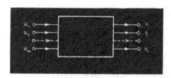

L. A. Zadeh
Associate Professor
Electrical Engineering

Fig. 1. Part of the title page of Zadeh's article [1].

rather than physical objects: "The distinguishing characteristic of system theory is its abstractness" [1, p. 16]. In this and later articles, Zadeh quoted the definition of a system from Webster's dictionary; a system is "an aggregation or assemblage of objects united by some form of interaction or interdependence" (Fig. 2). System theorists deal with abstract systems, "that is, systems whose elements have no particular physical identity" [1, p. 16]—they deal with black boxes. Zadeh believed that it was only a matter of time before system theory would be accepted. It turns out that he was right, and he became a protagonist of the system-theoretical approach in Electrical Engineering.

2 Zadeh's System Theory in the 1960s

After 1958, when Zadeh was a professor of electrical engineering at the University of California, Berkeley (UCB) he published papers on system theory and two well-known books with colleagues at his new department: Together with Charles Desoer, he published the volume *Linear System Theory: The State Space Approach* [2], which became a standard textbook. In 1969, together with Elijah Lucian Polak (born 1931), he edited the collection volume *System Theory* [3]. In his contribution to *System Theory*, "The Concepts of System, Aggregate, and State in System Theory," Zadeh presented the so-called state space approach. This general notion of *state* was the "new view" on system theory that Zadeh also presented in 1963 at the Second Systems Symposium at the Case Institute of Technology in Cleveland, Ohio. The conference proceedings were published in the next year by Mihajlo D. Mesarovic under the title *Views on General Systems Theory* [4]. This book does contain some very diverse approaches. In the foreword, however, Mesarovic wrote: "First of all, some of the participants took a definite stand, venturing to define a system and then discussing the consequences of such a definition. A second group of participants argued that the general systems theory should not be formalized since this very act will limit its generating power and make it more or less specific. A third group proposed to consider systems theory as a view point taken when one approaches the solution of a given (practical) problem. Finally, it was expressed that a broad-enough collection of powerful methods for the synthesis (design) of systems of diverse kinds should be considered as constituting the sought-for theory and any further integration was unnecessary. There were also participants that shared the viewpoints of more than one of the above groups" [4, p. xiv].

The economist and one of the founding members of the Society for General Systems Research, Kenneth E. Boulding, found himself inspired during some of the presentations to compose little poems, which were printed in the proceedings as introductions to the papers. Boulding treated Zadeh's contribution thusly:

"A system is a big black box

Of which we can't unlock he locks,

And all we can find out about

Is what goes in and what goes out.

Perceiving input-output pairs,

Related by parameters,

Permits us, sometimes, to relate

An input, output, and a state.

If this relation's good and stable

Then to predict we may be able,

But if this fails us – heaven forbid!

We'll be compelled to force the lid

K.B." [5, p. 39]

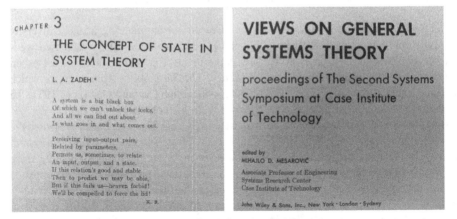

Fig. 2. Left: first page of Zadeh's article [5] with Boulding's poem, in [4] (right).

Zadeh based his abstract remarks on the concepts of the input, output and state of a system. While input and output were not expected to offer any difficulties, the concept of state appeared problematic. Zadeh noted that the idea of states had played an important role in the physical disciplines for a long time. It referred to a set of numbers that contain all of the information about a system's past and that determine its future behavior. The names Poincaré, Birkhoff, Markov, Nemytskii and Pontryagin stood for developments of more precise definitions of the concept of state in the fields where it was applied, such as dynamic systems and optimal control. However, the growing complexity and variety of the systems studied in the sciences had led to the necessity of framing the concept of state more broadly so that it could also be utilized for those systems that were not described by differential equations. Coming up with a general state concept was surely difficult, maybe even impossible, and so Zadeh limited himself "to sketch[ing] an approach that seems to be more natural as well as more general than those employed heretofore, but still falls short of complete generality" [5, p. 40].

As Boulding's poem predicted, Zadeh presented a proposal that was certainly difficult for the non-mathematical system scientists to understand. It is mentioned here because it serves as the bridge between Zadeh's earlier scientific works and the theory of fuzzy sets he was soon to devise.

40 years later Zadeh recalled in one of my interviews "System theory became grown up, but then computers took over. In other words, the center of attention shifted...So, before that, there were some universities that started departments of system sciences, departments of system engineering, something like that, but then they all went down. They all went down because Computer Science took over" [6].

Already in May 1962, Zadeh had contributed the paper "From Circuit Theory to System Theory" to the anniversary edition of the Proceedings of the IRE to mark the 50th anniversary of the Institute of Radio Engineers (IRE). In it he described problems and applications of system theory and its relations to network theory, control theory, and information theory: "It has been brought about, largely within the past two decades, by the great progress in our understanding of both inanimate and animate systems— progress which resulted, on the one hand, from a vast expansion in the scientific and technological activities directed toward the development of highly complex systems for such purposes as automatic control, pattern recognition, data processing, commu- nication, and machine computation, and, on the other hand, by attempts at quantitative analyses of the extremely complex animate and man–machine systems which are encoun- tered in biology, neurophysiology, econometrics, operations research and other fields" [7, p. 856f].

Then the following paragraph followed, that motivated Zadeh's later creation of fuzzy set theory: "In fact, there is a fairly wide gap between what might be regarded as «animate» system theorists and «inanimate» system theorists at the present time, and it is not at all certain that this gap will be narrowed, much less closed, in the near future. There are some who feel that this gap reflects the fundamental inadequacy of the con- ventional mathematics – the mathematics of precisely defined points, functions, sets, probability measures, etc.– for coping with the analysis of biological systems, and that to deal effectively with such systems, which are generally orders of magnitude more complex than man-made systems, we need a radically different kind of mathematics, the mathematics of fuzzy or cloudy quantities which are not describable in terms of probability distributions" [7, p. 867]. About two years later, when Zadeh became chair of the Department of EE at UCB in 1963 (Fig. 3, left), he experienced shifts of attention very intensively, for it was during his five-year tenure in this position that his department was renamed the Department of Electrical Engineering and Computer Science (EECS). During his chairmanship, in the year 1965, he tried to bridge this gap by introducing the new mathematical theory of fuzzy sets [8–11] which is based on the concept of "impre- cisely defined «classes»", which "play an important role in human thinking, particularly in the domains of pattern recognition, communication of information, and abstraction"

[8, p. 1]. A fuzzy set is "a notion which extends the concept of membership in a set to situations in which there are many, possibly a continuum of, grades of membership." [8, p. 1]. In his seminal article (Fig. 3, right), he wrote that these "fuzzy sets", as he named these "classes", "do not constitute classes or sets in the usual mathematical sense of these terms" [9, p. 338]. In that paper, Zadeh presented the framework for the new mathematical theory. He defined fuzzy sets, empty fuzzy sets, equal fuzzy sets, the complement and the containment of a fuzzy set. He also defined the union and intersection of fuzzy sets as the fuzzy sets that have membership functions that are the maximum or minimum, resp., of their membership values.[1]

U.C.
ELECTRONICS
NEWSLETTER

Published four times a year at Berkeley, California — Allan J. Lichtenberg, Editor

Vol. 5, No. 3 — Spring 1963

LOTFI A. ZADEH APPOINTED EE DEPARTMENT CHAIRMAN

Lotfi A. Zadeh, Professor of Electrical Engineering at the University of California in Berkeley since 1959, has been appointed chairman of the Department of Electrical Engineering. Professor Zadeh takes over the position from Professor Robert M. Saunders, who is returning to his full-time academic work, after having held the chairmanship for four years. The policy of appointing department chairmen for limited terms allows faculty members who are actively engaged in teaching and research to take the position.

Professor Zadeh is a graduate of an American missionary school in Teheran, Iran. He obtained the B.S. degree from the University of Teheran in 1942, the S.M. from M.I.T. in 1946, and the Ph.D. from Columbia in 1949. He joined the staff of the Department of Electrical Engineering at Columbia University in

L. A. Zadeh

INFORMATION AND CONTROL 8, 338-353 (1965)

Fuzzy Sets*

L. A. Zadeh

Department of Electrical Engineering and Electronics Research Laboratory, University of California, Berkeley, California

A fuzzy set is a class of objects with a continuum of grades of membership. Such a set is characterized by a membership (characteristic) function which assigns to each object a grade of membership ranging between zero and one. The notions of inclusion, union, intersection, complement, relation, convexity, etc., are extended to such sets, and various properties of these notions in the context of fuzzy sets are established. In particular, a separation theorem for convex fuzzy sets is proved without requiring that the fuzzy sets be disjoint.

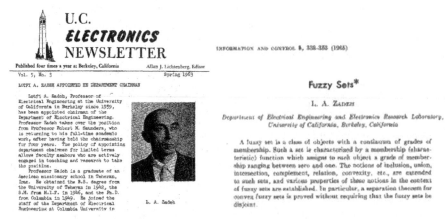

Fig. 3. Left: Zadeh became chair of the department; right: title page of Zadeh's paper [9].

Also in the year 1965, he offered a further "new view" on system theory: Within the theory of fuzzy sets, he was able to start establishing a theory of fuzzy systems: Given a system S with input $u(t)$, output $y(t)$ and state $x(t)$, it is a fuzzy system if input or out-put or state or any combination of them includes fuzzy sets, and, in Zadeh's view, only fuzzy systems can adequately cope with complex man–made systems and living systems. He explained that fuzzy systems relate to situations in which the source of imprecision is not a random variable or a stochastic process but rather a class or classes which do not possess sharply defined boundaries [10, p. 29] (Fig. 4).

FUZZY SETS AND SYSTEMS*

L. A. Zadeh

Department of Electrical Engineering, University of California, Berkeley, California

The notion of fuzziness as defined in this paper relates to situations in which the source of imprecision is not a random variable or a stochastic process, but rather a class or classes which do not possess sharply defined boundaries, e.g., the "class of bald men," or the "class of numbers which are much greater than 10," or the "class of adaptive systems," etc.

Fig. 4. Part of the title page of Zadeh's paper [10] to establish a "new view" on system theory.

[1] For basics on the theory of fuzzy sets see the paper "Lotfi Zadeh: Fuzzy Sets and Systems" in this volume. For a detailed study on that history see [12].

3 From Fuzzy Logic to Computing with Words

In an interview that Zadeh gave in 1994, he mentioned his surprise that Fuzzy Logic was "embraced by engineers" and "used in industrial process controls and in 'smart' consumer products such as hand-held camcorders that cancel out jittering and microwaves that cook your food perfectly at the touch of a single button." In that interview he also said that he had "expected people in the social sciences – economics, psychology, philosophy, linguistics, politics, sociology, religion and numerous other areas to pick up on it [Fuzzy Logic]. It's been somewhat of a mystery to me, why even to this day, so few social scientists have discovered how useful it could be." [13] see also [14].

In the 1970s, Zadeh was interested in applying fuzzy sets in linguistics. This idea led to interdisciplinary scientific exchange on the UCB-campus between him and the mathematicians Joseph Goguen and Hans-Joachim Bremermann, the psychologist Eleanor Rosch (Heider) and the linguist George Lakoff. During the 1970's Rosch developed her prototype theory on the basis of empirical studies. This theory assumes that people perceive objects in the real world by comparing them to prototypes and then ordering them accordingly. In this way, according to Rosch, word meanings are formed from prototypical details and scenes and then incorporated into lexical contexts depending on the context or situation. Rosch hypothesized that different societies process perceptions differently depending on how they go about solving problems [15]. Lakoff heard about Rosch's experiments and also about Zadeh'idea of linking English words to membership functions and establishing fuzzy categories in this way. Lakoff and Zadeh discussed this idea in 1971/72 and also the idea of fuzzy logic, after which Lakoff wrote his paper "Hedges: A Study in Meaning Criteria and the Logic of Fuzzy Concepts" [16]. In this work, Lakoff employed "hedges" (meaning barriers) to categorize linguistic expressions and he invented the term "fuzzy logic". Based on his later research, however, Lakoff decided that fuzzy logic was not an appropriate logic for linguistics, but: "Inspired and influenced by many discussions with Professor G. Lakoff concerning the meaning of hedges and their interpretation in terms of fuzzy sets," Zadeh had also written an article in 1972 in which he contemplated "linguistic operators", which he called "hedges": "A Fuzzy Set-Theoretic Interpretation of Hedges". Here he wrote: "A basic idea suggested in this paper is that a linguistic hedge such as very, more, more or less, much, essentially, slightly etc. may be viewed as an operator which acts on the fuzzy set representing the meaning of its operand" [17].

In the 1970s Zadeh expected that his theory of Fuzzy Sets "provides an approximate and yet effective means of describing the behavior of systems which are too complex or too ill-defined to admit of precise mathematical analysis" [18, p. 28]. He expected that "even at its present stage of development" his new fuzzy method "can be applied rather effectively to the formulation and approximate solution of a wide variety of practical problems, particularly in such fields as economics, management science, psychology, linguistics, taxonomy, artificial intelligence, information retrieval, medicine and biology. This is particularly true of those problem areas in these fields in which fuzzy algorithms can be drawn upon to provide a means of description of ill-defined concepts, relations, and decision rules" [18, p. 44].

For Zadeh, fuzzy logic was the basis for "computing with words" instead of "computing with numbers" [19] In his 1996 article "Fuzzy Logic = Computing with Words", he said that "the main contribution of fuzzy logic is a methodology for computing with words. No other methodology serves this purpose" [20, p. 103].

4   An Outlook: Perception-Based System Modelling

For the year 2001, Zadeh published a proposal "A New Direction in AI. Toward a Computational Theory of Perceptions". This theory was inspired by the remarkable human capability to operate on, and reason with, perception-based information, "to perform a wide variety of physical and mental tasks without any measurements and any computations (e.g. are parking a car, driving in city traffic, playing golf, cooking a meal, and summarizing a story). He assumed "that progress has been, and continues to be, slow in those areas where a methodology is needed in which the objects of computation are perceptions - perceptions of time, distance, form, and other attributes of physical and mental objects" [21, p. 73]. Zadeh compared the strategies of problem solving by computers, on the one hand, and by humans, on the other hand. Already in 1970 he had called it a paradox that the human brain is always solving problems by manipulating "fuzzy concepts" and "multidimensional fuzzy sensory inputs", whereas "the computing power of the most powerful, the most sophisticated digital computer in existence" is not able to do this. Therefore, he stated, "in many instances, the solution to a problem need not be exact", so that a considerable measure of fuzziness in its formulation and results may be tolerable" [22, p. 132]. For this purpose, the machine's ability "to compute with numbers" should be supplemented by an additional ability that is similar to human thinking: computing with words and perceptions (Fig. 5).

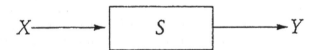

Input: $X_1, X_2, ...$
Output: $Y_1, Y_2, ...$
States: $S_1, S_2, ...$
State-Transition Function: $S_{t+1} = f(S_t, X_t), t = 1, 2,...$
Output Function: $Y_t = g(S_t, X_t)$

Fig. 5. Perception-based system modelling, [23].

In 1994 he presented perception-based system modeling: "A system, S, is assumed to be associated with temporal sequences of input $X_1, X_2, ...$; output $Y_1, Y_2, ...$; and states $S_1, S_2, ... S_t$ is defined by state-transition function f with $S_{t+1} = f(S_t, X_t)$, and output function g with $Y_t = g(S_t, X_t)$, for $t = 0, 1, 2, ...$ In perception-based system modelling, inputs, outputs and states are assumed to be perceptions, as state-transition function, f, and output function, g" [23, p. 77].

References

1. Zadeh, L.A.: System theory. Columbia Eng. Q. **8**, 16–19, and 34 (1954)
2. Zadeh, L.A., Desoer, Ch.A.: Linear System Theory: The State Space Approach. McGraw-Hill Book Company, New York (1963)
3. Zadeh, L.A., Polak, E. (eds.): System Theory. McGraw-Hill, Bombay (1969)
4. Mesarovic, M.D.: Views on general systems theory. In: Proceedings of the Second Systems Symposium at Case Institute of Technology. Robert E. Krieger Publ. Comp., Huntington, New York (1964)
5. Zadeh, L.A.: The Concept of State in System Theory, in [4], pp. 39–50
6. Zadeh, L.A.: In an interview with the author on July 26, 2000, Soda Hall, University of California, Berkeley, CA, USA
7. Zadeh, L.A.: From circuit theory to system theory. Proc. IRE **50**(5), 856–865 (1962)
8. Bellman, R., Kalaba, R., Zadeh, L.A.: Abstraction and Pattern Classification, Memo RM-4307-PR, US Air Force Project Rand. The Rand Corporation, Santa Monica (1964)
9. Zadeh, L.A.: Fuzzy sets. Inf. Control **8**, 338–353 (1965)
10. Zadeh, L.A.: Fuzzy sets and systems. In: Fox, J. (ed.) System Theory. Microwave Research Institute Symposia Series XV, pp. 29–37. Polytechnic Press, Brooklyn (1965)
11. Zadeh, L.A.: Shadows of Fuzzy Sets. In: Problemy peredachi informatsii, Akademija Nauk SSSR Moskva, vol. 2, no. 1, pp. 37–44 (1966). Engl. transl. in: Problems of Information Transmission: A Publication of the Academy of Sciences of the USSR; the Faraday Press coverto-cover transl., New York, NY, 2 1966
12. Seising, R.: The Fuzzification of Systems. The Genesis of Fuzzy Set Theory and Its Initial Applications – Its Development to the 1970s. Springer, Heidelberg (2007). https://doi.org/10.1007/978-3-540-71795-9
13. Zadeh, L.A.: Interview with Lotfi Zadeh, Creator of Fuzzy Logic by Betty Blair. Azerbaijada Int. **2**(4), 47–52 (1994)
14. Seising, R.: The experimenter and the theoretician – linguistic synthesis to tell machines what to do. In: Trillas, E., Bonissone, P., Magdalena, L., Kacprycz, J. (eds.) Combining Experimentation and Theory – A Homage to Abe Mamdani, pp. 329–358. Springer, Berlin (2012). https://doi.org/10.1007/978-3-642-24666-1_23
15. Rosch, E.: Natural categories. Cogn. Psychol. **4**, 328–350 (1973)
16. Lakoff, G.: Hedges a study in meaning criteria and the logic of fuzzy concepts. J. Philos. Log. **2**, 458–508 (1973)
17. Zadeh, L.A.: A fuzzy-set-theoretic interpretation of linguistic hedges. J. Cybern. **2**, 4–34 (1972)
18. Zadeh, L.A.: Outline of a new approach to the analysis of complex systems and decision processes. IEEE Trans. Syst. Man Cybern. SMC **3**(1), 28–44 (1973)
19. Zadeh, L.A.: From computing with numbers to computing with words – from manipulation of measurements to manipulation of perceptions. IEEE Trans. Circuits Syst. I Fundam. Theory Appl. **45**(1), 105–119 (1999)
20. Zadeh, L.A.: Fuzzy Logic = Computing with Words. IEEE Trans. Fuzzy Syst. **4**(2), 103–111 (1996)
21. Zadeh, L.A.: A new direction in AI. Toward a computational theory of perceptions. AI Mag. **22**(1), 73–84 (2001)
22. Zadeh, L.A.: Fuzzy languages and their relation to human and machine intelligence. In: Proceedings of the International Conference on Man and Computer, Bordeaux, pp. 13–165. Karger, Basel (1970)
23. Zadeh, L.A.: Fuzzy logic, neural networks, and soft computing. Commun. ACM **37**(3), 73–84 (1994)

A Latent Variable Model
in Conflict Research

Uwe M. Borghoff[2,3]([✉]), Sean Matthews[1], Holger Prüßing[2],
Christian T. Schäfer[1], and Oliver Stuke[1]

[1] Deloitte Analytics Institute, Berlin, Germany
[2] Center for Intelligence and Security Studies (CISS), Munich, Germany
ciss@unibw.de,info@casc.de
[3] Campus Advanced Studies Center (CASC),
Bundeswehr University Munich, Neubiberg, Germany
https://www.unibw.de/ciss,https://www.unibw.de/casc

Abstract. One of the focuses of interest for conflict research is crisis forecasting. While the approach often favored by the media and public approaches this challenge qualitatively with the help of pundits illustrating effects in a narrative form, quantitative models based on empirical data have been shown to be able to also provide valuable insights into multidimensional observations.

For these quantitative models, Bayes networks perform well on this kind of data. Both approaches arguably fail to meaningfully include all relevant aspects as

- expert knowledge is difficult to formalize over a complex multidimensional space and often limited to few variables (e.g. more A will lead to less B)
- empirical data can only tell us about things that are easily measurable and can only show correlations (in contrast to causalities that would be important for forecasting)

In this paper we will develop a method for combining empirical time series data with expert knowledge about causalities and "hidden variables" (nodes that belong to variables that are not directly observable), thereby bridging the gap between model design and fitting.

We build a toolset to use operationalized knowledge to build and extend a Bayes network for conflict prediction and, model unobservable probability distributions. Based on expert input from political scientists and military analysts and empirical data from the UN, Worldbank and other openly available and established sources we use our toolset to build an early-warning-system combining data and expert beliefs and evaluate its predictive performance against recordings of past conflicts.

Keywords: Statistical model · Expert scenario feedback · Bayes network

© Springer Nature Switzerland AG 2020
R. Moreno-Díaz et al. (Eds.): EUROCAST 2019, LNCS 12013, pp. 36–43, 2020.
https://doi.org/10.1007/978-3-030-45093-9_5

1   Traditional Approaches on Available Quantitative Data

Typical approaches seek to identify patterns and correlations in data to predict future developments. Approaches to conflict research, hence, rely on data representing the past and succeed best in predicting events that have already occurred. Readily available public datasets contain various observable measures and are, thus, applied to numerous questions in conflict research. Countries or other geographical entities are described by macroeconomic and social parameters such as gross domestic product or ethnic diversity [6].

Event data, on the other hand, contains information on event dates, involved actors and event type and is often used to operationalize the dependent variable [10].

2   Extension of the Data Base to Measure Additional Information

Given that not all potentially relevant parameters can be included in available data sets, researchers are including other sources to fill the blind spots. Natural language processing approaches, for instance, allow for the structuring of text as well as for social media and news data to be joined with other data in order to find relevant correlations and maximize the usefulness of each separate input beyond the accumulation of their respective contributions.

We suggest to follow the same path in including data that is derived from qualitative information in order to factor in dependencies that would otherwise be neglected.

2.1   Example for the Meaningful Recording of Structure, Learned from Data

While those approaches factor in both historic events and correlating indicators, they arguably fall short of valuing (hidden = not easily quantifiable) structural dependencies like relations between political decisions and the onset of crises or political links between actors that are otherwise not connected.

Mahoney et al. [8] apply Bayesian network structure learning with political science variables to produce meaningful priors and estimate the likelihood of *Events of Interest (EOI)*, such as Rebellion, International Crisis, Insurgency, Ethnic Religious Violence and Domestic political crisis.

The Bayesian aggregator outperforms all other compared models for every EOI [8].

2.2   Solution Spaces in Consideration of Qualitative Information

In public perception, the most visible forecasters of political events are experts whose opinions and predictions attract media attention [2]. Experts often provide valuable insights into a region or the relations between countries.

We find relevant work i.e. by Johansen [5], who has also applied scenario modelling to questions of conflict research. He chose a top down approach to model a consistent solution space with four parameters ("Actor", "Goal", "Method" and "Means"), where each parameter is defined in terms of an exhaustive set of possible states or values. He thereby constructs a model that derives structure solely from qualitative sources [5].

However, expert knowledge alone shows only mediocre results with regards to predictive power [12]. Nevertheless, expert opinions are valuable. Tetlock and Gardner [13] claim that a careful iterative selection of human predictors, for example via a tournament, has shown that they can produce consistently better forecasts than a wide range of competitors – including some of the most sophisticated algorithms [13].

2.3 Bayes Network Combined with Expert-Specified Nodes that Cannot Be Found in Data

Therefore, we hypothesize that the practical use of conflict prediction models needs a structural causally interpretable component and that not all relevant variables are (a) measurable or (b) available in data. We propose that the modelling of such latent variables contributes to both model performance and interpretability.

At the very least, the decision process in a crisis situation needs to assess the potential impact of proposed actions and consider the causal chain from action to impact.

3 Hybrid Approach

Bayes networks are already well established for a combination of expert and empirical observations as data into a prediction system. Current methods however limit the influence of expert knowledge to a-priori distributions and model design (see Fig. 1).

A hybrid approach for identifying the structure of a Bayesian network model, for instance, is presented in [4] by Huang et al. The authors propose a Bayesian network model for assessing threats of mass protests, using historical case data.

For a review of inference algorithms for hybrid Bayesian networks refer to [11].

The proposed research extends existing empirical approaches with qualitative expert knowledge by modeling a causal graph using expert opinion (structure + nodes) and learning the model using available data for all nodes plus combining this with hybrid virtual data based on expert scenario feedback. This is achieved by using a combination of empirical data and *Monte-Carlo scenario generated data* for a-postiori updates. To quantify multidimensional probability distributions implicitly encoded in qualitative expert knowledge, we develop a planning tool that allows for the likelihood weighting of different scenarios. We assume that experts are unable to easily quantify multidimensional distributions, but

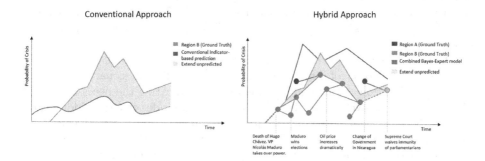

Fig. 1. Advantages of the new approach.

Disclaimer: The following data may not be sound in all detail and are certainly not complete. They are added for illustration purposes only.

7.10.2012: Death of Hugo Chávez. VP Nicolás Maduro takes over power. The legality of this action is disputed.

14.04.2013: Maduro won the elections by 50.6% Opposition claimed election irregularities.

February & March 2014: Violent protests against Maduro government. More than 40 fatalities and over 700 injured. Venezuela accuses Panama of conspiracy with the USA and breaks off all diplomatic and economic relations.

since 2014: Beginning of supply crisis. People are afraid of food shortage. Government accuses "extreme right opposition" and arrests Caracas mayor Antonio Ledezma.

6.12.2015: Parliamentary Elections. Opposition "Mesa de la Unidad Democrática" (MUD) wins by 56.3% (socialist Party of Maduro "Partido Socialista Unido de Venezuela" 40.9%)

since February 2016: Oil price increases dramatically; mineral water and medicines are scarce. Due to the high national debt, foreign countries no longer supply grain.

July 2016: Venezuela opens the border to Colombia for 12 hours to let people buy food and medicine.

September 2016: Maduro prevents a referendum on new elections. Massive protests follow.

9 January 2017: Parliament declares the President dismissed in the hope of new elections.

9 March 2017: the Supreme Court waived the immunity of all parliamentarians, withdrew all powers from parliament and transferred them to itself. Massive protests with several fatalities.

February 2018: Maduro orders a major manoeuvre, calling in one million military, militia and government officials for an alleged threat from Colombia.

20 May 2018. Presidential elections, which were internationally (EU, USA and several latin American countries) not recognized.

January 2019: Juan Guaidó declares himself interim president. USA, Ecuador, Colombia, Brasil are the first to recognize Guaidó as president.

have a strong intuition on the plausibility of claims. We present experts with scenario claims and receive a probability weighted feedback from the expert (e.g. "I'm sure this is a likely cause of events" or "I guess it might work in that direction").

We collect expert know-how about observed and un-observed variables each with a selected sub-set of dimensions (variables) in different scenarios (this will give us a calibration of the qualitative categories indicating the necessary probability weighting). We include directly connected and indirectly (a "hidden variable" in between) connected observations, thus allowing for the joint collection of virtual observations for non-connected nodes/variables. Based on an initial set of empirical observations and expert feedback we use active-learning methods to determine what aspects need to be further probed. This results in a Bayes network encoding the combination of empirical observations and expert knowledge. We then evaluate the performance of the model against past conflicts.

4   Data Sources

Since the suggested method relies on the combination of the approaches that the quantitative and qualitative viewpoint respectively excel in, data from both schools will be factored in.

Relevant data sources regarding quantitative data are event data found in ACLED as well as indicator data that is derived from UN, World Bank and other established sources. These sources provide data on individual events (e.g. assault, demonstration, cattle theft, ...) and data on Economic, cultural and socio-political indicators (e.g. child mortality, political representation, food prices, availability of drinking water, ethnic distribution) that influence crises (see also Fig. 2).

Experts on the matter of interest will be drawn from the *Center for Intelligence and Security Studies (CISS)*. CISS has a wide network of political scientists, military analysts and computer scientists, especially in the field of algorithms in cyber space and artificial intelligence. CISS also integrates its current research and findings into the academic master's programme MISS [1,3]; see also https://www.unibw.de/ciss/miss.

5   Implementation

Deloitte currently has an in-house tool for Bayesian-graphical scenario modelling, which satisfies most (but not quite all) of the requirements for our project. For instance we can set up 'earthquake' model that is often discussed in the AI literature, as follows:[1]

---

[1] We have experimented with various syntaxes for various purposes; this is one designed for interfacing to a graphical front end. We can easily change this as appropriate.

Fig. 2. Crisis management based on different and highly heterogeneous data sources.

```
Initialize;

DefineNode Earthquake;

AssignValues Earthquake quake no_quake;
P quake 1 / 1000;

DefineNode Burglar;

AssignValues Burglar burglar no_burglar;
P burglar 0.001;

DefineNode Radio;

AssignValues Radio announce no_announce;

DefineParents Radio Earthquake;
P announce quake 1.0;
P announce no_quake 0.0;

DefineNode Alarm;
AssignValues Alarm alarm no_alarm;

DefineParents Alarm Earthquake Burglar;
```

```
P alarm     burglar    quake 0.9901099;
P alarm no_burglar     quake 0.01099  ;
P alarm no_burglar no_quake 0.001     ;
P alarm     burglar no_quake 0.99001  ;

DefineNode Phonecall ;
AssignValues Phonecall phonecall no_phonecall;
DefineParents Phonecall Alarm;

P phonecall    alarm 1.0;
P phonecall no_alarm 0.0;
```

Currently, we can evaluate conditional and unconditional marginal queries of the model of the form

> Given that Fred has been told that the burglar alarm has gone off, what is the probability that there is a burglar,

or

> Given that Fred has been told that the burglar alarm has gone off, and the radio reports an earthquake, what is the probability that there is a burglar,

and even evaluate interventions (see [9])

> Given that Fred himself has deliberately triggered the burglar alarm, what is the probability that there is a burglar (!).

We are also able to complete models using maximum entropy.[2] What our package lacks, at the moment, is a facility to attach historical data, which can condition the probabilities of the various nodes. Fortunately, we have implemented the package in an advanced and flexible development environment that supports rapid and error-free modification. This means that experimental extensions are easy to program. We are thus confident that, at a technical level, the required new functionality is a straightforward matter of software engineering: powerful and general Gibbs samplers appropriate for this sort of problem (see [7]) are available as open source (e.g. Jags), so we can simply bind them to our system, no particular effort is required of us.

In fact we have previous implementation experience we can draw on with *ad hoc*-models of this sort (in financial risk). This leaves us free to deal with the real challenge of the project, which is not technical in this implementation sense: to develop an effective method and specific concepts/notation with which to bind historical information to the model in a way that most effectively supports our planned risk analysis as proposed in this paper.

---

[2] In an experimental extension!

Acknowledgement. We thank the anonymous reviewers for greatly improving the paper. Likewise, we thank Jonas Andrulis and Samuel Weinbach for influential discussions on the topic and their contributions to an earlier version of the paper.

References

1. Borghoff, U.M., Dietrich, J.-H.: Intelligence and security studies. In: Bode, A., Broy, M., Bungartz, H.-J., Matthes, F. (eds.) 50 Jahre Universitäts-Informatik in München, pp. 113–121. Springer, Heidelberg (2017). https://doi.org/10.1007/978-3-662-54712-0_9
2. Chadefaux, T.: Conflict forecasting and its limits. Data Sci. **1**, 7–17 (2017)
3. Corvaja, A.S., Jeraj, B., Borghoff, U.M.: The rise of intelligence studies: a model for Germany? Connect. Q. J. **15**(1), 79–106 (2016). https://doi.org/10.11610/Connections.15.1.06
4. Huang, L., Cai, G., Yuan, H., Chen, J.: A hybrid approach for identifying the structure of a Bayesian network model. Expert Syst. Appl. **131**, 308–320 (2019)
5. Johansen, I.: Scenario modelling with morphological analysis. Technol. Forecast. Soc. Chang. **126**, 116–125 (2018)
6. Kim, H.: Patterns of economic development in the world. J. Glob. Econ. **2**(113), 1–8 (2014)
7. Koller, D., Friedman, N.: Probabilistic Graphical Models: Principles and Techniques. The MIT Press, Cambridge (2009)
8. Mahoney, S., Comstock, E., de Blois, B., Darcy, S.: Aggregating forecasts using a learned Bayesian network. In: Proceedings of the 24th International Florida Artificial Intelligence Research Society Conference, pp. 626–631 (2011)
9. Pearl, J.: Causality: Models, Reasoning and Inference. Cambridge University Press, New York (2009)
10. Raleigh, C., Linke, A., Hegre, H., Karlsen, J.: Introducing ACLED-armed conflict location and event data. J. Peace Res. **47**(5), 651–660 (2010)
11. Salmeron, A., Rumi, R., Langseth, H., Madsen, A.L., Nielsen, T.D.: A review of inference algorithms for hybrid Bayesian networks. J. Artif. Intell. Res. **62**, 799–828 (2018)
12. Tetlock, P.E.: Expert Political Judgment. Princeton University Press, Princeton (2005)
13. Tetlock, P.E., Gardner, D.: Superforecasting. The Art and Science of Prediction. Random House, New York (2016)

Model Based Design of Inductive Components - Thermal Simulation and Parameter Determination

Simon Merschak[1][(✉)] , Mario Jungwirth[1] , Daniel Hofinger[2] ,
and Günter Ritzberger[2]

[1] University of Applied Sciences Upper Austria, Wels Campus,
Stelzhamerstrasse 23, 4600 Wels, Austria
{simon.merschak,mario.jungwirth}@fh-wels.at
[2] Fronius International GmbH, Research and Development,
Günter Fronius Strasse 1, 4600 Wels-Thalheim, Austria
{hofinger.daniel,ritzberger.guenter}@fronius.com
https://www.fh-ooe.at
https://www.fronius.com

Abstract. To ensure a faultless operation of an inductive component
and to optimize the efficiency, the thermal behavior of the component
has to be considered in an early design phase. In a previous project, a
computer aided design tool for inductors was developed. In this tool thermal network models are used to calculate the thermal behavior. Therefore thermal resistance values of all winding materials are required and
have to be measured. A measurement setup for the measurement of thermal resistances was developed, built and tested. In this paper the used
measurement principle and the consideration of systematic measurement
errors are described. A finite element method (FEM) simulation of the
measurement setup is used to verify the calculation of the thermal resistance values.

Keywords: Thermal resistance measurement · Thermal network
model · Thermal losses

1 Introduction

Inductive components like chokes and transformers are essential components of
modern power electronic systems. In order to ensure a faultless operation of
the system components and in order to optimize their efficiency, the thermal
behavior of these components has to be considered in an early design phase. To
reduce the time to market of such systems, a computer aided design tool for
inductors was developed in a previous project. This easy to use tool is able to
predict the physical behavior of inductors in operation and enables the designer
to optimize all significant design parameters in a short time. The tool can also
be used to simulate the thermal behavior of the component by use of thermal

© Springer Nature Switzerland AG 2020
R. Moreno-Díaz et al. (Eds.): EUROCAST 2019, LNCS 12013, pp. 44–51, 2020.
https://doi.org/10.1007/978-3-030-45093-9_6

network models. The input parameters for the thermal evaluation have to be determined through measurements. Therefore a novel measurement setup for the determination of thermal resistances in inductive components was constructed, simulated and tested.

2   Modeling Process

Various design requirements and an extensive variety of assembly components make inductor design a complex process. In previous publications an easy-to-use software tool was presented to determine suitable component parameters for optimized inductor designs [1]. Main features of the tool are a comprehensive database of all inductor assembly parts, a detailed calculation routine for the determination of core and winding losses, including a finite element algorithm to simulate eddy current losses in massive and litz wire windings and an extensive post processing package [2]. The thermal behavior of inductive components can be modeled by use of thermal network models with different levels of complexity. Currently a 2D thermal network simulation model is implemented in the design tool. Precise thermal resistance values for various winding structures are required to conduct the simulation.

2.1   Thermal Network Model

Thermal network models can be used to simulate the heat transfer in materials, which is the conduction, and the heat transfer between different materials, which is radiation and convection. Simple thermal network models are displayed in Fig. 1. As displayed in (1) and (2), a thermal network model is equivalent to an electric system. The heat flux $\dot{Q}$ is equivalent to the electric current I, the temperature difference is equivalent to the voltage and the thermal resistance is equivalent to the ohmic resistance [3]. Therefore software which is able to simulate electric circuits can also be used to simulate a thermal network model. Different winding materials have different thermal resistances. The exact value of the thermal resistance of the winding material is required to simulate the thermal network. These values are not included in the datasheet of the winding material and therefore they have to be measured.

$$\dot{Q} = \frac{\Delta T}{R_{th}} \tag{1}$$

$$I = \frac{U_{el}}{R_{el}} \tag{2}$$

2.2   Calculation of Thermal Resistance and Thermal Conductivity

For a cubic sample there are two ways to calculate the thermal resistance. It can be calculated with the temperature difference and the heat flux or it can

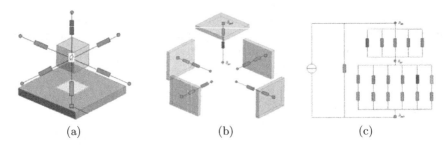

Fig. 1. Examples of simple thermal network models.

be calculated with the geometry parameters and the thermal conductivity (3). If both equations are combined, an equation for the calculation of the thermal conductivity can be formed (4).

$$R_{th} = \frac{\Delta T}{\dot{Q}} = \frac{l}{\lambda \cdot A} \tag{3}$$

$$\lambda = \frac{\dot{Q} \cdot l}{\Delta T \cdot A} \tag{4}$$

In the same way the thermal conductivity of a hollow cylinder can be calculated. With the used measurement setup, the winding has the form of a hollow cylinder and therefore (5) can be used for the calculation of the thermal conductivity of the winding [4]. For the calculation, the inner and the outer radius of the winding is needed. Further the heat flux provided from the heating cartridges has to be determined and the temperature difference between the inside and the outside of the winding has to be measured. With these values the thermal conductivity of the winding material can be calculated (Fig. 2).

$$\lambda = \frac{\ln \frac{R_a}{R_i}}{2 \cdot \pi \cdot h \cdot \frac{\Delta T}{Q}} \tag{5}$$

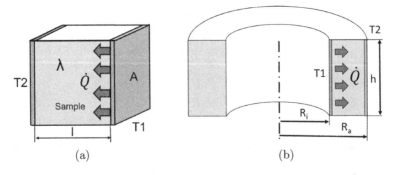

Fig. 2. Heat transfer through cubic sample (a) and hollow cylinder (b).

3   Measurement of Winding Parameters

In order to build up a database of thermal resistances for different winding materials and winding structures a novel measurement setup [5] was constructed. The measurement setup serves to measure the thermal resistance in the radial direction of wound inductors. Test materials can be various electrical conductors or insulation materials. The setup consists of a cylindric aluminium core with heating cartridges inside. At the top and bottom of the aluminium cylinder insulating plates are mounted to guide the main heat flux in radial direction through the winding. If the introduced heat flux $\dot{Q}$ is known and the temperature is measured at various points inside the winding structure, the thermal resistance R_{th} and subsequently the equivalent heat conductivity can be determined according to (6).

$$R_{th} = \frac{\Delta T}{\dot{Q}} = \frac{T_1 - T_2}{\dot{Q}} \tag{6}$$

3.1   Measurement Setup

A measurement setup for the determination of the thermal resistance of winding materials was built. It is displayed in Fig. 3. This measurement setup consists of an aluminum core which contains two 12 V heating cartridges and a bimetal switch to prevent the measurement setup from overheating. If the temperature value is rising above 150 °C, the bimetal switch deactivates the heating cartridges. At the top and at the bottom of the aluminum cylinder there are insulation plates made from PES which has a very low thermal conductivity. The winding material is wound on the aluminum core between the two PES insulation plates. One of the PES insulation plates provides a connector which allows to place the measurement setup in a wire-winding machine to speed up the winding process.

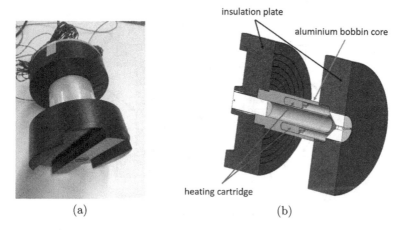

(a)   (b)

Fig. 3. Measurement setup (a) and position of the heating cartridges (b).

The schematic figure of the whole measurement setup is displayed in Fig. 4. A 12 V DC power supply is used to power two heating cartridges. The heating cartridges lead to a radial heat flux from the center of the winding to the outside of the winding. Inside the winding, thermocouples are placed. They are used to measure the temperature at different positions inside the winding structure. So a temperature gradient from the inside of the winding to the outside can be measured. A USB data acquisition card is used to record the temperature values. Finally, the thermal conductivity of the winding material is calculated from the temperature gradient in a LabVIEW program. A user interface was developed to set input parameters like the dimensions of the winding and to show the results.

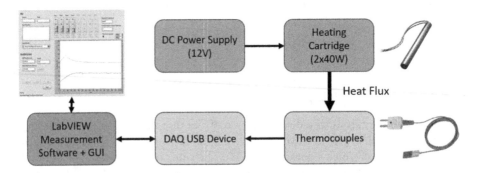

Fig. 4. Components of the thermal measurement setup.

3.2 Consideration of Heat Losses

In (6), it is only allowed to consider the radial heat flux $\dot{Q}_{rad}$ through the winding material. If the provided heat flux from the heating cartridges $\dot{Q}_{total}$ would be used for the calculation of the thermal resistance, the result would not be exact because there are heat flux losses $\dot{Q}_{axi}$ in axial direction.

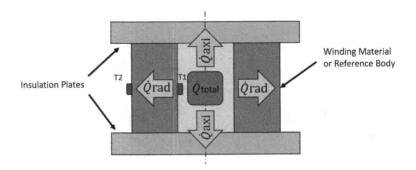

Fig. 5. Heat losses of the measurement setup.

In Fig. 5 the heat losses at the bottom and at the top of the heated core can be seen. It is important to mention that the measurement of the temperature difference is not started until the heating process is finished and the steady state is reached. To get the right values for the thermal conductivity, it is important to compensate the systematic measurement error due to the heat losses in axial direction. In order to determine the heat flux losses in axial direction, a reference body with known thermal conductivity λ is used instead of the winding material. The thermal conductivity of the reference body should be close to the thermal conductivity of the real winding material.

The temperature difference between the inner and the outer radius of the reference body is measured and the total heat flux from the heating cartridges is also known. So the heat flux losses in axial direction can be calculated with (8) as the difference between the total heat flux and the measured radial heat flux (7).

$$\dot{Q}_{rad} = \frac{\lambda \cdot 2 \cdot \pi \cdot h \cdot \Delta T}{\ln \frac{R_a}{R_i}} \tag{7}$$

$$\dot{Q}_{axi} = \dot{Q}_{total} - \dot{Q}_{rad} \tag{8}$$

3.3   FEM Simulation

In order to get an estimation for the axial heat flux loss, the heat transfer in the measurement setup was simulated with the FEM simulation software Comsol Multiphysics. The 3D geometry of the simulation model is displayed in Fig. 6.

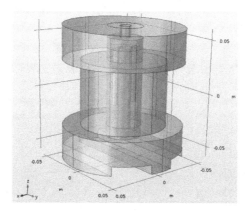

Fig. 6. Simulation model in Comsol Multiphysics.

In this stationary simulation, two heating cartridges are modeled as heat sources with a total heating rate of 10 W. The heat transfer from the measurement setup to the surrounding air is modeled as heat flux boundary condition.

In a first evaluation, the temperature distribution along the blue line in Fig. 7(a), inside the reference body, is evaluated for different thermal conductivity values. The resulting temperature distributions are displayed in Fig. 7(b). The heating rate of 10 W is kept constant for all simulations. The correlation between the temperature drop inside the winding and the value of the thermal conductivity can be seen in Fig. 7(b). This correlation is described by (5).

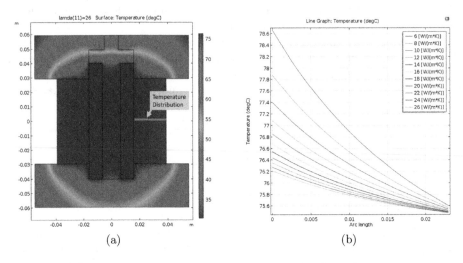

(a)   (b)

Fig. 7. Temperature distribution in the reference body for thermal conductivity ranging from 6 [W/(m·K)] to 26 [W/(m·K)].

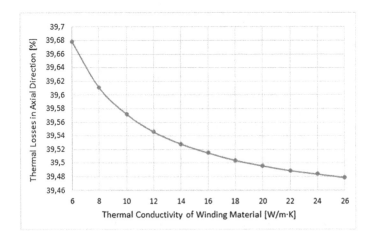

Fig. 8. Dependency of the axial heat flux losses on the thermal conductivity of the winding material.

The influence of the reference body's thermal conductivity on the axial heat flux losses is displayed in Fig. 8. In a range from 10 [W/(m·K)] to 20 [W/(m·K)], which applies for most winding structures, the variation of the axial heat flux losses due to the thermal conductivity change is only 0.076% and can be neglected.

4 Conclusion and Future Prospects

A novel measurement setup for the determination of thermal resistances of winding materials was developed, built and tested. It will be used to create a database of thermal resistances. In order to obtain correct measurement results, the influence of axial heat flux losses in the measurement setup had to be considered. A 3D FEM simulation was conducted to determine the axial heat flux losses and their dependency on the thermal conductivity of the winding material. For the analyzed measurement setup, the heat flux losses in axial direction were about 39.5% of the total supplied heat flux and the influence of thermal conductivity variations on the heat flux losses is negligible.

In a next step, measurements of a reference body with known thermal conductivity will be carried out to validate the simulation results. In further consequence an optimized layout of inductive components regarding thermal behavior can be archived by use of the new measurement data.

References

1. Jungwirth, M., Hofinger, D., Eder, A., Ritzberger, G.: Model based design of inductive components - a comparision between measurement and simulation. In: Moreno-Díaz, R., Pichler, F., Quesada-Arencibia, A. (eds.) EUROCAST 2015. LNCS, vol. 9520, pp. 546–551. Springer, Cham (2015). https://doi.org/10.1007/978-3-319-27340-2_68
2. Rossmanith, H., Albach, M., Patz, J., Stadler A.: Improved characterization of the magnetic properties of hexagonally packed wires. In: Proceedings of the 14th European Conference on Power Electronics and Applications, Birmingham, pp. 1–9. IEEE (2011)
3. Stephan, P., Kabelac, S., Kind, M., Mewes, D., Schaber, K., Wetzel, T. (eds.): VDI-Wärmeatlas. SRT. Springer, Heidelberg (2019). https://doi.org/10.1007/978-3-662-52989-8
4. Herwig, H., Moschallski, A.: Wärmeübertragung. Springer, Wiesbaden (2014). https://doi.org/10.1007/978-3-658-06208-8
5. Jaritz, M., Biela, J.: Analytical model for the thermal resistance of windings consisting of solid or litz wire. In: 15th European Conference on Power Electronics and Applications (EPE), Lille, pp. 1–10 (2013). https://doi.org/10.1109/EPE.2013.6634624

Predictive Model of Births and Deaths
with ARIMA

Diana Lancheros-Cuesta[1]([envelope]) [ORCID], Cristian Duney Bermudez[1] [ORCID],
Sergio Bermudez[1] [ORCID], and Geovanny Marulanda[2] [ORCID]

[1] Universidad Cooperativa de Colombia, Bogotá -Colombia Grupo de Investigación
Automatización Industrial, Bogotá, Colombia
{diana.lancheros,cristian.bermudez,sergio.bermudez}campusucc.edu.co
[2] Universidad Pontificia Comillas, Madrid, Spain
geovanny.marulanda@iit.comillas.edu

Abstract. The paper show the application of the ARIMA (Autoregressive integrated moving average) prediction model is made, which consists of the use of statistical data (in this case, birth and deaths in Colombia) to formulate a system in which an approximation of future data is obtained, this thanks to the help of a statistical software that allows us to interact with the variables of this model to observe a behavior as accurately as possible, the research source was extracted from the statistical data provided by the national statistical department, articles about cases in the that this method was used, and statistical texts. The development of the research was carried out observing the statistics provided by national statistical department, creating a database of births and deaths in Colombia per year, taking into account total figures at the national and departmental levels. Thanks to these data, a tool such as software and prior knowledge of the predictive model ARIMA (Autoregressive integrated moving average) is achieved to make an approximate prediction of what could happen in a given time, thus taking the measures required by each department, solving possible problems in the country.

Keywords: ARIMA · Prediction · Statistics · Correlation · Time series

1 Introduction

Nowadays, all information that leads to a solution of a social problem, where resources utilization is improved, needs to be considered. In this case, the health of Colombian people from different parts of the country. The paper show a results of research project that consists a prediction using an ARIMA [3] For this project, it was vital to have a previous knowledge of the ARIMA (Autoregressive integrated moving average) model, due to the importance in the selection of the adjustment variables for the prediction. The next step was to obtain

Supported by Universidad Cooperativa de Colombia.

R. Moreno-Díaz et al. (Eds.): EUROCAST 2019, LNCS 12013, pp. 52–58, 2020.
https://doi.org/10.1007/978-3-030-45093-9_7

the data from the last 10 years of births and non-fetal deaths by departments, so an organized data-based could be created. This data was downloaded from the website of National Bureau of Statistics [1], which are completely updated. Then, a software was required to interact with the variables of the statistical model and the data. The data obtained from the website of National Bureau of Statistics [1] and the data obtained as an output from the software were stored properly. The difference between them were observed, making emphasis in the errors of the outputs, which were in the minimum and maximum range for the model to be valid, in order to make a guess. These predicted values allow to solve problems that depend on calculations, such as resources management and places where these are required the most. This fact is very important because the analyzed data were sorted as national and by departments as well, being more relevant so the problematic could be identified and addresses quickly. Close approximations in the years of study were obtained from the general data. ARIMA (Auto-regressive integrated moving average) system allows to make a prediction, in a very accurate way, with statistical data that depend of a time series [3]. The larger the time series, the larger is the amount of information collected and it is easier to learn from the errors while analyzing the data. The best strategy to analyze a time series is to have a very specific data. When the data is generalized, errors are not taken into account, and if there are big errors, the model may tend to be inaccurate. The system allows to interfere the data, predicting aspects that may become a change factor for the model. In previous researches, the ARIMA (Auto regressive integrated moving average) model is applicable for cases that are related with health issues [2,4] and cases that require an optimal forecasting for a resources management [5,7,8] or for population control, as seen in this project. This model fits properly with the data used in this research, and shows a valid prediction that presents a low percentage error, so the decision making is viable from now on.

2   Method

It is common that time series evolve easily due to the influence of past facts over present ones, creating a dependency between their present values and past values. It can be said that when observations from an evolving variable through time are taken, there is a relation between those observations in different moments of time. The analysis, characterization and use of the dependency between observations in different moments of time makes possible the development of prediction models based on the past of the studied variable [3,6]. The ARIMA model is an autoregressive (AR) integrated (I) moving average (MA), where each of these components has a role to generate a prediction in a time series. The AR component means that the random perturbation of a model may have a behavior that depends on a previous time [3].

$$y_t = \mu + \theta_1 y_{t-1} + a_t \tag{1}$$

where y_t is the studied variable (births, deaths, etc.), depends on μ (independent variable) plus a parameter times $\theta_1 y_{t-1}$ (the variable from 1 period ago) plus another random perturbation. The MA component is where the mean and the observed errors within previous periods are taken into account.

$$y_t = \mu + a_t + \theta_1 a_{t-1} \tag{2}$$

where μ is a constant, a_t the observed errors and θ the moving average parameter. The required data for the development of the project, births and deaths in Colombia in the last few years, are collected. It is important to mention that the website gives the option to choose which data to download by year. A database is created in Excel with all the downloaded information from the website of DANE, sorting the data by year and department, considering the total national. To develop the statistical model, a previous study of the composition and behavior of the model is made. With this information, the necessary conditions for the model to forecast accurate predictions are determined. In addition, it is required that the ARIMA model adapts itself to the data without any problem. The chosen software (XLSTAT) allows to implement a large amount of tools to perform a statistical model with views, analysis and data modelling. It offers an adjustable interface, so the model parameters can be changed depending on what needs to be observed, and the possibility to compare the variables before and after applying the statistical model.

3   Results

The data obtained from the website of DANE and the data obtained as an output from the software were stored properly. The difference between them were observed, making emphasis in the errors of the outputs, which were in the minimum and maximum range for the model to be valid, in order to make a guess. These predicted values allow to solve problems that depend on calculations, such as resources management and places where these are required the most. This fact is very important because the analyzed data were sorted as national and by departments as well, being more relevant so the problematic could be identified and addresses quickly. Close approximations in the years of study were obtained from the general data. Here is where the sectioned data is appropriate for the analysis of the country, due to a precise behavior and sometimes an accurate or exact data in births and non-fetal deaths. For each of the departments of predictive analysis model data is performed. Figure 1 shows an example of data analysis of a department in Colombia called Vichada.

It becomes apparent that the behavior of the data from the departments can be predicted in a very accurate way, close to the real data. This is because the ARIMA model has a component that learns from errors through time. Also, these data are not combined with other information that may change the perspective of the analysis and observations. The Fig. 2 show data comparison chart between real births and data ARIMA outputs for 2016 for eleven departments en Colombia. The errors between the real data and the ARIMA model are show Table 1.

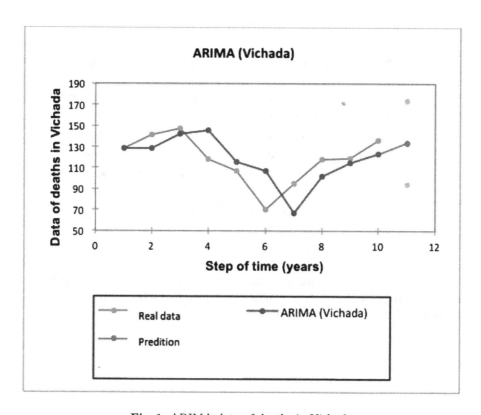

Fig. 1. ARIMA data of deaths in Vichada

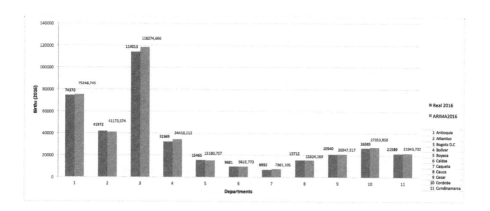

Fig. 2. Comparison real date of births - ARIMA

Table 1. Table Errors between the real data and the ARIMA model - births

Real data (2016)	Date ARIMA (2016)	Department	Error
74370	75248,745	Antioquia	−1,182
41972	41173,574	Atlantico	1,902
114013	118274,666	Bogota D.C	−3,738
32369	34418,112	Bolivar	−6,330
15465	15180,757	Boyaca	1,838
9681	9615,773	Caldas	0,674
6932	7361,105	Caqueta	−6,190
15712	15624,169	Cauca	0,559
20540	20947,517	Cesar	−1,984
26385	27353,850	Cordoba	−3,672
21589	21943,732	Cundinamarca	−1,643

The Fig. 3 show data comparison chart between real deaths and data ARIMA outputs for 2016 for eleven departments en Colombia. The errors between the real data and the ARIMA model are show Table 2.

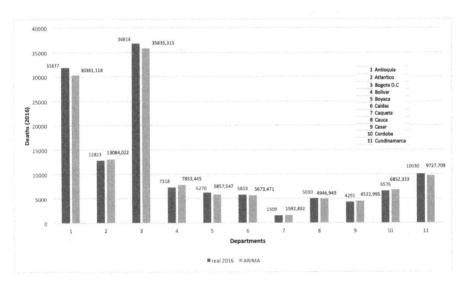

Fig. 3. Comparison real date of deaths - ARIMA

Table 2. Table Errors between the real data and the ARIMA model- deaths

Real data (2016)	Date ARIMA (2016)	Department	Error
31837	30361,118	Antioquia	4,635745617
12823	13084,022	Atlantico	−2,035574924
36814	35835,315	Bogota D.C	2,658458614
7318	7853,445	Bolivar	−7,316827426
6270	5857,547	Boyaca	6,578201186
5833	5673,471	Caldas	2,734941524
1509	1592,832	Caqueta	−5,555466835
5030	4946,949	Cauca	1,651120778
4291	4532,995	Cesar	−5,639601106
6576	6852,333	Cordoba	−4,202149733
10030	9727,709	Cundinamarca	3,013870339

4   Conclusions and Future Work

ARIMA model allows to make a prediction, in a very accurate way, with statistical data that depend of a time series. The larger the time series, the larger is the amount of information collected and it is easier to learn from the errors while analyzing the data. The best strategy to analyze a time series is to have a very specific data. When the data is generalized, errors are not taken into account, and if there are big errors, the model may tend to be inaccurate. The system allows to interfere the data, predicting aspects that may become a change factor for the model. Like in previous researches, the ARIMA model is applicable for cases that are related with health issues and cases that require an optimal forecasting for a resources management or for population control, as seen in this project. This model fits properly with the data used in this research, and shows a valid prediction that presents a low percentage error, so the decision making is viable from now on. As future work, validations and comparisons are made with other prediction models.

References

1. de la Republica de Colombia, P.: Dirección Nacional de Estadística. http://www.dane.gov.co/
2. Cortes, F., et al.: Time series analysis of dengue surveillance data in two Brazilian cities. Acta Tropica **182**, 190–197 (2018). https://doi.org/10.1016/j.actatropica.2018.03.006
3. Brillinger, D.R.: Time Series - an overview — ScienceDirect Topics. Elsevier (2001). https://www.sciencedirect.com/topics/economics-econometrics-and-finance/time-series

4. He, Z., Tao, H.: Epidemiology and ARIMA model of positive-rate of influenzaviruses among children in Wuhan, China: a nine-year retrospectivestudy. Int. J. Infect. Dis. **74**, 61–70 (2018). https://doi.org/10.1016/j.ijid.2018.07.003. www.sciencedirect.com/science/article/pii/S1201971218344618

5. Moon, T., Shin, D.H.: Forecasting construction cost index using interrupted time-series. KSCE J. Civil Eng. **22**(5), 1626–1633 (2018). https://doi.org/10.1007/s12205-017-0452-x

6. Nury, A.H., Hasan, K., Alam, M.J.B.: Comparative study of wavelet-ARIMA and wavelet-ANN models for temperature time series data in northeastern Bangladesh. J. King Saud Univ. - Sci. **29**(1), 47–61 (2017). https://doi.org/10.1016/j.jksus.2015.12.002

7. Torbat, S., Khashei, M., Bijari, M.: A hybrid probabilistic fuzzy ARIMA model for consumption forecasting in commodity markets. Econ. Anal. Policy **58**, 22–31 (2018). https://doi.org/10.1016/j.eap.2017.12.003. http://www.sciencedirect.com/science/article/pii/S031359261730067X

8. Wang, C.C.: A comparison study between fuzzy time series model and ARIMA model for forecasting Taiwan export. Exp. Syst. Appl. **38**(8), 9296–9304 (2011). https://doi.org/10.1016/j.eswa.2011.01.015. http://www.sciencedirect.com/science/article/pii/S0957417411000352

Automated Load Classification in Smart Micro-grid Systems

Thomas Schlechter[(✉)]

University of Applied Sciences Upper Austria, 4600 Wels, Austria
thomas.schlechter@ieee.org

1 Introduction

The ever evolving idea of smart grid substantially relies on the usage of renewable energies. The latter, however, have to cope with a variety of problems. E.g., the production of renewable energies like photovoltaic and wind energy depends on the weather, leading to a time varying energy output. Additionally, it is a complex task to save surplus energy. If the geographical position allows the installation of a power station using a reservoir or similar, the problem can be solved easily. However, in most cases this is not possible. Concepts have to be developed fulfilling this task.

Furthermore, one of the hardest challenges of any power system is to dynamically balance the demand and the supply of energy at any given point in time. In this context evolves the next major problem in smart microgrids, which is the task of load balancing. This issue is more challenging in smart microgrids than in main distribution grids. Opposite to smart microgrids, main distribution grids are typically significantly over-dimensioned regarding both the transmission capacity and their energy production flexibility. Therefore, big load changes can easier be compensated for in such systems. Contrary, load changes in smart microgrids can become huge relative to the overall system performance. It is a challenging task to handle this problem knowing certain physical implications to valid load changes [1].

At this point of load balancing, an additional problem comes into the play. To perform an efficient load balancing cycle, a method has to be found, which is able to decide, what type of device is connected to a certain plug within the smart microgrid - either manually or automatically. In this context it can be observed, that each device has a characteristic power consumption profile, both in a short-time window approach and also in a long-time window approach. The main focus of this work is to derive appropriate approaches fulfilling those requirements. Additionally, existing approaches will be rated against the proposed ones.

The paper is structured as follows. After a more deep problem statement description in Sect. 2, a summarized proposal for algorithms to be applied will be given in Sect. 3. Finally, Sect. 4 concludes the paper.

© Springer Nature Switzerland AG 2020
R. Moreno-Díaz et al. (Eds.): EUROCAST 2019, LNCS 12013, pp. 59–66, 2020.
https://doi.org/10.1007/978-3-030-45093-9_8

2 Problem Statement

2.1 Origin of the Problem

One issue coming in line with the mentioned problem is, that future smart grids will not only have to deal with the production of energy. Moreover, due to the decentralized structure of the distributed renewable energy sources, the integration of information and communication technologies is inevitable for the management and control of the whole system [2]. The starting point of data to be collected is localized within everyone's household. For this purpose smart meters will be installed in every consumer environment. The latter may result in individual consumers becoming aware of their power consumption and potentially sensitized for a more responsible usage behavior. However, the collected data has to be submitted to higher order instances in the smart grid and further processed there. Based on an evaluation, certain actions can be undertaken. To break down this management problem to less complex sub-problems, in [3] the introduction of a sub-system is envisioned, called a smart microgrid. Those consist of energy producers and consumers at a small scale, which are able to manage themselves in certain limits. From an external point of view, the smart microgrid can be considered a black box. To reduce the data exchange demand to the outside world, only important information, like energy surplus or lack of energy within the microgrid is reported. If the microgrid is connected to a high-voltage grid, then energy can be exchanged. If the microgrid is running in the socalled "island mode", then energy deficiencies need to be handled within the individual microgrid. Dynamical islanding is one of the main solutions to overcome faults and voltage sags, as described in [4].

2.2 Consumer Integration and Home Automation

One of the hardest challenges of any power system is to dynamically balance the demand and the supply of energy at any given point in time. From a technical point of view, the installation of devices, being able to provide a collection of data concerning current energy demand and supply and also communicating them amongst additional such devices in a complex network, is the key point.

Technical Implementation Challenges. Typically, there are two possibilities for points of view on this problem:

1. Let the consumer be an active part of the system.
2. Automate the power supply system in each household.

The two are not mutually exclusive. For example, the consumer provides information about his/her desires (heating on at 5 am, electrical car ready at 7 am, washing machine finished by 10 pm etc.). Based on those preferences, the home power supply system acts accordingly, being able to exchange information and also power with other smart microgrids within the main grid.

In literature, this technique is called Demand Response (DR) mechanism. The classic DR technique requires an active feedback from the individual consumer. The inputs given by the consumer might be based on personal preferences, market price observations provided by the installed system, etc. DR can, for example, be used to forecast future demands of a consumer by tracking his/her behaviour, the weather, etc., over a long period of time [5]. From the tracked data, consumption patterns can be developed, adopting the system behaviour to the consumer's demand. In this case, DR is used to automate the underlying system, providing home automation functionality. For future developments, the authors consider to integrate a real-time feedback option by the consumer. However, they admit that the learning algorithm then might become very complex. Either way, the consumers need to be able to use the individual appliances without any restriction [6]. As interface between the consumer and the system smart meters, sensors, digital control units and analytic tools will be used [7].

Most recent works combine the home automation approach with the classic DR approach to increase benefit for the consumers. For example, in [8] a concept is introduced, which introduces a priority list from of the user. The tasks desired by the user, like having a warm shower and providing warm water for the dishwasher, are listed according to personal preferences. Of course, a hot shower is required right now, the dishes can wait during this time. Therefore, if not both actions (heating water for the shower, providing warm water for the dishwasher) are supported at a certain point in time due to energy consumption peaks, the dishwasher will automatically be delayed in this example. More consumer-centric approaches are given in [9,10]. In those approaches, the user is iteratively informed about current possibilities to save energy, energy cost or both. Based on this information, the consumer might react accordingly. A by product of this approach is that the users can learn from their own behaviour and, in future, automatically adopt their actions, without considering the system's request.

Either way, the client feedback should be as simple as possible as to not overburden the consumer. Imaginable approaches might allow the user to specify certain possible time slots for specific tasks. The system takes care that all defined tasks will be finished within the given intervals, while optimal power usage over time is aimed as well. Mechanisms supporting this idea are, for example, introduced in [7,11]. Such approaches ensure that consumers keep control of their appliances and therefore the acceptance of the system increases. Suboptimal energy costs might be achieved only, however.

All the mentioned techniques and combinations of the latter have one thing in common: Each algorithm has to solve an optimisation problem based on the energy price on the market, the consumers demands (including a rating of the individual priority), the structure of the household and the available power supply. This optimization problem might become very complex. Efficient, intelligent algorithms need to be used.

The next major problem in smart microgrids is the task of load balancing. This issue is more challenging in smart microgrids than in main distribution grids. The reasons are twofold.

First, opposite to smart microgrids, main distribution grids are typically significantly over-dimensioned regarding both the transmission capacity and their energy production flexibility. Therefore, big load changes can easier be compensated for in such systems. Contrary, load changes in smart microgrids can become huge relative to the overall system performance. It is a challenging task to handle this problem howing certain physical implications to valid load changes [1]. In smart microgrids, typically mainly renewable energy sources will be used, which cannot be scaled in power generation very easily. While a spatial distribution of possible power generation processes can be named as a pro, the output of the energy source heavily depends on uncontrollable external conditions, like wind and daylight. Therefore, a smart microgrid will have to employ a load control strategy, which includes load shedding. To be able to implement such a load control strategy, a significant amount of the whole installation needs to be "smart" [12]. "Smart" here names the property to support load shedding or dispatching of their operation time. The most common techniques for load balancing can be identified as:

- Generator power control, where the output voltage of the generator is kept constant by applying intentional losses to droop the voltage [1],
- Demand dispatch, where the actual power consumption is shifted along the time axes based on the consumer preferences and desires [13],
- Load shedding, which is related to the demand dispatch method. However, this approach is used in more urgent cases, where low priority devices are switched of or lowered in performance, despite consumer preferences. Finding a valid set of devices affected by this method can be modeled by the Knapsack problem [14].
- Storage, where the energy is intermediately stored and can be used later on, when peaks in the net-load occur. However, storage in general is expensive (battery, hot water reservoirs, etc.) and might not be applicable today in many situations.

At this point of load balancing, an additional problem comes into the play. So far it has been assumed, that the connected devices are known by the smart device, for example, a fridge, dishwasher, water heating device. However, in a real scenario this is not the case. Old devices are not "smart" in a way needed to make the described approaches work. Still, they are required to cooperate with newly installed smart microgrid systems in an efficient way. Consumers cannot be forced to exchange all installed devices by new ones. For this reason, a method has to be found, which is able to decide, what type of device is connected to a certain plug within the smart microgrid. In this context it can be observed, that each device has a characteristic power consumption profile, both in a short-time window approach and also in a long-time window approach. For example, when a conventional light bulb is switched on, then the current flowing after a few microseconds is significantly lower, as the filament is heated and increases its resistance (short-time window). On the other hand, a dishwasher has cyclic behavior concerning energy consumption (for example, washing the dishes vs. drying the dishes). In the feature extraction models as approaches to this issue given in [15,16], the authors use FFT based implementations to identify the spectral distribution of the load's behavior by analyzing two main aspects of load switching behavior. First, low-frequency behaviour is studied, detecting significant observable load jumps in a system. Those might be caused by heaters, ovens etc. The second investigated behaviour is based on a harmonic analysis of the transients during a switching process. Using the latter observations allows for distinction between, for example, a light bulb and a computer monitor. Combining both methods allows a more precise detection of certain devices in a specific smart microgrid. However, the author of this paper is of the opinion, that this approach is not an appropriate one. The reason is, that the Fourier Transform (FT) has some major drawbacks. First, the magnitude spectrum of the FT does not deliver any information about the presence of certain frequency components over time. In the frequency domain representation a sinusoid, containing very high power but only being present for a very short time in the signal, cannot be distinguished from a sinusoid, containing low power but present over the whole observation period. The time dependency might be derived from the phase spectrum, but the computational effort for this approach is too high for real applications. Second, given a certain number of input samples, the resolution of the spectral estimate is always linear over the whole observation window. Investigating the spectrum of loads in a microgrid, however, should not show a time-invariant behaviour of the instantaneous spectral allocation of the load's frequency response. This effect occurs due to physical changes of the device as soon it is powered on (e.g., a light bulb becomes hot and changes its relative resistance continuously).

Social and Ethical Problems. So far, only the technical aspects of the main problem have been investigated in this document. However, whether the whole system and idea is successful or not also depends on the perception of the individual consumers. According to a recently published study of the Institute of Technology Assessment at the Austrian Academy of Science in cooperation with E-commerce Monitoring GmbH, power consumers want to be better informed about the potential and functionality of smart meters [17]. Furthermore, they want to have more influence on the actual behaviour of the system [18]. The problem is that recent discussions focused on what can technically be done, but the best system does not work properly if the users do not cooperate. In the example of smart grids, on the one hand consumers do not want a fully automated solution, as they feel a loss of freedom in their home. On the other hand, due to the fact that smart meters will not be installed voluntarily, according to the legal situation in Europe, most likely consumer driven actions within in the smart grid as described in the previous section might not occur as frequently as forecasted by the developers.

One big issue, which additionally decreases the acceptance of smart meters among the citizens is the uncertain situation concerning data safety in smart grid systems [19]. Consumers want to be sure that the data transmitted by the smart meters in their homes is safe. Otherwise, it might be possible to track a power consumption profile of a household, which might give hints in whether someone is home or not, so creating potential for burglaries. Furthermore, tracking the power consumption in detail, the idea of "Big brother is watching you" becomes more real than ever. Therefore, besides the technical possibility basic ethical rules have to be considered. If this is the case from a early stage in a project, the acceptance of the system by the consumer increases. This again enables the opportunity to a more efficient system, as the consumer becomes an active part of the system with own responsibility and possibility to decide.

From the author's point of view, integrating the consumer into the system can make the system more efficient than improving the basic technical implementation.

3   Load Classification Algorithm

A model implementing a load classification algorithm should be developed. The algorithm should be able to reliably detect specific devices within a smart microgrid network. The detection may be based on a time-frequency analyzes schema, implementing a feature detection approach. One potential approach towards feature extraction, which has been applied on spectrum sensing in LTE systems, has been presented in [20–22]. Contrary to the approach given in [15,16], the proposed approach will not be limited to the detection of harmonics for high-frequency feature extraction.

The idea relies on the following two methods. First, the power consumption of a specific part of the smart grid will be tracked and further processed. Second, based on the information gained from the feature extraction process, the device is classified and decisions concerning further actions are made. In [23] a simple prototype trying to achieve this approach is presented. In this work the proposed ideas shall be further developed and improved.

For the classification of certain loads within the smart microgrid, a database has to be developed, which includes characteristic profiles for each possible device. Existing databases, which might be reused and extended, can be found in [24] and [25] These include traditional light bulbs, energy saving light bulbs, ovens, washing machines, TV sets, etc. During the first part of this project, existing profile databases have to be extended by new profiles, which have to be developed based on measurements in real smart grids. The established AGH Smart Campus is a perfect example to be studied in this context.

A first test database might include only few different characteristic loads, while later on it might be extended step by step to further verify the developed detection algorithms.

The aforementioned methods will be used to classify the individual loads expected to be present in the smart microgrid. Further efficiency improvements can be achieved, if some smart high-level detection approach is chosen. The idea is, that any device present is not detected at once extracting all information present in the underlying signal, but in a step-by-step wise approach. First, a very rough estimation is performed, delivering the information of an upper class of device (e.g., heating device, lighting device, computer). This yields in a very quick and low cost rough overview in the situation in the system, allowing first basic adoptions and decisions within the system. After this, less critical decisions can be done, based on a fine estimation. Still, this will not be a full signal scan, but a fine feature extraction based on the rough estimate delivered before. Therefore, fine tuning of the smart grid, which already is adopted to the basic environmental conditions, can be done on this second step.

During research of this field it will become clear, how many levels of estimation are reasonable. A keen pre-estimation would be, that two or three levels of abstraction will provide good overall performance. The number of levels, for sure, will correlate with the actual environment. If there are many huge loads (for example, heating device) and only few small loads (for example, light bulbs) the number of levels will be different, as it would be for the inverse scenario. The optimum might be learned by the system itself during runtime.

4   Conclusion

In this paper we presented potential candidates for automated load classification in smart micro-grid systems. The proposed methods show major advantages compared to existing solutions. Those advantages as well as the necessity for improvements in that field have been illustrated.

References

1. Vandoorn, B., Renders, T., Degroote, L., Meersman, B., Vandevelde, L.: Active load control in islanded microgrids based on the grid voltage. Control **2**(1), 139–151 (2011)
2. Farhangi, H.: The path of the smart grid. IEEE Power Energy Mag. **8**(1), 18–28 (2010)
3. Mohn, T., Piasecki, R.: A smarter grid enables communal microgrids. In: Proceedings of the IEEE Green Technologies Conference (IEEE-Green), pp. 1–6, April 2011
4. Lasseter, R.H.: Smart distribution: coupled microgrids. Proc. IEEE **99**(6), 1074–1082 (2011)
5. Luh, P.B., Michel, L.D., Friedland, P.: Load forecast and demand response. In: Proceedings of the IEEE PES General Meeting, pp. 1–3, July 2010
6. Albadi, M., El-Saadany, E.F.: A summary of demand response in electricity markets. Electric Power Syst. Res. **78**(11), 1989–1996 (2008)
7. Jiang, B., Fei, Y.: Dynamic residential demand response and distributed generation management in smart microgrid with hierarchical agents. Energy Proc. **12**, 76–90 (2011)

8. Morganti, G., Perdon, A.M., Conte, G.: Optimising home automation systems: a comparative study on tabu search and evolutionary algorithms. In: Proceedings of the Mediterranean Conference on Control & Automation, pp. 1044–1049 (2009)

9. Pedrasa, T.D., Spooner, M.A., MacGill, I.F.: Coordinated scheduling of residential distributed energy resources to optimize smart home energy services. IEEE Trans. Smart Grid **1**(2), 134–143 (2010)

10. Negenborn, R., Houwing, M., De Schutter, B., Hellendoorn, H.: Adaptive prediction model accuracy in the control of residential energy resources. In: Proceedings of the IEEE International Conference on Control Applications, pp. 311–316 (2008)

11. Zhang, D., Papageorgiou, L.G.: Optimal scheduling of smart homes energy consumption with microgrid. Energy First **1**, 70–75 (2011)

12. Elmenreich, W., Egarter, D.: Design guidelines for smart appliances. In: Proceedings of the 10th International Workshop on Intelligent Solutions in Embedded Systems, July 2012

13. Lu, E., Reicher, D., Spirakis, C., Weihl, B.: Demand dispatch. IEEE Power Energy Mag. **8**(3), 20–29 (2010)

14. Martello, S., Toth, P.: Knapsack Problems: Algorithms and Computer Implementation. John Wiley, Hoboken (1990)

15. Zeifmann, M., Roth, K.: Nonintrusive appliance load monitoring: Review and outlook. IEEE Trans. Consum. Electron. **57**(1), 76–84 (2011)

16. Liang, J., Ng, S., Kendall, G., Cheng, J.: Load signature study iV part I: basic concept, structure and methodology. In: Proceedings of the IEEE Power and Energy Society General Meeting, p. 1, July 2010

17. IFZ Graz: Smart Meter - KonsumentInnen wollen selbst entscheiden, May 2012. http://www.ifz.aau.at/Media/Dateien/Downloads-IFZ/Energie-und-Klima/Smart-New-World/Presseinformation

18. IFZ Graz: Das Projekt 'Smart New World?, May 2012. http://www.ifz.aau.at/Media/Dateien/Downloads-IFZ/Energie-und-Klima/Smart-New-World/Fact-Sheet

19. Unabhängiges Landeszentrum für Datenschutz Schleswig-Holstein: Bundestag will aus Datenschutzsicht 'gefährlichen Unsinn' zu Smart Metern regeln, June 2011. https://www.datenschutzzentrum.de/presse/20110628-smartmeter.htm

20. Schlechter, T., Huemer, M.: Overview on blocker detection in LTE systems. In: Proceedings of Austrochip 2010, Villach, Austria, pp. 99–104, October 2010

21. Schlechter, T., Huemer, M.: Advanced filter bank based approach for blocker detection in LTE systems. In: Proceedings of IEEE International Symposium Circuits and System (ISCAS 2011), Rio De Janeiro, Brazil, pp. 2189–2192, May 2011

22. Schlechter, T., Juritsch, C., Huemer, M.: Spectral estimation for long-term evolution transceivers using low-complex filter banks. J. Eng. **2014**(6), 265–274 (2014)

23. Patel, S.N., Robertson, T., Kientz, J.A., Reynolds, M.S., Abowd, G.D.: At the flick of a switch: detecting and classifying unique electrical events on the residential power line (nominated for the best paper award). In: Krumm, J., Abowd, G.D., Seneviratne, A., Strang, T. (eds.) UbiComp 2007. LNCS, vol. 4717, pp. 271–288. Springer, Heidelberg (2007). https://doi.org/10.1007/978-3-540-74853-3_16

24. Anderson, K., Ocneanu, A., Benitez, D., Carlson, D., Rowe, A., Berges, M.: BLUED: a fully labeled public dataset for event-based non-intrusive load monitoring research. In: Proceedings of the 2nd KDD Workshop on Data Mining Applications in Sustainability (SustKDD), Beijing, China, August 2012

25. Zico Kolter, A., Johnson, M.J.: REDD: a public data set for energy disaggregation research. In: Proceedings of the 1st KDD Workshop on Data Mining Applications in Sustainability (SustKDD) (2011)

An Overview About the Physical Layer of the VHF Data Exchange System (VDES)

N. Molina[1]($\boxtimes$)(iD), F. Cabrera[1]($\boxtimes$)(iD), Victor Araña[1]($\boxtimes$)(iD),
and M. Tichavska[2]($\boxtimes$)(iD)

[1] University of Las Palmas de Gran Canaria, Las Palmas, Spain
{nicolas.molina,francisco.cabrera,victor.arana}@ulpgc.es
[2] MarineTraffic, Athens, Greece
miluse.tichavska@marinetraffic.com

Abstract. The AIS (*Automatic Identification System*) has become one of the most popular maritime communications systems in the world, allowing the data exchange and identification among vessels to avoid collisions at sea. In spite of the uncountable virtues of this system, security and channel saturation challenges have motivated the development of a new version of the AIS, called VDES (*VHF Data Exchange System*). This system improves the AIS robustness and includes additional services for a wider range of applications. In this paper, a description of the physical layer of the VDES is presented and the most relevant differences between the AIS and VDES are analysed. Moreover, the main challenges that the VDES has to face in the next years are also described and analysed, including a vision about the future applications that will be supported by this system.

Keywords: AIS (*Automatic Identification System*) · ACM (*Adaptive Coding and Modulation*) · VDES (*VHF Data Exchange System*)

1 A New Age for Maritime Communications

1.1 Description of the AIS

In the mid 90's the AIS (*Automatic Identification System*) was proposed by the IMO (*International Maritime Organization*) to improve marine safety and avoid vessel collisions in the oceans [1]. This system operates in VHF (*Very High Frequency*) band on two frequencies, AIS-1 (161.975 MHz) and AIS-2 (162.025 MHz), using a GMSK (*Gaussian Minimum Shift Keying*) modulation scheme combined with a NRZI (*Non Return to Zero Inverted*) coding scheme, and reaching a maximum baud rate of 9600 bps [2]. Moreover, this system uses a time division access scheme called SOTDMA (*Self-Organized Time Division Multiple Access*), which allows to transmit AIS messages according to the availability of the channel at all times.

There are 27 types of AIS messages, used by the different applications in this system. These applications include two types of AIS equipment, called class

R. Moreno-Díaz et al. (Eds.): EUROCAST 2019, LNCS 12013, pp. 67–74, 2020.
https://doi.org/10.1007/978-3-030-45093-9_9

A and class B. The AIS class A equipment is used by IMO regulated vessels, transmitting a maximum power of 12.5 W and shorter time intervals between messages [3]. On the other hand, the AIS class B equipment is used by any type of vessel, regardless of the IMO regulation, with a transmitted power limited to 2 W and greater time intervals between messages. Other AIS applications are AtoN (*Aid-to-Navigation*), the AIS base stations and SAR (*Search and Rescue*), which have specific types of messages.

The AIS has numerous advantages in terms of coverage and efficiency. This system can reach distances of around 40–60 km [4], also including global satellite coverage even in the Poles. Moreover, and in the face of adverse weather and orographic conditions, the AIS presents a great robustness [5]. The price of the AIS receivers and transceivers is accessible when compared to other systems with similar purposes, and devices are of easy installation and use [6].

1.2   AIS Security and Channel Saturation

The AIS presents limitations in terms of security and data transmission, which takes place with a limited bandwidth.

On the one hand, AIS data may be subject to manipulation. AIS transceivers can be manually turned off by the deck officers, incorrect GPS positions may override AIS information, or the officer of a vessel may choose not to enter their destination port into the AIS equipment for transmission. On the other hand, and while widely implemented, the AIS usage is not mandatory for all vessels. This is a fact that enables a comprehensive, yet partial view of overall vessel activity. Lastly, the AIS is an open system. In other words, the AIS does not incorporate encryption methods or other type of mechanisms that may guarantee this very relevant security aspect. Because of the later, and within the scope of illegal operations, the AIS may be subject to misusage [7,8].

Nevertheless, the industry may overcome the drawbacks of relying on AIS technology. This may include a team of expert analysts dedicated to the curation of AIS data inconsistencies and errors, using intelligent networks to record events and verify port callings and departures, and providing context and clarity on the last known vessel position. In this respect, even though the AIS can be an useful source of vessel tracking intelligence, it is necessary to understand the limitations and drawbacks of this technology, mainly for business and companies that use the AIS information to accomplish their strategies. Relying on AIS information alone may represent a level of inaccuracy and confusion due to technical limitations, a possible manipulation of the system and/or insufficient human/digital intelligence and verification.

Because the AIS operates only on two frequencies with a channel bandwidth of 25 KHz and has a maximum baud rate of 9600 bps, channel saturation may occur in sea areas with a great confluence of vessels, such as the ones shown in Fig. 1.

The AIS channel saturation depends on the number of vessels which transmit in a same coverage zone, but also on the vessels' speed. Indeed, and due to an AIS design focused on collision avoidance, vessels that navigate at higher speeds

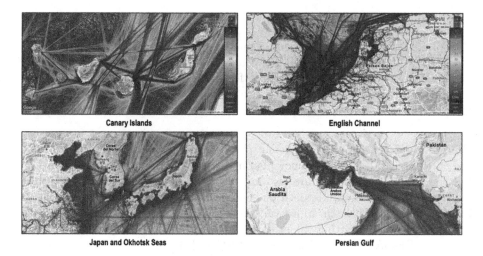

Fig. 1. Annual vessel traffic density in regions with AIS channel overloading danger. Source: MarineTraffic vessel traffic intelligence (https://www.marinetraffic.com)

transmit more messages, while those at anchor or lower speeds transmit a lower volume of messages. Channel saturation has become one of the most relevant challenges identified within the AIS. And, for instance, is also one of the main reason why an improvement of the AIS has been proposed [9,10].

2   VHF Data Exchange System (VDES)

2.1   Description of the VDES

In 2013, with the aim of addressing AIS channel saturation, the IALA and other organizations proposed a new version of AIS called VDES (*VHF Data Exchange System*). This system is conceived as an evolution of the AIS, including new services, improving the efficiency and increasing the transmission rates [11,12].

In addition to the AIS, which has the highest priority in VDES, the following services with different features and operation modes have been included:

– LR-AIS (*Long Range AIS*). *Ship-to-satellite* service used to stablish AIS satellite links.
– ASM (*Application Specific Messages*). *Ship-to-ship*, *ship-to-shore* and *ship-to-satellite* service used to create personalised messages with information about many types of applications.
– VDE (*VHF Data Exchange*). Service that allows the exchange of high data rates for different applications. There are two modalities for this service:
  • VDE-TER (*VDE for Terrestrial Links*). *Ship-to-ship* and *ship-to-shore* service.
  • VDE-SAT (*VDE for Satellite Links*). *Ship-to-satellite* service. Its technical features have not been published yet.

2.2 Physical Layer in VDES

The VDES implies an improvement of many technical aspects of the AIS. First, the VDES will increase the number of channels, also including new services. The new VDES services will require better features, so that new modulation and coding schemes will be added, which also supposes an increase of the maximum bit rate used.

Frequency Plan. A new frequency plan is proposed for VDES [13], and is presented in Fig. 2. This frequency plan maintains AIS-1 and AIS-2 frequencies, but the new VDES services are also included. LR-AIS and ASM were types of messages in the AIS standardisation, but now there are four channels with a bandwidth of 25 KHz each. Moreover, the VDE-TER service can use a variable channel bandwidth, getting values of 25, 50 or 100 KHz. Finally, the frequency plan for VDE-SAT will be discussed in WRC-19 [14].

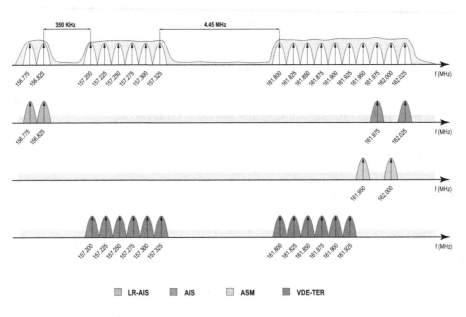

Fig. 2. VDES frequency plan proposed

Modulation and Coding Schemes. The new services included in VDES use different coding and modulation schemes depending on the radio link conditions. This technique is known as ACM (*Adaptive Coding and Modulation*) [15], and is based on the continuous monitoring of link quality parameters such as BER (*Bit Error Rate*) and SNR (*Signal to Noise Ratio*) to know the channel conditions. These parameters are exchanged between the communication equipment using

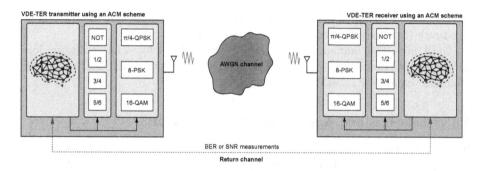

Fig. 3. ACM scheme for a VDE-TER communication

a return channel, allowing to decide which modulation is more efficient at that moment, as it is presented in the Fig. 3.

So far, the VDES standardisation shows the inclusion of three new modulation schemes: 8-PSK, π/4-QPSK and 16-QAM. In the case of the AIS and LR-AIS services, a GMSK modulation with NRZI coding scheme is kept, that it is also used by the ASM service, which also includes the 8-PSK modulation. For VDE-TER service, the three new types of modulations are used, but the modulation and coding scheme used by the AIS and LR-AIS is not included in this service [16].

Baud Rate. The baud rate of a communication system is conditioned by its modulation scheme and the channel bandwidth. For this reason, the integrated services in the VDES have different baud rates, as shown in Fig. 4, where are

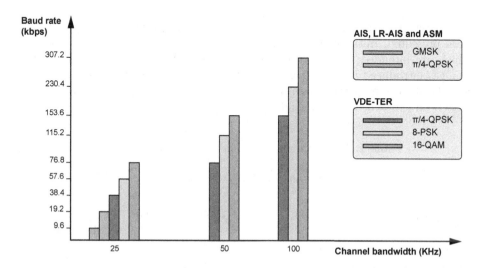

Fig. 4. Maximum baud rates for VDES services

represented the channel bandwidths (in KHz) on the X-axis and the maximum baud rate (in kbps) for each service in the Y-axis.

Until now, the maximum baud rate is 307.2 kbps for VDE-TER service, with a 16-QAM modulation scheme and 100 KHz of channel bandwidth, and the minimum baud rate is 9600 bps for the AIS.

2.3   Challenges and Applications Using the VDES

In the next years, the VDES system is expected to be fully operational and include all the services discussed above [17]. This will imply important challenges for maritime applications, as the ASM and VDE services that offer new uses regardless of user motivations.

Nowadays, new uses with the AIS have been developed. This has opened a new range of possibilities, despite the limited benefits of the AIS, but also contributing to the AIS channel saturation and consequently, the definition of the VDES as a solution for this problem. In relation to the environmental sciences, the AIS can be used to estimate the effects of vessel emissions [18,19], for the transmission of meteorological parameters [20], and even to analyse the effects of tsunamis in port zones [21]. However, the concept of IoV (*Internet of Vessels*) has appeared in the last years [22], closely related to the intelligent navigation or e-Navigation [23], which aims to interconnect different maritime services and equipment to improve marine navigation and reduce the problems derived from human errors. In relation to these trends, the authors consider that these will be the first application fields of the VDES, since performance and focus are improved considerably towards specific and non-limited applications.

3   Conclusions

In this paper, a review of the main aspects of the physical layer in VDES has been presented. Initially, a brief overview of the AIS and its main limitations, which are intended to be solved with the VDES, has been given. Then, the changes introduced by the VDES (except for VDE-SAT) have been explained. Specifically, the frequency plan, modulation and coding schemes and baud rate have been exposed, which are summarised in Table 1.

Table 1. Table captions should be placed above the tables.

	AIS and LR-AIS	ASM	VDE-TER
Number of channels	2	2	12
Modulation and coding schemes	GMSK	GMSK, $\pi/4$-QPSK	$\pi/4$-QPSK, 8-PSK, 16-QAM
Maximum baud rate (kbps)	9.6	19.2	307.2

Finally, some of the recent uses and applications of the AIS have been reviewed to justify the need of the VDES development, and also, to foretell the VDES application trends in the coming years.

References

1. Essaadali, R., Jebali, C., Grati, K., Kouki, A.: AIS data exchange protocol study and embedded software development for maritime navigation. In: 2015 IEEE 28th Canadian Conference on Electrical and Computer Engineering (CCECE), Canada, pp. 1594–1599 (2015)
2. ITU R: M.1371-5: Technical characteristics for an automatic identification system using time-division multiple access in the VHF maritime mobile band. International Telecommunication Union (2014)
3. Cervera, M.A., Ginesi, A.: On the performance analysis of a satellite-based AIS system. In: 2008 10th International Workshop on Signal Processing for Space Communications, Greece, pp. 1–8 (2008)
4. Plass, S., Poehlmann, R., Hermenier, R., Dammann, A.: Global maritime surveillance by airliner-based AIS detection: preliminary analysis. J. Navig. **68**(6), 1195–1209 (2015)
5. Kerbiriou, R., Lévêque, L., Rajabi, A., Serry, A.: The Automatic Identification System (AIS) as a data source for studying maritime traffic. In: International Maritime Science Conference, Croatia (2017)
6. Cabrera, F., Molina, N., Tichavska, M., Araña, V.: Automatic Identification System modular receiver for academic purposes. Radio Sci. **51**(7), 1038–1047 (2016)
7. Hall, J., Lee, J., Benin, J., Armstrong, C., Owen, H.: IEEE 1609 influenced automatic identification system (AIS). In: 2015 IEEE 81st Vehicular Technology Conference (VTC Spring), pp. 1–5. IEEE, Scotland (2015)
8. Goudossis, A., Katsikas, S.K.: Towards a secure automatic identification system (AIS). J. Marine Sci. Technol. **24**(2), 410–423 (2018)
9. Lee, S.J., Jeong, J.S., Kim, M.Y., Park, G.K.: A study on real-time message analysis for AIS VDL load management. J. Korean Inst. Intell. Syst. **23**(3), 256–261 (2013)
10. ITU R: Report M.2287-0: Automatic identification system VHF data link loading. International Telecommunication Union (2013)
11. Šafár, J., Hargreaves, C., Ward, N.: The VHF data exchange system. In: Antennas, Propagation & RF Technology for Transport and Autonomous Platforms 2017, Birmingham, United Kingdom, pp. 1–8 (2017)
12. Proud, R., Browning, P., Kocak, D.M.: AIS-based mobile satellite service expands opportunities for affordable global ocean observing and monitoring. In: OCEANS 2016 MTS/IEEE Monterey, Monterey, California, USA, pp. 1–8 (2016)
13. ITU R: M.2092-0: Technical characteristics for a VHF data exchange system in the VHF maritime mobile band. International Telecommunication Union (2015)
14. Eriksen, T., Braten, L.E., Haugli, H.C., Storesund, F.A.: VDESAT-A new maritime communications system. In: Proceedings of the Small Satellites, System & Services Symposium (4S), Malta, vol. 30, pp. 1–12 (2016)
15. Oien, G.E., Holm, H., Hole, K.J.: Impact of channel prediction on adaptive coded modulation performance in Rayleigh fading. IEEE Trans. Veh. Technol. **53**(3), 758–769 (2004)

16. Lázaro, F., Raulefs, R., Wang, W., Clazzer, F., Plass, S.: VHF Data Exchange System (VDES): an enabling technology for maritime communications. CEAS Space J. **11**(1), 55–63 (2019)
17. IALA Guideline G1117: VHF Data Exchange System (VDES) overview. IALA (2016)
18. Liu, T.K., Sheu, H.Y., Chen, Y.T.: Utilization of vessel automatic identification system (AIS) to estimate the emission of air pollutant from merchant vessels in a Port Area. In: OCEANS 2015-Genova, Genova, Italy, pp. 1–5 (2015)
19. Tichavska, M., Tovar, B.: Port-city exhaust emission model: an application to cruise and ferry operations in Las Palmas Port. Transp. Res. Part A: Policy Pract. **78**, 347–360 (2015)
20. Chang, S.J., Huang, C.H., Hsu, G., Chang, S.M.: Implementation of AIS-based marine meteorological applications. In: OCEANS 2014-TAIPEI, Taipei, Taiwan, pp. 1–4 (2014)
21. Makino, H., Gao, X., Furusho, M.: Grasp of a latent risk of ship evacuation during a tsunami using AIS data analysis. In: 2015 International Association of Institutes of Navigation World Congress (IAIN), Prague, Czech Republic, pp. 1–5 (2015)
22. Tian, Z., Liu, F., Li, Z., Malekian, R., Xie, Y.: The development of key technologies in applications of vessels connected to the Internet. Symmetry **9**(10), 211 (2017)
23. Amato, F., Fiorini, M., Gallone, S., Golino, G.: e-Navigation and future trend in navigation. TransNav Int. J. Mar. Navig. Saf. Sea Transp. **5**(1), 11–14 (2011)

A Job-Seeking Advisor Bot Based in Data Mining

J. C. Rodríguez-Rodríguez, Gabriele S. de Blasio📵, C. R. García📵,
and A. Quesada-Arencibia(✉)📵

Institute for Cybernetic Science and Technology,
University of Las Palmas de Gran Canaria, Las Palmas, Spain
jrodriguez@iuctc.ulpgc.es, alexis.quesada@ulpgc.es

Abstract. Promentor is a solution that advises job seekers how to effectively improve their chances of getting a job in a certain area of interest by focusing on what, at least historically, seems to work best. To this end, Promentor first analyzes previous selection processes, trying to quantitatively evaluate the effective value of the characteristics that the candidates put into play in the selection. With this evaluation, Promentor can estimate the value of a profile of the job seeker who requests advice based on the characteristics that make it up. Promentor then makes a simulation by applying each of the suggestions on the job seeker's profile exhaustively and evaluating the modified profile. In this way, Promentor identifies which suggestions offer the greatest increase in qualification, and are therefore more recommendable.

Promentor is a module of the employment web portal GetaJob.es, which has been developed in parallel and equipped with specific capabilities for collecting the data required by Promentor.

Keywords: Data mining · Job search · Employment portal · Training

1 Introduction

Data mining is the process of discovering potentially useful, interesting and previously unknown patterns in a large collection of data [1, 2]. It is part of the knowledge discovery process that attempts to infer patterns and trends that exist in the data. Usually these patterns cannot be easily detected with traditional data mining because the volume of data is too large or the relationships between the data are too complex.

Promentor is a system that makes recommendations aimed at improving a job seeker's chances of getting a job in a certain area of interest. It does this by applying data mining techniques to identify patterns (combinations of characteristics) of success or failure. It searches for discriminating patterns by analyzing candidates in the completed selection processes of an area of interest during the learning phase.

The discriminant table that is constructed during the learning phase allows Promentor to score a collection of characteristics (candidate profile). The higher the score, the more positive discrimination patterns (success discriminants) are contained in the collection of characteristics.

© Springer Nature Switzerland AG 2020
R. Moreno-Díaz et al. (Eds.): EUROCAST 2019, LNCS 12013, pp. 75–82, 2020.
https://doi.org/10.1007/978-3-030-45093-9_10

This discriminant table allows us to evaluate job seekers' profiles and how they are affected when job seekers follow suggestions. Consequently, it allows to estimate the effectiveness of the suggestions, and thus to build recommendations.

Promentor originates as an automatic personal assistance module on the web for job searches getajob.es. It is interesting to note that it does not try to recommend jobs like the systems described in [3, 4].

2 Profile Information

2.1 Input

The data entry takes place on the employment web portal getajob.es. In this portal job seekers register and fill in their profiles. In many cases this process is supervised by an employment agency, where a human assistant supervises the data entry. The portal provides forms designed from the experience of the employment agencies. It is clear that the design of forms or the choice of features play a significant role in the placement of the candidate, and thus, in the recommendations that can be given. However, from the very beginning, there was an imposition on Promentor to be agnostic about this. Consequently, Promentor will simply operate on the data supplied to it, without consideration of how it has been chosen or presented.

2.2 Pre-processing of Input

From the data entry, Promentor obtains a set of characteristics that describe the profile. The first step is to pre-process this set of characteristics to make the valuable information in the set more explicit. The result is the extended set of characteristics. Several inference rules are applied in succession to obtain this set: seniority and length of service, promotions, and merit accumulations.

Finally, the extended feature set is converted to a numerical representation for further processing efficiency (each feature is represented by a unique numerical identifier).

Seniority and Length of Service
The dates listed in the profile are replaced by the age and duration of periods.

The age of a date is defined as the time between that date and a reference date, which could be the date of the request for recommendation or, the date on which the selective process on which the data is considered takes place.

The length of a period is defined as the time between two dates. For example, the duration between the start and end dates of employment. Seniority and length of service are measured in years and months.

Promotions
It is possible to infer from a single characteristic one or more new characteristics that are simply contained in the first one, and which we can make explicit. This is called "promoting".

For example, from the feature {[PhD in Computer Science]} we can infer the new feature [PhD], so we would have the set: {[PhD], [PhD in Computer Science]}. It would also be possible to infer [Computer science], obtaining the set: {[PhD], [Computer science], [Computer science PhD]} being all the characteristics true.

Accumulations of Merits

Some characteristics are cumulative. This facilitates comparisons of merit in profiles.

For example: if we have three characteristics such as [Master in Computer Science], [Master in Electrical Engineering] and [Master in Mathematics], we can infer that in addition to those three characteristics you have the characteristic [Has three Master]. Also, if you have the characteristic [Has three Master], you automatically get the characteristics [Has one Master] and [Has two Master].

These new features are true regarding the profile description. With this operation we try to make them explicit in the set by promoting them to features from their implicit feature status embedded in features.

Numerical Labeling

Numerical labeling is the translation of characteristics into numerical identifiers to facilitate the manipulation of the feature sets. Each numerical identifier is unique, and if necessary, new identifiers are created as new features are introduced. In the process, the sets of characteristics become unreadable to humans.

3 Selection and Learning Processes

In order to achieve the goal, Promentor learns from past employee selection processes that have been recorded in the system.

Companies carry out selection processes to cover one or more jobs. A job is associated with an area of interest.

An area of interest is a way of categorizing professions based on what, how, where, … a job seeker would like to work. Examples of this can be: administrative, customer service, electricians, plumbers, software developers, … At Promentor a selection process is associated with a single area of interest.

3.1 The Job Offer

As far as Promentor is concerned, the first step in a selection process is the publication of the job offer. Every job offer specifies its conditions and the area of interest to which it is subscribed. It also indicates the number of vacancies available.

Job seekers are proposed for the job offered. By doing so, they send their job seeker profiles to be evaluated by the recruiter. The job seeker profile can be defined as a collection of characteristics such as academic careers taken, courses taken or experience accumulated.

3.2 Selection Process

The candidate undergoes the company's selection process: CV check (contained in the job seeker profile), interview… The result of the evaluation is the classification of the candidate in categories: (candidate's status). All candidate statuses can be terminal.

The pattern categories are constructed from the states and stages (see Fig. 1).

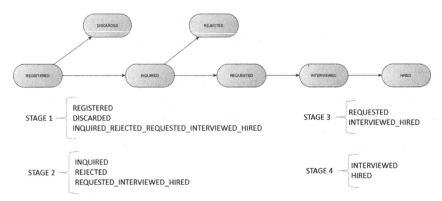

Fig. 1. The selection process is dissected into a series of states (7 states), which take place in a series of stages (4 stages). All states are terminal. That is, a candidate can eventually end up in any of these states. The categories (10 categories) can be made up of one or more states. This is the case, for example, with the category INQUIRED_REJECTED_REQUESTED_INTERVIEWED_HIRED. All the candidates who have remained in one of these states are part of that category.

The description of the states is straightforward:

– Registered: Candidates registered in the selection process.
– Discarded: Candidates discarded in the preliminary review.
– Inquired: Candidates examined in more detail.
– Rejected: Candidates rejected in the more detailed examination.
– Requested: Contact is made for interview.
– Interviewed: The candidate is interviewed.
– Hired: The candidate is hired.

The portal offers a wizard of the selection process in order to monitor those decisions. This way, it registers profiles of job seekers and how these profiles have been classified in multiple selection processes.

First Decision
The web portal presents recruiters with a small file with the Curriculum Vitae of all registered candidates. Based on this very general information, the recruiter discards candidates (Discarded) or decides to study some candidates more in detail (Inquired_Rejected_Requested_Interviewed_Hired). The recruiter may also not make any decisions about some candidates (Registered). We have here the categories Registered, Discarded, Inquired_Rejected_Requested_Interviewed_Hired.

Second Decision
Now, with more specific information about the candidates examined, the recruiter can reject candidates (Rejected) and request their contact information (Requested_Interviewed_Hired). Again, it could be that the recruiter does not decide any

action on some candidates (Inquired). We have here the categories Inquired, Rejected, Requested_Interviewed_Hired.

Third Decision
Among those who have been asked for contact information, some candidates will conduct the interview (Interviewed_Hired). Others will not (Requested). We have here the categories Requested, Interviewed_Hired.

Fourth and Last Decision
Some of the candidates interviewed are hired (Hired) or not (Interviewed). We have here the categories Interviewed, Hired.

3.3 Learning Phase

The purpose of the learning phase is to compute the pattern score table.

1. The selection processes associated with each area of interest are considered together.
2. For each selection process, the patterns common to all candidates evaluated within a stage (recruiter's decision) are not considered (the value assigned is 0). That is; they are considered indifferent (non-discriminatory) in the evaluation within the stage.
3. All patterns are grouped into sets according to the category. A pattern may appear in several sets. There are ten sets, corresponding to each of the ten categories (Registered, Discarded, Inquired_Rejected_Requested_Hired, Rejected, Inquired, Requested_Interviewed_Hired, Requested, Interviewed_Hired, Interviewed, Hired).
4. In each set, the most popular/common patterns are searched using a data mining technique. Searching for the most common patterns is a major problem in data mining. There are several well-established algorithms. The current implementation of Promentor uses FP-growth [5]. From now on, only the most common localized patterns will be operated.
5. For each pattern, the positive discriminant force in each category is calculated by comparing where the pattern YES appears and where it does NOT (see Table 1). The total positive discriminative force is the result of the weighted sum of the positive discriminative forces in each category (see Eqs. 1, 2, 3, 4).

$$\text{positive dominance} = \frac{\text{Number of positive candidates with the pattern}}{\text{Number of positive candidates}} \tag{1}$$

$$\text{negative dominance} = \frac{\text{Number of negative candidates with the pattern}}{\text{Number of negative candidates}} \tag{2}$$

If 'positive dominance' is greater than 'negative dominance',

$$\text{discriminance} = \frac{\text{Number of positive candidates with the pattern}}{\text{Number of positive candidates with the pattern} + \text{Number of negative candidates with the pattern}} \tag{3}$$

Otherwise,

$$\text{discriminance} = 0.0 \tag{4}$$

Table 1. The discriminant force of the pattern in each category is evaluated by contrasting the number of times it appears in "positive" patterns versus the number of times it appears in negative patterns.

K	Positive	Negative
REGISTERED	REGISTERED	DISCARDED, INQUIRED_REJECTED _REQUESTED_INTERVIEWED_HIRED
DISCARDED	DISCARDED	REGISTERED, INQUIRED_REJECTED _REQUESTED_INTERVIEWED_HIRED
INQUIRED_REJECTED_REQUESTED _INTERVIEWED_HIRED	INQUIRED_REJECTED_REQUESTED _INTERVIEWED_HIRED	REGISTERED, DISCARDED
INQUIRED	INQUIRED	REJECTED, REQUESTED_INTERVIEWED_HIRED
REJECTED	REJECTED	INQUIRED, REQUESTED_INTERVIEWED_HIRED
REQUESTED_INTERVIEWED_HIRED	REQUESTED_INTERVIEWED_HIRED	REJECTED, INQUIRED
REQUESTED	REQUESTED	INTERVIEWED_HIRED
INTERVIEWED_HIRED	INTERVIEWED_HIRED	REQUESTED
INTERVIEWED	INTERVIEWED	HIRED
HIRED	HIRED	INTERVIEWED

See example in Fig. 2, for the calculation of state discrimination for a given pattern.

The number of times the pattern appears at each location is counted and the ratios are calculated. Positive dominance is defined as the ratio of positive candidates to the pattern and the total number of positive candidates.

Fig. 2. The positive dominance is $\frac{2}{5}$. The negative dominance is $\frac{1}{6}$. $\frac{2}{5}$ is greater than $\frac{1}{6}$ so discriminance $= \frac{2}{2+1} = \frac{2}{3}$ for pattern A in DISCARDED category.

Negative dominance is defined as the ratio of negative candidates to the pattern against the total number of negative candidates.

If positive dominance is greater than negative dominance then the discrimination value has the value of the ratio of the number of positive candidates to the pattern to the total number of candidates. Otherwise the discrimination value is 0.

The value is "corrected" by applying a modifier. In this case "discarded" is an undesired state, therefore, the modifier penalizes the discrimination.

```
FOR EACH area_of_interest r:
  selective_process_set=get_selective_process_set(r)
  pattern_set=calculate_most_popular_patterns(selective_process_set)
  FOR EACH pattern i IN pattern_set:
    global_bonus=0
    FOR EACH selective_process j IN selective_process_set:
```

```
local_bonus=0
FOR EACH category k:
local_bonus=local_bonus+calculate_discriminace(i,j,k)x weight(k)
global_bonus=global_bonus+local_bonus
score_table(r,i)=global_bonus/size(selective_process_set)
```

Algorithm 1. This is the general algorithm that constructs the score table for each significant pattern in each area of interest. If the pattern is not significant then its value is zero. In other words, zero is the default value.

4   Recommendations and Suggestions

Promentor makes recommendations to job seekers. When a candidate makes a suggestion, the suggestion modifies his collection of characteristics. Examples of suggestions may be doing a degree, taking a course, gaining experience...

A recommendation is a list of suggestions ordered according to the potential benefit (suggestion ranking). The recommendation is made to maximize the chances of passing a selection process in an area of interest.

4.1   The Recommendation Request

For a job seeker profile, n profile variants can be generated with n being the number of suggestions available. The idea is to compare the different alternative candidate profiles generated in this way to find out which are the most promising. Thus, for each variant, your numerical score is calculated using the score table computed during the learning process for each area of interest. This is done by adding up the scores of the patterns found in the variant.

Scoring is the key for building recommendations. Finally, the suggestions are ranked from highest to lowest positive impact based on the study described. Suggestions in higher positions in the ranking will make up the recommendation.

5   Conclusions

Promentor has been introduced. Promentor aims to help improve the profile of job seekers. To do this, it tries to identify patterns of success discovered in previous selection processes, using a simple and direct approach with low computer costs.

Promentor is limited to the database of selection processes, so that everything it knows is restricted to this database. There are factors affecting selection that are not contained in the CV. For example, the circumstances surrounding the interview.

Promentor adapts, without modification, to the existing working methodology of employment agencies, and their standard forms. Being able to act on other areas could significantly improve the quality of Promentor's entry field.

Promentor is currently in the integration phase. No statistics are yet available. A suitable methodology for measuring the performance and effectiveness of Promentor has yet to be decided.

References

1. Sahu, H.B., Sharma, S., Gondhalakar, S.: A Brief Overview on Data Mining Survey (2011)
2. Agrawal, R., Srikant, R.: Fast algorithms for mining association rules in large databases. In: Proceedings of the 20th International Conference on Very Large Data Bases, VLDB 1994, pp. 487–499 (1994)
3. Al-Otaibi, S.T., Ykhlef, M.: A survey of job recommender systems. Int. J. Phys. Sci. 7(29), 5127–5142 (2012)
4. Aken, A., Litecky, C., Ahmad, A., Nelson, J.: Mining for computing jobs. IEEE Softw. 27(1), 78–85 (2010)
5. Han, J., Pei, J., Yin, Y., Mao, R.: Mining frequent patterns without candidate generation: a frequent-pattern tree approach. Data Min. Knowl. Disc. 8(1), 53–87 (2004). https://doi.org/10.1023/B:DAMI.0000005258.31418.83

Bluetooth Low Energy Technology Applied to Indoor Positioning Systems: An Overview

Gabriele S. de Blasio[(✉)][iD], A. Quesada-Arencibia[iD], Carmelo R. García[iD], and José Carlos Rodríguez-Rodríguez

Instituto Universitario de Ciencias y Tecnologías Cibernéticas (IUCTC),
Universidad de Las Palmas de Gran Canaria (ULPGC),
Las Palmas, Spain
{gabriel.deblasio,alexis.quesada,ruben.garcia}@ulpgc.es,
jcarlos@iuctc.ulpgc.es

Abstract. Indoor Positioning Systems (IPS) are an alternative to Global Positioning System (GPS) in those environments where its signal is attenuated. This is one of the main reasons why IPS has been the subject of much research over the last two decades and where different technologies and methods have been used. Among the technologies used in IPS are those that use radio frequency (RF) signals, such as Bluetooth Low Energy (BLE) or Wi-Fi. BLE is widely used in ubiquitous computing and in many applications of the Internet of Things (IoT) mainly due to its low power consumption and because it can provide advanced services to users. The aim of this paper is to provide an overview of the state of the art of BLE-based IPS including its main methods and algorithms.

Keywords: Indoor Positioning System · Bluetooth Low Energy

1 Introduction

Indoor Positioning Systems (IPS) can be classified depending on the type of signal employed into: radio frequency (RF), light, sound and magnetic fields IPS [9]. Among those technologies which use RF signals we can find Bluetooth Low Energy (BLE) which is widely used in Ubiquitous Computing and in many Internet of Things (IoT) applications [4].

There are several advantages offered by BLE: their emitters or beacons are portable, battery-powered, small, lightweight, easily deployable, have low-energy consumption and Received Signal Strength (RSS) readings, that is, the measurements of the power present in a received radio signal, are relatively easy to collect, producing positioning results with high accuracy and precision [14]. BLE technology can also provide location-based services (LBS) which can deliver location-triggered information to users and that are directly linked to indoor positioning and especially useful in many places, e.g. public transport stations: in such environments, these services can provide users with station guides, ticket

R. Moreno-Díaz et al. (Eds.): EUROCAST 2019, LNCS 12013, pp. 83–90, 2020.
https://doi.org/10.1007/978-3-030-45093-9_11

sales or online information [3]. Different challenges arise when working in indoor positioning using BLE technology, such as that the signal is susceptible to path loss, RSS readings suffer large fluctuations and degradation due to dynamical environments or multipath fading [11]. Another important challenge associated with BLE indoor positioning is the fact that mobile devices do not distinguish between the three primary advertising channels in which the beacons emit, so using the aggregation of these three signals combining their RSS values, may lead to a reduced positioning accuracy [4, 22].

Although there are several requirements for evaluating the performance of an indoor positioning system, the two most important are accuracy and precision: it is common to express positioning accuracy by means of the mean error and its precision by means of the cumulative probability function (CDF), which in practice is expressed in percentile format [18].

1.1 IPS Technologies

The main technologies used in IP can be classified as [9]:

- **Optical**, such as those that use visible light.
- **Sound-based**, such as those that use ultrasound.
- **RF-based**, such as BLE or Wi-Fi.
- Those that rely on **naturally occurring signals**, such as those that use the Earth's natural magnetic field.

1.2 Bluetooth Low Energy

BLE was introduced in 2009 as an extension to the Bluetooth Classic (4.0) and was designed to support the IoT [5, 8]. BLE use 40 2-Megahertz-wide channels in the 2.4 GHz band, divided into 3 primary advertising channels and 37 secondary advertising and data channels. BLE is used when it is not necessary to exchange a lot of data continuously in connections of about 1 millisecond, which leads to a lower consumption, compared to Bluetooth Classic.

Versions 4.1 and 4.2 were introduced later, with higher data range, higher packet capacity and much higher-strength secure connections. BLE 5 and BLE 5.1 were introduced recently (BLE 5.1 in January 2019): the former provides up to 4× the range, 2× the speed and 8× the broadcasting message capacity, along with enhancements that increase functionality for the IoT and the major feature of the latter is the possibility of absolute positioning in three-dimensional space through the angle of arrival (AoA) and angle of departure (AoD), which offer precision of direction in addition to the distance-only information that RSS traditionally brought [6, 7].

BLE emitters or beacons are portable, battery-powered, small, lightweight, have low energy consumption, give a high positioning accuracy and are easily deployable at low cost, transmitting data packets which are slightly different for some standards or protocols, such as iBeacon (Apple) and Eddystone (Google). Also, BLE beacons can exchange data with other devices in two modes, although the really interesting one, from a location point of view, is the advertising mode, that is, sending message packets regularly to other listening devices. In advertising mode, messages hop between the 3 narrow primary advertising channels, each of which has different RSS values, to gain redundancy and in order to reduce interference with other wireless technologies. Finally, BLE beacons can provide Location Based Services (LBS) and Proximity Based Services (PBS) that are very useful in IoT applications.

1.3   Position Calculation Techniques

According to how the person or object position is calculated, the main position calculation techniques used in BLE-based IPS can be classified as [10,18,25]:

- **Proximity:** Where is assumed that the person/object is located at the coordinates of the nearest beacon.
- **Triangulation:** These techniques use distance (multilateration) or angular (multiangulation) measurements between the person/object and the beacons to estimate their position.
- **Centroids:** In these techniques, the person/object position is estimated calculating the centroid of plane geometric figures.
- **Received Signal Strength Indicator (RSSI):** RSSI is a relative measure of the RSS with arbitrary units which, in combination with a path-loss propagation model, allows to calculate the distance between the beacon and the receiver.
- **Fingerprinting or Scene Analysis:** based on the signal power strength received at the receiver (RSS). Its process basically consists of collecting the signal from the beacons and associating it with a particular position.
- **Hybrid:** This technique is based on the combination of two or more of the above.

In the following sections, the important aspects of the main techniques will be discussed and some of its relevant works will be mentioned, but we will focus mainly on Fingerprinting, since it is currently the most used technique and the one that produces the best results [19,20,25].

2   Proximity

In this technique it is assumed that the closest beacon is the one whose signal reaches the person/object with the highest RSS, which is not always true in nonline-of sight (NLOS) propagation conditions [10].

Before the emergence of BLE, there were some attempts to use this technique but with unpromising results in terms of location accuracy [1]. It is usually used nowadays mainly in PBS, which focuses on using a user's proximity to a device to locate where and what they are nearby in an certain area [23]. In BLE-based IPS where high accuracy is sought, it is common to use position calculation techniques other than proximity [19].

3   Triangulation and Centroids

Multilateration is one type of triangulation-based positioning, which employ distance measurements between the target and beacons to obtain a position estimate [10]. The distances can be estimated through Time-of-Flight (TOF), Time Difference of Arrival (TDA), or using RF propagation loss.

Multiangulation is another type of triangulation-based positioning: it uses Angle of Arrival (AOA) to estimate the user/object location. AOA relies on a very simple geometric principle: determining the bearing of two fixed beacons in relation to user that is to be localized.

Among the most important works using those techniques are those of Feldmann et al. [15] and Wang et al. [24]. In the first work, authors present an experimental evaluation of a Bluetooth-based positioning system implemented in a Bluetooth-capable handheld device realizing empirical tests of the developed positioning system in different indoor scenarios. Triangulation combined with least square estimation is used to predict the position of the terminal obtaining a position estimation of a Personal Digital Assistant (PDA) with a precision (Root Mean Square, RMS) of 2.08 m. In the second work, authors analysed different indoor wireless positioning methods, especially the RSS based Bluetooth positioning. A proper propagation model was found, and a RSS and triangulation based positioning scheme was defined and three distance based triangulation algorithms were implemented: Least Square Estimation (LSE), Three-border and Centroid Method. Authors concluded that all three algorithms perform well in terms of positioning accuracy, however, LSE yields slightly better results.

4   Received Signal Strength Indicator (RSSI)

In this technique, the RSSI is used in combination with a path-loss propagation model to estimate the distance, d, between a beacon and the receiver through the relationship:

$$RSSI = -10n \, log_{10}(d) + A \tag{1}$$

where n is the path-loss exponent and A is the RSSI value at a certain reference distance from the receiver [25]. The relation between the RSSI and d may be affected by attenuation and interference.

Some important works in which in some way this technique is used are those of Zhu et al. [27] and Neburka et al. [21]. In the first, authors show that the probability of locating error less than 1.S meter is higher than 80%, and in the second authors shown that RSSI values are inaccurate and highly depending on features of the used BLE module.

5   Fingerprinting

Fingerprinting is a very popular measuring principle due to its simplicity [1, 22]. It consists of two phases: the calibration, training or offline phase and the positioning or online phase.

In the Calibration Phase (CP) or Training Phase (TP), a site inspection or survey is carried out, collecting the received signal strengths (RSS) from the different beacons at spatial points of known coordinates, called Reference Points (RP). Each RP is then characterized by a signal pattern or fingerprint. The set of fingerprints associated with all the RPs is stored in the so-called RP-database (RP-DB) or radio map. In the Positioning Phase (PP), a user situated at a some spatial points of unknown coordinates, called Test Points (TP), measures the signals coming from the different beacons and comparing these signals with those obtained in the CP through some matching algorithm, ultimately obtaining his position. The set of fingerprints associated with all the TPs is stored in the TP-database (TP-DB).

The most common fingerprinting matching algorithms can be classified as [13]:

(a) **Deterministic.** The most common algorithm is the Nearest Neighbor (NN) and its variants: k-Nearest Neighbor (kNN) and Weighted k-Nearest Neighbor (WkNN).
   In this type of matching, one or several survey locations stored in the RP-database are found that best match the observed RSS values stored in the TP-database. The difference between those algorithms is that, in the NN, the coordinates of the nearest RP with best signal match is assigned to the person/object, while in kNN and WkNN, location is estimated using respectively the average or weighted average of the coordinate's k-nearest best signal matches. For the WKNN algorithm, the estimated coordinates (x_e, y_e) of the TP are calculated using the equation:

$$(x_e, y_e) = \frac{\sum_{i=1}^{k}(x_i, y_i) \cdot w_i}{\sum_{i=1}^{k} w_i} \tag{2}$$

where (x_i, y_i) are the coordinates of the $i-$RP and w_i are the corresponding weights. One type of weight commonly used is:

$$w_i = \frac{1}{d_i} \tag{3}$$

where d_i is the Euclidean distance (or another distance or similarity metrics) in the signal space, that is, a multi-dimensional space in which, for a particular RP or TP, the RSS of all beacons are represented as a single point.

(b) **Probabilistic.** Those algorithms are based on statistical Bayesian inference. A set of training data is used that searches for the position of the user with the maximum likelihood: the signal strength values are represented as a probability distribution, and the algorithm calculates the probability of a user's location based on the online measurements and the RP-database. In terms of the position space and observations, Bayes rule can be written [2]:

$$P(s_i|o) = \frac{P(o|s_i) \cdot P(s_i)}{\sum_{i=1}^{k} P(o|s_i) \cdot P(s_i)} \tag{4}$$

where o is the real-time observation made by the person, s_i is a state vector which contains the coordinates of the location from RSS data collected at the time of the survey. The Eq. 4 represents the a posteriori probability that the person, making an observation o, is in state s_i. The a posteriori probability $P(s_i|o)$ is calculated for all s_i, and the state s_i for which this probability is maximum is the most likely position state of the person.

(c) **Machine Learning-Based.** The main algorithms employed so far are Neural Networks (NN) and Support Vector Machine (SVM), both usually applying a supervised approach [20,25]. A NN is composed of interconnecting artificial neurons were each neuron is weighted and used to compute the output. For the specific problem of localization, the NN has to be trained using the RSS values and the corresponding RP-coordinates [25]. Once the NN is trained, it can be used to obtain the user location based on the PP-RSS measurements. SVM is an supervised learning algorithm which can be used for both classification or regression, although it is mostly used in classification. Each data item is plotted as a point in n-dimensional space, where n is the number of features present and the value of each feature is the value of a particular coordinate. The SVM algorithm classifies data by finding the best hyperplane that separates all data points of one class from those of the other [13,26].

Some important works using the Fingerprinting techniques mentioned above are shown in Table 1.

Table 1. Some works using Fingerprinting techniques. The performance of the positioning technique (in metres) is expressed by its Accuracy (A) or Median Accuracy (MA) or Precision (P).

Ref.	Technique	Testbed area (m^2)	Beacon density $(1beacon/x\ m^2)$	Performance (m)	Comments
[14]	Probabilistic	600	30	P: <2.6 95.0%	—
[16]	Deterministic	163.8	7.4	P: <0.8 96.6%	—
[28]	Hybrid	2400	120	P: <2.57 90.0%	—
[17]	Deterministic	2236	131.5	MA: 0.77	BLE + Wi-Fi
[12]	Machine learning	60.5	12.1	A: 1.93	—

References

1. Anastasi, G., Bandelloni, R., Conti, M., Delmastro, F., Gregori, E., Mainetto, G.: Experimenting an indoor Bluetooth-based positioning service. In: 23rd International Conference on Distributed Computing Systems Workshops (ICDCS 2003 Workshops), 19–22 May 2003, Providence, RI, USA, pp. 480–483 (2003). https://doi.org/10.1109/ICDCSW.2003.1203598
2. Bensky, A.: Wireless Positioning Technologies and Applications. Artech House, Boston, London (2008)

3. de Blasio, G., Quesada-Arencibia, A., García, C.R., Molina-Gil, J.M., Caballero-Gil, C.: Study on an indoor positioning system for harsh environments based on Wi-Fi and Bluetooth Low Energy. Sensors **17**(6), 1299 (2017). https://doi.org/10.3390/s17061299. http://www.mdpi.com/1424-8220/17/6/1299

4. de Blasio, G., Quesada-Arencibia, A., García, C.R., Rodríguez-Rodríguez, J.C., Moreno-Díaz, R.: A protocol-channel-based indoor positioning performance study for Bluetooth Low Energy. IEEE Access **6**, 33440–33450 (2018). https://doi.org/10.1109/ACCESS.2018.2837497

5. Bluetooth SIG Proprietary: Bluetooth 4.0 core specification. https://www.bluetooth.org/docman/handlers/downloaddoc.ashx?doc_id=456433

6. Bluetooth SIG Proprietary: Bluetooth 5 core specification. https://www.bluetooth.org/docman/handlers/DownloadDoc.ashx?doc_id=421043

7. Bluetooth SIG Proprietary: Bluetooth 5 core specification. https://www.bluetooth.org/docman/handlers/downloaddoc.ashx?doc_id=457080

8. Bluetooth Special Interest Group (SIG): Bluetooth low energy. https://www.bluetooth.com/what-is-bluetooth-technology/bluetooth-technology-basics/low-energy

9. Brena, R.F., García-Vázquez, J., Galván-Tejada, C.E., Rodríguez, D.M., Rosales, C.V., Fangmeyer Jr., F.: Evolution of indoor positioning technologies: a survey. J. Sensors **2017**, 2630413:1–2630413:21 (2017). https://doi.org/10.1155/2017/2630413

10. Campos, R.S., Lovisolo, L.: RF Positioning: Fundamentals, Applications and Tools. Artech House, Boston, London (2015)

11. Cantón-Paterna, V., Calveras-Augé, A., Paradells-Aspas, J., Pérez-Bullones, M.A.: A Bluetooth Low Energy indoor positioning system with channel diversity, weighted trilateration and Kalman filtering. Sensors **17**(12), 2927 (2017). https://doi.org/10.3390/s17122927

12. Castillo-Cara, M., Lovón-Melgarejo, J., Rocca, G.B., Orozco-Barbosa, L., García-Varea, I.: An analysis of multiple criteria and setups for Bluetooth smartphone-based indoor localization mechanism. J. Sens. **2017**, 1928578:1–1928578:22 (2017). https://doi.org/10.1155/2017/1928578

13. Davidson, P., Piché, R.: A survey of selected indoor positioning methods for smartphones. IEEE Commun. Surv. Tutor. **19**, 1347–1370 (2017). https://doi.org/10.1109/COMST.2016.2637663

14. Faragher, R., Harle, R.: Location fingerprinting with Bluetooth Low Energy beacons. IEEE J. Sel. Areas Commun. **33**(11), 2418–2428 (2015). https://doi.org/10.1109/JSAC.2015.2430281

15. Feldmann, S., Kyamakya, K., Zapater, A., Lue, Z.: An indoor Bluetooth-based positioning system: concept, implementation and experimental evaluation. In: Proceedings of the International Conference on Wireless Networks, ICWN 2003, Las Vegas, NV, USA, 23–26 June, pp. 109–113. CSREA Press (2003)

16. Kajioka, S., Mori, T., Uchiya, T., Takumi, I., Matsuo, H.: Experiment of indoor position presumption based on RSSI of Bluetooth LE beacon. In: IEEE 3rd Global Conference on Consumer Electronics, Tokyo, Japan, 7–10 October 2014, pp. 337–339 (2014)

17. Kriz, P., Maly, F., Kozel, T.: Improving indoor localization using Bluetooth Low Energy beacons. Mob. Inf. Syst. **2016**, 2083094:1–2083094:11 (2016). https://doi.org/10.1155/2016/2083094

18. Liu, H., Darabi, H., Banerjee, P.P., Liu, J.: Survey of wireless indoor positioning techniques and systems. IEEE Trans. Syst. Man Cybern. Part C **37**(6), 1067–1080 (2007). https://doi.org/10.1109/TSMCC.2007.905750

19. Martín Mendoza-Silva, G., Matey-Sanz, M., Torres-Sospedra, J., Huerta, J.: BLE RSS measurements dataset for research on accurate indoor positioning. Data 4(1), 12 (2019). https://doi.org/10.3390/data4010012
20. Martín Mendoza-Silva, G., Torres-Sospedra, J., Huerta, J.: A meta-review of indoor positioning systems. Sensors 19(20), 4507 (2019). https://doi.org/10.3390/s19204507
21. Neburka, J., et al.: Study of the performance of RSSI based Bluetooth smart indoor positioning. In: 2016 26th International Conference Radioelektronika (RADIOELEKTRONIKA), pp. 121–125, April 2016. https://doi.org/10.1109/RADIOELEK.2016.7477344
22. Powar, J., Gao, C., Harle, R.: Assessing the impact of multi-channel BLE beacons on fingerprint-based positioning. In: 2017 International Conference on Indoor Positioning and Indoor Navigation, IPIN 2017, Sapporo, Japan, 18–21 September 2017, pp. 1–8 (2017). https://doi.org/10.1109/IPIN.2017.8115871
23. Sadowski, S., Spachos, P.: Optimization of BLE beacon density for RSSI-based indoor localization. In: 17th IEEE International Conference on Communications Workshops, ICC Workshops 2019, Shanghai, China, 20–24 May 2019, pp. 1–6 (2019). https://doi.org/10.1109/ICCW.2019.8756989
24. Wang, Y., Yang, X., Zhao, Y., Liu, Y., Cuthbert, L.: Bluetooth positioning using RSSI and triangulation methods. In: 10th IEEE Consumer Communications and Networking Conference, CCNC 2013, Las Vegas, NV, USA, 11–14 January, pp. 837–842 (2013). https://doi.org/10.1109/CCNC.2013.6488558
25. Zafari, F., Gkelias, A., Leung, K.K.: A survey of indoor localization systems and technologies. IEEE Commun. Surv. Tutor. 21(3), 2568–2599 (2019). https://doi.org/10.1109/COMST.2019.2911558
26. Zhang, L., Liu, X., Song, J., Gurrin, C., Zhu, Z.: A comprehensive study of Bluetooth fingerprinting-based algorithms for localization. In: 27th International Conference on Advanced Information Networking and Applications Workshops, WAINA 2013, Barcelona, Spain, 25–28 March 2013, pp. 300–305 (2013). https://doi.org/10.1109/WAINA.2013.205
27. Zhu, J., Luo, H., Chen, Z., Li, Z.: RSSI based Bluetooth Low Energy indoor positioning. In: 2014 International Conference on Indoor Positioning and Indoor Navigation, IPIN 2014, Busan, South Korea, 27–30 October 2014, pp. 526–533 (2014). https://doi.org/10.1109/IPIN.2014.7275525
28. Zhuang, Y., Yang, J., Li, J., Qi, L., El-Sheimy, N.: Smartphone-based indoor localization with Bluetooth Low Energy beacons. Sensors 16, 596 (2011). https://doi.org/10.3390/s16050596

Pioneers and Landmarks in the Development of Information and Communication Technologies

The Origin, Evolution and Applications of Visual Bio-cybernetics Concepts Included in the Original MIT-NASA Reports for Designing a Mars Rover (1965–1985)

R. Moreno-Díaz Jr.[1]([✉]), R. Moreno-Díaz[1], A. Moreno-Díaz[2], and A. Moreno-Martel[3]

[1] Universidad de Las Palmas de Gran Canaria, Las Palmas, Spain
roberto.morenodiaz@ulpgc.es
[2] Universidad Politécnica de Madrid, Madrid, Spain
[3] University of Aberdeen, Scotland, UK

Abstract. In this paper we will initially make a review of the ample and rich set of ideas found in [2, 3], but in the core of it we shall focus in the revision of concepts related to the bio-cybernetics of visual processes, artificial and machine vision and their consequences in the following 20 years, that is, from 1965 to 1985. In both cited references it is worthy to note not only the industrial description of the designs for autonomous vehicles, the vision acquisition, preprocessing and transmission systems, but at the same time the conceptual exuberance together with a very didactic exposition of details, far from the pre-assumed obscurity of a restricted-access technological report, but much closer to an educational effort of clarity oriented to the academic audience. Quite possibly, the scientific and technical impact of these reports would have been significantly increased at the time, should the legal obligation of keeping them of restricted access for a period of time not had been included. They can be downloaded from the NTRS – NASA Technical Reports Server (https://ntrs.nasa.gov/).

Keywords: Mars exploration systems · Cybernetics · Artificial vision · AI

1 The Environment

In the three-year period 1965–1967 a fruitful collaboration between NASA and the Instrumentation Laboratory Group at the Massachusetts Institute of Technology (IL-MIT), commanded by Dr. Warren McCulloch, took place. A series of reports, developed under a contract intended to build robotic devices to be "junket" (using McCulloch's own terminology [1]) to Mars in search of possible signs of life, became a rich source of ideas, concepts, models and designs that would deeply influence later research in a no smaller number of disciplines. The general context in which this research took place is that of a NASA involved in a time race against the USSR for "conquering" the Red Planet. The goal to place an un-manned robot on Mars' surface, which eventually was accomplished in 1976, came preceded by a number of orbital ships and by the technological development to build a remote controlled robot able to explore the planet's surface.

R. Moreno-Díaz et al. (Eds.): EUROCAST 2019, LNCS 12013, pp. 93–100, 2020.
https://doi.org/10.1007/978-3-030-45093-9_12

Several contracts between NASA and IL-MIT allowed McCulloch's research group and external consultants to join the effort and implement their previous ideas and models on information processing in natural nervous systems in a practical, focused, engineering Project. In retrospect, it is very illustrative to find in both reports [2, 3], of around 100 pages each including detailed schemes, illustrations and references, such a combination of concepts ranging from machine multisensorial integration, neural computation and visual processing in animals to image processing and modeling of planetary environments. Quite possibly, the scientific and technical impact of these reports would have been significantly increased at the time, should the legal obligation of keeping them of restricted access for a period of time not had been included. They can be downloaded from the NTRS – NASA Technical Reports Server (https://ntrs.nasa.gov/). On the formal side of the contracts that gave birth to these reports, it is important to note that the OSRD, Office for Scientific Research and Development (predecessor of the National Science Foundation and directed by Vannevar Bush) set the model of contracting with universities and industry during WWII - that was used for many years afterwards and is still used (with legal variations).

2   The Instrumentation Laboratory MIT Report R-548

This report, titled "Sensory, decision and control systems" [3], is probably the richest of the two reports in terms of concepts, and can be regarded as an actual compendium of applied engineering based on the formal models that W.S. McCulloch's group previously developed, including:

(a) Artificial vision: design and implementation of camera-computer systems; non-(or very slightly)-supervised vision mechanisms for robots and autonomous vehicles; visual and tactile sensory integration.
(b) Artificial Intelligence: Neural networks theory; non-supervised decision making; visual data interpretation and semantic content in nets; behavior and control of actuation in autonomous vehicles.
(c) Bio-cybernetics and models of information processing: formal models and simulations of structure and function of the reticular formation of vertebrate brain; probabilistic models of decision making within the reticular formation of vertebrates; models of neural facilitation and lateral inhibition in the optic tectum of amphibians; modelling of the Group II ganglion cells ("bug detectors") of the frogs retina.

The staff included Warren McCulloch, William Kilmer and Carl Sagan as Consultants, Louis Sutro as assistant director, Richard Catchpole and David Peterson as research assistants, and Jay Blum, Jerome Krasner, Roberto Moreno-Diaz, David Tweed and Joseph Convers as Staff members.

Being the main goal the determination of the presence of life in Mars by visual means, the report devotes some contents to the description of the physical rover, which was a variation of the ones already designed for Moon exploration. On this rover, a complete set of cameras and computers will wander on the planet's surface. Cameras will sensor

Martian landscapes and computers will preprocess the images taken by the cameras and decide on whether the images would be sent to the Earth control to be further analyzed, and the actions to be performed by the rover, like turning, advancing, controlling the robotic arm and hand, the attitude of wheels and other internal monitoring.

The visual subsystems consisted of two cameras, to allow stereoscopic vision whose outputs feed the visual first stage computer in charge of describing the images in terms of basic elements like edges, shadows and some movement analysis. These in turn are fed to the visual decision computer, whose 15 analog output lines wield yield degrees to which certain properties are present on the images. The decision subsystem represents actually the basic intelligence of the robot. Such a computer should be capable of either interpreting what is the object that has been viewed by the visual system in terms of a small number of possible corresponding acts or deciding what is the most likely act that the robot should perform in response to not only visual but to all inputs. Eventually two computers may be developed to handle each of the two requirements, although in the repot only one is being designed that is sufficiently general to handle the needs of both.

For designing its architecture, they took inspiration from the structure of the reticular formation of the Tectum in amphibians, a place in the nervous system that was described to be a decision maker – from visual signals - in those animals. Basically, the Tectum computes velocity and size of moving objects out of signals that originate in the retina in 4 types of cells that feed different processing layers of it. Tectal cells properties are best described via four basic interaction processes: Facilitation, Adaptation, Maximum activity selection and Distribution. These processes were implemented in the artificial version of the Tectum for the Mars rover's computer.

The decision subsystem, as designed in the report, is a probabilistic computer, whose operation resembles the one of a fuzzy system. A concept of importance in the decision system is that of the probability vector (PV). Probability vectors are associated with a decision for which there are n possible responses. Thus, a PV is an n-component vector $H = (p1, p2, \ldots pn)$. Its jth component, pj, represents the probability that the jth response is the appropriate one. For example, in the probability vector $P1 = (0.35, 0.1, 0.15, 0.4)$ the probability is 0.15 that response 3 is the appropriate response. A normalized probability vector is one representing mutually exclusive and collectively exhaustive responses; the components of such a vector thus sum to 1.

Probability vectors are calculated by each computation module (equivalent to a layer of cells in the Tectum) from an initial state (initial guess), communicated to the rest of modules, combined, refined in the values of their components (in an iterative fashion) until the modules reach general agreement or convergence on the decision the robot has to take. It is, in fact, a consensus machine.

The physical structure of the decision system resembles the one of the actual Tectum: it is a stack of layers or modules very similar to "poker chips" (using their own words) in which an intercrossing of parallel and cooperative computation takes place: Each module computes directly from the input information it receives and makes a best guess as to what the corresponding act probably should be. After the initial guess, the modules communicate their decisions to each other over low capacity information channels. Then each module takes the information (from all other modules) and combines it in a nonlinear fashion with the information coming directly into it to arrive at a mixed guess as to what act should be performed. This is in turn communicated to the subset of modules to which it is connected above and below.

Modules are decomposed in two parts: The A part operates on the module's input information. The B part operates on information from above, below and from the A part. The A part with five binary input variables and four analog output variable outputs, is a nonlinear probability transformation network.

3   The Instrumentation Laboratory MIT Report R-548

This second Report titled, "Development of visual contact and decision subsystem for a Mars rover" [2], describes in detail the construction of autonomous vehicles with sensory and actuation capacities, including local decision making – to some degree independent from remote operations from Earth. The main technological topics in this report include:

(a) Navigation and Exploration: Building of 3-D environment models of Mars surface; binocular vision; multisensorial integration.
(b) Sensory data processing: Design and implementation of real-time visual data pre-processing software (including segmentation, object extraction, reconstruction of damaged scenes); data compression systems for environment modelling and data transmission.
(c) Decision-making mechanisms: Non supervised machine learning and decision-making; computational models of brain processing; bioinspired models of visual processing; layered computation; biological learning.

Thus, it explains the detailed implementation of mechanisms and concepts from the previous report, down to an engineering plan. Although the line between research and development cannot be sharply drawn in the text, it can be divided into two broad groups of sections:

- Development includes: "camera computer chains" (binocular vision digital image processing), "representation of the environment" (3D world modelling) and "tying together of the subsystems" (integration).
- Research results include: "Nervous systems as interconnected computers" (biocybernetics), "mathematical concepts of visual processing" (image processing algorithms for retinal simulation), "animal learning" for "adaptation to changes in its environment" (learning, adaptation, intelligence).

Two of the detailed contents are worth mentioning: First, since the rover has to navigate in a – then – practically unknown environment, there arises the need for building a "model" of the physical environment on whatever information (mainly visual) was available. We think this is the earliest effort in robotics to - and the root of today's research of – providing an autonomous robot with an adaptive environment model (that changes, and whose features were fairly mysterious) to interact with.

Second, the neurocybernetic schema, which is a constant in many writings of McCulloch, yields one of the most detailed plan of a general view of the nervous systems as interconnected computers: the engineers model of the neuroscientists descriptions known to that day.

They also show an extension of the decision subsystem developed in the previous report, adding some learning capabilities by increasing the number of memory registers and feedback in the stack modules. This new features makes the whole subsystem resemble a by then not yet defined Perceptron, quite closely.

4 Subsequent Influence

As for subsequent influence of both reports, the concepts and ideas that can be found in the sections dedicated to visual processing were later developed in a very fruitful way by Moreno-Díaz and collaborators.

Thus, for example, following a chronological thread, there is a clear path of development of models in biocybernetics that starts with the functional description of the neurophysiology of the visual system of the frog [4], which is taken as the experimental basis for the construction of a successful formal model [5]. This is in turn the starting point for the proposals of both a generalization of the retinal process (oriented towards a global theory of its functioning) and of models of the data process in the retina of other vertebrates:

First, then, and starting from Section 8 of [2], a proposal is based for a general mathematical model of data processing in the retina based on integral transformations [6, 7], an extension for non-linear data processing mechanisms in the retina [8], the software implementation of such retinal models [9] and the computer simulation of non-linear retinal processes [10].

Second another line of work includes the extension to other species of the models that were initially developed on retinal neurophysiological data of amphibians. Thus, we have a first spatiotemporal model of retinal cells in cats [11] and a layered computation model of the visual process in the retina of birds [12]. On the other hand, a more global and complete formal proposal about the visual process in the frog's retina is found in [13]. The extension of the results of the work developed in [5] and those mentioned in this paragraph gave natural way to the exposition of the foundations of a general theory of data processing in the retina of vertebrates that we can find in [14] and [15].

Another interesting line of work is the one that emanates from some ideas implicit in [3]. In the papers mentioned so far, the interest is focused on the construction of models of neuronal behavior seen both as independent elements and as elements that interact and are organized in layers and processing channels, elements that perform a series of operations and transformations on the input data that allow us to extract in terms of basic descriptors (lines, contrasts, objects, textures, movement) visual information of interest for the global system whose correct functioning is serving. Starting from the frog's Tectum model in [3] a framework for interpreting the semantic content of the neuronal signal at higher levels and the possibility of modelling in neuropsychology are proposed in [16, 17].

At the same time, and having as a common root the first computer implementations of the visual models of inferior vertebrates that we can find in Section 4 of [2], proposals for computational implementations of retinal mechanisms for image processing are developed with a more practical perspective as we find in [18–21]. These theoretical developments will give rise to applied industrial implementations of artificial vision and vision for robots through various research projects and technological innovation in later years, with a summary of these contributions in image processing and form recognition in [22].

Finally, it is natural, following the study of information coding and its transmission between successive layers of neurons in the visual system and deeper into the nervous system (with or without feedback, [23]), that ideas on the integration of neuronal signals of different origins in the visual path that are included in the formal model of the frog tectum developed in [3], would result in more complex theoretical constructs about multisensory integration and cooperation within, and between, nervous subsystems. A group of papers that develop both theoretical proposals and possible practical applications in visual robotics are referenced in [24–27].

5 Conclusions

We like to refer to these two reports as two "modern old" reports – they are fresh in concepts, methods and practicalities and philosophically they follow the ideas within the structure-and-function neuronal school of "What the frog's eye tells the frog's brain" (Lettvin, McCulloch, Maturana, Pitts) [4].

Being reports on a project meant for reaching specific goals, they include solid theory-developing and implementation-ready proposals together with actual realizations. The staff was, quite probably, the top research group in Vision (in a broader sense that includes – brain theory – biocybernetics and camera-computer chains for modelling and implementation) in the US Academia at the time. The texts present some extra current value: the academic/educational value for systems theory and cybernetics subjects (especially in topics related to model construction, interplay between natural and artificial computation, bioinspiration, systems design, goal oriented research etc.) that could be used once the reports were cleared by the Defense Office.

Seen with "todays eyes" something could be missing: computational power in the implementations. This is of course due to a comparative lack of technological development at the time. In the artificial vision system, the deep changes in technology are evident. In the 60s last century, they used a PDP-9 computer for visual interpretation and commanding motion; IBM 360 Model 75 for the decision subsystem; and a third computer telephone connected to the PDP for environment modelling. That is, state-of-the-art machinery of those days.

References

1. McCulloch, W.S.: Logic and closed loops for a computer junket to mars. In: Caianiello, E.R. (ed.) Neural Networks. Springer, Heidelberg (1968). https://doi.org/10.1007/978-3-642-87596-0_7
2. Sutro, L., Moreno-Díaz, R., Kilmer, W., McCulloch, W.S.: Development of visual, contact and decision subsystems for a Mars rover. Instrumentation Laboratory MIT Report R-565 (1967)

3. Sutro, L., et al.: Sensory, decision and control systems. Instrumentation Laboratory MIT Report R-548 (1966)
4. Lettvin, J., Maturana, H., McCulloch, W.S., Pitts, W.: What the frog's eye tells the frog's brain. Proc. Inst. Radio Engr. **47**(11), 1940–1951 (1959)
5. Moreno-Diaz, R.: An analytical model of the group 2 ganglion cell in the frog's retina. Report, Instrumentation Laboratory, MIT, E-1858 (1965)
6. Moreno-Díaz, R., Rubio-Royo, F., Mira-Mira, J.: Aplicación de las Transformaciones integrales al proceso de datos en la retina. Revista de Automática **2**(5), 7–17 (1969)
7. Moreno-Díaz, R.: Models of retinal processes. Brain Theory Newsl. **3**(1), 7–10 (1977)
8. Moreno-Díaz, R., Rubio, E., Rubio-Royo, F.: A generalized model for non-linear retinal processing. In: Rose, J. (ed.) Current Topics in Cybernetics and Systems. Springer, Heidelberg (1978). https://doi.org/10.1007/978-3-642-93104-8_124
9. Moreno-Díaz, R., Santana, O.: Computer programs to implement retinal models. Int. J. Biomed. Comput. **10**, 217–229 (1979)
10. Santana, O., Moreno-Díaz, R.: Simulación de procesos retinales no lineales. Rev Real Academia de Ciencias (J. Span. R. Acad. Sci.) **LXXIV**, 2 (1980)
11. Fernández-Escartín, V., Moreno-Díaz, R.: A spatio-temporal model of cat's retinal cells. Biol. Cybern. **30**(15), 16–22 (1978). https://doi.org/10.1007/BF00365479
12. Moreno-Díaz, R., Rubio Royo, E., Núñez, A.: A layered model for visual processing in Avian retina. Biol. Cybern. **38**(85), 85–89 (1980). https://doi.org/10.1007/BF00356034
13. Moreno-Díaz, R., Rubio-Royo, F., Rubio-Royo, E.: A theoretical proposal to account for visual computation in a frog's retina. Int. J. Biomed. Comput. **11**, 415–426 (1980)
14. Moreno-Díaz, R., Rubio-Royo, F.: Towards a theory of visual processing in vertebrate retina. Appl. Syst. Cybern. **IV**, 1859–1864 (1981)
15. Moreno-Díaz, R., Santana, O., Rubio Royo, E., Núñez, A.: Bases de una Teoría del proceso de datos en la Retina. Biocibernética: Implicaciones en biología, medicina y tecnología. Siglo XXI Editores (1984)
16. Mira-Mira, J., Moreno-Díaz, R.: Un marco teórico para interpretar la función neuronal a altos niveles. Biocibernética: implicaciones en biología, medicina y tecnología. Siglo XXI Eds. (1984)
17. Mira-Mira, J., Moreno-Díaz, R., Delgado-García, A.: Modelos implícitos en neuropsicología. Revista de informática y automática **11**(38), 30–37 (1978). ISSN 0210-8712
18. Rubio-Royo, E., Núñez, A., Moreno-Díaz, R.: Generalización del concepto centro-periferia en una teoría de la retina. Span. R. Acad. Sci. J. **LXXV**(1), 290–292 (1981)
19. Santana, O., Méndez, A., Moreno-Díaz, R.: Momentos normalizados para el proceso de datos visuales. Span. R. Acad. Sci. J. **LXXV**(1), 287–289 (1981)
20. Núñez, A., Candela Solá, S., Muñoz-Blanco, J.A., Moreno-Díaz, R.: Mecanismos mínimos de operación espacio-temporal en la retina de los vertebrados. Span. R. Acad. Sci. J. **LXXV**(4), 915–931 (1981)
21. González Rodríguez, M., Moreno Díaz, R.: Modelo de procesos retinales del color a nivel analítico. In: Proceedings II Simposium Nacional de Ing Biomédica, Madrid (1983)
22. Fortes, J., Moreno-Díaz, R.: Robótica visual. Bit 8(37) (1985)
23. Núñez, A., Moreno-Díaz, R.: Realimentación en el Sistema multivariable retina-centros visuales superiores. In: Proc. II Simposium Nacional de Ing Biomédica (1983)
24. Mira-Mira, J., Moreno-Díaz, R., Delgado-García, A.: A theoretical proposal to embody cooperative decision in the nervous system. In: Proceedings of the Conference on General Systems Research (1983)
25. Moreno-Díaz, R., Fortes, J., Mira, J., Delgado, A.: Proceso de datos visuales e integración multisensorial en robots. In: Proc. Jornadas de Robótica y Fabricación Flexible (1983)

26. Araujo, C.P.S., Moreno-Díaz, R.: Intersensorial transformations: general systems description and implications. In: Pichler, F., Díaz, R.M., Albrecht, R. (eds.) EUROCAST 1995. LNCS, vol. 1030, pp. 73–87. Springer, Heidelberg (1996). https://doi.org/10.1007/BFb0034751
27. Mira, J., Delgado, A., Moreno-Díaz, R.: The fuzzy paradigm for knowledge representation in cerebral dynamics. In: Proceedings of the First International Fuzzy Systems Association Congress, vol. I (1985)

Lotfi Zadeh: Fuzzy Sets and Systems

Rudolf Seising[(✉)] [iD]

Deutsches Museum, Munich, Germany
r.seising@deutsches-museum.de

Abstract. We give an outline of the history of the origin of the theory of Fuzzy Sets and Systems in the life of its founder Lotfi A. Zadeh. We present roots of this theory in the developments of computers, systems and information theory.

Keywords: Lotfi A. Zadeh · Thinking machines · System theory · Fuzzy sets · Fuzzy systems

1 Introduction

"Fuzzy Set and Systems" was the title of one of the first texts that the electrical engineer and Berkeley-professor Lotfi A. Zadeh (1921–2017) wrote about the theory of fuzzy sets [1]. He published these first ideas on a mathematical theory of fuzzy quantities by the mid-1960s [1–4]. In this contribution, we outline the prehistory up to then. In the next section, we show how Zadeh's thinking as an assistant professor at the Columbia University in New York leading to the theory of fuzzy sets was rooted in the developments of computers, systems theory and information theory. Some other aspects on the history of fuzzy sets and systems the reader will find in the paper "From Linear Systems to Fuzzy Systems to Perception-based Systems" in this volume[1].

2 Thinking Machines

About two years ago, on 6 September 2017, Lotfi A. Zadeh, the founder of the theory of fuzzy sets and systems passed away. He was the son of an Azerbaijani father with Iranian roots, and a Russian mother. He was born in 1921 in Baku, Azerbaijan and the family chose to move back to Tehran, Iran in 1931 due to Stalin's harsh immigration policies. He studied electrical engineering in Tehran, where the University awarded him the BSc. degree in 1942. In 1943, he immigrated to the USA where he worked for the *International Electronic Laboratories* in New York. In the following year, he went to Boston to continue his studies at the *Massachusetts Institute of Technology* (MIT). Amongst others, he attended the lectures of Norbert Wiener (18941964) and Ernst A. Guillemin (1898–1970). In 1946, Zadeh had earned a Master of Science degree. His thesis was "An Investigation of Current Distribution and Radiation Field of a Solenoidal

---

[1] For other aspects of the history of fuzzy sets and systems see the paper "From Linear Systems to Fuzzy Systems" in this volume. For a detailed study on that history see [5].

© Springer Nature Switzerland AG 2020
R. Moreno-Díaz et al. (Eds.): EUROCAST 2019, LNCS 12013, pp. 101–108, 2020.
https://doi.org/10.1007/978-3-030-45093-9_13

Antenna". He changed to the Columbia University in New York and he acquired a position as an instructor in the Department of Electrical Engineering. He was responsible for teaching the theories of circuits and electromagnetism. In 1950, his department at Columbia awarded him a PhD for his thesis "Frequency Analysis of Variable Networks" [6].

In 1950, the same year when Alan M. Turing published his famous article "Computing Machinery and Intelligence", in which he proposed the "Imitation game", later named "Turing Test" [7], Zadeh wrote an article entitled "Thinking Machines – A New Field in Electrical Engineering" (Fig. 1) for the *Columbia Engineering Quarterly* [8]. He was interested in "the principles and organization of machines which behave like a human brain." Such machines were then variously referred to as "thinking machines", "electronic brains", "thinking robots", and similar names. He mentioned that the "same names are frequently ascribed to devices which are not «thinking machines» in the sense used in this article"; he therefore made the following distinction: "The distinguishing characteristic of thinking machines is the ability to make logical decisions and to follow these, if necessary, by executive action." [8, p. 12]

Fig. 1. Heading of Zadeh's article [8].

In this paper, he jumped to a conclusion that he retracted more than half a century later: "In 1950, I wrote a paper entitled 'Thinking Machines – A New Field in Electrical Engineering'. Like many others, I had greatly underestimated the difficulty of designing machines that can approximate to the remarkable human ability to reason and make decisions in an environment of uncertainty and imprecision". [9, p. 96] In this early paper [8] he said "that a thinking machine is a device which arrives at a certain decision or answer through the process of evaluation and selection. Despite its simplicity, Haufe's machine is typical in that it possesses a means for arriving at a logical decision based on evaluation of a number of alternatives. Thus, M.I.T.'s differential analyser is not a thinking machine, for it can not make any decisions, except trivial ones, on its own initiative. However, the recently built large-scale digital computers, UNIVAC and BINAC, are endowed with the ability to make certain non-trivial decisions and hence can be classified as thinking machines. The modern large-scale digital computers are relatively narrow thinkers". However, he was right in a different respect: "In fact, such machines may be commonplace in anywhere from ten to twenty years hence. Furthermore, it is absolutely certain that thinking machines will play a major role in any armed conflict that may arise in the future" [8].

3 Information Theory

In one of many interviews Zadeh told me that he "was very much influenced by Shannon's talk that he gave in New York in 1946 in which he described his information theory". [10] Zadeh began to deliver lectures on automata theory, and in 1949, he organized and moderated a discussion meeting on digital computers at Columbia University in New York, in which Claude E. Shannon, Edmund Berkeley and Francis J. Murray took part. It was probably the first public debate on this subject ever! [11].

In the years before Shannon was engaged in developing a general theory of communication of information. The parallel work on secret binary codes and ways to improve them obviously led him to employ statistical considerations. He included "new factors" to the theory of telegraphy by Ralph Vinton Lyon Hartley [12], in particular the effect of noise in the communication channel, and the savings possible due to the statistical structure of the original message and due to the nature of the final destination of the information" [13, p. 379]. To illustrate this mathematical theory, Shannon had drawn a diagram of a general communication model (Fig. 2).

In his famous paper, Shannon formulated the essential difficulty of the communication process: "The fundamental problem of communication is that of reproducing at one point either exactly or approximately a message selected at another point" [13, p. 379]. Obviously, he conceived of communication purely as the transmission of messages – completely detached from the meaning of the symbols.

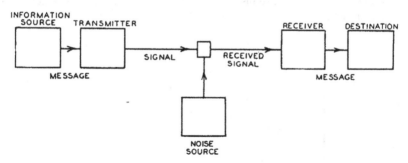

Fig. 2. Shannon's communication model [13].

Referring to this "fundamental problem", Zadeh presented in February 1952 "Some Basic Problems in Communication of Information" at the meeting of the Section of Mathematics and Engineering of the New York Academy of Sciences. Here, we sketch one of these problems that deals with the recovery process of transmitted signals. In the proceedings of this meeting, Zadeh wrote: "Let $X = \{x(t)\}$ be a set of signals. An arbitrarily selected member of this set, say $x(t)$, is transmitted through a noisy channel C and is received as $y(t)$. As a result of the noise and distortion introduced by Γ, the received signal $y(t)$ is, in general, quite different from $x(t)$. Nevertheless, under certain conditions it is possible to recover $x(t)$ – or rather a time-delayed replica of it – from the received signal $y(t)$" [14, p. 201].

In this paper, he didn't examine the case where $\{x(t)\}$ is an ensemble; he restricted his view to the problem to recover $x(t)$ from $y(t)$ "irrespective of the statistical character of $\{x(t)\}$" [14, p. 201]. Corresponding to the relation $y = \Gamma x$ between $x(t)$ and $y(t)$ he represented the recovery process of $y(t)$ from $x(t)$ by $x = \Gamma^{-1}y$, where $\Gamma^{-1}y$ is the inverse of Γ, if it exists, over $\{y(t)\}$. Zadeh represented signals as ordered pairs of points in a signal space S, which is imbedded in a function space with a delta-function basis, and to measure the disparity between $x(t)$ and $y(t)$ he attached a distance function $d(x,y)$ with the usual properties of a metric. Then he considered the special case in which it is possible to achieve a perfect recovery of the transmitted signal $x(\mathbf{t})$ from the received signal $y(t)$. He supposed that "$X = \{x(t)\}$ consist of a finite number of discrete signals $x_1(t)$, $x_2(t)$,..., $x_n(t)$, which play the roles of symbols or sequences of symbols. The replicas of all these signals are assumed to be available at the receiving end of the system. Suppose that a transmitted signal x_k is received as y. To recover the transmitted signal from y, the receiver evaluates the 'distance' between y and all possible transmitted signals x_1, x_2,. .., x_n, by the use of a suitable distance function $d(x,y)$, and then selects that signal which is 'nearest' to y in terms of this distance function (Fig. 3). In other words, the transmitted signal is taken to be the one that results in the smallest value of $d(x, y)$. This in brief, is the basis of the reception process." [14, p. 201]. By this process the received signal x_k is always 'nearer' – in terms of the distance functional $d(x, y)$ – to the transmitted signal $y(t)$ than to any other possible signal xi, i.e. $d(x_k, y) < d(x_i, y)$, $i \neq k$, for all k and i. At the end of his reflection of this problem Zadeh conceded that "in many practical situations it is inconvenient, or even impossible, to define a quantitative measure, such as a distance function, of the disparity between two signals. In such cases we may use instead the concept of neighborhood, which is basic to the theory of topological spaces" [14, p. 202].

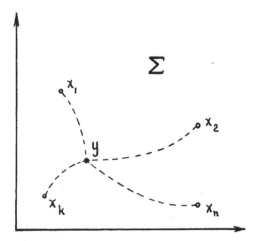

Fig. 3. Zadeh's illustration: "Recovery of the input signal by means of the comparison of the distances between the received signal y and all possible transmitted signals" [14].

Spaces such as these, Zadeh surmised, could be very interesting with respect to applications in communication engineering. Therefore, this problem of the recovery process of transmitted signals which is a special case of Shannon's fundamental problem of communication, the problem "of reproducing at one point either exactly or approximately a message selected at another point" [12], was one of the problems that initiated Zadeh's thoughts about not precisely specified quantitative measures, i.e. or cloudy or fuzzy quantities. About 15 years later, he proposed his new 'concept of neighborhood', which is now basic to the theory of fuzzy systems!

4 Fuzzy Sets

Since the late 1950s, Zadeh had criticized the relationship between mathematics and his own scientific–technical discipline of electrical engineering. The tools offered by mathematics were not appropriate for the problems that needed to be handled. Information and communication technology had led to the construction and design of high-grade complex systems. To measure and analyze these systems it took much more effort than had been the case just a few years before. Methods that are much more exact were now required to identify, classify, or characterize such systems or to evaluate and compare them in terms of their performance or additivity.

To provide a mathematically exact expression of experimental research with real systems, it was necessary to employ meticulous case differentiations, differentiated terminology, and definitions that were adapted to the actual circumstances, things for which the language normally used in mathematics could not account. The circumstances observed in reality could no longer simply be described using the available mathematical means.

In the summer of 1964, Zadeh was thinking about pattern recognition problems and grades of membership of an object to be an element of a class.[2] Almost 50 years later, he recalled how he was thinking at the time: "While I was serving as chair, I continued to do a lot of thinking about basic issues in systems analysis, especially the issue of unsharpness of class boundaries. In July 1964, I was attending a conference in New York and was staying at the home of my parents. They were away. I had a dinner engagement, but it had to be canceled. I was alone in the apartment. My thoughts turned to the unsharpness of class boundaries. It was at that point that the simple concept of a fuzzy set occurred to me. It did not take me long to put my thoughts together and write a paper on the subject." [18, p. 7]

Zadeh submitted his paper "Fuzzy Sets" to the editors of the journal *Information and Control* in November 1964, and it was published the following June [2]. He introduced new mathematical entities as classes or sets that "are not classes or sets in the usual sense of these terms, since they do not dichotomize all objects into those that belong to the class and those that do not". He established "the concept of a fuzzy set, that is a class in which there may be a continuous infinity of grades of membership, with the grade of membership of an object x in a fuzzy set A represented by a number $f_A(x)$ in the interval [0, 1]" [1].

Zadeh generalized various concepts, union of sets, intersection of sets, and so forth. He defined equality, containment, complementation, intersection, and union relating to fuzzy sets A, B in any universe of discourse X as follows (for all $x \in X$) (Fig. 4):

---

[2] For a detailed history of the theory of fuzzy sets see [5, 15–17].

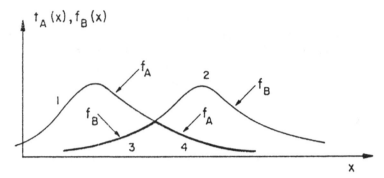

Fig. 4. Zadeh's figure of fuzzy sets [1].

- $A = B$ if and only if $f_A(x) = f_B(x)$
- $A \subseteq B$ if and only if $f_A(x) \leq f_B(x)$
- A' is the complement of A, if and only if $f_{A'}(x) = 1 - f_A(x)$
- $A \cup B$ if and only if $f_{A \cup B}(x) = \max(f_A(x), f_B(x))$
- $A \cap B$ if and only if $f_{A \cap B}(x) = \min(f_A(x), f_B(x))$

In these definitions of fuzzy operators for the union and intersection, we discover the influence of Zadeh's earlier works on electrical filters. Zadeh regarded his theory of fuzzy systems as a general system theory that he planned to use to cope with the so-called input–output analysis of systems. His interpretation goes back to Shannon's discovery of the use of electrical circuits to model logical statements. In the case of conventional sets, every set C from sets $A_1, \ldots, A_i, \ldots A_n$ can be combined with one another using the conjunctions $\cup$ and $\cap$ such that it represents a network of circuits $a_1, \ldots, a_n$ $(i, j = 1, \ldots, n)$. By this logic, $A_1 \cup A_j$ and $A_1 \cap A_j$ are, respectively, series and parallel combinations of the circuits A_1 and A_j. For the analogous interpretation in the case of fuzzy sets, Zadeh employed the concept of the sieve. He provided the membership function $f_i(x)$ of A_i at x with a sieve $S_i(x)$ with mesh size $f_i(x)$. This interpretation results in immediate and clearly evident correlations of the parallel combinations of sieves $S_i(x)$ and $S_j(x)$ with $f_i(x) \vee f_j(x)$ and of their series combinations with $f_i(x) \wedge f_j(x)$ (Fig. 5). Considering that the term sieve connotes the meaning of a filter, then one may catch the analogy of fuzzy sets and electrical filters.

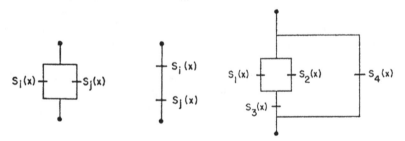

Fig. 5. Zadeh's illustrations of parallel (left) and serial combination (middle) of sieves and a network of sieves (right) [1, p. 14].

Looking for applications of his theory Zadeh compared the strategies of problem-solving by computers, on the one hand, and by humans, on the other hand. In a conference paper in 1969, he called it a paradox that the human brain is always solving problems by manipulating "fuzzy concepts" and "multidimensional fuzzy sensory inputs", whereas "the computing power of the most powerful, the most sophisticated digital computer in existence" is not able to do this. Therefore, he stated, "in many instances, the solution to a problem need not be exact", so that a considerable measure of fuzziness in its formulation and results may be tolerable" [19]. A decade later Zadeh wrote the article "Making Computers Think like People" [20]. In his view, computers do not think like humans. For this purpose, the machine's ability "to compute with numbers" should be supplemented by an additional ability that is similar to human thinking: computing with words and perceptions (Fig. 6).

Fig. 6. Part of the title page of Zadeh's article [20].

References

1. Zadeh, L.A.: Fuzzy sets and systems. In: Fox, J. (ed.) System Theory. Microwave Research Institute Symposia Series XV. Polytechnic Press, Brooklyn, pp. 29–37 (1965)
2. Zadeh, L.A.: Fuzzy sets. Inf. Control **8**(3), 338–353 (1965)
3. Bellman, R.E., Kalaba, R., Zadeh, L.A.: Abstraction and pattern classification. J. Math. Anal. Appl. **13**, 1–7 (1966)
4. Zadeh, L.A.: Shadows of fuzzy sets. In: Problemy peredachi informatsii, Akademija Nauk SSSR Moskva, **2**(1), 37–44 (1966). Engl. transl. in: Problems of Information Transmission: A Publication of the Academy of Sciences of the USSR; the Faraday Press coverto-cover transl., New York, NY, 2 (1966)
5. Seising, R.: The Fuzzification of Systems. The Genesis of Fuzzy Set Theory and Its Initial Applications—Developments up to the 1970 s. Studies in Fuzziness and Soft Computing 216. Springer, Heidelberg (2007). https://doi.org/10.1007/978-3-540-71795-9

6. Zadeh, L.A.: Frequency analysis of variable networks. In: Proceedings of the IRE, vol. 38, pp. 291–198, March 1950

7. Turing, A.M.: Computing Machinery and Intelligence. In: Mind, LIX, pp. 433–460 (1950)

8. Zadeh, L.A.: Thinking Machines–A new field in electrical engineering. Columbia Eng. Q. **12–13**, 30–31 (1950)

9. Zadeh, L.A.: The evolution of systems analysis and control: a personal perspective. IEEE Control Syst. Mag. **16**, 95–98 (1996)

10. Zadeh, L.A.: In an interview with the author on 8 September 1999, in Zittau, at the margin of the 7th Zittau Fuzzy Colloquium at the University Zittau/Görlitz

11. Zadeh, L.A.: In an interview with the author on 15 June 2001, University of California, Berkeley, CA, Soda Hall

12. Hartley, R.V.L.: Transmission of information. Bell Syst. Tech. J. **8**(3), 535–563 (1928)

13. Shannon, C.E.: A mathematical theory of communication. Bell Syst. Tech. J. **27**, 379–423 and 623-656 (1948)

14. Zadeh, L.A.: Some basic problems in communication of information. The New York Academy of Sciences, Series II **14**(5), 201–204 (1952)

15. Seising, R.: When computer science emerged and fuzzy sets appeared. The contributions of Lotfi A. Zadeh and other pioneers. IEEE Syst. Man Cybern. Mag. **1**(3), 36–53 (2015)

16. Seising, R., Lotfi, A.: Zadeh (1921–2017) – his life and work from the perspective of a historian of science. Int. J. Fuzzy Sets Syst. **331**, 3–11 (2018)

17. Seising, R.: Fuzzy mathware and soft computing: obituary for a visionary scientist: Lotfi Aliasker Zadeh (1921–2017). Mathware Soft Comput. Mag. **24**(2), 6–15 (2017)

18. Zadeh, L.A.: My life and work—A retrospective view. Appl. Comp. Math. **10**(1), 4–9 (2011)

19. Zadeh, L.A.: Fuzzy languages and their relation to human and machine intelligence. In: Marois, M. (ed.) Proceedings of International Conference Man and Computer, Bordeaux, 22–26 June 1970, p. 132 (1970)

20. Zadeh, L.A.: Making computers think like people. IEEE Spectr. **8**, 26–32 (1984)

Mathematical Reconstruction of the Enigma

Otmar Moritsch[(⊠)]

Technisches Museum, Mariahilfer Straße 212, 1140 Vienna, Austria
otmar.moritsch@tmw.at

Abstract. This paper demonstrates the mathematical reconstruction of the three-rotor German cyphering machine Enigma by Polish cryptologists around Marian Rejewski in 1932 [1].

Keywords: Cryptography · Cyphering machine · Enigma · Radio message · Telegraphy · History of information and communication

1 The Enigma Machine and Its Operation Procedure

Around 1930, the standard cyphering machine of the German Wehrmacht was the three-rotor Enigma, which looks from the outside like a typewriter in a wooden box. The electro-mechanical parts of the Enigma, which are essential for the encryption process, are the keyboard (*Tastatur*), the plugboard with plug connections (*Steckerverbindungen*), three interchangeable code rotors or wheels (*Walzen*) with adjustable rings (*Ringe*), the reflector drum (*Umkehrwalze*) and the lamp panel (*Lampenfeld*).

When a key is pressed, the electrical circuit closes, with the current flowing via the plugboard (S), then through the three code rotors (W_1, W_2, W_3) as well as the reflector drum (U) and from there back through the three code rotors and the plugboard to the lamp panel. After a keypress, the rotor located in the first position (on the right) rotates forward by one rotational step before the electrical circuit closes as the mechanical force applied to the key is used to move the first rotor one step forward. As long as the key is pressed, the electrical circuit remains closed and the allocated cipher is illuminated by electricity from a battery inside. The three rotors have different wirings inside and a turnover notch on different rotating positions. On an average, the middle rotor moves forward after 26 and the left rotor after 676 keystrokes.

Printed codebooks were delivered to the troops by courier and defined the daily changes in settings for a month in advance. According to these instructions, before using an Enigma machine every encoder had to carry out the following settings: he had to establish six specific plug connections (*Steckerverbindungen*) on the plugboard, insert the rotors into the machine in the stated wheel order (*Walzenlage*) and rotate the rings into the stated position (*Ringstellungen*). After closing the lid of the machine, he had to rotate the three rotors until the required basic position (*Grundstellung*) – for example the digits 02-03-01 for B-C-A – were visible. Ring settings, the basic setting of the rotors and plug connections changed daily, while the order of the rotors was changed in 1932 every three months. However, these settings of the daily keying element (*Tagesschlüssel*) on

R. Moreno-Díaz et al. (Eds.): EUROCAST 2019, LNCS 12013, pp. 109–115, 2020.
https://doi.org/10.1007/978-3-030-45093-9_14

their own could not ensure secure radio traffic. Enemy cryptologists would have found it relatively easy, from an adequate amount of intercepted encrypted radio messages that originated from the same code, to deduce this code by mathematical means and decrypt the radio messages. In addition, the loss of a single code book – which could be expected in times of war – would have compromised all the radio traffic for a month. For this reason, a further encryption level was provided: that of the message key (*Spruchschlüssel*) – a further rotor setting for the encryption of the actual message.

The rotor setting of the message key was not in the code book, but had to be selected freely by the encoder for each individual radio message. He typed the selected three letters – for example A-D-K – two times in succession on his machine which had been preset for the daily keying element. The thus resulting six ciphers were written down. Then he set the rotors in accordance with the three letters of his message key and encoded the message with this setting; the ciphers illuminating in each case were also jotted down. The ciphers were then sent by the radio operator according to the generally used Morse alphabet: first the six ciphers of the message key which was typed twice and then sent to avoid any transmission errors that regularly occur in radio communication as a result of atmospheric interference. Following the message key the code of the message was sent. On the receiving end the operators worked on the same Enigma machine with identical daily keying elements. The received Morse signals were subsequently decoded beginning with the six ciphers of the message key, which has to be appear again twice, for example A-D-K-A-D-K. Then the receiver had to set the three rotors to the starting position A-D-K and the remaining radio message had to be typed into the Enigma.

The operators had the free choice of three letters for the message key. They often chose trivial combinations, like A-A-A or A-B-C etc. In the case of frequently occurring message keys which were encoded with an identical daily keying element, the plain letters chosen by the operator could sometimes be guessed and the so-called six characteristic transformations of a given day could be reconstructed. Normally 60 to 80 radio messages a day were enough to find out the six characteristic transformations of the day – the six letters of the message key.

2   The Enigma Equation and the Reconstruction of the Internal Wirings of the Three Rotors and the Reflector Drum

The respective transformation of a plain letter to its corresponding cipher in any arbitrary place could be written as follows (both rotors W_2 and W_3 can be assumed as constant in most cases – at least for the six positions (*Stelle*) of the message key:

$$\text{Stelle} = S\,D\,W_1 D^{-1} W_2 W_3 U\,W_3^{-1} W_2^{-1} D\,W_1^{-1} D^{-1} S^{-1} \tag{1}$$

The operator D represents one rotational step (*Drehschritt*) forward of the right rotor W_1 and is equal to a simple alphabet permutation.

Assuming that the middle and left rotor do no rotate momentarily, thus do not execute a turnover, they can be combined together with the reflector drum to a constant block U_C. This enables the derivation of six equations for the six characteristic transformations of a day, which show S, W_1 and U_C as three unknowns on one side:

$$\text{Stelle}_1 = SDW_1 D^{-1} U_C DW_1^{-1} D^{-1} S^{-1} \tag{2}$$

$$\text{Stelle}_2 = SD^2W_1D^{-2}U_CD^2W_1^{-1}D^{-2}S^{-1} \tag{3}$$

$$\text{Stelle}_3 = SD^3W_1D^{-3}U_CD^3W_1^{-1}D^{-3}S^{-1} \tag{4}$$

$$\text{Stelle}_4 = SD^4W_1D^{-4}U_CD^4W_1^{-1}D^{-4}S^{-1} \tag{5}$$

$$\text{Stelle}_5 = SD^5W_1D^{-5}U_CD^5W_1^{-1}D^{-5}S^{-1} \tag{6}$$

$$\text{Stelle}_6 = SD^6W_1D^{-6}U_CD^6W_1^{-1}D^{-6}S^{-1} \tag{7}$$

The constant block contains the reflector drum and two rotors:

$$U_C = W_2W_3U\ W_3^{-1}W_2^{-1} \tag{8}$$

After transforming Eqs. (2), (3), (4), (5) and (6) one finds:

$$T_1 = D^{-1}S^{-1}\text{Stelle}_1 SD = W_1D^{-1}U_CD\ W_1^{-1} \tag{9}$$

$$T_2 = D^{-2}S^{-1}\text{Stelle}_2 SD^2 = W_1D^{-2}U_CD^2W_1^{-1} \tag{10}$$

$$T_3 = D^{-3}S^{-1}\text{Stelle}_3 SD^3 = W_1D^{-3}U_CD^3W_1^{-1} \tag{11}$$

$$T_4 = D^{-4}S^{-1}\text{Stelle}_4 SD^4 = W_1D^{-4}U_CD^4W_1^{-1} \tag{12}$$

$$T_5 = D^{-5}S^{-1}\text{Stelle}_5 SD^5 = W_1D^{-5}U_CD^5W_1^{-1} \tag{13}$$

$$T_6 = D^{-6}S^{-1}\text{Stelle}_6 SD^6 = W_1D^{-6}U_CD^6W_1^{-1} \tag{14}$$

The goal for the next calculation steps is to eliminate the constant terms in the equations for the characteristic transformations. The related products of T read:

$$T_1T_2 = W_1D^{-1}U_CDW_1^{-1}W_1D^{-2}U_CD^2W_1^{-1} = W_1D^{-1}\left(U_CD^{-1}U_CD\right)D\ W_1^{-1} \tag{15}$$

$$T_2T_3 = W_1D^{-2}U_CD^2W_1^{-1}W_1D^{-3}U_CD^3W_1^{-1} = W_1D^{-2}\left(U_CD^{-1}U_CD\right)D^2W_1^{-1} \tag{16}$$

$$T_3T_4 = W_1D^{-3}U_CD^3W_1^{-1}W_1D^{-4}U_CD^4W_1^{-1} = W_1D^{-3}\left(U_CD^{-1}U_CD\right)D^3W_1^{-1} \tag{17}$$

Inserting (16) into (15) and respective (17) into (16) one gets:

$$T_1T_2 = \left(W_1D\ W_1^{-1}\right)T_2T_3\left(W_1D\ W_1^{-1}\right)^{-1} \tag{18}$$

$$T_2T_3 = \left(W_1D\ W_1^{-1}\right)T_3T_4\left(W_1D\ W_1^{-1}\right)^{-1} \tag{19}$$

Since the transformations T and their products originate from the same starting position, namely the basic position of the daily keying element, their corresponding cycles must have the same structure.

At the end of 1932, Marian Rejewski got also several spy documents listing daily keying elements with the secret rotor orders and plugboard connections for the months of September and October. Now he was able to reconstruct the missing elements: the wirings of the rotors and the reflector drum. E.g. for a day in September, the following daily keying elements were given:

> wheel order (*Walzenlage*) III-II-I
> ring position (*Ringstellung*) 01-01-01
> basic position (*Grundstellung*) 01-01-01
> plug connections (*Steckerverbindungen*) (BE)(RS)(KD)(WM)(CX)(PQ)

Inserting the six characteristic transformations of a day, one can find with the help of the Eqs. (9), (10), (11) and (12) and (15), (16) and (17) the following cycles for the several products of T:

```
T₁ T₂ = (ATP)(BHIQWOMRC)(DNZGXVJLF)(EYU)(K)(S)
T₂ T₃ = (ASZURPFQB)(CGW)(DHV)(EXNOYJTIK)(L)(M)
T₃ T₄ = (AYK)(BWIVPOEDL)(CSQJGUHNM)(FXZ)(R)(T)
```

If one rearranges the cycles – cyclic permutations of the letters inside a cycle is allowed – one can find a consistent solution:

```
T₁ T₂ = (K)(CBHIQWOMR)(EYU)(XVJLFDNZG)(TPA)(S)
T₂ T₃ = (M)(KEXNOYJTI)(CGW)(SZURPFQBA)(HVD)(L)
T₃ T₄ = (T)(MCSQJGUHN)(KAY)(LBWIVPOED)(XZF)(R)
```

With (18) one finally gets the wiring pattern of the first rotor:

$$W_I \ D \ W_I^{-1} = \text{(KMTHXSLRINQOJUWYGADFPVZBEC)}$$

The correct wiring of rotor W_I corresponds to one of the 26 possible permutations of the alphabet.

In the secret spy documents for the month October were e.g. the following daily keying elements given:

> wheel order (*Walzenlage*) I-III-II
> ring position (*Ringstellung*) 01-01-01
> basic position (*Grundstellung*) 01-01-06
> plug connections (*Steckerverbindungen*) (AB)(CD)(EF)(GH)(IJ)(KL)

According to the previous calculation for the wiring pattern of rotor W_I, one gets for the wiring pattern of rotor W_{II}:

$$W_{II} \ D \ W_{II}^{-1} = \text{(AJPCZWRLFBDKOTYUQGENHXMIVS)}$$

In order to find the correct wirings for both rotors W_I and W_{II}, one has to look at two days with the same wheel order and identical basic position for the rotor W_{III} but with arbitrary basic positions for the rotors W_I and W_{II}. Assuming that the following keying elements are given for the month September:

wheel order (*Walzenlage*) III-II-I
ring position (*Ringstellung*) 01-01-01

day T1: basic position (*Grundstellung*) 01-01-01
plug connections (*Steckerverbindungen*) (BE)(RS)(KD)(WM)(CX)(PQ)

day T2: basic position (*Grundstellung*) 01-02-02
plug connections (*Steckerverbindungen*) (AB)(CD)(EF)(GH)(IJ)(KL)

The corresponding equations for the two characteristic transformations of a day are:

$$\text{Stelle}_{1\_T1} = S_1 D W_I D^{-1} W_{II} W_{III} U \, W_{III}^{-1} W_{II}^{-1} D \, W_I^{-1} D^{-1} S_1^{-1} \tag{20}$$

$$\text{Stelle}_{1\_T2} = S_2 D W_I D^{-1} W_{II} W_{III} U \, W_{III}^{-1} W_{II}^{-1} D \, W_I^{-1} D^{-1} S_2^{-1} \tag{21}$$

After rearranging the equations one gets:

$$W_{III} U W_{III}^{-1} = W_{II}^{-1} D W_I^{-1} D^{-1} S_1^{-1} \text{Stelle}_{1\_T1} S_1 D \, W_I D^{-1} W_{II} \tag{21}$$

$$W_{III} U W_{III}^{-1} = W_{II}^{-1} D W_I^{-1} D^{-1} S_2^{-1} \text{Stelle}_{1\_T2} S_2 D \, W_I D^{-1} W_{II} \tag{22}$$

Since the left-hand side of the equations above has to be equal, one can find the correct wiring of the rotors W_I and W_{II} with the following permutations:

```
W_I  = (EKMFLGDQVZNTOWYHXUSPAIBRCJ)
W_II = (AJDKSIRUXBLHWTMCQGZNPYFVOE)
```

In order to reconstruct the third rotor, Marian Rejewski compared four days with the third rotor at third (left-hand) position and equidistant basic positions of rotor W_{III} but with arbitrary basic positions for the rotors W_I and W_{II}. Assuming following keying elements for September are listed in the secret spy documents:

wheel order (*Walzenlage*) III-II-I
ring position (*Ringstellung*) 01-01-01

day T1: basic position (*Grundstellung*) 01-01-01
plug connections (*Steckerverbindungen*) (BE)(RS)(KD)(WM)(CX)(PQ)

day T2: basic position (*Grundstellung*) 02-02-02
plug connections (*Steckerverbindungen*) (AB)(CD)(EF)(GH)(IJ)(KL)

day T3: basic position (*Grundstellung*) 03-03-03
plug connections (*Steckerverbindungen*) (MN)(OP)(QR)(ST)(UV)(WX)

day T4: basic position (*Grundstellung*) 04-04-04
plug connections (*Steckerverbindungen*) (CR)(DJ)(EK)(FZ)(GP)(HW)

This enables the derivation of the corresponding four equations for the characteristic transformation of a day:

$$\text{Stelle}_{1\_T1} = S_1 DW_I D^{-1} W_{II} W_{III} U\, W_{III}^{-1} W_{II}^{-1} D\, W_I^{-1} D^{-1} S_1^{-1} \tag{23}$$

$$\text{Stelle}_{1\_T2} = S_2 DW_I D^{-1} W_{II} W_{III} U\, W_{III}^{-1} W_{II}^{-1} D\, W_I^{-1} D^{-1} S_2^{-1} \tag{21}$$

$$\text{Stelle}_{1\_T3} = S_3 DW_I D^{-1} W_{II} W_{III} U\, W_{III}^{-1} W_{II}^{-1} D\, W_I^{-1} D^{-1} S_3^{-1} \tag{22}$$

$$\text{Stelle}_{1\_T4} = S_4 DW_I D^{-1} W_{II} W_{III} U\, W_{III}^{-1} W_{II}^{-1} D\, W_I^{-1} D^{-1} S_4^{-1} \tag{24}$$

After rearranging the equations one gets:

$$Z_1 = W_{III} U W_{III}^{-1} = W_{II}^{-1} DW_I^{-1} D^{-1} S_1^{-1} \text{Stelle}_{1\_T1} S_1 D\, W_I D^{-1} W_{II} \tag{25}$$

$$Z_2 = W_{III} U\, W_{III}^{-1} = W_{II}^{-1} D\, W_I^{-1} D^{-1} S_2^{-1} \text{Stelle}_{1\_T2} S_2 DW_I D^{-1} W_{II} \tag{26}$$

$$Z_3 = W_{III} U W_{III}^{-1} = W_{II}^{-1} DW_I^{-1} D^{-1} S_3^{-1} \text{Stelle}_{1\_T3} S_3 D\, W_I D^{-1} W_{II} \tag{27}$$

$$Z_4 = W_{III} U\, W_{III}^{-1} = W_{II}^{-1} D\, W_I^{-1} D^{-1} S_4^{-1} \text{Stelle}_{1\_T4} S_4 DW_I D^{-1} W_{II} \tag{28}$$

Similar to (18) and (19) one can build the products of the transformations Z and eliminate the reflector drum U:

$$Z_1 Z_2 = \left(W_{III} D\, W_{III}^{-1} \right) Z_2 Z_3 \left(W_{III} D\, W_{III}^{-1} \right)^{-1} \tag{29}$$

$$Z_2 Z_3 = \left(W_{III} D\, W_{III}^{-1} \right) Z_3 Z_4 \left(W_{III} D\, W_{III}^{-1} \right)^{-1} \tag{30}$$

Inserting the characteristic transformations of the four days into (26) and (27) one can derive the following cycles for the corresponding products of Z (here the cycles are already rearranged in the consistent cyclic order):

```
Z₁ Z₂ = (QNOWXIMSZDLKY)(RBVFJACGTUHPE)
Z₂ Z₃ = (EZMLJXTDHQROI)(KPNVWGSBAFYCU)
Z₃ Z₄ = (UHTRWJAQYEKMX)(OCZNLBDPGVISF)
```

One finally gets the wiring pattern of the third rotor W_{III}:

```
WIII D W⁻¹III = (QEUFVNZHYIXJWLRKOMTAGBPCSD)
```

Again, the correct wiring of rotor W_{III} corresponds to one of the 26 possible permutations of the alphabet.

By using two different days in September and October with identical basic position for rotor W_{III} but arbitrary basic positions for the other rotors, one can reconstruct the correct wiring of the third rotor W_{III} and the reflector drum U:

day T1: wheel order (*Walzenlage*) III-II-I
ring position (*Ringstellung*) 01-01-01
basic position (*Grundstellung*) 01-01-01
plug connections (*Steckerverbindungen*) (BE)(RS)(KD)(WM)(CX)(PQ)

day T2: wheel order (*Walzenlage*) I-III-II
ring position (*Ringstellung*) 01-01-01
basic position (*Grundstellung*) 01-01-06
plug connections (*Steckerverbindungen*) (AB)(CD)(EF)(GH)(IJ)(KL)

The reflector drum is then given by:

$$U = W_{III}^{-1}W_{II}^{-1}D\,W_I^{-1}D^{-1}S_I^{-1}\text{Stelle}_{1\_T1}S_1DW_ID^{-1}W_{II}W_{III} \tag{31}$$

$$U = W_{III}^{-1}W_{II}^{-1}D^7W_I^{-1}D^{-7}S_2^{-1}\text{Stelle}_{1\_T2}S_2D^7W_ID^{-7}W_{II}W_{III} \tag{32}$$

Since the left hand side of the equations above has to be equal, one can find the correct wiring of the reflector drum (in the calculation above the reflector drum of 1937 was used):

```
U = (YRUHQSLDPXNGOKMIEBFZCWVJAT)
```

For the last rotor W_{III} one finally gets:

```
WIII = (BDFHJLCPRTXVZNYEIWGAKMUSQO)
```

Although there was no data given concerning the wiring layouts of the three rotors, as would have been required for producing a replica of the Enigma, Marian Rejewski was finally able to accomplish the calculations of the missing wiring layouts with the help of the available spy documents, various mathematical theorems as well as ciphers of intercepted German radio messages.

Reference

1. Rejewski, M.: How Polish mathematicians deciphered the Enigma. Ann. Hist. Comput. **3**(3), 213–234 (1981)

Heinz von Foerster and Early Research in the Field of Pattern Recognition at the Biological Computer Laboratory

Jan Müggenburg[✉] [iD]

Leuphana Universität Lüneburg, Universitätsallee 1, 21335 Lüneburg, Germany
mueggenburg@leuphana.de

Abstract. This paper deals with the early phase of research in the field of pattern recognition in the 1960s using Heinz von Foerster's Biological Computer Laboratory (BCL) as an example. In a first step, the foundation, rise and decline of the BCL are outlined. Subsequently, three machines will be presented that were expected to model the natural phenomenon of pattern. Finally, the suggestion is made to refer to these biological computers as "lively artifacts" and to understand their observation as an integral aspect of their mode of operation.

Keywords: Heinz von Foerster · Biological Computer Laboratory · Biological computing · Pattern recognition · Cybernetics

1 Introduction: Numa-rete Would Count No More

On a cold winter's day in January 1965, two physicists from Urbana-Champaign, Illinois travelled to New York City. Heinz von Foerster (1911–2002) and his doctoral student Paul Weston (*1935) followed an invitation from television broadcaster CBS, to present one of the machines they had designed as part of their research on pattern recognition in a new science show. The show was hosted by America's most popular newscaster, Walter Cronkite (1916–2009), who had achieved worldwide fame a good year earlier with his live coverage of the assassination attempt on President Kennedy.

The machine that Foerster and Weston brought to New York didn't look very spectacular at first glance. With its angular shape and numeric display module, it resembled an electronic cash register where the keyboard had been replaced with a pane of glass. However, if a random number of differently shaped objects were placed on the surface of the machine, it could perform an amazing trick: In a fraction of a second, the correct number of objects appeared on its two-digit display module. Weston had named his invention Numa-rete, because with its central component, a network of photo and computing cells, the machine was supposed to imitate the function of the retina in the animal and human eye and perceive patterns in its environment.

On that very day in the winter of 1965, however, the Numa-rete had suddenly turned blind. About an hour before the recording of the show Foerster and Weston had to find out that the machine had been damaged during its journey to New York. In vain Weston

R. Moreno-Díaz et al. (Eds.): EUROCAST 2019, LNCS 12013, pp. 116–122, 2020.
https://doi.org/10.1007/978-3-030-45093-9_15

tried to repair his machine. Helplessly, he had to watch as the greatest conceivable embarrassment for a young scientist came closer and closer. It was Foerster who at the last moment had the decisive idea. Finally America's television audience was able to witness their favorite newscaster placing small objects on the surface of a biological computer, excited that it was displaying the right number. What the audience couldn't see, however, was that Foerster secretly controlled the numerical display of the biological computer using a cable remote control hidden under the table. At the very last moment, Foerster had succeeded in saving the day by replacing the artificial intelligence of the brain-dead Numa-rete with his own [12, 18].

The research institution whose projects Foerster and Weston presented in Cronkite's show was called Biological Computer Laboratory (BCL) and was one of the few research facilities in the US were actual cybernetic research took place in the 1960s [7]. While in the early stage of cybernetics the computer had only served as a metaphor and model, research with actual special-purpose-computer was conducted at the BCL. Its research focus was the construction of biological computers that were supposed to be modeled after examples from nature. This resulted in artificial sensory organs, neural networks, and self-organizing automata; a remarkable prefiguration of contemporary approaches to Artificial Intelligence Research (AI) and Computer Science. At the center of cybernetic research at the BCL was the problem of pattern recognition. Based on recent neuroscientific research, it was assumed that the sensory organs of vertebrates are pre-interpreting visual, acoustic, tactile or other sensory stimuli in the form of patterns. Foerster and his team at the BCL were working on a whole series of different special-purpose-computers that would imitate this natural form of biological computation. They hoped that experimenting with these machines would ultimately result in new solutions in the field of automated image and sound recognition.

2   A Very Brief History of the Biological Computer Laboratory

The history of the BCL is inseparably linked to the history of its director, the Austrian physicist and cybernetician Heinz von Foerster [14]. After he emigrated to the USA in 1949, he got to know the first generation of cyberneticians around Norbert Wiener (1894–1964) and Warren McCulloch (1898–1969) as participant of the famous Macy conferences [16]. In the 1950s, Foerster worked as a professor of electrical engineering at the Department for Electrical Engineering (DEE) of the University of Illinois in Urbana-Champaign. Foerster proved his skills in the area of grantsmanship and reformulated the cybernetic program as research in biological computing. In this way in 1958 he succeeded in convincing a number of military sponsors such as the Office of Naval Research to fund a laboratory at the DEE dedicated to this new field of research.

2.1   Growth Phase of the BCL

During its growth phase in the first half of the 1960s, Foerster was able to bring together some protagonists from the early stage of cybernetics, such as the neuropsychiatrist Ross Ashby (1903–1972) or the psychologist Gordon Pask (1928–1996), as permanent staff or visiting researchers for his new laboratory [20]. Researchers such as the German philosopher Gotthard Günther (1900–1984) or the Chilean biologist Humberto Maturana (*1928) became acquainted with cybernetics through their research at the BCL and later contributed to its theoretical advancement. The research group was completed by several doctoral students who were enthusiastic about Foerster's interdisciplinary program and were highly involved in the research. In order to fund his growing department, Foerster had to further develop the BCL's research agenda and convince potential sponsors that cybernetics was ideally suited and well-prepared to explore topics such as AI, self-organization or bionics. In the early 1960s the BCL thus grew into a well-networked and renowned institution in the field of AI-research.

2.2 Decline of the BCL

In the second half of the decade, however, Foerster found it increasingly difficult to guarantee the flow of further research funds. While around 1960 Foerster had been able to benefit from the status of cybernetics as an interdisciplinary and future-oriented science, the situation changed dramatically now. Rival approaches such as Symbolic AI pushed cybernetics, which continued to insist on the neuronal approach, out of its traditional fields of research and application. Moreover, Foerster could no longer rely on the first generation of cyberneticists that had supported him in founding his laboratory. Staff personnel changes in Washington and a new law that prohibited military funding of research that lacked a direct relationship to specific military function in 1970 resulted in Foerster's good connections to the military research offices gradually being cut off [14]. While the BCL had to struggle with financial problems in the second half of the decade, the research carried out there shifted away from the design of electronic machines to more theoretically oriented work. Foerster and his collaborators were increasingly concerned with social problems such as human language and began to look at social applications of cybernetics in various fields of society [14]. Small projects organized by students as well as teaching formats inspired by the American counterculture now shaped the BCL and led to disputes with the university administration [5]. After Foerster applied for retirement in the summer of 1974, his department was phased out at the DEE and finally closed in 1976 [13].

2.3 Departure from the Mainstream of Research

Thus, while the first years of the BCL were characterized by the design of machines, its proximity to the "military-industrial-academic complex" [8] and a high degree of independence from the faculty and university administration, the second half of the laboratory's life was marked by a "departure from the mainstream of research" [13], dwindling support from sponsors in Washington, as well as a focus on teaching. It was precisely in these difficult times, however, that numerous papers were written, which today are generally perceived as the most influential products of the research carried out at Foerster's laboratory. The heterogeneous strands of BCL research were combined with Foerster's theory of a "cybernetics of cybernetics" or "second-order cybernetics" and Maturana's "biology of cognition" to form an interdisciplinary epistemology [17].

Eventually this recursive application of cybernetics to itself influenced a variety of scientific and non-scientific fields such as sociology, pedagogics, media theory, psycho therapy and others.

3 Pattern Recognition Research at the BCL

At the center of cybernetic research at the BCL was the problem of pattern recognition. Based on McCulloch's "Experimental Epistemology" [10] and on recent neuroscientific research [4, 6], it was assumed that the sensory organs of vertebrates pre-interpret visual, acoustic, tactile or other sensory stimuli in the form of patterns. Following Kant McCulloch firmly believed in the existence of a physiological a priori [1]. Foerster and his team were working on a whole series of different special-purpose-computers that would imitate this natural form of biological computation. They hoped that experimenting with these machines would ultimately result in new solutions in the field of automated image and sound recognition. They were corresponding with other pioneers in the field of Pattern Recognition such as Oliver Selfridge (1926–2008) and Frank Rosenblatt (1928–1971) and were involved in the emerging field of AI research [15]. Three of the biological computers designed at BCL are briefly presented below.

3.1 The Adaptive Reorganizing Automaton

The first biological computer designed at the BCL was an artificial neural network called "Adaptive Reorganizing Automaton" (ARA) [2]. Its designer, the engineer Murray Babcock (1924–1999), was inspired by contemporary writings of prominent neurophysiologists from the 1950s, such as those of the later Nobel Prize winner John Carew Eccles (1903–1997) [6]. Babcock's goal was to construct a computer that could be aware of its past and present states in relation to its environment, so that this knowledge would enable it to behave accordingly in reaction to future stimuli. In contrast to the famous Von Neumann architecture, the machine did not consist of specialized functional segments, but of a sufficiently large number of structurally and functionally identical basic units - the artificial neurons.

By using a combination of Top Down and Bottom Up-approaches Babcock constructed three different components that collectively imitated the complex functionality of the biological neuron. The machine consisted of 28 "energy transducer units", 112 "facilitator units" and 28 "autonomous components" [2]. Together they formed a network through which pulse modulated signals could propagate. The energy transducers simulated the ability of a neuron to convert multiple input signals into one output signal and had individually adjustable thresholds. The facilitator units acted as the artificial synapses of the machine and raised or lowered the frequency of the output signal depending on the number of previously passing signals. The machine's Autonomous Components randomly emitted pulses as soon as a sensor detected that the machine was inactive for a certain period of time. The machine could be used in two ways: First, as a special purpose computer, if one treated the initial states of the energy transducer units and the facilitator units as the input of the machine and the final state of the facilitator units as its output. Secondly, as an artificial organism, if one observed and documented its emerging patterns of behavior.

3.2 The Numa-rete

The Numa-rete was modeled after the retina of a frog and it was designed to reproduce the biological phenomenon of pattern recognition. The design of the machine was inspired by the neurophysiological frog experiments by Humberto Maturana and Jerome Lettvin (1920–2011), who, in the late 1950s, published an essay with the central claim that for vertebrates a lateral connectivity of neurons in the retina already prompts an intelligent pre-interpretation and computation of visual stimuli before they reach the brain [9].

The design of the Numa-Rete allegedly imitated the underlying natural algorithm of this mechanism and, according to Foerster, operated according to the same, simple fundamental principle of retinal connectivity. Just as the frog-eye from the experiment could recognize small, agile points ('flies') and large shadows ('birds') immediately, the Numa-rete could display the number of objects placed on its artificial retina, thus baffling visitors of the BCL because of this seemingly intelligent capacity. Its designer Paul Weston had arranged 400 photocells in a square of 20×20 cells and connected them with corresponding computing units on an underlying level [19]. Weston had inserted this double network into a 60 cm high housing with a square base of 90 cm edge length and covered it with a pane of glass. If several objects were placed on this artificial retina, so that for each object a group of photocells was covered underneath, the underlying network of computing units changed its state in accordance to the electric currents from the photocells. A two-digit display on top of the machine would then display the number of objects. As Weston later recalled, his machine was not an exact copy of its natural prototype: "[T]he way I built the NumaRete was rather the Rube Goldberg-scheme for doing it." [12].

3.3 The Dynamic Signal Analyzer

Another major group of BCL-projects was related to speech recognition and the general functioning of the mammalian auditory system. Following the work of the Hungarian physiologist Georg von Békésy (1899–1972) who had claimed that the essential parts of the mammalian ear could be described as a signal analyzer a machine was build to model this mechanism [4]. Designed by Murray Babcock, the "Dynamic Signal Analyzer" (DSA) was built on the assumption that the analysis of the travelling waves in the basilar membrane of the inner ear was computationally equivalent to a Fourier transform being performed on the acoustic wave [3].

The DSA was a room-filling machine, composed of several hundred transistors and semiconductor diodes, three oscilloscopes, a control console, an amplifier, a microphone and many other electronic components. Within this complex setup the precomputation of the acoustic signal was realized through a series of spectrum analyses performed on an electrical current produced by a conventional microphone. By comparing the input of phonetic elements spoken into the microphone with the output signals displayed on three oscilloscopes it was hoped to find invariants of human speech and how these invariants were distinguished by the human ear. The design of the DSA thus followed the idea that the mammalian auditory system operated similar to an electromechanical signal processor and could be described and understood by means of well known engineering principles.

4 Conclusion: Lively Artifacts

Borrowing a term from McCulloch [11] I propose that we refer to the pattern recognition machines built at the BCL as lively artifacts. The moniker captures the essential nature of these machines by addressing three different qualifications for these cybernetic machines. First, they are lively in the sense of lifelike. BCL engineers wanted to apply biological organizational and structural principles to the design of their machines and thus tried to model their artifacts on natural prototypes. For example, if they were planning to design artificial neurons, their first step would be to look at nature, which usually meant catching up with the latest findings in neurophysiology and bio-chemistry. Second, the term artifact references the fact that these machines are still objects made by human engineers. Consequently, the machine's designs involved a great deal of tinkering and even, often enough, a hint of trickery.

However, it is the third qualification that lively artifacts addresses that makes it the best term for these machines. Lively, as a synonym for spirited and active, infers that these machines appear to exhibit a life of their own. Whereas the first two qualifications focus on design and experimentation processes of cybernetic machines, liveliness speaks to the performativity and aesthetics of these machines. A machine's liveliness describes the nature of the interaction between the finished artifact and an individual, an individual who can be, but does not necessarily need to be, the machine's designer. When analyzing the liveliness of cybernetic artifacts, we have to deal with the trinity of the object, the creator of the object, and the subject interacting with the object.

References

1. Abraham, T.: Rebel Genius. Warren S. McCullochs Transdisciplinary Life in Science. MIT Press, Cambridge (2016)
2. Babcock, M.L.: Reorganization by adaptive automation, Doctoral thesis. University of Illinois (1960)
3. Babcock, M.L. Erickson, R.J., Neill, D.M.: A dynamic signal analyzer, BCL Technical report No. 3–1. Wright-Patterson Airforce Base, Ohio (1962)
4. von Békésy, G.: Experiments in Hearing. McGraw-Hill, New York (1960)
5. Clarke, B.: From information to cognition: the systems counterculture, Heinz von Foerster's pedagogy, and second-order cybernetics. Constructivist Found. **7**(3), 194–205 (2012). http://constructivist.info/7/3/196
6. Eccles, J.C.: The Neurophysiological Basis of Mind. Oxford Press, Oxford (1953)
7. Kline, R.: The Cybernetics Moment. Or Why We Call Our Age the Information Age. John Hopkins University Press, Baltimore (2015)
8. Leslie, S.W.: The Cold War and American Science: The Military-Industrial-Academic Complex. MIT and Stanford, New York (1993)
9. Lettvin, J., Maturana, H.R., McCulloch, W.S., Pitts, W.H.: What the frog's eye tells the frog's brain. Proc. Inst. Radio Eng. **47**, 1940–1951 (1959)
10. McCulloch, W.S. (ed.): Embodiments of Mind. MIT Press, Cambridge Mass. (1988)
11. McCulloch, W.S.: Living models for lively artifacts. In: Arm, D.L. (ed.) Science in the Sixties. The Tenth Anniversary AFOSR Scientific Seminar, Albuquerque, pp. 73–83 (1965)
12. Müggenburg, J., Hutchinson, J.A.: Kybernetik in Urbana. Ein Gespräch zwischen Paul Weston, Jan Müggenburg und James Andrew Hutchinson, Österreichische Zeitschrift für Geschichtswissenschaften **19**(4), 126–139 (2008)

13. Müller, A: The end of the biological computer laboratory. In: Müller, A., Müller, K.H. (eds.) An Unfinished Revolution? Heinz von Foerster and the Biological Computer Laboratory BCL 1958–1976, Vienna, pp. 314–317 (2007)
14. Müller. A.: A brief history of the BCL. In: Müller, A, Müller, K.H. (eds.) An Unfinished Revolution? Heinz von Foerster and the Biological Computer Laboratory BCL 1958–1976, Vienna, pp. 279–302 (2007)
15. Nilsson, N.J.: The Quest for Artificial Intelligence. Cambridge University Press, Cambridge (2009)
16. Pias, C. (ed.): Cybernetics. The Macy Conferences 1946–1953. The Complete Transactions, Chicago (2016)
17. Scott, B.: Second-order cybernetics: an historical introduction. Kybernetes **33**(9–10), 1365–1378 (2004). https://doi.org/10.1108/03684920410556007
18. Weston, P.: Mere Tuna. In: Bröcker, M., Riegler, A. (eds.) Human Becoming–Becoming Human. Heinz von Foerster Festschrift (2001). https://www.univie.ac.at/constructivism/HvF/festschrift/weston.html
19. Weston, P.: Photocell fields counts random objects. Electronics **34**(38), 46–47 (1961)
20. Wilson, K.L.: The work of visiting cyberneticians in the biological computer laboratory. Cybern. Forum **9**(2), 36–39 (1979)

Remarks on the Design of First Digital Computers in Japan - Contributions of Yasuo Komamiya

Radomir S. Stanković[1(✉)], Tsutomu Sasao[2], Jaakko T. Astola[3], and Akihiko Yamada[4]

[1] Mathematical Institute of SASA, Belgrade, Serbia
Radomir.Stankovic@gmail.com
[2] Department of Computer Science, Meiji University,
Kawasaki, Kanagawa 214-8571, Japan
[3] Department of Signal Processing, Tampere University of Technology,
Tampere, Finland
[4] Computer Systems and Media Laboratory, Tokyo, Japan

Abstract. This paper presents some less known details about the work of Yasuo Komamiya in development of the first relay computers using the theory of computing networks that is based on the former work of Oohashi Kan-ichi and Mochiori Goto at the Electrotechnical Laboratory (ETL) of Agency of Industrial Science and Technology, Tokyo, Japan. The work at ETL in the same direction was performed under guidance of Mochinori Goto.

Keywords: Digital computers · Relay-based computers · Parametron computers · Transistorised computers · History · Arithmetic circuits

1 Introduction

In the first half of the 20th century, many useful algorithms were developed to solve various problems in different areas of human activity. However, most of them require intensive computations, due to which their applications, especially wide applications, have been suppressed by the lack of the corresponding computing devices. Even before that, already in late thirties, it was clear that discrete and digital devices are more appropriate for such applications that require complex computations. Therefore, in fifties of the 20th century, the work towards development of digital computers was a central subject of research at many important national level institutions. This research was performed equally in all technology leading countries all over the world, notably USA, Europe, and Japan.

In this paper, we briefly discuss the work in this area performed in Japan at the Electrotechnical Laboratory (ETL), Agency of Industrial Science and Technology, conducted by Mochinori Goto, and in particular we present some

© Springer Nature Switzerland AG 2020
R. Moreno-Díaz et al. (Eds.): EUROCAST 2019, LNCS 12013, pp. 123–130, 2020.
https://doi.org/10.1007/978-3-030-45093-9_16

less known details about the work of Yasuo Komamiya and his contributions in the development of the first relay computers based on the theory of computing networks [4–6] by elaborating the results of Mochinori Goto and Oohashi Kan-ichi, the actual at that time and the former director of ETL.

2   Professional Biography of Yasuo Komamiya

In September 1944, Yasuo Komamiya graduated from the School of Engineering of Tokyo Imperial University, Tokyo, Japan in the area of electrical engineering. During his study, he attended courses of Mochinori Goto, who was at the same time the Director of the Electrotechnical Laboratory (ETL) of Agency of Industrial Science and Technology, Japan, which was important for his further professional work and development as will be discussed below. Within this course, Komamiya got the subject for his research leading to a theory of computing networks due to which Komamiya was awarded by the degree of PhD in engineering from Tokyo Imperial University.

In November 1944, Komamiya joined the Basic Research Division of the ETL as an engineer. Starting from January 1957, he spent a year at the Computation Laboratory of Harvard University, which was at that time led by Howard Aiken who was championing the application of arithmetic expressions for Boolean functions to describe and design arithmetic circuits. By expanding his research interest, from September 1962, he stayed for a year at the Digital Computation Laboratory, University of Illinois working in coding theory.

Komamiya stayed with ETL until 1980, by holding different positions including the Head of Applied Mathematics, Department of Physics Division, the Division Director of Control Systems Research, Director of Electronic Components (Devices) Research. From 1980 to 1986 he was a Professor at the Graduate School of Integrated Science and Technology, Kyushu University in Fukuoka. After retirement in 1986, Komamiya was appointed as a Professor of Meiji University in Kanagawa where he worked until 1993.

Figure 1 shows a photo of Komamiya (central figure) with associates.

3   Theory of Computing Networks

The work towards the first computers in Japan is strictly related to the Electrotechnical Laboratory (ETL) of the Japanese government that was established by the Ministry of Communication of Japan in 1891. In 1948, ETL was reorganised into two units, the ETL conducting research on power engineering and power electronic, and the Electrical Communication Laboratory performing research in communication and electronic engineering including design of computing devices and development of the first computers. The underlined theoretical framework can be traced as follows. Oohashi Kan-ichi, who served as the Director of ETL until 1945, developed a theory of electric relay circuits by taking into account the delay in functioning of electric relays. This theory required solving a functional equation that depends on time as an explicit variable. The problem of solving this

Fig. 1. From left to the right Tsutomu Sasao, Teruhiko Yamada, Yasuo Komamiua, Yoshihiro Iwadare, and an associate whose name is not recorded.

functional equation was explored further by Mochinori Goto, who was appointed as the Director of Power Engineering Research Division, and further promoted into the General Director of ETL in 1952. Goto solved the problem initiated by Oohashi by expanding the Boolean algebra by modeling the relay delay time as a logic function, which enables modeling behaviour of relay circuits as a function depending on time by solving an equation including an unknown logic function [1–3]. The theory called by Goto *Logical Mathematics* permitted to analyze and design relay circuits via calculations. The work of Goto was reviewed by Alonzo Church in *The Journal of Symbolic Logic*, Vol. 20, No. 3, 1955, 285–286.

This theory implemented by Yasuo Komamiya become a mathematical foundation for constructing computing networks [4,6]. In several publications from 1951 to 1959, Yasuo Komamiya discussed the theoretical foundations for converting the design of computing networks from an engineering discipline into a subdiscipline of applied mathematics [4–10].

In the case of adders, the chief idea is to represent the sum of natural numbers in terms of binary values as

$$A_1 + A_2 + \cdots + A_n = d_m 2^m + d_{m-1} 2^{m-1} + \cdots + d_1 2 + d_0,$$

where $A_i, d_i \in \{0, 1\}$ and the addition is the integer sum.

Various kinds of adders can be derived as special cases.

The arithmetic and Reed-Muller expressions are viewed as particular examples of Fourier series-like expressions over different fields, the complex field C and the Galois field $GF(2)$. These expressions are used to describe arithmetic circuits as adders and multipliers. The first publications on that subject Komamiya prepared in 1951 when he discussed the conversion of decimal to binary systems [4], but it is reasonable to assume that he did a research on that subject earlier. Note that related approaches were promoted by Aiken and his research group, and this joint interests possibly explain the visit of Komamiya to the Computation Laboratory of Harvard University in 1957.

The theory of computing networks by Komamiya [6] was reviewed by Calvin C. Elgot, in *The Journal of Symbolic Logic*, Vol. 23, No. 3, 1958, 366.

The work of Komamiya presented in [6] was reported later also in [12] and pointed out in [13].

In [10], it is explicitly stated the relationship between the canonical sum-of-product expressions F for Boolean functions and their positive polarity Reed-Muller forms f in the following manner. If all logical OR symbols in F are replaced by exclusive-or symbols and if negated variables appearing in the fundamental product are omitted, the resulting expression is f. The constant term 1 will appear in f if there exists fundamental product $\overline{x}_1\overline{x}_2\overline{x}_3\cdots\overline{x}_n$.

Conversely, F can be obtained from f if exclusive-or symbols are replaced by logical OR symbols and if each product is multiplied by the missing variables, each in negated form. If there exists the constant term 1 in f, the fundamental product $\overline{x}_1\overline{x}_2\overline{x}_3\cdots\overline{x}_n$ should appear in F. So, it can be seen that F and f has an $1-1$ correspondence, where f is the Reed-Muller transform of F.

Another important and useful result in converting decimal to binary numbers is the observation that a decimal number $r_{10} = x_1+x_2+x_3+\cdots+x_n$, $x_i \in \{0,1\}$, can be expressed as $r_{10} = (y_k, y_{k-1}, y_{k-2}, \ldots, y_1, y_0)$, with $y_i = SB(n, 2^i)$, where $SB(n, k)$ are n-variable symmetric functions defined as the exclusive-or of all the products of binary variables x_i consisting of k literals, with by definition $SB(n, 0) = 1$. Therefore, $SB(n, 0) = 1$, $SB(n, 1) = \bigoplus x_i$, $SB(n, 2) = \bigoplus_{i<j} x_i x_j$, $SB(n, 3) = \bigoplus_{i<j<k} x_i x_j x_k$, ..., and finally $SB(n, n) = x_1 x_2 x_3 \cdots x_n$.

4   First Relay Computers in Japan

The work towards designing of a non stored-program computer working in the asynchronous mode started in November 1944 by Komamiya and Ryouta Suekane. The prototype was completed in 1952 and named by Prof. Mochinori Goto, the Director of ETL at that time, as ETL Mark I [17]. This prototype served as a pilot version for designing an entirely functional computer Mark II based on the same principles.

The team engaged in designing of ETL Mark I and ETL Mark II consisted of the following members of the Mathematical Research Group of ETL. Mochinori Goto, the Director of ETL, served as the Director of the Construction Committee, working at the same time as a Professor of the University of Tokyo. Yasuo Komamiya, the Chief of the Mathematics Research Group of ETL, served as the Chief Designer and the Sub-Director of the Construction Committee. Ryouta Suekane was a member of the Construction Committee and served as an Assistant Designer in charge for logical design. Masahide Takagi served as Assistant Designers in charge for circuit designs. Shigeru Kuwabara was another Assistant Designer. They all were members of the Mathematical Research Group of ETL.

To develop a fast, reliable and low-power machine, they specially designed three different types of relays as basic elements. The S type relay for storage: They are fast, have a few contacts, and low-power. The C type relay for control and arithmetic: They are fast, have many contacts, but high-power. The G type relay for gates: They have many contacts, low-power, but slow. In the combinational part, they used double-rail logic: Every bit is represented by a pair of signals $(A, \overline{A})$, which makes asynchronous self-checking operation possible. On the other hand, in the storage part, they used single-rail logic. The C type relay has main and holding coils, and works as a Set-Reset flip-flop. The ETL Mark II is the floating point parallel machine with 200 words consisting of 22,253 relays [11]. It was the largest and fastest relay computer in the world at that time.

Figure 2 shows the ETL Mark II computer exhibited in the National Museum of Nature and Science in Tokyo. ETL Mark II was built by Fuji Communication Apparatus Manufacturing Co. (now Fujitsu), while the company Shinko-Seisaku-Sho made the input equipment, i.e., tape readers and perforators. Mr. K. Noda, Chief of Electromagnetic Machinery Section of ETL designed motor generator for the power supply of the computer. Mr. T. Okabe Chief of the Temperature Control Research Group of ETL designed the air-conditioning process of the room where the computer was located [11].

The construction and functioning of Mark II was described with many details in the book *Theory and Structure of The Automatic Relay Computer ETL Mark II* by the designer staff of ETL. Final editing was done by Y. Komamiya and S. Kuwabara. This book was published by the ETL, Agency of Industrial Science and Technology, Japan, in the series *Researches of The Electrotechnical Laboratory* as the volume No. 556 and printed by the International Academic Printing Co., Ltd. (Kokusai Bunken Insatsusha) in Fujimi-cho, Chiyoda-ku, Tokyo, Japan for 500 samples in September 1956. This makes that in the antiquarian book shops its price presently achieves around USA $1,250.

In Synopsis of this book, Prof. Mochinori Goto wrote

Since the end of war, the Mathematics Research Group of Electrotechnical Laboratory has been studying logical mathematics, especially its applications to relay networks. As an outcome of this study, the theory of relay networks considering the time lag of switching elements was completed by one of the authors, M. Goto, and the theory of computing networks by Y. Komamiya in 1951. Applying the results of these studies, the Mathematical research Group have been engaged in the construction of the Plot Model of the automatic relay computer ETL Mark I completed in 1952, and subsequently have been engaged in the construction of the automatic relay computer ETL Mark II for practical use completed in November 1955. These computers, therefore, have been built on the same principles. The features of these computers are as follows. The computers are controlled without electric clock pulses, each relay being energized by its preceding relay in the same fashion as a row of falling nine-pins. As a result,

The calculating speed of the computers are 4 to 5 times as fast as any other relay computer, and

These computers perform self-checking simultaneously with calculation, if the computer misses a calculation, the unit causing the trouble performs automatically the 2nd trial. Intermittent troubles are omitted automatically and the computer stops only in the event of an essential trouble. Therefore, there can be no trouble as long as the computers are running.

These features were developed from a new computer design based on our research in logical mathematics.

This book was review by Calvin C. Elgot in *The Journal of Symbolic Logic*, Vol. 23, No. 1, 1958, 60.

In 1956, the cost of the ETL Mark II was about Japanese 36,000,000 Yen or about USA $100,000. The exchange rate of 1.00 USA $ was about 360.00 Yen. For a comparison the salary of a bank worker who just graduate a university was 10,000 Yen per month at that time.

Fig. 2. The ETL Mark II relay computer, photo by T. Sasao.

5   Summary of the Work on First Computers in Japan

In this section, we give a brief summary of the development of first computers in Japan. For more details, we refer to [18].

Transistor was invented at Bell Laboratories in 1948. However, at that time ETL people could not research on transistors formally. They started an informal study in a series of meetings on transistors. Electrical Communication Laboratory succeeded in manufacturing germanium transistors in 1951.

As Electrical Communication Laboratory became a part of the public corporation in 1952, ETL established Electronics Department in it in 1954 and started the research on transistors and transistor computers formally.

Eiichi Goto invented Parametron in 1954. Then, Electrical Communication Laboratory decided to develop Parametron computers. Notice that at that time the reliability of transistors was low while parametrons were more reliable and less expensive. ETL developed ETL Mark III transistor computer using point-contact transistors in July, 1956. This is the first transistor computer in Japan. Electrical Communication Laboratory developed the first Parametron computer MUSASINO-1 in March 1957.

Figure 3 shows a photo of Eiichi Goto (seating) and Hidetoshi Takahasi. Recall that a parametron is a resonant circuit containing a non-linear reactive element which oscillates at half the driving frequency.

ETL developed ETL Mark IV transistor computer using junction-type transistors in November, 1957. Hidetoshi Takahasi and Eiichi Goto of University of Tokyo developed another Parametron computer PC-1 in March, 1958. This is a binary, single-address computer with magnetic core memory. Later the advantage was given to transistor computers mainly because of the speed. Several commercial transistor computers were developed based on first ETL transistor computers [18]. Moreover, Dr. Mochinori Goto, General Director of ETL named the ETL's first transistor computer as ETL Mark III following the relay computers ETL Mark I and Mark II which Dr. Komamiya developed. Recently, there is a renewed interest in basic principles of parametron devices related to quantum computing and also nanomechanical computers. *D-Wave Systems* from Canada adopted superconductivity quantum bits (qubits) and quantum annealing system. To amplify magnetic flux, they introduced the quantum flux Parametron invented by Eiichi Goto in 1986. Quantum annealing system was proposed by Prof. Nishimori of Tokyo Institute of Technology, who was awarded NEC C&C Prize on Nov. 28, 2018. It might be said that the combination of quantum flux parametron and quantum annealing system created a new quantum computer. Nanomechanical computers based on the principle of parametron were proposed in 2008 at the NTT Research Labs in Kanagawa.

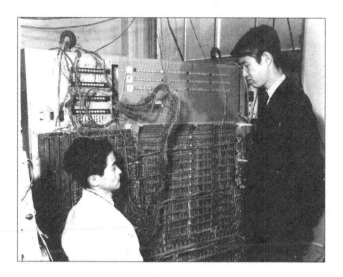

Fig. 3. Eiichi Goto (seating) and Hidetoshi Takahasi.

References

1. Goto, M.: Application of logical mathematics to the theory of relay contact networks. Electr. Soc. Jpn. **69**, April 1949. 125 pages
2. Goto, M.: Application of logical mathematics to the theory of relay networks. Jpn. Sci. Rev. **1**(3), 35–42 (1950)
3. IPSJ Computer Museum, page for Goto Mochinori. http://museum.ipsj.or.jp/en/results.html?cx=005960164513048960843
4. Komamiya, Y.: Theory of relay networks for the transformation between the decimal system and the binary system. Bull. E.T.L. **15**(8), 188–197 (1951)
5. Komamiya, Y.: Theory of computing networks. Research of E.T.L., No. 526, November 1951
6. Komamiya, Y.: Theory of computing networks. In: Proceedings of the First Japan National Congress for Applied Mechanics, pp. 527–532 (1952)
7. Komamiya, Y.: Logical representation of algebraic condition in computing networks. In: Symposium of Switching Circuits and Automation, Spring Joint Meeting of Institute of Electrical Engineers of Japan, May, pp. 1–9 (1958)
8. Komamiya, Y.: Some properties of Boolean function. In: Autumn Meeting of Mathematical Society of Japan (Applied Mathematics Branch), pp. 27–31, October 1958
9. Komamiya, Y., Theory of Computing Networks. Researchers of the Applied Mathematics Section of Electrotechnical Laboratory in Japanese Government, 2 Chome, Nagata-cho, Chiyodaku, Tokyo, p. 40, 10 July 1959
10. Komamiya, Y., Theory of Computing Networks. Researchers of the Applied Mathematics Section of Electrotechnical Laboratory in Japanese Government, 2 Chome, Nagata-cho, Chiyodaku, Tokyo, No. 580, September 1959
11. Komamiya, Y.: Automatic relay computers ETL Mark I and Mark II, Joho Shori. IPSJ Mag. **7**(6), 513–520 (1976)
12. Nozaki, A.: Switching Theory. Kyoritsu Shuppan, Tokyo (1972). (in Japanese)
13. Sasao, T.: Switching Theory for Logic Synthesis. Kluwer Academic Publishers, Boston (1999)
14. Sasao, T., Stanković, R.S., Astola, J.T.: Reprints from the early days of information sciences, contributions of Yasuo Komamiya to switching theory. TICSP #65 (2012). ISBN 978-952-15-3453-9
15. Stanković, R.S., Sasao, T.: A discussion on the history of research in arithmetic and Reed-Muller expressions. IEEE Trans. Comput. Aided Des. Integr. Circuits Syst. **20**(9), 1177–1179 (2001)
16. Stanković, R.S., Sasao, T., Astola, J.T.: Contributions of Yasuo Komamiya to switching theory. In: Proceedings of the 2015 Reed-Muller Workshop, Waterloo, Canada, 21 May 2015
17. Suekane, R.: Description of the pilot model of automatic relay computer. Bull. E.T.L., **19**(4), 252–260. (in Japanese)
18. Yamada, A.: Research and development of transistor computers at Electrotechnical Laboratory, History of Electrical Engineering Workshop of the Institute of Electrical Engineering of Japan (IEEJ), Tokyo, Japan, 1 October 2018. (in Japanese with the abstract in English)

From the Discovery of Electro-Magnetism and Electro-Magnetic Induction to the Maxwell Equations

Franz Pichler[✉]

Johannes Kepler University Linz, Linz, Austria
`telegraph.pichler@aon.at`

1 Introduction

This lecture discusses, in its first part, the discovery of the physical phenomenon of electro-magnetism, as found in 1820 by the Danish professor of physics Hans Christian Ørsted and the discovery of the effects of electro-dynamics due to the French mathematician André Marie Ampère in 1820–1823. The main result shows in principle that an electrical current is able to move a magnet. In the second part, the important discovery in 1831 of the English scientist Michael Faraday of the phenomenon of electro-magnetic induction has our interest. By this is proven, to state it in simple words, that a moving magnet generates an electrical current in a closed circuit of conductive wire. Both discoveries are important milestones in the development of electricity and the associated electrical technology. In the final third part, we discuss the important contribution given in 1864 by the Scottish professor of physics at King's College, London, James Clerk Maxwell[1], in providing a mathematical framework which establishes a bidirectional relation of electro-dynamics with electro-magnetic induction. Maxwell was able to give the concept of electrical forces of Faraday a mathematical form. Both the effects of electro-dynamics of Ampère and the effects of electro-magnetic induction of Faraday can be derived by Maxwell from a set of equations, known today as the "Maxwell equations". As a result they allow the hypothesis of the existence of electromagnetic waves, which travel with the speed of light. Maxwell considered also light as electromagnetic waves. More than twenty years later, in 1887, the German physics professor Heinrich Hertz proved by experiments the existence of electromagnetic waves (radio). Today the system of Maxwell equations are together with the law of gravitation, as found by Isaac Newton, cornerstones of classical physics.

Our presentation should be considered as a tutorial contribution to history. Many scientists have already succeeded in the past to trace the evolution of

---

[1] Maxwell was born in Edinburgh in 1831; he graduated from Cambridge University in 1855; worked on electro-magnetism at King's College, London, 1860–1865; and after an interruption, started up the Cavendish Laboratory of experimental physics as professor at Cambridge in 1871–1879.

© Springer Nature Switzerland AG 2020
R. Moreno-Díaz et al. (Eds.): EUROCAST 2019, LNCS 12013, pp. 131–140, 2020.
https://doi.org/10.1007/978-3-030-45093-9_17

Maxwell equations in detail and to make it easier for the non-specialist to understand their importance [9]. We mention here the papers of Lindell [10], of Helsinki University of Technology, and the paper of Bucci [2], of the University of Naples. Our presentation follows a similar pedagogical goal as the writing of the late German professor of physics and collaborator of the Deutsches Museum München, Gerlach [6], namely to try to give a contribution to the education of scientists and engineers in the philosophy of science.

2   Electro-Magnetism and Electro-Dynamics

It must have been a special "Sternstunde der Menschheit", to use an expression of the Austrian writer Stefan Zweig, when the Danish professor of physics Hans Christian Ørsted announced, in his 1819/20 lecture on "Galvanismus" at the University of Copenhagen that a "Schliessungsdraht" (the short-circuit wire on a Voltaic column) made the needle of a nearby magnetic compass turn. Ørsted called this an "electric conflict" which moves in circles along the electric wire. At that time, he did not imagine a magnetic force. In a short notice of July 21, 1820 [17], written in Latin, Ørsted informed a number of colleagues and institutions about his discovery. As reported in the book of E.H. Pfaff on the history of electro-magnetism, the echo was tremendeous [16]. His experiments were immediately repeated and confirmed at many places. Additional new results were discovered. The Estonian physicist Seebeck found that the force, generated by the "Schliessungsdraht", was constant on circles around. Of special importance were the findings of the French mathematician André Marie Ampère. By the experiments done at his private laboratory in Paris, Ampère identified the observed forces as being magnetical. If the "Schliessungsdraht" was wound in a spiral, this "solenoid" showed the behavior of a usual natural magnet. However, the most important contribution of Ampère was that he showed by his "Fundamental Law of Electro-Dynamics" how the force K between two differential parts of length ds of the wires, which carry an electric current of values i and i', respectively, can be computed

$$K = \frac{i\,ds \times i'\,ds}{r^2} \left(cos\epsilon - \frac{3}{2} \times cos\delta \times \cos\delta' \right) \tag{1}$$

where r denotes the distance between the two differential parts ds of the wire, δ, δ' and *epsilon* are angles between the differential parts $i\,ds$, $i'\,ds$ and the two wires, respectively. In case of parallel wires and if $i\,ds$ and $i'\,ds$ are normally aligned so that δ and δ' are at 90° to each other, Ampère's fundamental law reduces to

$$K = \frac{i\,ds \times i'\,ds}{r^2} \tag{2}$$

Equation 2 has the form of Coulomb's law in Electrostatics. If r is understood as a function $r(s, s')$ where s and s' denote the coordinates of $i\,ds$ and $i'\,ds$, respectively, then Ampère's fundamental law can also be written as

$$K = \frac{i\,ds \times i'\,ds}{r^2} \left(\frac{1}{2} \times \frac{dr}{ds} \times \frac{dr}{ds'} - r\frac{d^2r}{dsds'} \right) \tag{3}$$

André Marie Ampère can be considered as the founder of electro-dynamics. His lectures of 1820, 1822, 1823 and 1825 are the subject of his important publication of 1827 [1]. James Clerk Maxwell praised the contribution of Ampère in his "treatise" with the following words:

> *The experimental investigation by which Ampère established the laws of the mechanical action between electrical currents is one of the most brilliant achievements in science. The whole, theory and experiments, seems as if it had leaped, full grown and full armed, from the brain of the "Newton of electricity". It is perfect in form, and unassailable in accuracy, and it is summed up in a formula from which all the phenomena may be deduced, and which always remain the cardinal formula of electro-dynamics.*

The scientific results of Ørsted, Ampère and all other physicist which were involved in the field of electro-magnetism at the time provided the basis for many important practical inventions and developments which followed. Samuel F.B. Morse used electro-magnets in his system of electrical telegraphy (1837, 1844); electro-magnets were also the important component in the telephone of Alexander Graham Bell (1876). The first electrical motors (Jacobi (1834), Page (1838), Froment (1844)) used electro-magnets to convert electrical power into mechanical power. Electro-magnetic switches, commonly called "relays" (from the French relais), have been for a long time the most important components for automated telephone switches (Strowger 1885) and also for computers (Zuse 1931, Aiken 1944). These in turn were partly replaced by the invention of the electronic tube (DeForest 1906, von Lieben 1906). Today transistors (Bardeen and Brattain 1948) and integrated circuits (Kilby 1962) have replaced such applications of electro-magnets.

3   Electro-Magnetic Induction

With Ørsted's discovery, the question arose if it would be possible to generate electricity by magnetism. Michael Faraday, professor at the Royal Institution[2], devoted nine years work to get an answer to this question until finally in 1831 he found a solution by the following two experiments:

In the first experiment he put on an iron ring two windings of electrical wire. One of the windings, let it called to be the "primary winding" P, could be connected to a strong galvanic battery, called the "calorimeter" by Faraday. The secondary winding, called S, got short-circuited by a piece of wire, a kind of "Schließungsdraht" in the sense of Ørsted. Next to this wire, Faraday put a magnetic compass, called "galvanometer" by him. Faraday performed now the following experiment: by connecting P to the calorimeter, the needle of the galvanometer showed a short reaction. Disconnecting P from the Calorimeter, the

---

[2] The Royal Institution was founded by Benjamin Thompson in 1799 to further scientific research, loosely associated with the Royal Society. Maxwell's King's College, was near the Royal Institution, and Maxwell heard lectures as well as visiting Faraday there.

needle reacted again, however in the opposite direction. Faraday had discovered an effect on which all transformers are based today.

In the second experiment Faraday made a winding of electrical wires, a solenoid, and connected it to a galvanometer. By moving the solenoid towards the poles of a natural magnet the galvanometer showed a short reaction. Moving the solenoid backwards the galvanometer reacted again, however in the opposite direction. With this experiment, Faraday discovered "magneto-electric induction", on which all "magnet-inductors" depend today which generate electricity from mechanical power.

Faraday published his results in a series of papers in the Journal of the Royal Society of 1832–1833. James Clerk Maxwell described, in his famous "Treatise" of 1873, electro-magnetic induction in the following terms:

The total electromotive force acting round a circuit at any instant is measured by the rate of decrease of the number of lines of magnetic force which pass through it [12].

The discovery of electro-magnetic induction together with electro-magnetism can be considered as the most important milestones in the history of electricity. This discovery motivated a number of scientists on continental Europe to investigate this effect in terms of already existing theoretical results. Franz Neumann, a well known professor of physics in Königsberg, Prussia, was able to derive the law of electro-magnetic induction in terms of his theory of potentials. Professor Helmholtz and also professor William Thomson showed that by the assumption of the preservation of energy, electro-magnetic induction follows directly from Ampère's universal law of electro-dynamics.

One of the most important physicists in Germany of that time was Wilhelm Weber (1804–1891), professor at Göttingen and Leipzig. In his fundamental work "Elektrodynamische Maßbestimmungen" of 1846, Weber confirmed the results of Ampère. Weber's goal was, however, to extend the results of Ampère to include also Faraday's law of electro-magnetic induction. To reach this Weber replaced the differential electrical current elements of Ampère by differential electrical charges. From his colleague, professor Gustav Theodor Fechner (1801–1887), he took the idea of modeling an electrical current by an "electrical liquid" which moves electrical charges with a certain velocity trough the wires. The result of Weber was the following "universal law of electro-dynamics" which allows us to compute the force K between moving electrical charges e and e' with time-variable distance r as

$$K = \frac{e \times e'}{r^2} \left(1 - \frac{1}{c^2} \left(\frac{dr}{dt} \right)^2 + \frac{2r}{c^2} \times \frac{d^2r}{dt^2} \right) \qquad (4)$$

The constant c in 4 has the dimension of a velocity. Its actual value was determined in 1856 by Weber and Kohlrausch to be close to the velocity of light, a fact which was later confirmed by Maxwell. In case e and e' are stationary, Weber's law 4 reduces to Coulomb's law. Wilhelm Weber showed in his work that his law not only implies Ampère's law of electro-dynamics but also the law of electro-magnetic induction of Faraday. From Maxwell we have in this regard in his "Treatise" the following comment:

*After deducing from Ampère's formula for the action between the elements
of currents, his own formula for the action between moving electrical particles, Weber proceeded to apply his formula to the explanation of the production of electric currents by magneto-electric induction. In this he was
eminently successful, and we shall indicate the method by which the laws
of induced currents may be deduced from Weber's formula.* [13]

The law of electro-magnetic induction has found many practical applications.
Transformers are still today important components for the construction of high-
voltage power lines. Magneto-electric machines driven by steam engines produced
in history the necessary electric power to give electrical light to cities and to
light houses at the sea. They found also use in railway signalling and for the
electrical ignition of dynamite in mining. Another historic application was the
generation of electrical pulses for medical treatments. By the discovery of the
"dynamo-electric principle" by Werner von Siemens and Charles Wheatstone in
1867 dynamo-electric machines, called "dynamos" for short, replaced in many
applications the former magneto-electric machines. This led also to effective
electrical motors for the first time. Again, the electro-magnetic induction of
Faraday gave the basis for that.

4   Modeling Electrical Phenomena by "Lines of Force"

Michael Faraday had, as we know, only little knowledge of mathematical for-
malisms. This is also shown in his important book on his researches in electricity [3–5]. However with his concept of "lines of forces" he nevertheless made an
important contribution which allowed Maxwell to model electrical phenomena
by mathematical means. Maxwell set to himself the goal, to develop, by using
Faraday's ideas, a mathematical theory of electricity which was to be different
from the already existing results of Ampère and Weber. Maxwell demonstrated
his results in a series of lectures in 1855/56. As a mathematical analogy he
used methods from the theory of hydromechanics in modelling the ideas of Faraday concerning his "lines of forces". The scientific importance of these lectures
of Maxwell can be seen by the fact that Ludwig Boltzmann edited them in a
special volume of "Ostwald's Klassiker der Exakten Wissenschaften" [14]. For
Maxwell the results of the 1855/56 lectures on Faraday's lines of forces however
seemed not to attain his final satisfaction. In a series of papers on "On Physical
Lines of Force" in 1861/62 he took a more speculative point of view and this
time used, to model the electro-magnetical and electro-inductive phenomena,
mathematical methods as provided in theoretical mechanics by the concept of
molecular vortices. Maxwell got the idea of using this method from a paper by
William Thomson. This work of Maxwell again got edited by Ludwig Boltzmann
in "Ostwald's Klassiker der Exakten Wissenschaften" [15]. In the Addendum to
this work Boltzmann made the following statements:

*The translation of the lectures published in 1861/62 contain all Maxwell
equations for electro-magnetism, including the equations for moving bodies.*

[Furthermore:] The results of the series of lectures translated here has to be considered as one of the most important achievements of the theory of physics.

Boltzmann believes, as he states, that the mechanical models, elaborated in detail by Maxwell found little interest among his colleagues in Germany. Concerning the importance of the work he wrote the following:

Nobody will see a verification of the correctness of Maxwell equation system by the mechanical models of this series of lectures. They will also be of no use to cancel the arguments of Heinrich Hertz on the later work of Maxwell, which does not try to derive the equations but considers them solely as a phenomenological description of reality. However, the mechanical models proved to be important means for Maxwell's discoveries. They allowed him to describe the electro-magnetic phenomena as a result of near actions, contrary to the actions at distance of Ampère, Weber et al., which follow the example of the gravitational forces of Newton. The equations of Maxwell led also the way to the experiments which finally proved the existence of electro-magnetic waves. They certainly are the most interesting mathematical formulas which the history of physics can offer.

5   On Maxwell's Equation System

Although, according to Boltzmann, the paper "On Physical Lines of Force" contains already the results of the later system of equations, today known as the "Maxwell equations", their original presentation was quite different. In his later work, "A Dynamical Theory of the Electromagnetic Field" of 1864, published in 1865 in the Transactions of the Royal Society [11], Maxwell neglected all mechanical arguments and elaborated the resulting electrical facts by giving a set of 20 equations involving 20 variables. These equations summarized all the knowledge on electricity and its relation to magnetism which was known at this time. They contain also the radical concept of the "Displacement Current" to make Ampère's equation for the generation of electro-magnetism valid also for "open circuits" [18]. In the introduction to this work Maxwell made in point (20) the following important remark:

The general equations are next applied to the case of magnetic disturbance propagated through a non-conducting field, and it is shown that the only disturbances which can be so propagated are those which are traverse to the direction of propagation, and that the velocity v, found from experiments such as those of Weber, which expresses the number of electrostatic units of electricity which are contained in one electromagnetic unit. This velocity is so nearly that of light, that it seems we have strong reasons to conclude that light itself (including radiant heat, and other radiations if any) is an electromagnetic disturbance in the form of waves propagated through the electromagnetic field according to electromagnetic law.

We know that Heinrich Hertz proved by his experiments in 1886 that Maxwell was right to assume that light has the nature of electromagnetic waves. The results derived in his paper of 1865 motivated Maxwell to publish his most famous book with title "A Treatise on Electricity and Magnetism" in 1873. The book covers in detail all aspects of the existing theoretical knowledge of electricity of the time. Volume I deals with Electrostatics and Electrokinetics, Volume II with Magnetism. Maxwell emphasizes in the preface of the book that he followed Faraday's "Experimental Researches in Electricity" by putting the "mathematics" of Faraday in the conventional form of mathematical symbols. Starting with "Faraday's lines of Forces" (1855/56) followed by his work "On Physical Lines of Force" (1861/62) and finally publishing his "A Dynamical Theory of the Electromagnetic Field" (1865) and his "A Treatise on Electricity and Magnetism" (1873) he found finally the set of equations which describe in a fundamental manner the interplay between electricity and magnetism. Today this set of equations, commonly called the "Maxwell equations", as we will see, are not presented in the form as originally derived by Maxwell. They appear today in a form as later formulated by Heaviside [7] and by Hertz [8].

6   Deriving Maxwell Equations from the Fundamental Laws of Faraday and Ampère

In this paper we are not able to show the different steps which Maxwell had to take to develop his equations. It would be a very complicated story. Our way to derive the first two of the equations consists of the following steps: We start with the fundamental results of Faraday concerning electro-magnetic induction and Ampère's law of electro-magnetism. Using the concept of Faraday's lines of force we try to give a verbal formulation of Faraday's law of electro-magnetic induction, from which we derive their mathematical integral form. After the modification of Ampère's law by adding Maxwell's displacement current we have already again the equation in integral form.

Michael Faraday found, as we have seen before, that the change of the magnetic flux density $\mathbf{B}$ on a plane A generates, in an electrical wire wound around A, an electro-motoric force $\mathbf{U}$ given by

$$\mathbf{U} = -\frac{d}{dt}\int_A \mathbf{B}dA \qquad \text{(Faraday's law of electro-magnetic induction)} \qquad (5)$$

The minus sign comes from the fact that, according to the rule of Lenz, the induced electro-motoric force $\mathbf{U}$ tries to equalize the electro-magnetic flux. (Bold letters represent vector quantities.)

The results of Ampère concerning his law of electromagnetism can be expressed as follows: The electrical current of density $\mathbf{I}$ through a plane A generates, on any closed curve R which goes around A, a magnetic potential which is the sum of all locally given magnetic forces $\mathbf{H}$ on R. In mathematical form, this can be expressed by the following equation

$$\oint_R \mathbf{H}ds = \int_A \mathbf{I}dA \qquad \text{(Ampère's law of electromagnetism)} \qquad (6)$$

Maxwell, in analogy to the case of magnetic flux, made the assumption that every electric current has to be closed. In consequence, he had to assume that electrical charges changing in time induce also a change of the induced displacement $\mathbf{D}$. The associated electric current has then to be given by $d\mathbf{D}/dt$. Following this assumption of Maxwell in 6 and expressing in 5 the electromotoric force $\mathbf{U}$ by summing up the electric force $\mathbf{E}$ on a curve R around A we get

$$\oint_R \mathbf{E}ds = -\frac{d}{dt}\int_A \mathbf{B}dA \qquad \text{(Maxwell's equation 1)} \qquad (7)$$

$$\oint_R \mathbf{H}ds = \int_A \left(\mathbf{I} + \frac{d\mathbf{D}}{dt}\right)dA \qquad \text{(Maxwell's equation 2)} \qquad (8)$$

Summing up, the displacement $\mathbf{D}$ on A gives the charge q of A, since any magnet has necessarily two poles; which means for magnetism that monopoles do not exist (independent magnetic charges do not exist), therefore the magnetic flux density $\mathbf{B}$ on any plane A computes to zero. In mathematical form we thus have

$$\int_A \mathbf{D}dA = q \qquad (9)$$

$$\int_A \mathbf{B}dA = 0 \qquad (10)$$

Equations 7, 8 and also 9, 10 are integral equations. They reflect in mathematical form Faraday's concept of "line of force" to describe the electrical laws for the interplay between electrical currents and generated magnetic effects and vice versa in the 3-dimensional space. Maxwell in his paper on "physical line of force", when he found for the first time a version of his important equations, but also in his later work, did not use integral equations. The application of the theory of molecular vortices of hydromechanics led to solutions given by differential equations. This type of equation is generally desired in physics, since it allows in an easier way to include additional conditions which are required. We are in the position to transform the integral equations, as given by 7–10 into differential equations. The application of the theorem of Stokes to 7 and 8 gives

$$rot\,\mathbf{E} = -\frac{d\mathbf{B}}{dt} \qquad \text{(Maxwell's equation 1)} \qquad (11)$$

$$rot\,\mathbf{H} = \mathbf{I} + \frac{d\mathbf{D}}{dt} \qquad \text{(Maxwell's equation 2)} \qquad (12)$$

Applying the theorem of Gauß to 9 and 10 yields

$$div\,\mathbf{D} = \rho \qquad (13)$$

$$div\,\mathbf{B} = 0 \qquad (14)$$

The Eqs. 11–14 are linear partial differential equations. They differ from the equations which can be found in Maxwell's work, including his final book the "Treatise". It was Oliver Heaviside who published for the first time this form of the Maxwell equations [7].

To complete the picture we have to add the following "material" equations

$$\mathbf{D} = \varepsilon\,\mathbf{E} \tag{15}$$

$$\mathbf{B} = \mu\mathbf{H} \tag{16}$$

$$\mathbf{I} = \sigma\mathbf{E} \quad \text{(Ohm's Law)} \tag{17}$$

7 Concluding Remarks

This paper tries to guide the reader from the time of the discovery of electro-magnetism, electro-dynamics and electro-magnetic induction as accomplished by Ørsted, Ampère and Faraday, up to the development of Maxwell's equations. As pointed out, Maxwell was not convinced that the theoretical work of the physicists on the European continent, such as Ampère, Weber and Neumann, to name the most active ones, had reached a final form in modeling electrical phenomena. They followed the concept that electrically generated forces act instantaneously at a distance. The force of gravitation was a important example for it. Maxwell studied electricity and magnetism carefully, using Faraday's "Experimental Researches in Electricity and Magnetism". By Faraday's method of the "Lines of Force" Maxwell got the idea to develop a theory of electricity and magnetism which generate forces which act "in time" at a distance, the force being carried by a medium between physical objects. From his equations, he found the result that in a vacuum, which means that the carrier has the properties of the ether, the speed of electro-magnetic action is equal to the speed of light. This made him speculate that light could be considered as a electromagnetic wave.

References

1. Ampère, M.: Sur la théorie mathématique de phenoménes électrodynamiques uniquement déduite de léxpériences. In: Mémoires de l'Académie Royale des Sciences de l'Institut de France, Année 1823, Paris, Hamburg, vol. II, Chapt. III, 528, pp. 175–388 (1827)
2. Bucci, O.M.: Evolution of electromagnetics in the nineteenth century. In: Sarkar, T.K., et al. (eds.) History of Wireless, pp. 189–214. Wiley, Hoboken (2006)
3. Faraday, M.: Experimental Researches in Electricity, vol. I. Taylor & Francis, London (1839)
4. Faraday, M.: Experimental Researches in Electricity, vol. II. Taylor & Francis, London (1844)
5. Faraday, M.: Experimental Researches in Electricity, vol. III. Taylor & Francis, London (1855)

6. Gerlach, W.: Was ist und wozu dient die Elektrodynamik? vol. Heft 1, pp. 5–33. Deutsches Museum, Abhandlungen und Berichte (1966)

7. Heaviside, O.: Electromagnetic Theory, vol. I, II. Taylor & Francis, London (1893)

8. Hertz, H.: Ueber die Grundgleichungen der Elektrodynamik für ruhende Körper. In: Untersuchungen über die Ausbreitung der elektrischen Kraft, 13, Kapitel, Leipzig (1892)

9. Hunt, B.J.: The Maxwellians. Cornell University Press, Ithaca (1991)

10. Lindell, I.: Evolution of electromagnetics in the nineteenth century. In: Sarkar, T.K., et al. (eds.) History of Wireless, pp. 165–188. Wiley, Hoboken (2006)

11. Maxwell, J.C.: A dynamical theory of the electromagnetic field. Philos. Trans. R. Soc. London **155**, 459–512 (1865)

12. Maxwell, J.C.: Lehrbuch der Electricität und des Magnetismus. German translation by Dr. B. Weinstein, vol. II, 541, Berlin (1873)

13. Maxwell, J.C.: A Treatise on Electricity and Magnetism, vol. I, II. Clarendon Press, Oxford (1873)

14. Maxwell, J.C.: Über Faraday's Kraftlinien. Herausgegeben von L. Boltzmann. Ostwald's Klassiker der Exakten Wissenschaften, vol. 69. Theil, Leipzig (1895)

15. Maxwell, J.C.: Über Faraday's Kraftlinien. Herausgegeben von L. Boltzmann. Ostwald's Klassiker der Exakten Wissenschaften, vol. 102. Theil, Leipzig (1898)

16. Pfaff, C.: Der Elektro-Magnetismus, eine historisch-kritische Darstellung der bisherigen Entdeckungen nebst eigenthümlichen Versuchen, Hamburg (1824)

17. Ørsted, H.C.: Versuche über die Wirkung des elektrischen Conflicts auf die Magnetnadel. German translation in: Annalen der Physik und Chemie, pp. 295–304. Sechster Band, Leipzig, S. (1820)

18. Sengupta, D.L., Sarkar, T.K.: Maxwell, Hertz, the Maxwellians and the early history of electromagnetic waves. In: Sarkar, T.K., et al. (eds.) History of Wireless, pp. 215–228. Wiley, Hoboken (2006)

Stochastic Models and Applications to Natural, Social and Technical Systems

Bounds on the Rate of Convergence for Nonstationary $M^X/M_n/1$ Queue with Catastrophes and State-Dependent Control at Idle Time

Alexander Zeifman[1]([✉]), Yacov Satin[2], Ksenia Kiseleva[2], Tatiana Panfilova[2], Anastasia Kryukova[2], Galina Shilova[2], Alexander Sipin[2], and Elena Fokicheva[2]

[1] Institute of Informatics Problems, Federal Research Center "Informatics and Control" of RAS, Vologda Research Center of RAS, Vologda State University, Vologda, Russia
a_zeifman@mail.ru
[2] Vologda State University, Vologda, Russia

Abstract. Inhomogeneous continuous-time Markov chain with a special structure of infinitesimal matrix is considered as the queue-length process for the corresponding queueing model with possible batch arrivals, possible catastrophes and state-dependent control at idle time. For two wide classes of such processes we suppose an approach is proposed for obtaining explicit bounds on the rate of convergence to the limiting characteristics.

1   Introduction

We consider the general nonstationary Markovian queueing model with possible batch arrivals, possible catastrophes and state-dependent control at idle time. The previous investigations in this area deal with different particular classes of this general model, see, for instance, [1,2,4,7,8,13,14]. Detailed discussion and references one can find in [7]. We obtain bounds on the rate of convergence for such models and study some specific situations. The general approach is closely connected with the notion of the logarithmic norm and the corresponding bounds for the Cauchy matrix, and it has been described in [15].

Let

$b_k(t)$ be intensity of arrival of group of k customers to the non-empty queue, which does not depend on the current size of the length of queue;

$\mu_k(t)$ be intensity of service of a customer in the queue, if the current size of the length of queue equals k;

$\beta_k(t)$ be disaster (catastrophe) intensity, if the current size of the length of queue equals k;

$h_k(t)$ be intensity of transition from zero to k (resurrection in terms of [7], or mass arrivals in terms of [1]).

© Springer Nature Switzerland AG 2020
R. Moreno-Díaz et al. (Eds.): EUROCAST 2019, LNCS 12013, pp. 143–149, 2020.
https://doi.org/10.1007/978-3-030-45093-9_18

Consider the corresponding queue-length process $X(t)$. Then the intensity matrix $Q(t) = (q_{ij}(t))_{i,j=0}^\infty$ for $X(t)$ takes the following form:

$$Q(t) = \begin{pmatrix} q_{00}(t) & h_1(t) & h_2(t) & h_3(t) & h_4(t) & \cdots & \cdots \\ \beta_1(t) + \mu_1(t) & q_{11}(t) & b_1(t) & b_2(t) & \cdots & & \cdots & \cdots \\ \beta_2(t) & \mu_2(t) & q_{22}(t) & b_1(t) & b_2(t) & \cdots & \cdots \\ \cdots & \cdots & \cdots & \cdots & \cdots & \cdots & \cdots \\ \beta_j(t) & 0 & \cdots & \mu_j(t) & q_{jj}(t) & b_1(t) & \cdots \\ \vdots & \vdots & \vdots & \vdots & \vdots & \vdots & \ddots \end{pmatrix},$$

where $q_{ii}(t)$ are such that all row sums of the matrix equal to zero for any $t \geq 0$.

Then, applying the modified combined approach of [15,17] we can obtain bounds on the rate of convergence of the queue-length process to its limiting characteristics and compute them. We separately consider the important special cases.

2   Basic Notions

Denote by $p_{ij}(s,t) = P\{X(t) = j\,|\,X(s) = i\}$, $i,j \geq 0$, $0 \leq s \leq t$ the transition probabilities of $X(t)$ and by $p_i(t) = P\{X(t) = i\}$ – the probability that $X(t)$ is in state i at time t. Let $\mathbf{p}(t) = (p_0(t), p_1(t), \dots)^T$ be probability distribution vector at instant t. Applying the standard approach (see for instance [14,16]) we assume that all the intensity functions $q_{ij}(t)$ are locally integrable on $[0, \infty)$. Henceforth it is assumed that the intensity matrix $Q(t)$ is essentially bounded, i.e. $\sup_i |q_{ii}(t)| = L(t) \leq L < \infty$, for almost all $t \geq 0$.

Probabilistic dynamics of the process $\{X(t),\ t \geq 0\}$ is given by the forward Kolmogorov system

$$\frac{d\mathbf{p}(t)}{dt} = A(t)\mathbf{p}(t), \tag{1}$$

where $A(t) = Q^T(t) = (a_{ij}(t))_{i,j=0}^\infty = (q_{ji}(t))_{i,j=0}^\infty$ is the transposed intensity matrix.

Throughout the paper by $\|\cdot\|$ we denote the l_1-norm, i. e. $\|\mathbf{p}(t)\| = \sum_k |p_k(t)|$, and $\|A(t)\| = \sup_j \sum_i |a_{ij}|$. Let Ω be a set all stochastic vectors, i.e. l_1 vectors with non-negative coordinates and unit norm. Hence we have $\|A(t)\| = 2\sup_k |q_{kk}(t)| \leq 2L$ for almost all $t \geq 0$. Hence the operator function $A(t)$ from l_1 into itself is bounded for almost all $t \geq 0$ and locally integrable on $[0; \infty)$. Therefore we can consider (1) as a differential equation in the space l_1 with bounded operator.

It is well known (see [3]) that the Cauchy problem for differential equation (1) has a unique solutions for an arbitrary initial condition, and $\mathbf{p}(s) \in \Omega$ implies $\mathbf{p}(t) \in \Omega$ for $t \geq s \geq 0$.

Denote by $E(t,k) = E(X(t)|X(0) = k)$ the conditional expected number of customers in the system at instant t, provided that initially (at instant $t = 0$) k customers were present in the system.

Recall that a Markov chain $\{X(t),\ t \geq 0\}$ is called *weakly ergodic*, if $\|\mathbf{p}^*(t) - \mathbf{p}^{**}(t)\| \to 0$ as $t \to \infty$ for any initial conditions $\mathbf{p}^*(0)$ and $\mathbf{p}^{**}(0)$, where $\mathbf{p}^*(t)$ and $\mathbf{p}^{**}(t)$ are the corresponding solutions of (1). A Markov chain $\{X(t),\ t \geq 0\}$ has the limiting mean $\varphi(t)$, if $\lim_{t\to\infty}(\varphi(t) - E(t,k)) = 0$ for any k.

3   Nonstationary $M^X/M_n/1$ Queue Without Catastrophes with the Special Resurrection Intensities

In this section we study as in [7] the queueing model without catastrophes (i.e. all $\beta_j(t) = 0$) with the special resurrection rates $h_j(t) = b_j(t)$, for any j, t. In addition, we suppose in this section that $b_{k+1}(t) \leq b_k(t)$ for all k.

In accordance with these assumptions, we arrive at the model described in [16, 17] as queue with state-independent batch arrivals and state-dependent service intensities. Since $p_0(t) = 1 - \sum_{i=1}^{\infty} p_i(t)$ due to normalization condition, one can rewrite the system (1) as follows:

$$\frac{d\mathbf{z}(t)}{dt} = B(t)\mathbf{z}(t) + \mathbf{f}(t), \tag{2}$$

with $\mathbf{z}(t) = (p_1(t), p_2(t), \dots)^T$, $\mathbf{f}(t) = (a_{10}(t), a_{20}(t), \dots)^T$, and the corresponding $b_{ij}(t) = a_{ij}(t) - a_{i0}(t)$, $i, j \geq 1$.

Put now $\mathbf{y} = D(\mathbf{z}^* - \mathbf{z}^{**})$, where D is the upper triangular matrix:

$$D = \begin{pmatrix} d_1 & d_1 & d_1 & \cdots \\ 0 & d_2 & d_2 & \cdots \\ 0 & 0 & d_3 & \cdots \\ \vdots & \vdots & \vdots & \ddots \end{pmatrix}.$$

Then one has:

$$\frac{d\mathbf{y}(t)}{dt} = B_D(t)\mathbf{y}(t), \tag{3}$$

where $B_D(t) = DB(t)D^{-1} = b_{i,j,D}(t)$ with elements $b_{i,i+1,D}(t) = \frac{d_i}{d_{i+1}}\mu_i(t)$, $b_{i+k,i,D}(t) = \frac{d_{i+k}}{d_i}b_k(t)$, and all other off-diagonal elements equal zero for any $t \geq 0$. Hence matrix $B_D(t)$ is essentially nonnegative for any $t \geq 0$, therefore we can apply the notion of logarithmic norm of an operator function and related bounds.

Recall the respective notion and the corresponding bounds.

Namely, if $K(t)$, $t \geq 0$ is a one-parameter family of bounded linear operators on a Banach space $\mathcal{B}$, then

$$\gamma(K(t))_{\mathcal{B}} = \lim_{h \to +0} \frac{\|I + hK(t)\| - 1}{h} \tag{4}$$

is called the logarithmic norm of the operator $K(t)$. If $\mathcal{B} = l_1$ then the operator $K(t)$ is given by the matrix $K(t) = (k_{ij}(t))_{i,j=0}^{\infty}$, $t \geq 0$, and the logarithmic norm of $K(t)$ can be found explicitly:

$$\gamma\left(K\left(t\right)\right) = \sup_{j}\left(k_{jj}\left(t\right) + \sum_{i \neq j}|k_{ij}\left(t\right)|\right), \quad t \geq 0.$$

Then one has the following bound on the rate of convergence

$$\|\mathbf{x}(t)\| \leq e^{\int_0^t \gamma(K(\tau))d\tau}\|\mathbf{x}(0)\|,$$

if $\mathbf{x}(t)$ is an arbitrary solution of the differential equation

$$\frac{d\mathbf{x}(t)}{dt} = K(t)\mathbf{x}(t),$$

with the corresponding initial condition $\mathbf{x}(0)$.

Moreover, if the matrix $K(t)$ is essentially nonnegative then

$$\gamma\left(K\left(t\right)\right) = \sup_{j}\sum_{i}k_{ij}\left(t\right), \quad t \geq 0.$$

In this situation there is a number of classes of processes with possibility of sharp bounding the rate of convergence, see details in [15–17].

Put

$$\alpha_j(t) = \mu_j\left(t\right) - \frac{d_{j-1}}{d_j}\mu_{j-1}\left(t\right) + \sum_{i=1}^{\infty}\left(1 - \frac{d_{i+j}}{d_j}\right)b_i\left(t\right),$$

and $\alpha(t) = \inf \alpha_j(t)$.

Then we can compute

$$\gamma\left(B_D(t)\right) = \sup_{j}\sum_{i}b_{i,j,D}\left(t\right) = -\alpha(t).$$

Therefore the corresponding results of [16,17] give us the following statement.

Theorem 1. *Let there exist an increasing sequence $\{d_j,\ j \geq 1\}$ of positive numbers with $d_1 = 1$, such that*

$$\int_0^{\infty} \alpha(t)\,dt = +\infty. \tag{5}$$

Then the Markov chain $X(t)$ is weakly ergodic and the following bound holds:

$$\|\mathbf{y}\left(t\right)\| \leq e^{-\int_s^t \alpha(u)du}\|\mathbf{y}\left(s\right)\|, \tag{6}$$

for any initial conditions $s \geq 0$, $\mathbf{p}^*(s)$, $\mathbf{p}^{**}(s)$ and any $t \geq s$. Moreover, if $W = \inf_{i \geq 1}\frac{d_i}{i} > 0$, then $X(t)$ has the limiting mean and

$$|\varphi(t) - E(t,k)| \leq \frac{2}{W}e^{-\int_s^t \alpha(u)du}\|\mathbf{y}(s)\|, \tag{7}$$

for any $s \geq 0$, $t \geq s$, and any k.

Remark 1. One can apply all the bounds both for the ordinary nonstationary $M/M/S$ queue and for nonstationary $M^X/M/S$ queue with batch arrivals and S servers, the corresponding results one can find in [5,17]. Moreover, these bounds give us the way for obtaining the perturbation bounds and for computing of the limiting characteristics of the corresponding queue-length-process, see [9–12].

4   General Nonstationary $M^X/M_n/1$ Queue with Mass Arrivals and Catastrophes

Let now resurrection intensities $h_j(t)$ be arbitrary locally integrable functions such that $h_0(t) = \sum_{i \geq 1} h_j(t) \leq L$ in accordance with our general assumptions.

Let catastrophe intensities be *essential*, i.e. let

$$\int_0^\infty \beta_*(t)\, dt = +\infty, \tag{8}$$

where $\beta_*(t) = \inf_i \beta_i(t)$.

Rewrite the forward Kolmogorov system (1) as

$$\frac{d\mathbf{p}}{dt} = A^*(t)\,\mathbf{p} + \mathbf{g}(t), \quad t \geq 0. \tag{9}$$

Here $\mathbf{g}(t) = (\beta_*(t), 0, 0, \dots)^T$, $A^*(t) = \left(a_{ij}^*(t)\right)_{i,j=0}^\infty$, and

$$a_{ij}^*(t) = \begin{cases} a_{0j}(t) - \beta_*(t), & \text{if } i = 0, \\ a_{ij}(t), & \text{otherwise} . \end{cases} \tag{10}$$

Put now $\mathbf{w} = D(\mathbf{p}^* - \mathbf{p}^{**})$, where D be the diagonal matrix $D = diag(d_0, d_1, d_2, \dots)$ and $\{d_i\}$, $1 = d_0 \leq d_1 \leq \dots$ is a non-decreasing sequence of positive numbers.

Then one has:

$$\frac{d\mathbf{w}(t)}{dt} = A_D^*(t)\mathbf{w}(t), \tag{11}$$

where $A_D^*(t) = DA^*(t)D^{-1} = a_{i,j,D}^*(t)$ with the corresponding elements. Hence matrix $A_D^*(t)$ is essentially nonnegative for any $t \geq 0$, therefore we can apply the notion of logarithmic norm of an operator function and related bounds.

Put firstly all $d_j = 1$.

Then we have

$$\gamma(A_D^*(t)) = \sup_j \sum_i a_{i,j}^*(t) = -\beta_*(t).$$

Then the corresponding results of [13–15] give us the following statement.

Theorem 2. *Let catastrophe rates be essential, i.e. assumption (8) be fulfilled. Then the queue-length process $X(t)$ is weakly ergodic in the uniform operator topology and the following bound hold*

$$\|\mathbf{p}^*(t) - \mathbf{p}^{**}(t)\| \leq$$

$$\leq e^{-\int_0^t \beta_*(\tau)\, d\tau} \|\mathbf{p}^*(0) - \mathbf{p}^{**}(0)\| \leq 2e^{-\int_0^t \beta_*(\tau)\, d\tau}, \tag{12}$$

for any initial conditions $\mathbf{p}^*(0), \mathbf{p}^{**}(0)$ and any $t \geq 0$.

Let now, in the contrary numbers d_i grow fast enough.
Put

$$\beta_{**}(t) = \inf_i \left(|a_{ii}^*(t)| - \sum_{j \neq i} \frac{d_j}{d_i} a_{ji}^*(t) \right). \tag{13}$$

Then

$$\gamma(A_{\mathsf{D}}^*(t)) = \sup_i \sum_j \frac{d_j}{d_i} a_{j,i}^*(t) = -\beta_{**}(t),$$

and we obtain the following statement.

Theorem 3. *Let $\{d_i\}$, $1 = d_0 \leq d_1 \leq \dots$ be a non-decreasing sequence such that $W = \inf_{i \geq 1} \frac{d_i}{i} > 0$, and*

$$\int_0^\infty \beta_{**}(t)\, dt = +\infty. \tag{14}$$

Then $X(t)$ has the limiting mean, say $\phi(t) = E(t, 0)$, and the following bound holds:

$$|E(t, j) - E(t, 0)| \leq \frac{1 + d_j}{W} e^{-\int_0^t \beta_{**}(\tau)\, d\tau}, \tag{15}$$

for any j and any $t \geq 0$.

Remark 2. One can apply all the bounds for the specific classes of queueing models which are connected with nonstationary $M^X/M/S$ queue with batch arrivals, S servers and possible resurrections and catastrophes. The corresponding results for these models in the cases considered were firstly obtained in [13–15].

Remark 3. Such processes can be applied also for modeling of geoinformation systems, see for instance [6].

Acknowledgments. The work by Zeifman was supported by the State scientific grant of the Vologda region.

References

1. Chen, A., Renshaw, E.: The $M/M/1$ queue with mass exodus and mass arrivals when empty. J. Appl. Probab. **34**(1), 192–207 (1997)
2. Chen, A., Renshaw, E.: Markovian bulk-arriving queues with state-dependent control at idle time. Adv. Appl. Probab. **36**(2), 499–524 (2004)
3. Daleckii, J.L., Krein, M.G.: Stability of Solutions of Differential Equations in Banach Space, vol. 43. American Mathematical Society, Providence (2002)
4. Di Crescenzo, A., Giorno, V., Nobile, A.G., Ricciardi, L.M.: A note on birth-death processes with catastrophes. Stat. Probab. Lett. **78**(14), 2248–2257 (2008)
5. Granovsky, B.L., Zeifman, A.I.: Nonstationary queues: estimation of the rate of convergence. Queueing Syst. **46**, 363–388 (2004)
6. Ismaeel, A.G.: Effective technique for allocating servers to support cloud using GPS and GIS. In: IEEE 2013 Science and Information Conference, pp. 934–939 (2013)
7. Li, J., Zhang, L.: $M^X/M/c$ queue with catastrophes and state-dependent control at idle time. Front. Math. China **12**(6), 1427–1439 (2017)
8. Zhang, L., Li, J.: The M/M/c queue with mass exodus and mass arrivals when empty. J. Appl. Probab. **52**, 990–1002 (2015)
9. Zeifman, A., Satin, Ya., Korolev, V., Shorgin, S.: On truncations for weakly ergodic inhomogeneous birth and death processes. Int. J. Appl. Math. Comput. Sci. **24**, 503–518 (2014)
10. Zeifman, A., Korolev, V., Satin, Y., Korotysheva, A., Bening, V.: Perturbation bounds and truncations for a class of Markovian queues. Queueing Syst. **76**(2), 205–221 (2014). https://doi.org/10.1007/s11134-013-9388-0
11. Zeifman, A.I., Korolev, V.Y.: On perturbation bounds for continuous-time Markov chains. Stat. Probab. Lett. **88**, 66–72 (2014)
12. Zeifman, A.I., Korotysheva, A.V., Korolev, V.Yu., Satin, Ya.A: Truncation bounds for approximations of inhomogeneous continuous-time Markov chains. Theor. Probab. Appl. **61**, 513–520 (2017)
13. Zeifman, A.I., Satin, Y.A., Korotysheva, A.V., Korolev, V.Y., Bening, V.E.: On a class of Markovian queuing systems described by inhomogeneous birth-and-death processes with additional transitions. Doklady Math. **94**(2), 502–505 (2016). https://doi.org/10.1134/S1064562416040177
14. Zeifman, A., Korotysheva, A., Satin, Y., Razumchik, R., Korolev, V., Shorgin, S.: Ergodicity and truncation bounds for inhomogeneous birth and death processes with additional transitions from and to origin. Stoch. Models **33**, 598–616 (2017)
15. Zeifman, A., Korotysheva, A., Satin, Y., Kiseleva, K., Korolev, V., Shorgin, S.: Bounds for Markovian queues with possible catastrophes. In: Proceedings of 31st European Conference on Modelling and Simulation ECMS 2017, Digitaldruck Pirrot GmbHP Dudweiler, Germany, pp. 628–634 (2017)
16. Zeifman, A., et al.: On sharp bounds on the rate of convergence for finite continuous-time Markovian queueing models. In: Moreno-Díaz, R., Pichler, F., Quesada-Arencibia, A. (eds.) EUROCAST 2017. LNCS, vol. 10672, pp. 20–28. Springer, Cham (2018). https://doi.org/10.1007/978-3-319-74727-9_3
17. Zeifman, A., Razumchik, R., Satin, Y., Kiseleva, K., Korotysheva, A., Korolev, V.: Bounds on the rate of convergence for one class of inhomogeneous Markovian queueing models with possible batch arrivals and services. Int. J. Appl. Math. Comput. Sci. **28**(1), 141–154 (2018)

Some Remarks on the Prendiville Model in the Presence of Jumps

Virginia Giorno$^{(\boxtimes)}$, Amelia G. Nobile, and Serena Spina

Dipartimento di Informatica, Università di Salerno, Via Giovanni Paolo II, n. 132, 84084 Fisciano, SA, Italy
{giorno,nobile,sspina}@unisa.it

Abstract. The inhomogeneous Prendiville process in the presence of catastrophes at a generic state of the space of the states is considered. The transition probabilities and the moments are determined in closed form for the homogeneous case and when the intensities of the involved processes have the same time dependence.

1 Introduction

The Prendiville process is a birth-death (BD) process defined on a discrete finite space and characterized by state dependent rates: the births are favorite when the population size is low and the deaths are more frequent for large size of the populations. This process has been extensively used in biology, in physics, in population dynamic and in epidemiology. In [14], it was also considered as an adaptive queueing system model with finite capacity in which the customers are discouraged to joint the queue when the queue size is large and, at the same time, the server accelerates the service. A theoretical study of the time-non-homogeneous Prendiville process can be found in [22].

In the present paper, we study the Prendiville process under the effect of random jumps or catastrophes. Specifically, a catastrophe is an event that occurs at random times and produces an instantaneous variation of the state of the system by passing from the current state to a fixed state. The literature concerning the BD processes subject to catastrophes is very wide (see, for instance, [1–12, 15–21]). The catastrophes assume various meaning depending on the applicative context: for example, a catastrophe at zero state can be considered as the effect of a fault that clears the queue, while in a population dynamics a catastrophe can be interpreted as the effect of an epidemic or an extreme natural disaster (forest fire, flood,...).

Let $\widetilde{N}(t)$ be a Prendiville BD process defined in $\mathcal{S} = [L, H]$ with $0 \leq L < H < +\infty$ and let $\{N_k(t), \ t \geq 0\}$, with $k \in \mathcal{S}$, be the Prendiville process subject to jumps at the state k. We assume that the transition rate functions $r_{j,n}(t) = \lim_{h \downarrow 0} P[N_k(t + h) = n | N_k(t) = j]/h$ are defined as follows:

$$r_{n,n+1}(t) = \alpha_n(t) = \lambda(t)(H - n), \quad r_{n,n-1}(t) = \beta_n(t) = \mu(t)(n - L), \quad n \in \mathcal{S},$$
$$r_{n,k}(t) = \xi(t), \qquad \forall \ n \in \mathcal{S} \setminus \{k\}, \tag{1}$$

This work is partially supported by G.N.C.S.- INdAM.

R. Moreno-Díaz et al. (Eds.): EUROCAST 2019, LNCS 12013, pp. 150–157, 2020.
https://doi.org/10.1007/978-3-030-45093-9_19

where $\lambda(t) > 0$, $\mu(t) > 0$, $\xi(t) \geq 0$ are continuous functions of the time.

The outline of the paper is the following. To study the process $N_k(t)$, we consider the relations between the processes $\widetilde{N}(t)$ and $N_k(t)$; then, closed form results are obtained; finally, some numerical cases are discussed.

2   The Prendiville Process with Jumps

The transition probabilities $p_{j,n;k}(t|t_0) = \mathbb{P}\{N_k(t) = n \,|\, N_k(t_0) = j\}$ satisfy the following system:

$$\frac{d}{dt}p_{j,n;k}(t|t_0) = -[\alpha_n(t) + \beta_n(t) + \xi(t)]p_{j,n;k}(t|t_0) + \alpha_{n-1}(t)p_{j,n-1;k}(t|t_0)$$
$$+\beta_{n+1}(t)p_{j,n+1;k}(t|t_0), \qquad (j \in \mathcal{S}, n \in \mathcal{S} \setminus \{k\})$$
$$\frac{d}{dt}p_{j,k;k}(t|t_0) = -[\alpha_k(t) + \beta_k(t) + \xi(t)]p_{j,k;k}(t|t_0) + \alpha_{k-1}(t)p_{j,k-1;k}(t)$$
$$+\beta_{k+1}(t)p_{j,k+1;k}(t|t_0) + \xi(t)$$

and the initial condition $\lim_{t \downarrow t_0} p_{j,n;k}(t|t_0) = \delta_{j,n}$. Moreover, assuming that

$$\lim_{t \to +\infty} \int_{t_0}^{t} \xi(u) \, du = +\infty, \tag{2}$$

the probabilities $p_{j,n}(t|t_0)$ can be related to the transition probabilities $\widetilde{p}_{j,n}(t|t_0)$ of the process $\widetilde{N}(t)$. Indeed, conditioning on the time of the first catastrophe, we have:

$$p_{j,n;k}(t|t_0) = \varphi(t|t_0)\, \widetilde{p}_{j,n}(t|t_0) + \int_{t_0}^{t} \xi(\tau)\, \varphi(\tau|t_0)\, p_{k,n;k}(t|\tau) \, d\tau, \tag{3}$$

whereas, conditioning on the age of the catastrophes process, we obtain:

$$p_{j,n;k}(t|t_0) = \varphi(t|t_0)\, \widetilde{p}_{j,n}(t|t_0) + \int_{t_0}^{t} \xi(\tau)\, \varphi(t|\tau)\, \widetilde{p}_{k,n}(t|\tau) \, d\tau, \tag{4}$$

where $\varphi(t|t_0) = \exp\left\{ -\int_{t_0}^{t} \xi(u)du \right\}$. The ℓth-order moments of the processes $\widetilde{N}(t)$ and $N_k(t)$ satisfy the following relation:

$$M_\ell(t|j, t_0) = \varphi(t|t_0)\, \widetilde{M}_\ell(t|j, t_0) + \int_{t_0}^{t} \xi(\tau)\varphi(t|\tau)\, \widetilde{M}_\ell(t|k, \tau) \, d\tau.$$

Let

$$\mathcal{T}_{j,k}(t_0) = \inf\{t \geq t_0 : N_k(t) = k\}, \qquad N_k(t_0) = j$$

be the first visit time (FVT) of $N_k(t)$ to k starting from the state j; one has:

$$g_{j,k}(t|t_0) = \frac{d\mathbb{P}\{\mathcal{T}_{j,k}(t_0) \leq t\}}{dt}$$
$$= \varphi(t|t_0)\, \widetilde{g}_{j,k}(t|t_0) + \xi(t)\, \varphi(t|t_0)\left[1 - \int_{t_0}^{t} \widetilde{g}_{j,k}(\tau|t_0) \, d\tau\right], \tag{5}$$

where $\widetilde{g}_{j,k}(t|t_0)$ is the pdf of the first passage time (FPT) of $\widetilde{N}(t)$ from j to k. Note that $\int_{t_0}^{+\infty} g_{j,k}(\tau|t_0)\, d\tau = 1$, hence the FVT of $N_k(t)$ to k occurs with probability 1, whereas $\int_{t_0}^{+\infty} \widetilde{g}_{j,k}(\tau|t_0)\, d\tau \leq 1$.

2.1 The Prendiville Process

The transient behavior of the Prendiville process with jumps can be obtained by using the results of the Prendiville process in the absence of jumps. As proved in [22], the process $[\widetilde{N}(t)|\widetilde{N}(t_0) = j]$ is distributed as a shifted sum of independent binomial variables:

$$[\widetilde{N}(t)|\widetilde{N}(t_0) = j] \sim L + \mathrm{Bin}(j - L, b_1(t|t_0)) + \mathrm{Bin}(H - j, b_2(t|t_0)), \qquad (6)$$

where

$$b_1(t|t_0) = \exp\{-A(t|t_0)\}\left[1 + \int_{t_0}^{t} \lambda(u)\exp\{A(u|t_0)\}\, du\right],$$

$$b_2(t|t_0) = 1 - \exp\{-A(t|t_0)\}\left[1 + \int_{t_0}^{t} \mu(u)\exp\{A(u|t_0)\}\, du\right],$$

with

$$A(t|t_0) = \int_{t_0}^{t} [\lambda(u) + \mu(u)]\, du. \qquad (7)$$

Specifically, we have:

$$\widetilde{p}_{j,n}(t|t_0) = [b_1(t|t_0)]^{n-L}[1 - b_2(t|t_0)]^{H-j}[1 - b_1(t|t_0)]^{j-n}$$

$$\times \sum_{\ell=\max(0,n-j)}^{\min(H-j,n-L)} \binom{H-j}{\ell}\binom{j-L}{n-L-\ell}\left[\frac{b_2(t|t_0)}{b_1(t|t_0)}\frac{1 - b_1(t|t_0)}{1 - b_2(t|t_0)}\right]^{\ell} \quad j, n \in \mathcal{S}. \ (8)$$

Note that if the process starts from one of the end points of S, then one of the random variables in (6) becomes degenerate; in these cases one has:

$$[\widetilde{N}(t)|\widetilde{N}(t_0) = L] \sim L + \mathrm{Bin}(H - L, b_2(t|t_0)),$$

$$[\widetilde{N}(t)|\widetilde{N}(t_0) = H] \sim L + \mathrm{Bin}(H - L, b_1(t|t_0)).$$

Recalling (4) one can determine the transition probabilities via numerical procedures. Moreover, making use of (6), the moments generating function is

$$\widetilde{\mathcal{M}}(z,t|j,t_0) = e^{Lz}\left[1 - b_1(t|t_0) + b_1(t|t_0)e^{z}\right]^{j-L}\left[1 - b_2(t|t_0) + b_2(t|t_0)e^{z}\right]^{H-j}.$$

3 Closed Form Results

In this section, we consider some particular cases in which the transition probabilities $p_{j,n;k}(t|t_0)$ can be expressed in a closed form.

- *The homogeneous case.* Let

$$\alpha_n(t) = \lambda(H-n) \qquad \beta_n(t) = \mu(n-L), \qquad \xi(t) = \xi. \tag{9}$$

In this case, from (4), making use of (8) one has

$$p_{j,n;k}(t) = e^{-\xi t}\,\widetilde{p}_{j,n}(t) + \xi \int_0^t e^{-\xi \tau}\,\widetilde{p}_{k,n}(\tau)\,d\tau = e^{-\xi t}\,\widetilde{p}_{j,n}(t) + \frac{\xi \lambda^{H-k}\mu^{k-L}}{(\lambda+\mu)^{H-L}}$$

$$\times \sum_{i=\max(0,n-k)}^{\min(H-k,n-L)} \binom{H-k}{i}\binom{k-L}{n-L-i}\mathcal{I}_{k-n+2i,H-k-i,n-L-i}(\xi,t), \tag{10}$$

where $p_{j,n;k}(t) \equiv p_{j,n;k}(t|0)$ and

$$\mathcal{I}_{a,b,c}(\xi,t) = \int_0^t e^{-\xi\tau}\left[1 - e^{-(\lambda+\mu)\tau}\right]^a \left[\frac{\mu}{\lambda} + e^{-(\lambda+\mu)\tau}\right]^b \left[\frac{\lambda}{\mu} + e^{-(\lambda+\mu)\tau}\right]^c d\tau,$$

with a,b,c non negative integers. The integral $\mathcal{I}_{a,b,c}(\xi,t)$ can be evaluated by using the change of integration variable $y = e^{-(\lambda+\mu)\tau}$ and some series expansions. In this way one obtains

$$\mathcal{I}_{a,b,c}(\xi,t) = \frac{1}{\lambda+\mu}\int_{e^{-(\lambda+\mu)t}}^1 y^{\xi/(\lambda+\mu)-1}(1-y)^a \left(\frac{\mu}{\lambda}+y\right)^b \left(\frac{\lambda}{\mu}+y\right)^c dy$$

$$= \sum_{h=0}^a \binom{a}{h}(-1)^h \sum_{i=0}^b \binom{b}{i}\left(\frac{\mu}{\lambda}\right)^{b-i} \sum_{r=0}^c \binom{c}{r}\left(\frac{\lambda}{\mu}\right)^{c-r}\frac{1 - e^{-[\xi+(\lambda+\mu)(h+i+r)]t}}{\xi + (\lambda+\mu)(h+i+r)}. \tag{11}$$

Moreover, the moments can be calculated; in particular the mean is:

$$E[N_k(t)|N_k(0) = j]$$
$$= \frac{e^{-(\lambda+\mu+\xi)t}\left[\left(e^{(\lambda+\mu+\xi)t} - 1\right)(H\lambda + L\mu + k\xi) + j(\lambda+\mu+\xi)\right]}{\lambda+\mu+\xi}. \tag{12}$$

Let $q_{n;k}$ be the steady state distribution for $N_k(t)$. We have:

$$q_{n;k} = \lim_{t\to\infty} p_{j,n;k}(t) = \lim_{t\to\infty}\left[e^{-\xi t}\,\widetilde{p}_{j,n}(t) + \xi \int_0^t e^{-\xi\tau}\,\widetilde{p}_{k,n}(\tau)\,d\tau\right] \equiv \xi\widetilde{\Pi}_{k,n}(\xi),$$

where

$$\widetilde{\Pi}_{j,n}(s) = \int_0^{+\infty} e^{-st}\widetilde{p}_{j,n}(t)\,dt$$

$$= \frac{\lambda^{H-j}\mu^{j-L}}{(\lambda+\mu)^{H-L}}\sum_{\ell=\max(0,n-j)}^{\min(H-j,n-L)} \binom{H-j}{\ell}\binom{j-L}{n-L-\ell}I_{j-n-2\ell,H-j-\ell,n-L-\ell}(s)$$

is the Laplace transform (LT) of the transition probability $\widetilde{p}_{j,n}(t)$ with

$$I_{a,b,c}(s) = \sum_{h=0}^a \binom{a}{h}(-1)^h \sum_{i=0}^b \binom{b}{i}\left(\frac{\mu}{\lambda}\right)^{b-i} \sum_{r=0}^c \binom{c}{r}\left(\frac{\lambda}{\mu}\right)^{c-r}\frac{1}{s + (\lambda+\mu)(h+i+r)}.$$

The LT $\widetilde{\varPi}_{i,j}(s)$ plays a role also in the FVT problem. Indeed, due to (5), one can write

$$\gamma_{j,k}(s) = \int_0^\infty e^{-st} g_{j,k}(t)dt = \frac{\xi}{s+\xi} + \frac{s}{s+\xi}\widetilde{\gamma}_{j,k}(s) \qquad (j \neq k), \qquad (13)$$

where $\widetilde{\gamma}_{j,k}(s) = \widetilde{\varPi}_{j,k}(s+\xi)/\widetilde{\varPi}_{k,k}(s+\xi)$ is the LT of the FPT density. Moreover, from (13), one obtains $\mathbb{P}(T_{j,k} < \infty) = \int_0^\infty g_{j,k}(t)dt \equiv \gamma_{j,k}(0) = 1$, so the FVT is a certain event, the moments of the FVT can be evaluated as follows:

$$\mathbb{E}(T_{j,k}^l) = (-1)^l \frac{d^l \gamma_{j,k}(s)}{ds^l}\bigg|_{s=0}.$$

For the process (9), in Fig. 1 the transition probabilities $p_{0,n;0}(t)$ for $n = 0,1,2,3$ with $\lambda = 0.5$, $\mu = 1$ and $L = 0$, $H = 10$ are plotted for various choices of ξ. Note that in the asymptotic behavior the probabilities increase with ξ for $n = 0,1,2$ whereas they decrease for $n = 3$ as ξ increases, this last behavior is common to the probabilities for $n > 3$. For the same choices of the parameters in Fig. 2 the mean and the coefficient of the variation are plotted. As expected, due to the choice of $k = 0$, the mean decreases as ξ increases.

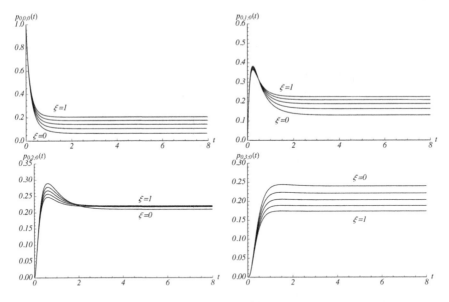

Fig. 1. For th process (9) the probabilities $p_{0,n;0}(t)$ for $n = 0,1,2,3$ with $\lambda = 0.5$, $\mu = 1$ and $L = 0$, $H = 10$ and $\xi = 0, 0.2, \ldots, 1$ are plotted as function of the time.

- *An inhomogeneous case.* Let

$$\alpha_n(t) = \lambda\psi(t)(H - n), \quad \beta_n(t) = \mu\psi(t)(n - L), \quad \xi(t) = \xi\psi(t), \qquad (14)$$

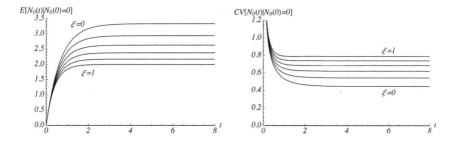

Fig. 2. Mean and coefficient of variation for $N_0(t)$ with $N_0(0) = 0$ for the same parameters of Fig. 1.

where $\psi(t)$ is a continuous function such that

$$\lim_{t \to \infty} \int_{t_0}^{t} \psi(u)\, du = +\infty.$$

Inhomogeneous BD processes with rates characterized by the same time dependence are treated also in [13]. In this case, the transformation $\Psi(t|t_0) = \int_{t_0}^{t} \psi(u)\, du$ $(t \geq t_0)$ allows to bring the transient analysis of the process $N_k(t)$ to the case of the Prendiville process with jumps obtained by setting $\psi(t) = 1$. In this way, we obtain:

$$p_{j,n;k}(t|t_0) = e^{-\xi\Psi(t|t_0)}\, \widetilde{p}_{j,n}(\Psi(t|t_0)) + \frac{\xi\lambda^{H-k}\mu^{k-L}}{(\lambda+\mu)^{H-L}} \sum_{i=\max(0,n-k)}^{\min(H-k,n-L)} \binom{H-k}{i}$$

$$\times \binom{k-L}{n-L-i} \mathcal{I}_{k-n+2i,H-k-i,n-L-i}(\xi, \Psi(t|t_0)),$$

with $\widetilde{p}_{j,n}(t)$ given in (10) and $\mathcal{I}_{a,b,c}(\xi,t)$ defined in (11). Moreover, the mean can be obtained from (12) by substituting t with $\Psi(t|t_0)$.

For the numerical analysis of the inhomogeneous case, we consider the following periodic function

$$\psi(t) = D + \left[1 - \cos\left(\frac{2\pi t}{Q}\right)\right]\left[A + B\sin\left(\frac{2\pi t}{Q}\right) + C\left(\frac{2\pi t}{Q}\right)\right]. \tag{15}$$

For the process (14), in Fig. 3 the transition probabilities $p_{0,n;0}(t|0)$ for $n = 0, 1, 2, 3$ are plotted; the parameters are chosen as follows: $\lambda = 0.5$, $\mu = 1$, $L = 0$, $H = 10$, $\psi(t)$ given in (15) with $A = 0.05$, $B = 0.04$, $C = 0.1$, $D = 1/4$ and period $Q = 2$. Due to the periodicity of the function $\psi(t)$, the transition probabilities admit the asymptotic behavior given by $q_{n;k}(t) = \lim_{\ell \to +\infty} p_{j,n;k}(t|t_0)$. For the same choices of the parameters in Fig. 4 the mean and the coefficient of the variation are plotted. Due to the choice of $k = 0$, the mean decreases as ξ increases. Note that the periodic behavior of $\psi(t)$ also affects the probabilities and the moments.

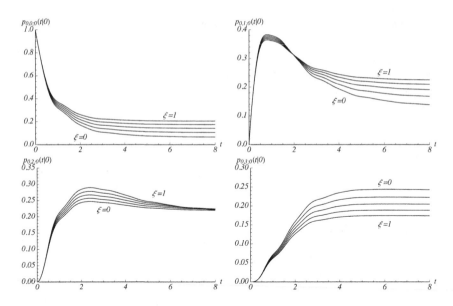

Fig. 3. For the process (14) the probabilities $p_{0,n;0}(t|0)$ for $n = 0, 1, 2, 3$ with $\lambda = 0.5$, $\mu = 1$ and $L = 0$, $H = 10$ and $\xi = 0, 0.2, \ldots, 1$; $\psi(t)$ is chosen as in (15) with $A = 0.05$, $B = 0.04$, $C = 0.1$, $D = 1/4$ and $Q = 2$.

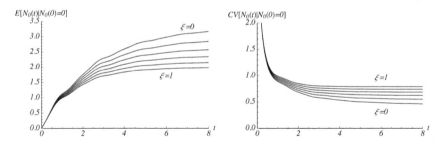

Fig. 4. Mean and coefficient of variation for $N_0(t)$ with $N_0(0) = 0$ for the same parameters of Fig. 3.

References

1. Brockwell, P.J., Gani, J., Resnick, S.I.: Birth, immigration and catastrophe processes. Adv. Appl. Prob. **14**, 709–731 (1982)
2. Brockwell, P.J.: The extinction time of a birth, death and catastrophe process and of a related diffusion model. Adv. Appl. Prob. **17**, 42–52 (1985)
3. Brockwell, P.J.: The extinction time of a general birth and death process with catastrophes. J. Appl. Prob. **23**, 851–858 (1986)
4. Chao, X., Zheng, Y.: Transient analysis of immigration birth-death processes with total catastrophes. Prob. Engin. Inform. Sci. **17**, 83–106 (2003)
5. Dharmaraja, S., Di Crescenzo, A., Giorno, V., Nobile, A.G.: A continuous-time Ehrenfest model with catastrophes and its jump-diffusion approximation. J. Stat. Phys. **16**, 326–345 (2015). https://doi.org/10.1007/s10955-015-1336-4

6. Di Crescenzo, A., Giorno, V., Nobile, A.G., Ricciardi, L.M.: On the M/M/1 queue with catastrophes and its continuous approximation. Queueing Syst. **43**, 329–347 (2003)
7. Di Crescenzo, A., Giorno, V., Nobile, A.G., Ricciardi, L.M.: A note on birth-death processes with catastrophes. Stat. Prob. Lett. **78**, 2248–2257 (2008)
8. Di Crescenzo, A., Giorno, V., Nobile, A.G., Krishna Kumar, B.: A double-ended queue with catastrophes and repairs: and a jump-diffusion approximation. Methodol. Comput. Appl. Prob. **14**, 937–954 (2012). https://doi.org/10.1007/s11009-011-9214-2
9. Di Crescenzo, A., Giorno, V., Krishna Kumar, B., Nobile, A.G.: A time-non-homogeneous double-ended queue with failures and repairs and its continuous approximation. Mathematics **6**(5), 81 (2018)
10. Economou, A., Fakinos, D.: A continuous-time Markov chain under the influence of a regulating point process and applications in stochastic models with catastrophes. Eur. J. Oper. Res. **149**, 625–640 (2003)
11. Giorno, V., Nobile, A.G., Saura, A.: Prendiville stochastic growth model in the presence of catastrophes. In: Trappl, R. (ed.) Cybernetics and Systems 2004, pp. 151–156. Austrian Society for Cybernetics Studies, Vienna (2004)
12. Giorno, V., Nobile, A.G.: On a bilateral linear birth and death process in the presence of catastrophes. In: Moreno-Díaz, R., Pichler, F., Quesada-Arencibia, A. (eds.) EUROCAST 2013. LNCS, vol. 8111, pp. 28–35. Springer, Heidelberg (2013). https://doi.org/10.1007/978-3-642-53856-8_4
13. Giorno, V., Nobile, A.G., Spina, S.: On some time non-homogeneous queueing systems with catastrophes. Appl. Math. Comput. **245**, 220–234 (2014)
14. Giorno, V., Negri, C., Nobile, A.G.: A solvable mode for a finite-capacity queueing system. J. Appl. Prob. **22**, 903–911 (1985)
15. Giorno, V., Spina, S.: Some remarks on stochastic diffusion processes with jumps. Lect. Notes Seminario Interdisciplinare di Matematica **12**, 161–168 (2015)
16. Kyriakidis, E.G.: Stationary probabilities for a simple immigration birth-death process under the influence of total catastrophes. Stat. Prob. Lett. **20**, 239–240 (1994)
17. Krishna Kumar, B., Arivudainambi, D.: Transient solution of an M/M/1 queue with catastrophes. Comput. Math. Appl. **40**, 1233–1240 (2000)
18. Pakes, A.G.: Killing and resurrection of Markov processes. Com. Stat. Stoch. Models **13**, 255–269 (1997)
19. Peng, N.F., Pearl, D.K., Chan, W., Bartoszynski, R.: Linear birth and death processes under the influence of disasters with time-dependent killing probabilities. Stoch. Proc. Appl. **45**, 243–258 (1993)
20. Sinitcina, A., et al.: On the bounds for a two-dimensional birth-death process with catastrophes. Mathematics **6**(5), 80 (2018)
21. Swift, R.J.: Transient probabilities for a simple birth-death-immigration process under the influence of total catastrophes. Int. J. Math. Math. Sci. **25**, 689–692 (2001)
22. Zheng, Q.: Note on the non-homogeneous Prendiville process. Math. Biosci. **148**, 1–5 (1998)

Continuous-Time Birth-Death Chains Generate by the Composition Method

Virginia Giorno and Amelia G. Nobile[(✉)]

Dipartimento di Informatica, Università di Salerno, Via Giovanni Paolo II, n. 132, 84084 Fisciano, SA, Italy
{giorno,nobile}@unisa.it

Abstract. Starting from a special birth-death process, obtained via the composition method applied to two double-ended systems, we consider a restricted birth-death process and construct the corresponding symmetric process with respect to zero state.

1 Introduction and Background

Continuous-time birth-death (BD) chains are frequently used in queueing systems, mathematical finance, reliability theory, populations growth, epidemiology and ecology to model the stochastic transient and asymptotic dynamics of real systems. In the literature, a relevant effort has been devoted to analyze BD processes also under the effect of catastrophes and/or with time dependent rates by focusing on the determination of *(i)* the transition probabilities for unrestricted and restricted BD chain; *(ii)* the first-passage time (FPT) densities in the presence of absorbing and reflecting boundaries; *(iii)* the symmetry relations between transition probabilities and FPT densities for different BD chains (cf., for instance, [1]–[14]).

In this paper, by starting from a double-ended BD process obtained via the composition method applied to two continuous-time BD chains, we obtain some restricted and unrestricted BD processes. Specifically, by assuming that $0 \leq \vartheta \leq 1$, $\lambda > 0$, $\mu > 0$ and $\varrho = \lambda/\mu$,

- In Sect. 2, we consider the BD process $\{R(t), t \geq 0\}$ on the nonnegative integers, with 0 reflecting boundary, characterized by birth and death rates:

$$\widetilde{\lambda}_n := \lambda \, \frac{\vartheta + (1-\vartheta)\varrho^{-n-1}}{\vartheta + (1-\vartheta)\varrho^{-n}}, \quad n \in \mathbb{N}_0, \quad \widetilde{\mu}_n := \mu \, \frac{\vartheta + (1-\vartheta)\varrho^{-n+1}}{\vartheta + (1-\vartheta)\varrho^{-n}}, \quad n \in \mathbb{N}; \quad (1)$$

- In Sect. 3, we focus on the symmetric BD chain $\{N(t), t \geq 0\}$ on $\mathbb{Z}$ characterized by rates:

$$
\begin{aligned}
\lambda_n = \mu_{-n} &= \lambda \, \frac{\vartheta + (1-\vartheta)\varrho^{-n-1}}{\vartheta + (1-\vartheta)\varrho^{-n}}, \quad n \in \mathbb{N}_0, \\[2mm]
\mu_n = \lambda_{-n} &= \mu \, \frac{\vartheta + (1-\vartheta)\varrho^{-n+1}}{\vartheta + (1-\vartheta)\varrho^{-n}}, \quad n \in \mathbb{N}.
\end{aligned}
\qquad (2)
$$

This paper is partially supported by G.N.C.S. - INdAM.

R. Moreno-Díaz et al. (Eds.): EUROCAST 2019, LNCS 12013, pp. 158–166, 2020.
https://doi.org/10.1007/978-3-030-45093-9_20

Let $\{\widetilde{M}_i(t), t \geq 0\}$ $(i = 1, 2)$ be a double-ended system conditioned to start from j at time 0, with birth rates $\lambda_{i,n}$ and death rates $\mu_{i,n}$ $(n \in \mathbb{Z})$. For $n \in \mathbb{Z}$, we set $\lambda_{1,n} = \lambda$, $\mu_{1,n} = \mu$ and $\lambda_{2,n} = \mu$, $\mu_{2,n} = \lambda$. The transition probabilities $\widetilde{f}_{j,n}^{(i)}(t) = P\{\widetilde{M}_i(t) = n | \widetilde{M}_i(0) = j\}$ of $\widetilde{M}_i(t)$ is (see, for instance, [1]):

$$\widetilde{f}_{j,n}^{(i)}(t) = \varrho^{(-1)^{i+1}(n-j)/2} e^{-(\lambda+\mu)t} I_{n-j}(2t\sqrt{\lambda\mu}), \qquad j, n \in \mathbb{Z}, \qquad (3)$$

where $I_\nu(z) = \sum_{m=0}^{\infty} (z/2)^{\nu+2m}/[m!(m+\nu)!]$ $(\nu \in \mathbb{N}_0)$ denotes the modified Bessel function of the first kind. We note that

$$\widetilde{f}_{j,n}^{(2)}(t) = \varrho^{-(n-j)} \widetilde{f}_{j,n}^{(1)}(t), \qquad j, n \in \mathbb{Z}. \qquad (4)$$

Let $\widetilde{T}_{j,n}^{(i)}$ be the first-passage time (FPT) from j to n for $\widetilde{M}^{(i)}(t)$ $(i = 1, 2)$ and let $\widetilde{g}_{j,n}^{(i)}(t)$ be its probability density function (pdf). One has:

$$\widetilde{g}_{j,n}^{(i)}(t) = \frac{|n-j|}{t} \widetilde{f}_{j,n}^{(i)}(t), \quad j, n \in \mathbb{Z}, j \neq n, \ i = 1, 2 \qquad (5)$$

and, from (4), it follows $\widetilde{g}_{j,n}^{(2)}(t) = \varrho^{-(n-j)} \widetilde{g}_{j,n}^{(1)}(t)$ for $j, n \in \mathbb{Z}, j \neq n$.

1.1 Composition Method

Let $0 \leq \vartheta \leq 1$ be a real number and, for any fixed $t \geq 0$, let $\widetilde{U}_j(t)$ be a random variable uniform in $(0, 1)$, independent on $\widetilde{M}_1(t)$ and $\widetilde{M}_2(t)$. We consider the discrete stochastic process $\{\widetilde{M}(t), t \geq 0\}$ obtained by the composition method:

$$\widetilde{M}(t) = \begin{cases} \widetilde{M}_1(t), \ 0 \leq \widetilde{U}_j(t) < \widetilde{Q}_j(\vartheta) \\ \widetilde{M}_2(t), \ \widetilde{Q}_j(\vartheta) \leq \widetilde{U}_j(t) < 1, \end{cases}$$

where $\widetilde{Q}_j(\vartheta) = \vartheta/[\vartheta + (1-\vartheta)\varrho^{-j}]$, $j \in \mathbb{Z}$. The transition probability $\widetilde{\xi}_{j,n}(t) = P\{\widetilde{M}(t) = n | \widetilde{M}(0) = j\}$ of $\widetilde{M}(t)$ is a mixture of the probabilities $\widetilde{f}_{j,n}^{(1)}(t)$ and $\widetilde{f}_{j,n}^{(2)}(t)$, so that from (4) it follows:

$$\widetilde{\xi}_{j,n}(t) = \widetilde{Q}_j(\vartheta) \widetilde{f}_{j,n}^{(1)}(t) + [1 - \widetilde{Q}_j(\vartheta)] \widetilde{f}_{j,n}^{(2)}(t) = \frac{\vartheta + (1-\vartheta)\varrho^{-n}}{\vartheta + (1-\vartheta)\varrho^{-j}} \widetilde{f}_{j,n}^{(1)}(t), \qquad (6)$$

for $j, n \in \mathbb{Z}$. The infinitesimal rates of the discrete process $\widetilde{M}(t)$ are:

$$\widetilde{\lambda}_n := \lim_{h \downarrow 0} \frac{\widetilde{\xi}_{n,n+1}(h)}{h} = \lambda \frac{\vartheta + (1-\vartheta)\varrho^{-n-1}}{\vartheta + (1-\vartheta)\varrho^{-n}}, \qquad n \in \mathbb{Z}$$

$$\widetilde{\mu}_n := \lim_{h \downarrow 0} \frac{\widetilde{\xi}_{n,n-1}(h)}{h} = \mu \frac{\vartheta + (1-\vartheta)\varrho^{-n+1}}{\vartheta + (1-\vartheta)\varrho^{-n}}, \qquad n \in \mathbb{Z}, \qquad (7)$$

so that $\widetilde{\lambda}_n + \widetilde{\mu}_n = \lambda + \mu$ for $n \in \mathbb{Z}$.

1.2   Birth-Death Chain $\widetilde{N}(t)$

Let $\{\widetilde{N}(t), t \geq 0\}$ be a BD chain on $\mathbb{Z}$ with rates given in (7). Some results on the transition probabilities and on FPT densities for $\widetilde{N}(t)$ are given in [8] and [9]. Moreover, one can prove that $\widetilde{N}(t) \overset{d}{=} \widetilde{M}(t)$ for all $t \geq 0$, i.e. the BD process $\widetilde{N}(t)$ is distributed as $\widetilde{M}(t)$. Hence, denoting by $\widetilde{f}_{j,n}(t) = P\{\widetilde{N}(t) = n | \widetilde{N}(0) = j\}$, from (6), it follows:

$$\widetilde{f}_{j,n}(t) = \frac{\vartheta + (1 - \vartheta)\varrho^{-n}}{\vartheta + (1 - \vartheta)\varrho^{-j}} \, e^{-(\lambda + \mu) t} \, \varrho^{(n-j)/2} \, I_{n-j}(2t\sqrt{\lambda\mu}), \quad j, n \in \mathbb{Z}. \quad (8)$$

For $j, n \in \mathbb{Z}$, we denote by

$$\widetilde{\Lambda}_{j,n}(t) = -\frac{d}{dt} P\{\widetilde{N}(t) < n | \widetilde{N}(0) = j\} = \widetilde{\lambda}_{n-1} \widetilde{f}_{j,n-1}(t) - \widetilde{\mu}_n \widetilde{f}_{j,n}(t) \quad (9)$$

the probability current in the state n at time t, conditioned upon $\widetilde{N}(0) = j$. Let $\widetilde{T}_{j,n}$ be the FPT for the BD process $\widetilde{N}(t)$ and let $\widetilde{g}_{j,n}(t)$ be its pdf. For $j, n \in \mathbb{Z}, n \neq j$ one has:

$$\widetilde{g}_{j,n}(t) = \frac{\vartheta + (1 - \vartheta)\varrho^{-n}}{\vartheta + (1 - \vartheta)\varrho^{-j}} \, \widetilde{g}^{(1)}_{j,n}(t) = \frac{\vartheta + (1 - \vartheta)\varrho^{-n}}{\vartheta + (1 - \vartheta)\varrho^{-j}} \, \frac{|n - j|}{t} \, \widetilde{f}^{(1)}_{j,n}(t). \quad (10)$$

We note that the Laplace transform (LT) of the transition probability $\widetilde{f}_{j,n}(t)$ is:

$$\widetilde{\varphi}_{j,n}(s) = \int_0^{+\infty} e^{-st} \widetilde{f}_{j,n}(t) \, dt = \frac{\vartheta + (1 - \vartheta)\varrho^{-n}}{\vartheta + (1 - \vartheta)\varrho^{-j}} \, \frac{[\psi_\epsilon(s)]^{n-j}}{\mu[\psi_1(s) - \psi_2(s)]}, \quad j, n \in \mathbb{Z}, \quad (11)$$

where $\epsilon = 1$ if $n \leq j$, $\epsilon = 2$ if $n > j$ and

$$\psi_1(s), \psi_2(s) = \frac{s + \lambda + \mu \pm \sqrt{(s + \lambda + \mu)^2 - 4\lambda\mu}}{2\mu}, \qquad \psi_1(s) > \psi_2(s).$$

Therefore, from (9), the LT of the probability current $\widetilde{\Lambda}_{j,n}(t)$ is:

$$\widetilde{\Omega}_{j,n}(s) = \int_0^\infty e^{-st} \widetilde{\Lambda}_{j,n}(t) \, dt = \frac{[\psi_\epsilon(s)]^{n-j}}{\mu[\psi_1(s) - \psi_2(s)]}$$
$$\times \frac{\mu \vartheta [\psi_{3-\epsilon}(s) - 1] + (1 - \vartheta) \varrho^{-n} [\mu \psi_{3-\epsilon}(s) - \lambda]}{\vartheta + (1 - \vartheta)\varrho^{-j}}, \quad j, n \in \mathbb{Z}. \quad (12)$$

In particular, for $j = n = 0$ one has:

$$\widetilde{\Omega}_{0,0}(s) = \frac{\mu \psi_2(s) - \mu \vartheta - \lambda (1 - \vartheta)}{\mu[\psi_1(s) - \psi_2(s)]}. \quad (13)$$

Furthermore, from (10) the LT of the FPT pdf $\widetilde{g}_{j,n}(t)$ is:

$$\widetilde{\gamma}_{j,n}(s) = \int_0^\infty e^{-st} \widetilde{g}_{j,n}(t) \, dt = \frac{\vartheta + (1 - \vartheta)\varrho^{-n}}{\vartheta + (1 - \vartheta)\varrho^{-j}} \, [\psi_\epsilon(s)]^{n-j}, \quad j, n \in \mathbb{Z}. \quad (14)$$

The LT's $\widetilde{\varphi}_{j,n}(s)$, $\widetilde{\Omega}_{j,n}(s)$ and $\widetilde{\gamma}_{j,n}(s)$, given in (11), (13) and (14), play an important role to determine the transition probabilities of the restricted BD process $R(t)$ and of the symmetric BD chain $N(t)$.

2   Restricted Birth-Death Process

Let $\{R(t), t \geq 0\}$ be a continuous-time BD process with state space $\mathbb{N}_0 = \{0, 1, \ldots\}$, 0 being a regular reflecting state, characterized by BD rates given in (1). The transition probabilities $p_{j,n}(t) = P\{R(t) = n | R(0) = j\}$ of $R(t)$ satisfy the following relation (cf. [10]):

$$p_{j,n}(t) = \widetilde{f}_{j,n}(t) - \int_0^t \widetilde{\Lambda}_{j,0}(u)\, p_{0,n}(t - u)\, du, \qquad j, n \in \mathbb{N}_0, \qquad (15)$$

where $\widetilde{f}_{j,n}(t)$ and $\widetilde{\Lambda}_{j,n}(t)$ are the transition probabilities and the probability current of the BD chain $\widetilde{N}(t)$. For $j, n \in \mathbb{N}_0$, let $\pi_{j,n}(s) = \int_0^{+\infty} e^{-st}\, p_{j,n}(t)\, dt$ be the LT of the transition probabilities $p_{j,n}(t)$. From (15), recalling (11) and (12), one is led to

$$\pi_{0,n}(s) = \frac{\widetilde{\varphi}_{0,n}(s)}{1 + \widetilde{\Omega}_{0,0}(s)} = \frac{\left[\vartheta + (1 - \vartheta)\, \varrho^{-n}\right] [\psi_2(s)]^n}{\mu\, \psi_1(s) - \mu\, \vartheta - \lambda\, (1 - \vartheta)}, \qquad n \in \mathbb{N}_0,$$

$$(16)$$

$$\pi_{j,n}(s) = \widetilde{\varphi}_{j,n}(s) - \widetilde{\Omega}_{j,0}(s)\, \widetilde{\pi}_{0,n}(s) = \frac{\vartheta + (1 - \vartheta)\, \varrho^{-n}}{\vartheta + (1 - \vartheta)\, \varrho^{-j}} \left\{ \frac{[\psi_\epsilon(s)]^{n-j}}{\mu\, [\psi_1(s) - \psi_2(s)]} \right.$$

$$\left. - \frac{\varrho^{-j}\, [\psi_\epsilon(s)]^{n+j}}{\mu\, [\psi_1(s) - \psi_2(s)]} + \frac{\varrho^{-j}\, [\psi_2(s)]^{n+j}}{\mu\, \psi_1(s) - \mu\, \vartheta - \lambda\, (1 - \vartheta)} \right\}, \qquad j, n \in \mathbb{N}_0$$

where $\epsilon = 1$ if $n \leq j$, $\epsilon = 2$ if $n > j$. Taking the inverse LT in (16), after some calculations, the transition probabilities $p_{j,n}(t)$ of $R(t)$ can be obtained:

$$p_{j,n}(t) = \frac{\vartheta + (1 - \vartheta)\, \varrho^{-n}}{\vartheta + (1 - \vartheta)\, \varrho^{-j}}\, e^{-(\lambda + \mu)t} \left\{ \varrho^{(n-j)/2}\, I_{n-j}(2t\sqrt{\lambda\mu}) \right.$$

$$+ \frac{\beta}{\lambda}\, \varrho^{(n-j+1)/2}\, I_{n+j+1}(2t\sqrt{\lambda\mu}) + \varrho^{-j} \left(\frac{\lambda}{\beta}\right)^{n+j} \left(1 - \frac{\lambda\mu}{\beta^2}\right)$$

$$\left. \times \sum_{k=n+j+2}^{+\infty} \left(\frac{\beta}{\lambda}\right)^k \varrho^{k/2}\, I_k(2t\sqrt{\lambda\mu}) \right\}, \qquad j, n \in \mathbb{N}_0. \qquad (17)$$

where $\beta = \mu\, \vartheta + \lambda\, (1 - \vartheta)$. In Figs. 1 and 2, the probabilities $p_{j,n}(t)$, given in (17), are plotted for $t = 10$, $\lambda = 1.0$, $\mu = 2.0$ and for some values of ϑ.

3   Symmetric BD Chain with Respect to Zero State

Let $\{N(t), t \geq 0\}$ be a BD chain with state-space $\mathbb{Z}$ characterized by BD rates given in (2). Since the chain $N(t)$ is symmetric with respect to zero state, the transition probabilities $f_{j,n}(t) = P\{N(t) = n | N(0) = j\}$ are such that $f_{j,n}(t) = f_{-j,-n}(t)$ for all $j, n \in \mathbb{Z}$, so that $f_{0,0}(t) + 2 \sum_{n=1}^{+\infty} f_{0,n}(t) = 1$. Hence, we have:

$$\sum_{n=1}^{+\infty} \varphi_{0,n}(s) = \frac{1 - s\, \varphi_{0,0}(s)}{2\, s},$$

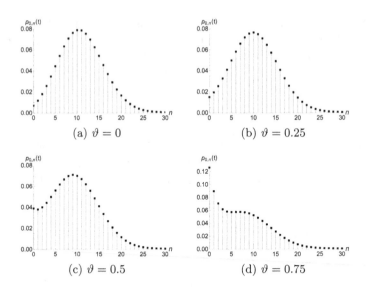

Fig. 1. The probabilities $p_{0,n}(t)$ are plotted for $\lambda = 1.0$, $\mu = 2.0$ and $t = 10$.

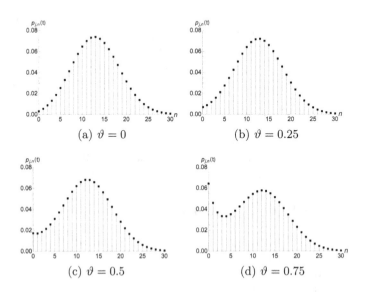

Fig. 2. The probabilities $p_{j,n}(t)$ are plotted for $j = 3$, $\lambda = 1.0$, $\mu = 2.0$ and $t = 10$.

where $\varphi_{j,n}(s)$ are the LT of the probabilities $f_{j,n}(t)$. Moreover, the LT $\Omega_{j,n}(s)$ of the probability current $\Lambda_{j,n}(t) = -\frac{d}{dt}P\{N(t) < n|N(0) = j\}$, satisfies the following relations:

$$\Omega_{0,0}(s) = [-1 + s\,\varphi_{0,0}(s)]/2, \qquad \Omega_{j,0}(s) = \gamma_{j,0}(s)\,\Omega_{0,0}(s), \quad j \in \mathbb{N},$$

where $\gamma_{j,0}(s)$ is the LT of the FPT pdf $g_{j,0}(t)$ from j to 0 for $N(t)$. The probabilities $f_{j,n}(t)$ for $N(t)$ can be expressed in terms of the probabilities $p_{j,n}(t)$ of the restricted process $R(t)$ and of the probability current $\Lambda_{j,0}(t)$ of $N(t)$:

$$f_{j,n}(t) = p_{j,n}(t) + \int_0^t \Lambda_{j,0}(u)\, p_{0,n}(t-u)\, du, \qquad j, n \in \mathbb{N}_0, \tag{18}$$

with $p_{j,n}(t)$ given in (17). Hence, taking the LT's in (18), one obtains:

$$\varphi_{0,n}(s) = \varphi_{0,-n}(s) = \frac{\pi_{0,n}(s)}{2 - s\,\pi_{0,0}(s)}, \qquad n \in \mathbb{N}_0 \tag{19}$$

$$\varphi_{j,n}(s) = \varphi_{-j,-n}(s) = \begin{cases} \pi_{j,n}(s) + \Omega_{j,0}(s)\,\pi_{0,n}(s), & j \in \mathbb{N}, n \in \mathbb{N}_0 \\ \gamma_{j,0}(s)\,\varphi_{0,n}(s), & n < 0 < j. \end{cases} \tag{20}$$

In order to obtain $f_{j,n}(t)$ for $j, n \in \mathbb{Z}$, we now consider the following cases:
Case $j = 0$. Substituting $\pi_{0,n}(s)$, given in (16), in (19), one has:

$$\varphi_{0,n}(s) = \varphi_{0,-n}(s) = \frac{\pi_{0,n}(s)}{2 - s\,\pi_{0,0}(s)} = \frac{[\vartheta + (1-\vartheta)\varrho^{-n}]\,[\psi_2(s)]^n}{2\,\mu\,\psi_1(s) - 2\,\mu\,\vartheta - 2\,\lambda\,(1-\vartheta) - s}$$

$$= -\frac{[\vartheta + (1-\vartheta)\varrho^{-n}]\,[\psi_2(s)]^{n+1}}{\mu\,\psi_2^2(s) + (\lambda - \mu)(1 - 2\vartheta)\,\psi_2(s) - \lambda}, \qquad n \in \mathbb{N}_0, \tag{21}$$

from which, taking the inverse LT, one obtains:

$$f_{0,n}(t) = f_{0,-n}(t) = \left[\vartheta + (1-\vartheta)\varrho^{-n}\right] e^{-(\lambda+\mu)t}\,\varrho^{n/2}\left[I_n(2t\sqrt{\lambda\mu}) + \frac{(\lambda-\mu)(2\vartheta-1)}{\mu(z_1 - z_2)}\right.$$

$$\left. \times \sum_{k=n+1}^{+\infty} (-1)^{k-n}\varrho^{-(k-n)/2} I_k(2t\sqrt{\lambda\mu})(z_1^{k-n} - z_2^{k-n})\right], \qquad n \in \mathbb{N}_0, \tag{22}$$

where z_1, z_2 are the roots of the second degree equation $\mu\,z^2 + (\lambda - \mu)\,(1 - 2\,\vartheta)\,z - \lambda = 0$, with $z_1 > z_2$. In Fig. 3 the probabilities $f_{0,n}(t)$, given in (22), are plotted for $\lambda = 1.0$, $\mu = 2.0$, $t = 10$ and some values of ϑ.

Cases $(n < 0 < j)$ and $(j \in \mathbb{N}, n \in \mathbb{N}_0)$. For $N(t)$ and $\widetilde{N}(t)$, we have $g_{j,0}(t) = \widetilde{g}_{j,0}(t)$, so that $\gamma_{j,0}(s) = \widetilde{\gamma}_{j,0}(s)$. The LT's of the transition probabilities of $N(t)$ are:

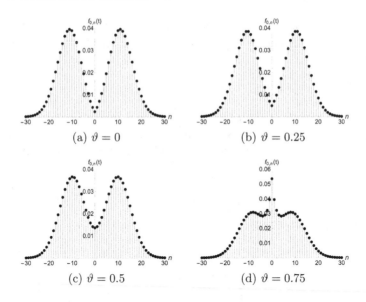

Fig. 3. The probabilities $f_{0,n}(t)$ are plotted for $\lambda = 1.0$, $\mu = 2.0$ and $t = 10$.

$$\varphi_{j,n}(s) = \varphi_{-j,-n}(s) = \gamma_{j,0}(s)\,\varphi_{0,n}(s)$$
$$= \frac{\vartheta + (1-\vartheta)\varrho^n}{\vartheta + (1-\vartheta)\varrho^{-j}}\,\frac{\varrho^{-j}\,\varphi_{0,j-n}(s)}{\vartheta + (1-\vartheta)\varrho^{n-j}}, \qquad n < 0 < j, \qquad (23)$$

$$\varphi_{j,n}(s) = \varphi_{-j,-n}(s) = \pi_{j,n}(s) + \Omega_{j,0}(s)\,\pi_{0,n}(s)$$
$$= \pi_{j,n}(s) + \frac{\vartheta + (1-\vartheta)\varrho^{-n}}{\vartheta + (1-\vartheta)\varrho^{-j}}\,\frac{\mu\vartheta - \lambda\vartheta - \mu - \mu\psi_2(s)}{\mu\psi_1(s) - \mu\vartheta - \lambda(1-\vartheta)}$$
$$\times \frac{\varrho^{-j}\,[\psi_2(s)]^{n+j}}{2\mu\,\psi_1(s) - 2\mu\,\vartheta - 2\lambda(1-\vartheta) - s}, \qquad j \in \mathbb{N}, n \in \mathbb{N}_0, \quad (24)$$

from which, taking the inverse LT, one obtains:

$$f_{j,n}(t) = f_{-j,-n}(t) = \frac{\vartheta + (1-\vartheta)\varrho^n}{\vartheta + (1-\vartheta)\varrho^{-j}}\,\frac{\varrho^{-j}\,f_{0,j-n}(t)}{\vartheta + (1-\vartheta)\varrho^{n-j}}, \qquad n < 0 < j, \qquad (25)$$

$$f_{j,n}(t) = f_{-j,-n}(t) = p_{j,n}(t) + \frac{\vartheta + (1-\vartheta)\varrho^{-n}}{\vartheta + (1-\vartheta)\varrho^{-j}}\,\varrho^{-j}$$
$$\times \left\{ \frac{\mu}{\beta} \sum_{k=1}^{+\infty}\left(\frac{\beta}{\lambda}\right)^k \frac{f_{0,k+n+j+1}(t)}{\vartheta + (1-\vartheta)\varrho^{-k-n-j-1}} \right.$$
$$\left. - \frac{\alpha}{\lambda}\sum_{k=0}^{+\infty}\left(\frac{\beta}{\lambda}\right)^k \frac{f_{0,k+n+j+1}(t)}{\vartheta + (1-\vartheta)\varrho^{-k-n-j-1}} \right\}, \qquad j \in \mathbb{N}, n \in \mathbb{N}_0, \quad (26)$$

where $\alpha = \lambda\vartheta + \mu(1-\vartheta)$ and $\beta = \mu\vartheta + \lambda(1-\vartheta)$. In Fig. 4 the probabilities $f_{j,n}(t)$, given (25) and (26), are plotted for $j = 3$, $t = 10$, $\lambda = 1.0$, $\mu = 2.0$ and some values of ϑ.

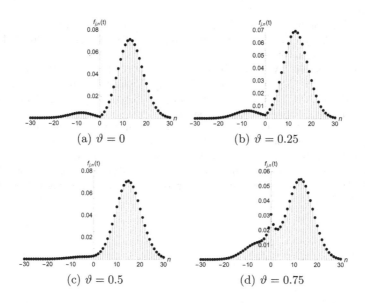

Fig. 4. The probabilities $f_{j,n}(t)$ are plotted for $j = 3$, $\lambda = 1.0$, $\mu = 2.0$ and $t = 10$.

References

1. Conolly, B.W.: On randomized random walks. SIAM Rev. **13**(1), 81–99 (1971)
2. Crawford, F.W., Suchard, M.A.: Transition probabilities for general birth-death processes with applications in ecology, genetics, and evolution. J. Math. Biol. **65**, 553–580 (2012)
3. Dharmaraja, S., Di Crescenzo, A., Giorno, V., Nobile, A.G.: A continuous-time Ehrenfest model with catastrophes and its jump-diffusion approximation. J. Stat. Phys. **161**, 326–345 (2015)
4. Di Crescenzo, A., Giorno, V., Krishna Kumar, B., Nobile, A.G.: A double-ended queue with catastrophes and repairs, and a jump-diffusion approximation. Method. Comput. Appl. Probab. **14**, 937–954 (2012)
5. Di Crescenzo, A., Giorno, V., Krishna Kumar, B., Nobile, A.G.: A time-non-homogeneous double-ended queue with failures and repairs and its continuous approximation. Mathematics **6**, 81 (2018)
6. Di Crescenzo, A., Giorno, V., Krishna Kumar, B., Nobile, A.G.: $M/M/1$ queue in two alternating environments and its heavy traffic approximation. J. Math. Anal. Appl. **458**, 973–1001 (2018)
7. Di Crescenzo, A., Giorno, V., Nobile, A.G.: Constructing transient birth-death processes by means of suitable transformations. Appl. Math. Comput. **465**, 152–171 (2016)

8. Di Crescenzo, A., Martinucci, B.: On a symmetry, nonlinear birth-death process with bimodal transition probabilities. Symmetry **1**, 201–214 (2009)
9. Di Crescenzo, A., Martinucci, B.: A review on symmetry properties of birth-death processes. Lecture Notes of Seminario Interdisciplinare di Matematica **12**, 81–96 (2015)
10. Giorno, V., Nobile, A.G.: First-passage times and related moments for continuous-time birth–death chains. Ricerche di Matematica **68**(2), 629–659 (2018). https://doi.org/10.1007/s11587-018-0430-8
11. Giorno, V., Nobile, A.G.: On a bilateral linear birth and death process in the presence of catastrophes. In: Moreno-Díaz, R., Pichler, F., Quesada-Arencibia, A. (eds.) EUROCAST 2013. LNCS, vol. 8111, pp. 28–35. Springer, Heidelberg (2013). https://doi.org/10.1007/978-3-642-53856-8_4
12. Giorno, V., Nobile, A.G., Spina, S.: On some time non-homogeneous queueing systems with catastrophes. Appl. Math. Comput. **245**, 220–234 (2014)
13. Giorno, V., Nobile, A.G., Pirozzi, E.: A state-dependent queueing system with asymptotic logarithmic distribution. J. Math. Anal. Appl. **458**, 949–966 (2018)
14. Pruitt, W.E.: Bilateral birth and death processes. Trans. Am. Math. Soc. **107**(3), 508–525 (1963)

Unimodularity of Network Representations of Time-Series for Unbiased Parameter Estimation

Florian Sobieczky[(✉)]

SCCH - Software Competence Center Hagenberg, Softwarepark 21,
4232 Hagenberg, Austria
florian.sobieczky@scch.at
https://www.scch.at

Abstract. For the problem of parameter estimation in time-series models of a stream of physical measurements under potential local quality impairments (indicated by an additional binary variable), a lamplighter-graph representation is introduced. This representation makes conditional estimation using large samples conditioned on specific degrees of quality impairment possible. Due to the unimodularity of the network representation, there are unbiased estimators.

Keywords: Unimodularity · Time-series parameter estimation · Production-process data analysis · Graphical models · Unbiased point estimation

1 Introduction

We start with describing the stochastic model of the specific production problem involving a production line, with its processes for the work piece and the state of the production machinery. For these processes we will assume a standard probabilistic setting with a sufficiently large probability space $(\Omega, \mathcal{F}, P)$, with respect to which all defined processes are measurable functions. Our notation includes $\chi_A(\omega)$ for the indicator of an event $A \in \mathcal{F}$, and $\langle ., \ldots, . \rangle$ for tuples, as in $G = \langle V, E \rangle$ for a graph with vertex-set V and edge-set E. $\mathbb{N}$ will denote the natural numbers including zero, and $\mathbb{Z}$ the set of all integers.

1.1 Problem Description: Parameter Estimation of Production Data

We consider the problem of unbiased point estimation (see the introduction of [1]), and related use of graph models (references given in [2]) describing the process variables in a production process in which an array of work items is sequentially passed through a fixed array of machines, like in the case of a conveyor belt. We consider a sequence of processes indexed over the same discrete

R. Moreno-Díaz et al. (Eds.): EUROCAST 2019, LNCS 12013, pp. 167–175, 2020.
https://doi.org/10.1007/978-3-030-45093-9_21

time-set, where each element of the sequence corresponds to the process of one work item that occupies a single position at a single instance. As the time variable is increased, the whole array is shifted by one step and each item is passed through the next machine in the line carrying out the same manufacturing step on the subsequent work item (Fig. 1).

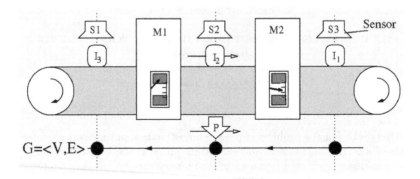

Fig. 1. Production line scenario of single items passing along an array of positions between which work is carried out by machines each of which may be working correctly or not. At a fixed time, measurements Y_n specific to the work item at the n-th position of the line are carried out between machines n and $n + 1$, which are in a (possibly operational) state indicated by η_n, and η_{n+1}, respectively. Note the direction of the arrows in the directed graphical model are pointed backwards, referring to the possible influence of a former item on a newly incoming one.

Furthermore, we consider the array of machines indexed by $\mathbb{N}$ being at the fixed considered time in a state $\eta \in 2^{\mathbb{N}} = \{f : \mathbb{N} \to \{0,1\}\}$, a two-valued function, the n-th value describing the n-th machine to be in mode 0 (e.g. functioning) or 1 (e.g. mal-functioning). The process environment to be modeled here is an array of machines each of which at a given instant may either be in a well-working mode, or a disfunctional one. In order to make the stochastic model as simple as possible, we only consider the sequence of processes *evaluated at specific times*. This still gives a set of random variables indexed by integers, however, referring to the positions of the sensors performing the measurements. When only considering this at a single instance in time we have the 'snapshot view':

Definition 1: A snapshot of a production line is the pair of processes Y_n and η_m indexed by the positions of the sensors $n \in \mathbb{Z}$ and machines $m \in \mathbb{Z}$ on $\langle \Omega_1, \mathcal{F}_1, \mu \rangle$, and $\langle \Omega_2, \mathcal{F}_2, \nu \rangle$, respectively, for which the joint distribution of Y, namely $\bar{\mu} : \mathcal{F}_1 \otimes \cdots_{n \text{ times}} \otimes \mathcal{F}_1 \to [0,1]$ is a Bayes network

$$\bar{\mu}(\{Y_i \in B_i\}_{i=1}^n) = \prod_{i=1}^{n} \mu(Y_i \in B_i | \{Y_k\}_{k=i+1}^n), \tag{1}$$

where $B_i \in \mathcal{F}1$, and each η_m is a Bernoulli process over (discrete) time.

This definition captures the idea that a work item coming in on a sequentially arranged (line) environment, at a later time, may not influence a work item having passed the current position, before.

1.2   Directed Graphical Models (DGM) for Production Lines

A directed graphical model is a graph $G = \langle V, E \rangle$ of which V represents a set of random variables. As is usually the case in a Bayesian network, the directed edge set E reflects on the underlying correlation structure of the joint distribution [13]. We will extend this view by letting the edges also represent random variables, namely the operational mode of the corresponding machine. Alternatively, one can pass to the bipartite graph with an extra vertex with the additional state variable dividing each edge into two edges.

We are now considering the typical situation that only work items having previously passed a certain position of the production line may influence (the measurement of) a work item that has newly arrived at that same position. This is suggestive of the directed graphical model sketched at the bottom of Fig. 1 with directed edges pointing backward, giving expression to this dependency-structure. In particular, condition (1) of the production line snapshot is represented by this DGM.

It is our aim to directly study point estimators of Y_n of the form

$$\widehat{Y}_J = \frac{1}{|J|} \sum_{j \in J} Y_j, \tag{2}$$

with J a subset of the vertices of the graphical model. While it is well known that the arithmetic mean of a finite sample of identically distributed variables is an unbiased estimator of their distribution's mean [12], we are going beyond this situation by also including the machine states η_m into the estimator.

(1). **Linear DGM:** The graphical model on the graph $G_0 = \langle V_0, E_0 \rangle$ with vertex set $V_0 = \mathbb{Z}$ and directed edge set $E_0 = \{\langle k, k-1 \rangle \mid k \in V_0\}$, measurement Y_k attached to $k \in V_0$, and state variable η_k to the edge $\langle k, k-1 \rangle$ will be called linear DGM.

(2). **Tree-like DGM:** Graph: The graph of this DGM is built on the graph G_0 of the linear DGM. Given a vertex k from V_0, construct a binary ('canopy') tree T_k in such a way that for each vertex $l \in V_0 = \mathbb{Z}$ on the semi-infinite ray $\{k, k-1, k-2, ...\} \subset V_0$ in G_0, a finite binary rooted tree of depth (height) $k - l$ is attached with the root being vertex l, and the finite tree's vertices of height i over this root being aligned over vertex $l + i$ in V_0 (see Fig. 2). Now, when passing from k to $k + 1$, it is realized that the canopy tree starting from k is a sub graph of that starting at $k + 1$. This implies that limsup and liminf coincide (the liminf is the union and the limsup is the intersection of these (equal) unions), so the set-theoretical limit exists.

Assigning to each edge the direction pointing towards its root on V_0, or, if it's an edge from E_0 assign the direction that points to the left, yields a bi-infinite directed binary tree $T = \langle V_T, E_T \rangle$. Note that by construction, G_0 is a sub graph of T.

Labeling: For each vertex $v \in V_T$, assign one of the two incoming edges the label "0", and label the other with "1". Make sure, that all the edges in E_0 are labeled with "0". Then this implies that all sequences of edge-labels starting with any edge in E_T and following a path in the directed tree in the direction of the adjoining edges finally merge into a sequence of constant "0"-s.

DGM: To turn this graph into a directed graphical model, we first adapt the labels for vertices in $V_0 \subset V_T$ from the linear DGM and then assign to each of the remaining vertices v the random variable Y_j, where $j = k + l$, k is the label of the closest vertex in V_0, and l its distance to the given (remaining) vertex.

For a given edge $e \in E_T$, assign to it the random variable given by the indicator function of the event in $\mathcal{F}_2$ that a given sequence of "0"-s and "1"-s is equal to the sequence of labels assigned to the edges of the path resulting in following the directions assigned.

Finally, to make this a DGM of the snapshot of the production line scenario, realize that the binary tree T projects onto G_0 in a natural way, by construction: Identify each sequence of edge labels along the directed edges in T with the sequence of states of the machine between the projected vertices in V_0. So, while the random variables of the line DGM only encode the states of single machines, the random variables given by the sequences of labels along the path in the tree from the given edge along the directed edges additionally encode the sequence of machine states at edges towards the left of the projected edge in G_0.

Thus, the tree-like DGM is a much more informative graphical model, than the linear DGM. We shall now *attempt* to define unbiased estimators of the form (2) for them.

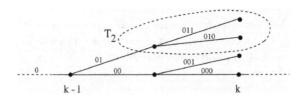

Fig. 2. Left: Part of infinite binary tree-like DGM. The bottom branch of the tree corresponds to the subgraph (the path) G_0. The sub tree T_2 is one of those used in the construction of T out of G_0. The edges are labeled by $\{0,1\}$-valued sequences, indicating the whole state of the array of machines towards the left up to the machine between the projected vertices in V_0. It is seen from this, that even though the tree has a direction which ultimately derives from the (non)-independence structure of the directed graphical network (Definition 1), the graph lacks left-right reflection symmetry, which the unimodularization provides (see Fig. 3).

1.3   Unimodularity

While the term 'unimodularity' originates from topological group theory, it has been extended to transitive graphs and networks. In discrete probability theory it has been used with much success in the generalization of results from the percolation theory on Euclidean lattices. It has been formulated in the form of a statement about the local discrete geometry of the graph under the name of the 'mass transport principle' [6–8].

In its simplest form, the principle says that a transitive graph $G = \langle V, E \rangle$ is unimodular if for all functions $f : V \times V \to \mathbb{R}$ with diagonal invariance $(\forall_{\gamma \in \text{AUT}(G)} f(x, y) = f(\gamma x, \gamma y)$ with $\text{AUT}(G)$ the automorphism group of the graph G), it holds that

$$\sum_{y \in V} f(x, y) = \sum_{y \in V} f(y, x). \tag{3}$$

If $f(x, y)$ represents mass being sent from x to y then unimodularity of G represents the principle of conservation of mass transferred through its vertices as indicated by the function $f(\cdot, \cdot)$.

1.4   Statement of the Result

As a direct application of the unimodularity of the network carrying data, consider the following Theorem, proven in [10] (see also [11]), of which we repeat the proof for clarity, below.

Theorem. *If $Z(x)$ is a random variable depending on vertices $x \in V$ of a unimodular graph $G = \langle V, E \rangle$. Let C_x be a finite subset of V, chosen randomly by a measure μ invariant under $AUT(G)$, such that $x \in V$. Then*

$$\mathbb{E}\left[\frac{1}{|C_x|} \sum_{y \in C_x} Z(y) \right] = \mathbb{E}[Z(x)], \tag{4}$$

where the expected value is with respect to the measure μ.

The result states that averaging over the subgraph induced by the finite subset C chosen from a measure invariant under $\text{AUT}(G)$, there is no bias introduced into the estimation of the mean. This was used in [10] to estimate random walk return probabilities on percolation sub graphs. If the underlying graph is transitive but fails to fulfill (3) then it may still be quasi-transitive, leading to a relation similar to (4) which, however, carries a correcting factor involving the then non-trivial modular function of the graph [8], leading to a more complicated form of the mass transport principle [9].

If the measurements Y_n and η_m each have identical marginal distributions (without having to be independent!), there may be invariance of the joint distribution with respect to the automorphism group of the graph. We now check whether the DGMs introduced in Sect. 1.2, which are both transitive graphs, do actually have a measure invariant under the automorphism group.

In the case of the line DGM, the group of shifts along the labels of the machine positions $\mathbb{Z}$ acts transitively, i.e. every vertex can be reached by it starting from any given vertex, and is a subgroup of the automorphism group. The stabilizer of a given vertex (subgroup leaving this vertex invariant) consists only of the trivial group (containing only the identity). By Lemma 1.29 in [5], this makes the group of shifts unimodular.

On the other hand, for the tree-like DGM, the transitive group of shifts preserving the direction of the tree given by the independence structure induced by the production line, does not contain the rotations of the tree around a single vertex. This makes the ratio of cardinalities of stabilizers of vertices x and y acting on y x, respectively, (compare with in Theorem 8.7 of [9]) unequal to one. In the language of topological groups this is expressed by saying that the modular function is not identically equal to one.

We arrive at the fact that we *can* apply unimodularity to show that estimators of the form (2) are unbiased (by using (4)), and we *cannot* in the case of the tree-like DGM. In the next section, yet another DGM will be considered, which can be considered a 'unimodularization' of the tree-like DGM of Sect. 2.1.

2 Unimodular Graphical Models

Instead of just presenting the 'unimodularized' tree-like DGM, we construct it on the basis of a calculation showing the lack of symmetry that graphical models have, if their representing graph is non-unimodular.

2.1 Proof of the Theorem

Due to the transitiveness of the graph, the function $f : V \times V \to \mathbb{R}$ given by $f(x, y) = \mathbb{E}[\frac{\chi_{C_x}(y)}{|C_x|} Z(x)]$ is diagonally invariant, that is $f(\gamma x, \gamma y) = f(x, y)$ for all transformations γ from $\text{AUT}(G)$. By unimodularity and (3), we have

$$\mathbb{E}[Z(x)] = \sum_{y \in V} f(x, y) = \sum_{y \in V} f(y, x) = \sum_{y \in V} \mathbb{E}\left[\frac{\chi_{C_y}(x)}{|C_y|} Z(y)\right] = \mathbb{E}[\frac{1}{|C_y|} \sum_{y \in C_x} Z(y)].$$

The last inequality is due to $y \in C_x$ being equivalent to $x \in C_y$. $\qquad\square$

Note that the last line of the proof highlights how the invariance of the measure with respect to transformations from $\text{AUT}(G)$ is relevant for the selection of C_x: It must be equally probable (according to μ) to pick y as a member of C_x as x being a member of C_y.

2.2 How to Unimodularize a Production Line

Our model, the "snapshot of a production line" has the symmetry of a lamplighter graph [4] (a lamp lit indicates a local quality-impairing event). The lamplighter graph is unimodular [3]. Therefore: If x represents an arbitrary state in the production line (i.e. a configuration of sensor readings indicating the working-state of each machine together with the position of a single work piece in the line), and $Z(x)$ some AUT(G)-invariantly distributed numerical characteristic of vertex x, then the arithmetic mean of Z across any finite subset C of other states chosen by an invariant percolation process (such as Bernoulli percolation) gives an unbiased estimator of the mean of Z.

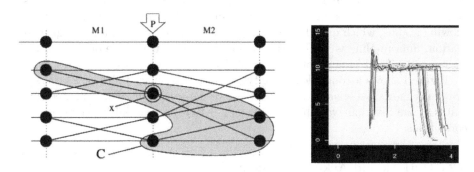

Fig. 3. Left: Unimodular network representation of a time-series of measurements with potential local quality impairments: Instead of the binary tree (non-unimodular) the lamplighter graph emerges as a unimodular network representation from which to draw samples (here: the randomly chosen vertex-set C) for unbiased point estimation. Right: Application for measurements of a controlled industrial process variable (Drill-speed over time in seconds). A finite set of different instances n during the process define the "stations" at which measured values from the sensor is obtained. In between the events of incoming measurements the control mechanism may fail to regulate the variable to be within predefined specification bounds. This defines a 'failure' ($\eta_n = 1$).

Here is how to arrive at this lamplighter graph (see [4] for details). We let $\mathrm{lab}(v)$ be the label of the projected vertex $v \in V_T$ onto $V_0 = \mathbb{Z}$. Starting with the tree-like DGM T, from Sect. 1.2, interpret this tree as the graph 'horocyclic' product of the tree with a bi-infinite path $P = \langle \mathbb{Z}, \text{Nearest Neighb.}(\mathbb{Z}) \rangle$. This is $H := T \times P = \langle V_H, E_H \rangle$ given by $V = \{\langle v, k \rangle \in V_T \times V_P \mid \mathrm{lab}(v) + k = 0\}$, and $E_H = \{\langle \langle e_1, e_2 \rangle, f \rangle \in E_T \times E_P \mid \langle \pi(e_1), \pi(e_2) \rangle = \langle k, k-1 \rangle\}$. Geometrically, this represents just an alignment with every infinite ray along the directed edges in T with the path P.

Now: Switch P to an infinite bifurcated path (v-shape): P_2, where the bifurcation opens towards the left (decreasing labels of T). Then the resulting horocyclic product $T \times P_2$ will have a bifurcated structure towards the left. It becomes a multitree (the directedness of the edges is inherited to the edges of the horocyclic product). Continue to attach ray-shaped branches to the increasingly bifurcated graph, until it becomes a binary tree (use the construction technique for the canopy tree). The result is the horocyclic product of T with itself (compare [3], Chap. 12.13). A subset is sketched on the left side of Fig. 3.

3 Applications

3.1 Controlled Processes

The practical value when making estimations (may be indeed from the momentarily available sensor data of a production line environment) lies in the ability to chose a reasonably sized sample set C of vertices of the DGM's graph and calculate the point estimator using only the data associated with these vertices. Note, that both, the tree-like DGM and its unimodularization have exponentially growing graphs, which quickly becomes unfeasible to process all for a single estimation. Subsampling is vital in this online-task, so that the use of an unbiased estimator is useful because of the potentially small sample size.

Figure 3 shows a controlled velocity process v_n with set point of 10 rot/sec, becoming more stable as more data of the response is available. Unbiased estimates of the invariant variable $v_n \cdot \chi_{>\epsilon}(|V_n - V_{SP}|)$ are useful for the control's corrections.

3.2 Outlook and Acknowledgments

For production environments with more sophisticated structures of interdependence of different measurements than that of the graphical models derived from Definition 1, it is desirable to construct unimodularizations. In particular, a production line environment, in which work items may jump by more than a single step, and leave out several machines is of interest in real applications.

Also, the approach to construct unbiased estimators using quasi-unimodular graphs (see Sect. 1.4), by using the (reciprocal) modular function represents an interesting challenge, whenever unimodular graphs such as the horocyclic product cannot be derived, easily.

The research in this paper has been supported by the Austrian Ministry for Transport, Innovation and Technology, the Federal Ministry of Science, Research and Economy, and the Province of Upper Austria in the frame of the COMET center SCCH.

References

1. Jacob, P.E., Thiery, A.H.: On non-negative unbiased estimators. Ann. Statist. **43**(2), 769–784 (2015)
2. Wiesel, A., Eldar, Y., Hero, A.: Covariance estimation in decomposable Gaussian models. IEEE Trans. Sig. Proc. **58**(3), 1482–1492 (2010)
3. Woess, W.: Lamplighters, Diestel-Leader graphs, random walks, and harmonic functions. Comb. Probab. Comput. **14**, 415–433 (2005)
4. Woess, W.: What is a horocyclic product, and how is it related to lamplighters? Internationale Mathematische Nachrichten **224**, 1–27 (2013)
5. Woess, W.: Random Walks on Infinite Graphs and Groups, CUP 2000
6. Häggström, O.: Infinite clusters in dependent automorphism invariant percolation on trees. Ann. Probab. **25**, 1423–1436 (1997)

 7. Benjamini, I., Lyons, R., Peres, Y., Schramm, O.: Group-invariant percolation on graphs. GAFA **9**, 20–66 (1999)
 8. Kaimanovich, V.: Invariance, Quasi-invariance, and unimodularity for random graphs. J. Math. Sci. **19**(5), 747–764 (2016)
 9. Lyons, R., Peres, Y.: Probability on Trees and Networks, Chap. 8. CUP, New York (2017)
10. Sobieczky, F.: An interlacing technique for spectra of random walks and its application to finite percolation clusters. J. Theor. Prob. **23**(3), 639–670 (2010)
11. Sobieczky, F.: Bounds for the annealed return probability on large finite percolation clusters. Electron. J. Prob. **17**(29), 1–17 (2012)
12. Johnson, R.A., Wichern, D.W.: Applied Multivariate Statistical Analysis, pp. 547–553. Pearson Prentice Hall, Upper Saddle River (2007). Result 3.1
13. Koller, D., Friedman, N.: Probabilistic Graphical Models, pp. 237–246. MIT Press, Cambridge (2009)

Volatility Modelling for Air Pollution Time Series

G. Albano[1(✉)], Michele La Rocca[2], and C. Perna[2]

[1] Department of Political and Social Studies, University of Salerno, Fisciano, Italy
pialbano@unisa.it
[2] Department of Economics and Statistics, University of Salerno, Fisciano, Italy
{larocca,perna}@unisa.it

Abstract. We consider a class of semiparametric models for univariate air pollutant time series which is able to incorporate some stylized facts usually observed in real data, such as missing data, trends, conditional heteroschedasticity and leverage effects. The inference is provided by a semiparametric approach in a two step procedure in which the cycle-trend component is firstly estimated by a local polynomial estimator and then the parametric component is chosen among several "candidate models" by the Model Confidence Set and estimated by standard procedures. An application to PM_{10} concentration in Torino area in the North-Italian region Piemonte is shown.

Keywords: GARCH-type models · ARMA models · PM_{10} time series

1 Introduction

In the last two decades several works have tried to study associations between health effects and air pollution. In particular day-to-day selected particulate air pollution has been identified as cause of an increased risk of various adverse health outcomes, including cardiopulmonary mortality (see, for example, [1] and references therein). Air quality monitoring becomes thus a fundamental issue in order to detect any significant pollutant concentrations which may have possible adverse effects on human health. Among the main indicators to measure air quality, PM_{10} plays a crucial role. It measures "particulate matter" with diameter less than 10 μm.

From a statistical perspective, it is important to accurately model stylized facts of air pollutant time series such as an often large presence of missing values, a trend-cycle component, reflecting the underlying levels of the series and the repetitive movement due to the seasons, the presence of significant autocorrelations among near observations, heteroschedastic effects tipically observed in form of fat tails, clustering of volatility and leverage effects. The aim of this paper is to propose a model able to capture the main characteristics of PM_{10} time series. In particular, we consider a semiparametric model in which the trend-cycle is estimated by using a local polynomial approach; such estimate is then used to

© Springer Nature Switzerland AG 2020
R. Moreno-Díaz et al. (Eds.): EUROCAST 2019, LNCS 12013, pp. 176–184, 2020.
https://doi.org/10.1007/978-3-030-45093-9_22

impute missing values if present in the data at hand. The detrended component is assumed to follow an ARMA-GARCH-type model which is able to take into account several stylized patterns of the PM_{10} time series.

The first choice allows to have flexible local structures which are not influenced by missing values outside the estimation window. The latter choice appears to be necessary since the detrended time series might show a dependence structure due both to the intrinsic characteristic of the data and to the cycle-trend estimation step. Indeed, the difficult task of selecting tuning parameters in that step might possibly induce neglected nonlinearities in the detrended series. Also, a GARCH-type model is able to capture heteroschedasticity in the data. The choice of the "best" model, among some possible specifications of the GARCH models, has been made by using the Model Confidence Set (MCS) (see [2]) which allows to discriminate among models looking at their forecasting performance.

The paper is organized as follows. Section 2 presents the proposed modelling strategy, focusing on several alternative specifications of the conditional volatility dynamics. In Sect. 3 the results of an application to real data are discussed. Some remarks close the paper.

2   The Model

Let Y_t be the one dimensional random process representing the daily average of an air pollutant at day t, modelled as:

$$Y_t = m(t) + \zeta_t, \qquad t \in \mathbb{N}, \tag{1}$$

where $m(t)$ represents the deterministic trend-cycle component and ζ_t represents the stochastic component. The stochastic term ζ_t is assumed to follow a stationary and invertible Autoregressive Moving Average (ARMA)-Generalized Autoregressive Conditional Heteroschedastic (GARCH) specification originally proposed in [3], described by the following equations:

$$\zeta_t = \Phi_1 \zeta_{t-1} + \Phi_2 \zeta_{t-2} + \ldots + \Phi_p \zeta_{t-p} + \varepsilon_t + \theta_1 \varepsilon_{t-1} + \ldots + \theta_q \varepsilon_{t-q} \tag{2}$$

$$\varepsilon_t = \sigma_t z_t, \tag{3}$$

$$\sigma_t^2 = h(\sigma_{t-1}^2, \ldots, \sigma_{t-q'}^2, \varepsilon_{t-1}^2, \ldots, \varepsilon_{t-p'}^2, \psi_\sigma), \tag{4}$$

where $\{z_t\}$ is a sequence of independent and identically distributed random variables such that $\mathbb{E}(z_t) = 0$ and $\mathbb{E}(z_t^2) = 1$, σ_t is the conditional standard deviation of ε_t; the function $h(\cdot)$ refers to one of the ARCH-type dynamics and the vector ψ_σ contains all the conditional variance dynamic parameters, its specification depending on the structure in the data, such as leverage effects and/or asymmetry. The ARMA component (2) in the model (1) allows to capture the stochastic behaviour of the process Y_t; the GARCH component (3) and (4) takes into account the heteroschedastic effects tipically observed in form of fat tails, clustering of volatility and leverage effects. We will assume that the trend-cycle component in (1) is a smooth function of time and that the ARMA process is stationary and ergodic. Moreover, to preserve the positivity and the weak ergodic stationarity of the conditional variance σ_t^2, suitable conditions on the parameters in (2)–(4) are also assumed.

Let

$$\mathbf{y} = (y_1, \ldots, y_n)'$$

be the time series of PM_{10} emissions observed at times $t_1 < t_2 < \ldots < t_n = T,$. We assume that it is daily observed, so $t_{j+1} - t_j = 1$ $(j = 1, \ldots, n-1)$. Further, some of y_j could be missing, due to some defaults in the machine. So, in the estimation of model (1)–(4), first of all we need to impute the missing values.

The approach we use is the imputed local polynomial smoother proposed in [5]. It is a two-step procedure in which a local polynomial smoother is used to estimate the missing observations of the response variable and then the same estimator is again applied to the complete sample to obtain an estimate of the function $m(\cdot)$ (see [4] for details).

Under general hypothesis, the asymptotic properties of the previous estimator can be obtained (see [5]) as well as expressions for the bias and the variance. Moreover the method seems to work better than other standard methods with time series with possibly high percentage of missing values ([4]). Finally, since it is essentially based on a local estimator approach, it is non parametric and so able to handle general trend structure in the data. The detrended time series $\widehat{\zeta}_i = y_i - \widehat{m}(t_i)$ is then modelled with an ARMA-GARCH-type model (2)–(4) focusing on some alternative possible specifications of the conditional volatility function $h(\cdot)$. The simplest conditional volatility dynamics is the GARCH(p', q') specification (SGARCH), introduced in [6] for analysing financial time series:

$$\sigma_t^2 = \omega + \sum_{i=1}^{p'} \alpha_i \varepsilon_{t-i}^2 + \sum_{j=1}^{q'} \beta_j \sigma_{t-j}^2. \tag{5}$$

Despite its popularity, the SGARCH specification is not able to account for time series exhibiting higher volatility after negative shocks than after positive ones as theorized by the leverage effect. To take into account these latter effects, along with possible asymmetries, several generalizations of this model have been proposed. Here, we focus on three of the most used generalizations of (5) which could be suitable to account for the stylized facts usual observed in PM_{10} series. The first one is the EGARCH(p', q') model (see [7]). It assumes that the conditional volatility dynamics follows the equation:

$$\log(\sigma_t^2) = \omega + \sum_{i=1}^{p'} \left[\eta_i z_{t-i} + \gamma_i (|z_{t-i}| - \mathbb{E}|z_{t-i}|) \right] + \sum_{j=1}^{q'} \varphi_j \log(\sigma_{t-j}^2),$$

where $z_t = \varepsilon_t / \sigma_t$. The second generalization is GJR-GARCH(p', q') model (see [8]) which assumes that the conditional volatility dynamics follows:

$$\sigma_t^2 = \omega + \sum_{i=1}^{p'} \left(\eta_i + \gamma_i \mathbf{I}_{\{\varepsilon_{t-i} < 0\}} \right) \varepsilon_{t-i}^2 + \sum_{j=1}^{q'} \beta_j \sigma_{t-j}^2,$$

where $\mathbb{I}_{\{\varepsilon_{t-i}<0\}}$ assumes value one if $\varepsilon_{t-i} < 0$ for $i = 1, \ldots, p'$ and zero otherwise. Finally, the last considered generalization is the Asymmetric-Power-ARCH(p', q') APARCH(p', q') ([9]), described by the following equation:

$$\sigma_t^{(\delta)} = \omega + \sum_{i=1}^{p'} \alpha_i \left(|\varepsilon_{t-i} - \gamma_i \varepsilon_{t-i}| \right)^\delta + \sum_{j=1}^{q'} \beta_j \sigma_{t-j}^{(\delta)}.$$

with $x^{(\delta)} = \frac{x^\delta - 1}{\delta}$. The APARCH specification results in a very flexible model which includes several of the most popular univariate ARCH parameterizations.

The choice among these different specifications and different error term distributions can be made by looking at their ability to predict future values of the series by means of the MCS proposed in [2]. Starting from a set of M competing models, this procedure allows to construct a Set of Superior Models (SSM) by using a sequence of tests in which the null hypothesis of equal predictive ability is not rejected at a fixed significance level α.

3   Application to PM_{10} Time Series from Torino Area

In order to evaluate the performance of the proposed approach for modelling PM_{10} time series, a real dataset has been considered. It consists of PM_{10} concentration (in $\mu g/m^3$) measured from five monitoring stations from 1 January 2015 to 19 October 2016 in the town of Torino - http://www.arpa.piemonte. gov.it. In Table 1 it is reported the typology of each station (urban-traffic or land-urban) together with the number of observations in each time series and the percentage of missing values. It ranges from 3.8% for Rubino station to 34% of Grassi station. In the estimation of the function $m(\cdot)$ in (1) we choose the degree of the polynomials equal to 1, the two kernel functions as the Epanechnikov kernel and the smoothing parameters by means of a least squares cross validation as suggested in [10] and implemented in the R package np. Figure 1 reports the local polynomial smoothers, from which the detrended time series are derived as:

$$\zeta_i^{(s)} = y_i^{(s)} - \widehat{m}^{(s)}(t) \tag{6}$$

for each station s = "Consolata", "Grassi", "Lingotto", "Rebaudengo" and "Rubino". By looking at Fig. 2, which reports the autocorrelation functions and the partial autocorrelation functions of the estimated detrended series, it is evident that a strong dependence structure is still evident. Although the autocorrelation plots can be used to identify the orders p and q of the ARMA model (2), an automatic selection criterion based on Bayesian Information Criterion has been used. The identified orders are shown in Table 2. We can observe that for the stations "Grassi", "Rebaudengo" and "Rubino" only an autoregressive structure with a small order p (1 and 2) is present. In order to verify if a conditional heteroschedasticity is present in the residuals of the ARMA models (2), we perform an Engle's ARCH test (see [3]). In this analysis we choose the lag for testing ARCH effects equal to 7 in order to look at the dependence of the data in a week. The test rejects the null hypothesis of no ARCH effects for all the stations, being all the p−values of the order of 10^{-4}. So, as discussed in the previous section, we consider the following ARCH-type specifications: SGARCH, EGARCH, GJR-GARCH and APARCH, with orders $p' = q' = 1$. For each specification we consider six different distributions of the error term z_t in (3): the standard normal $\mathcal{N}(0,1)$, the Student−t T_g with g degrees of freedom, the Generalized Error Distribution $Ged(0, 1, k)$ where k is the shape parameter, the skew-Normal, the skew Student-t and the Skew-GED. This choice covers a large class of asymmetric and heavy-tailed distributions.

The 24 resulting models are compared by means of MCS in which, as loss function, we choose the mean square error, frequently used to assess the validity of a forecasting scheme. Moreover, we fix $\alpha = 0.10$ and the forecast sample equal to the last 30 observations, corresponding to one month horizon. In Table 3, we show the composition of the SSM for the 24 considered models as well as the model ranking, the value of the test statistic T_{max}, the related probability (MCS p−value) and the value of the loss function.

It is evident that in four of the five stations, the SSM contains the simplest GARCH model with standard normal error distribution. Only for Rebaudengo station, the SSM includes only the more complex APARCH model. Such result allows to argue that the SGARCH model, despite its simplicity, is able to capture the heteroschedatic structure in the error term of model (1) in the sense that more complicated GARCH-type models have similar forecasting performance.

In Table 4, the estimates of the coefficients of the model (2)–(4), in which p and q are chosen as in Table 2 and $p' = q' = 1$, are shown for the five stations. In parenthesis the standard errors are reported.

Table 1. Typology, number of observations and percentage of missing values for PM_{10} time series in the Torino area.

Station	Typology	Obs.	% missing
Consolata	Urban-traffic	626	0.049
Grassi	Urban-traffic	434	0.340
Lingotto	Land-urban	599	0.090
Rebaudengo	Urban-traffic	605	0.080
Rubino	Land-urban	633	0.038

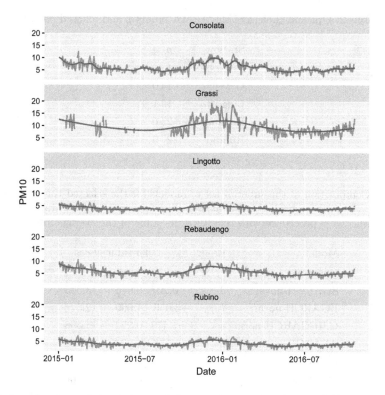

Fig. 1. Local polynomial estimates of the trend-cycle component $\widehat{m}(t)$ for the transformed PM_{10} concentration in the 5 considered stations.

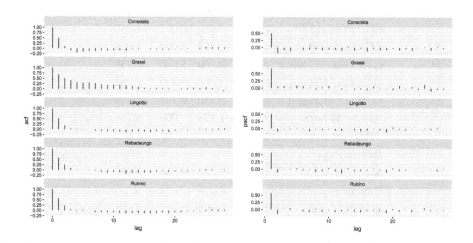

Fig. 2. Autocorrelation functions and partial autocorrelation functions of the detrended series (6) for all the stations.

Table 2. Optimal choices of the ARMA orders p and q based on BIC.

Station	p	q
Consolata	1	1
Grassi	1	0
Lingotto	1	1
Rebaudengo	2	0
Rubino	2	0

Table 3. Composition of the superior set of models for the selected times series. The entries are: the rank according to the average loss function; the test statistic and the p-value of the MCS procedure with $\alpha = 0.10$; the value of the average loss.

Model	Rank	T_{max}	p-value	Loss
Consolata				
SGARCH-norm	3	0.207	0.976	13.333
EGARCH-norm	1	−3.050	1.000	12.526
GJR-GARCH-norm	4	1.479	0.224	14.398
APARCH-norm	2	−2.176	1.000	12.734
Grassi				
SGARCH-norm	2	0.672	0.536	193.062
GJR-GARCH-norm	1	−0.672	1.000	189.191
Lingotto				
SGARCH-norm	8	1.617	0.157	1.264
EGARCH-norm	5	0.859	0.563	1.155
EGARCH-std	7	1.389	0.277	1.387
EGARCH-ged	4	−0.453	1.000	1.026
GJR-GARCH-norm	1	−2.700	1.000	0.739
GJR-GARCH-std	6	0.964	0.499	1.282
GJR-GARCH-ged	9	1.691	0.134	1.364
APARCH-norm	3	−1.774	1.000	0.788
APARCH-ged	2	−2.109	1.000	0.808
Rebaudengo				
apARCH-norm	1	−4.539	1.000	9.212
Rubino				
SGARCH-norm	3	0.468	0.804	3.871
EGARCH-norm	1	−2.828	1.000	3.537
GJR-GARCH-norm	2	−0.065	1.000	3.813
APARCH-norm	4	1.713	0.121	4.048

Table 4. Estimated coefficients of the ARMA-GARCH model (2)–(4). In parenthesis the standard errors are reported.

Station	Φ_1	Φ_2	θ_1	ω	α_1	β_1
Consolata	0.9941		−0.8175	0.3978	0.1268	0.8723
	(0.0011)		(0.0177)	(0.0854)	(0.0596)	(0.0284)
Grassi	0.9670			3.8186	0.1114	0.8393
	(0.0098)			(0.5730)	(0.0207)	(0.0174)
Lingotto	0.9798		−0.5787	0.5457	0.0732	0.9236
	(0.0037)		(0.0511)	(0.0831)	(0.0115)	(0.0077)
Rebaudengo	0.5404	0.3531		0.5562	0.0840	0.9150
	(0.0493)	(0.0495)		(0.1691)	(0.0122)	(0.0089)
Rubino	0.6420	0.2742		1.3995	0.2401	0.7589
	(0.0804)	(0.0800)		(0.2380)	(0.0289)	(0.0184)

Concluding Remarks

The paper provides a model for univariate environmental time series incorporating their main peculiarities, such as cycle-trend components, correlation among consecutive observations and conditional heteroschedasticity. The suggested procedure combines a non parametric local estimator for the trend-cycle and an ARMA-GARCH-type model to describe the detrended time series. This semiparametric approach allows to combine flexibility of local structures, which are not influenced by missing values outside the estimation window, and the standard estimation procedure of the involved models. Moreover, the ARMA-GARCH class of models is able to capture neglected dependence structures in the detrended time series, essentially due both to the intrinsic characteristic of the data and to the cycle-trend estimation step.

Four GARCH-type specifications and six different distributions of the error term have been considered leading to $6 \cdot 4 = 24$ different models. The MCS procedure has been suggested to choose among the models, by looking at their forecasting performances. An application to PM_{10} concentration in Torino is provided and for this dataset the MCS procedure shows that the simplest standard GARCH model with normal distributed error is able to capture the heteroschedatic structure in the data in the sense that more complicated GARCH-type models have similar forecasting performance.

References

1. Health—Particulate Matter—Air & Radiation—US EPA. Epa.gov. Accessed 17 Nov 2010
2. Hansen, P.R., Lunde, A., Nason, J.M.: The model confidence set. Econometrica **79**(2), 453–497 (2011)
3. Engle, R.F.: Autoregressive conditional heteroskedasticity with estimates of the variance of United Kingdom inflation. Econometrica **5**, 987–1007 (1982)

4. Albano, G., La Rocca, M., Perna, C.: On the imputation of missing values in univariate PM_{10} time series. In: Moreno-Díaz, R., Pichler, F., Quesada-Arencibia, A. (eds.) EUROCAST 2017. LNCS, vol. 10672, pp. 12–19. Springer, Cham (2018). https://doi.org/10.1007/978-3-319-74727-9_2

5. Pèrez, A., Vilar, J.M., Gonzàlez, W.: Asymptotic properties of local polynomial regression with missing data and correlated errors. Ann. Inst. Stat. Math. **61**, 85–109 (2009)

6. Bollerslev, T.: Generalized autoregressive conditional Heteroskedasticity. J. Econom. **31**, 307–32 (1986)

7. Nelson, D.B.: Conditional heteroschedasticity in asset returns: a new approach. Econometrica **59**(2), 217–235 (1991)

8. Glosten, L.R., Jagannathan, R., Runkle, D.E.: On the relation between the expected value and the volatility of the nominal excess return on stocks. J. Finance **48**(5), 1779–1801 (1993)

9. Ding, Z., Granger, C.W., Engle, R.F.: A long memory property of stock market returns and a new model. J. Empir. Finance **1**(1), 83–106 (1993)

10. Li, Q., Racine, J.: Cross-validated local linear nonparametric regression. Statistica Sinica **14**, 485–512 (2004)

HAR-type Models for Volatility Forecasting: An Empirical Investigation

G. Albano[1]([✉]) and D. De Gaetano[2,3]

[1] Department of Political and Social Studies, University of Salerno, Fisciano, Italy
pialbano@unisa.it
[2] Department of Economics, University of Roma Tre, Rome, Italy
[3] SOSE-Soluzioni per il Sistema Economico, Rome, Italy
ddegaetano@sose.it

Abstract. The paper addresses the problem of forecasting realized volatility in the context of HAR-type models. Some extensions of the basic HAR-RV model are discussed. The forecasting performance of the considerec HAR-type models are compared in terms of suitable loss functions, by using the Model Confidence Set procedure, on two real datasets.

Keywords: HAR-RV model · Model Confidence Set · Out-of-sample forecasts

1 Introduction

In the last two decades volatility has played an important role in financial and economic fields such as in pricing derivatives, in portfolio optimization and risk management. Several recent studies have recognized the properties of realized volatility constructed from high frequency data and, in this context, a wide range of volatility models have been developed for modelling its dynamics. Notably, Corsi in [1] proposed the so called heterogeneous autoregressive RV model (HAR-RV) which can well capture stylized facts in volatility, such as long memory and multi-scaling behaviour. HAR-RV has become the standard benchmark for analyzing and forecasting financial volatility dynamics (see, for example, [4]). Recently, many authors (see, e.g., [2] and the references therein) have also investigated the impact of jump components on forecasting accuracy, and several models generalizing HAR-RV to include jumps have been provided [3,4].

The aim of the paper is to compare several recent HAR-type models with the earlier specification proposed in [1]. The considered models, which explicitly introduce the decomposition of the realized volatility in a continuous component and in a jump component, are evaluated in term of fitting and forecasting on two real datasets. In particular, the forecasting performances of the models are compared by means of the Model Confidence Set (MCS) procedure, firstly proposed in [5]. Different loss functions have been considered in order to investigate their role in discriminating among models for realized volatility.

© Springer Nature Switzerland AG 2020
R. Moreno-Díaz et al. (Eds.): EUROCAST 2019, LNCS 12013, pp. 185–194, 2020.
https://doi.org/10.1007/978-3-030-45093-9_23

The paper is organized as follows. Section 2 sets up the notation and describes the eleven HAR-type models employed in the analysis. Section 3 provides an application to real data. Some concluding remarks close the paper.

2   Methodology

Let $p(t)$ be a logarithmic asset price descrited by the following jump diffusion process:

$$dp(t) = \mu(t)dt + \sigma(t)dW(t) + k(t)dq(t), \quad 0 \leq t \leq T. \tag{1}$$

In (1) $\mu(t)$ is a continuous and locally bounded variation process, $\sigma(t) > 0$ denotes the càdlàg instantaneous volatility, $W(t)$ is a standard Brownian Motion and $q(t)$ is a counting process with intensity $\lambda(t)$. The function $k(t) = p(t) - p(t-)$ is the size of the jump at time t.

The object of interest is the integrated variance (IV), i.e. the amount of variation of the function $\sigma^2(\cdot)$ accumulated over a twentyfour hour day $(t, t+1)$:

$$IV_t = \int_{t-1}^{t} \sigma^2(s)ds, \quad t = 1, \ldots, T.$$

Suppose there exist m intraday returns, and let $r_{t,i}$ be the i-th intraday return at day t, then the realized volatility RV_t at day t is defined as

$$RV_t = \sum_{i=1}^{m} r_{t,i}^2 \quad t = 1, \ldots, T.$$

In order to disentangle the continuous and the jump components of the realized volatility, the realized bi-power variation was introduced in [6]. It is defined as

$$BV_t = (2/\pi)^{-1} \sum_{i=1}^{m-1} |r_{t,i}||r_{t,i+1}|$$

Further, when the difference between RV_t and BV_t is statistically significant, the jump component can be estimated in the following way:

$$J_t = [RV_t - BV_t] \times I[ZJ_{BV}(t) > \Phi_\alpha] \tag{2}$$

where $ZJ_{BV}(t)$ is the adjusted jump ratio statistic defined in [7] to formally test the presence of jumps, and $I[\cdot]$ is the indicator function which identifies the significance of the $ZJ_{BV}(t)$ statistic in excess of a given critical value of the Gaussian distribution Φ_α. Analogously, the continuous jump component is:

$$C_t = BV_t \times I[ZJ_{BV}(t) > \Phi_\alpha] + RV_t \times I[ZJ_{BV}(t) \leq \Phi_\alpha] \tag{3}$$

We also consider the approach proposed in [8], which introduces estimators able to capture the variation due to negative or positive returns, i.e. the negative and positive realized semivariances, defined as

$$RS_t^- = \sum_{i=1}^{m} r_{t,i}^2 \times I[r_{t,i} < 0], \quad RS_t^+ = \sum_{i=1}^{m} r_{t,i}^2 \times I[r_{t,i} \geq 0]. \tag{4}$$

The difference between the realized semivariances is the so called signed jump variation, i.e.

$$\Delta J := RS^+ - RS^-. \tag{5}$$

The different decompositions of the realized variance can be introduced in the HAR-RV model to better analyze financial volatilty dynamics. In the following, we rely on the HAR-RV class of volatility models extensivelly used in econometric literature (see [9] and the references therein) referring, in particular, to the models proposed in [3]. They are essentially regression models in which the response variable is the average realized variance over the period $[t + 1, t + h]$, i.e.

$$RV_{t+1,t+h} = h^{-1} [RV_{t+1} + RV_{t+2} + \ldots + RV_{t+h}] \tag{6}$$

We consider the following eleven HAR-RV-type models.

Model HAR-RV. In this model, lags of RV are used at daily, weekly and monthly aggregated periods:

$$RV_{t+1,t+h} = \beta_0 + \beta_1 RV_t + \beta_5 RV_{t-1,t-4} + \beta_{22} RV_{t-5,t-21} + \epsilon_t \tag{7}$$

where ϵ_t is a random variable representing the error term and $RV_{t-1,t-4}$ and $RV_{t-5,t-21}$ are the weekly and monthly averages of the realized variance, respectively.

Model HAR-RV-J. In this model the jump component J_t defined in (2) is added as esplicative variable:

$$RV_{t+1,t+h} = \beta_0 + \beta_1 RV_t + \beta_5 RV_{t-1,t-4} + \beta_{22} RV_{t-5,t-21} + \beta_{J1} J_t + \epsilon_t. \tag{8}$$

Model HAR-RV-CJ. It is obtained from HAR-RV model by decomposing realized volatility into continuous sample path variation and discontinuous jump variation:

$$RV_{t+1,t+h} = \beta_0 + \beta_{C1} C_t + \beta_{J1} J_t + \beta_{C5} C_{t-1,t-4} + \beta_{J5} J_{t-1,t-4}$$
$$+ \beta_{C22} C_{t-5,t-21} + \beta_{J22} J_{t-5,t-21} + \epsilon_t \tag{9}$$

where C_t, J_t, $C_{t-1,t-4}$, $J_{t-1,t-4}$, $C_{t-5,t-21}$, $J_{t-5,t-21}$ are the daily, weekly and monthly continuous and discontinuous sample path variation as in (2) and (3).

Model PS. It is obtained decomposing the lagged realized variance by using realized semivariances:

$$RV_{t+1,t+h} = \beta_0 + \beta_1^- RS_t^- + \beta_1^+ RS_t^+ + \beta_5 RV_{t-1,t-4} + \beta_{22} RV_{t-5,t-21} + \epsilon_t \tag{10}$$

Model PSlev. It is similar to the previous model and it includes a possible inverse leverage effect:

$$RV_{t+1,t+h} = \beta_0 + \beta_1^- RS_t^- + \beta_1^+ RS_t^+ + \gamma RV_t I_{(r_t < 0)}$$
$$+ \beta_5 RV_{t-1,t-4} + \beta_{22} RV_{t-5,t-21} + \epsilon_t. \tag{11}$$

Model HAR-RS. It uses the heterogeneous structure of the basic HAR model over positive and negative realized semivariances:

$$
\begin{aligned}
RV_{t+1,t+h} &= \beta_0 + \beta_1^- RS_t^- + \beta_1^+ RS_t^+ + \beta_5^- RS_{t-1,t-4}^- \\
&+ \beta_5^+ RS_{t-1,t-4}^+ + \beta_{22}^- RS_{t-5,t-21}^- + \beta_{22}^+ RS_{t-5,t-21}^+ + \epsilon_t.
\end{aligned}
\tag{12}
$$

Model CG. This model is a version of the HAR-RS model that includes the lagged daily discontinuous jump variation:

$$
\begin{aligned}
RV_{t+1,t+h} &= \beta_0 + \beta_1^- RS_t^- + \beta_1^+ RS_t^+ + \beta_5^- RS_{t-1,t-4}^- + \beta_5^+ RS_{t-1,t-4}^+ \\
&+ \beta_{22}^- RS_{t-5,t-21}^- + \beta_{22}^+ RS_{t-5,t-21}^+ + \beta_{J1} J_t + \epsilon_t.
\end{aligned}
\tag{13}
$$

Model HAR-RV-SJ. It introduces as a esplicative variable the signed jump defined in Eq. (5):

$$
RV_{t+1,t+h} = \beta_0 + \beta_{\Delta 1} \Delta J_t + \beta_{C1} C_t + \beta_5 RV_{t-1,t-4} + \beta_{22} RV_{t-5,t-21} + \epsilon_t.
\tag{14}
$$

Model HAR-RV-CSJ. It considers signed jumps over a weekly, monthly and daily interval:

$$
\begin{aligned}
RV_{t+1,t+h} &= \beta_0 + \beta_{\Delta 1} \Delta J_t + \beta_{C1} C_t + \beta_{\Delta 5} \Delta J_{t-1,t-4} + \beta_{C5} C_{t-1,t-4} \\
&+ \beta_{\Delta 22} \Delta J_{t-5,t-21} + \beta_{C22} C_{t-5,t-21} + \epsilon_t
\end{aligned}
\tag{15}
$$

Model HAR-RV-Sjd. It discriminates between positive and negative signed jumps:

$$
\begin{aligned}
RV_{t+1,t+h} &= \beta_0 + \beta_{\Delta 1}^- \Delta J_t I_{(\Delta J_t < 0)} + \beta_{\Delta 1}^+ \Delta J_t I_{(\Delta J_t > 0)} + \beta_{C1} C_t \\
&+ \beta_5 RV_{t-1,t-4} + \beta_{22} RV_{t-5,t-21} + \epsilon_t.
\end{aligned}
\tag{16}
$$

Model HAR-RV-CSjd. It is an extension of the previous model in which the signed jumps are considered at daily, weekly and monthly horizons:

$$
\begin{aligned}
RV_{t+1,t+h} &= \beta_0 + \beta_{\Delta 1}^- \Delta J_t I_{(\Delta J_t < 0)} + \beta_{\Delta 1}^+ \Delta J_t I_{(\Delta J_t > 0)} + \beta_{C1} C_t \\
&+ \beta_{\Delta 5}^- \Delta J_{t-1,t-4} I_{(\Delta J_{t-1,t-4} < 0)} + \beta_{\Delta 22}^- \Delta J_{t-5,t-21} I_{(\Delta J_{t-5,t-21} < 0)} \\
&+ \beta_{\Delta 5}^+ \Delta J_{t-1,t-4} I_{(\Delta J_{t-1,t-4} > 0)} + \beta_{C5} C_{t-1,t-4} \\
&+ \beta_{\Delta 22}^+ \Delta J_{t-5,t-21} I_{(\Delta J_{t-5,t-21} > 0)} + \beta_{C22} C_{t-5,t-21} \epsilon_t.
\end{aligned}
\tag{17}
$$

3   Application to Real Data

This paper uses 5-min, high frequency transaction data from two German DAX component stocks: BMW ans Allianz. The sample period is from January 2, 2002 to December 27, 2012, for a total of 2791 daily observations. Realized volatility dynamics of the two series are plotted in Fig. 1. We first address the issue of comparing models in-sample. In the estimation of HAR-type models, to

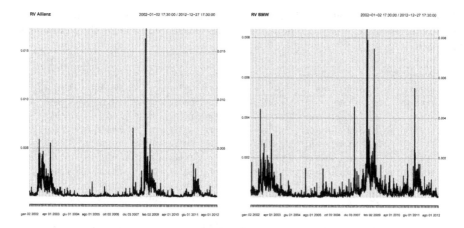

Fig. 1. Realized volatility dynamics of Allianz market (left) and BMW (right).

limit the effect on the parameters estimates of periods of high variance, we use Weighted Least Squares (WLS) in which the weights are obtained as the inverse of the fitted values from a preliminary Ordinary Least Square (OLS) regression. In the estimation of the jump components J_t and C_t, we fix $\alpha = 0.95$. The standard errors of the regression coefficient estimates have been adjusted by using the Newey-West/Bartlett heteroskedasticity consistent covariance matrix estimator with 5 lags. This correction allows for series correlation up to the order 5. Table 1 shows the estimated parameters for all models (7)–(17) together with the estimated standard errors and the Newey-West adjusted t-statistics. It is evident that, for both the series, all the coefficients in the basic HAR-RV model are significant, confirming the long memory behaviuor of volatility. In the HAR-RV-J model, the estimation results are in accordance with the findings in [10], i.e. the jump component is significant and its estimate is negative. In HAR-RV-CJ model, the lagged squared jump components are all not significant for both the series. Also this result is in line with the recent findings in [3] and [10] and confirms that, when the continuous component is introduced in the model, it incorporates most of the relevant information for predicting volatility nullfying the effect of the jump components. In the PS model in which the decomposition between positive and negative semivariances is included, both of them are highly significant. This result is at odds with the results in [8] but in line with those in [3]. By looking at the PSlev model, we find that the coefficient of the leverage effect is not significant for both the time series. In the HAR-RS model, all the coefficients of the positive and negative realized semivariances are significant except for the one month lagged positive semivariance. When a one day lagged jump is introduced as in the model CG, it results significant at 5% only for BMW time series and not significant for Allianz. In the HAR-RV-SJ model, all the coefficients, including the signed jump component, are highly significant for both the series. When signed jumps over a longer period are introduced, as in the HAR-RV-CSJ model, the one week and the one month signed jump become insignificant. In the HAR-RV-Sjd model, in which the signed jumps are included

Table 1. ALLIANZ (left) and BMW (right) estimation by WLS with fitted values of an OLS regression for weights. The t-statistics are calculated by using the Newey-West approach.

ALLIANZ					BMW		
Model	Coef.	Estimate	Std. Error	t−value	Estimate	Std. Error	t value
HAR-RV	β_0	$1.068e-05^{***}$	$2.745e-06$	3.892	$1.852e-05^{***}$	$4.311e-06$	4.296
	β_1	$5.477e-01^{***}$	$2.573e-02$	21.290	$4.533e-01^{***}$	$2.563e-02$	17.684
	β_5	$2.905e-01^{***}$	$2.877e-02$	10.099	$3.641e-01^{***}$	$2.997e-02$	12.150
	β_{22}	$1.431e-01^{***}$	$2.419e-02$	5.916	$1.411e-01^{***}$	2.650e-02	5.326
		Adjusted R^2: 0.5323			Adjusted R^2: 0.4991		
HAR-RV-J	β_0	$1.074e-05^{***}$	$2.743e-06$	3.917	$1.834e-05^{***}$	$4.310e-06$	4.254
	β_1	$5.498e-01^{***}$	$2.576e-02$	21.346	$4.588e-015^{***}$	$2.580e-02$	17.783
	β_5	$2.902e-01^{***}$	$2.876e-02$	10.090	$1.402e-01^{***}$	$2.649e-02$	5.292
	β_{22}	$1.437e-01^{***}$	$2.417e-02$	5.945	$1.402e-01^{***}$	$2.649e-02$	5.292
	β_{J1}	$-3.398e-01^{*}$	$1.371e-01$	-2.479	$-2.820e-01^{*}$	$1.555e-01$	-1.814
		AdjustedR^2:: 0.5328			Adjusted R^2: 0.4996		
HAR-RV-CJ	β_0	$1.091e-05^{***}$	$2.746e-06$	3.974	$1.806e-05^{***}$	$4.296e-06$	4.204
	β_{C1}	$5.470e-01^{***}$	$2.577e-02$	21.229	$4.530e-01^{***}$	$2.569e-02$	17.636
	β_{J1}	$2.206e-01$	$1.372e-01$	1.608	$1.662e-01$	$1.525e-01$	1.090
	β_{C5}	$2.982e-01^{***}$	$2.925e-02$	10.194	$3.728e-01^{***}$	$3.029e-02$	12.309
	β_{J5}	$-4.184e-02$	$1.585e-01$	-0.264	$-1.709e-01$	$1.886e-01$	-0.906
	β_{C22}	$1.414e-01^{***}$	$2.450e-02$	5.773	$1.425e-01^{***}$	$2.717e-02$	5.245
	β_{J22}	$8.185e-02$	$3.025e-01$	0.271	$2.020e-01$	$3.380e-01$	0.598
		AdjustedR^2: 0.5324			Adjusted R^2: 0.5004		
PS	β_0	$1.135e-05^{***}$	$2.695e-06$	4.211	$1.981e-05^{***}$	$4.264e-06$	4.647
	β_1^-	$7.749e-01^{***}$	$4.964e-02$	15.611	$6.573e-01^{***}$	$5.150e-02$	12.764
	β_1^+	$2.636e-01^{***}$	$5.425e-02$	4.860	$2.331e-01^{***}$	$5.164e-02$	4.514
	β_5	$3.058e-01^{***}$	$2.839e-02$	10.774	$3.674e-01^{***}$	$2.962e-02$	12.404
	β_{22}	$1.446e-01^{***}$	$2.372e-02$	6.097	$1.391e-01^{***}$	$2.626e-02$	5.296
		AdjustedR^2:: 0.5408			Adjusted R^2: 0.5025		
PSlev	β_0	$1.151e-05^{***}$	$2.693e-06$	4.273	$2.004e-05^{***}$	$4.239e-06$	4.726
	β_1^-	$6.613e-01^{***}$	$7.940e-02$	8.329	$6.156e-01^{***}$	$7.809e-02$	7.883
	β_1^+	$3.165e-01^{***}$	$6.196e-02$	5.108	$2.487e-01^{***}$	$6.055e-02$	4.107
	γ	$5.566e-02^{\cdot}$	$3.138e-02$	1.774	$1.639e-02$	$2.946e-02$	0.556
	β_5	$3.084e-01^{***}$	$2.838e-02$	10.864	$3.691e-01^{***}$	$2.956e-02$	12.486
	β_{22}	$1.439e-01^{***}$	$2.369e-02$	6.074	$1.412e-01^{***}$	$2.613e-02$	5.403
		AdjustedR^2:: 0.5412			Adjusted R^2: 0.5033		
HAR-RS	β_0	$1.200e-05^{***}$	$2.707e-06$	4.434	$2.188e-05^{***}$	$4.375e-06$	5.0016
	β_1^-	$7.698e-01^{***}$	$4.958e-02$	15.526	$6.365e-01^{***}$	$5.210e-02$	12.218
	β_1^+	$2.617e-01^{***}$	$5.408e-02$	4.839	$2.263e-01^{***}$	$5.188e-02$	4.362
	β_5^-	$3.787e-01^{***}$	$7.414e-02$	5.108	$4.052e-01^{***}$	$7.754e-02$	5.225
	β_5^+	$2.114e-01^{*}$	$8.354e-02$	2.530	$3.467e-01^{***}$	$6.981e-02$	4.966
	β_{22}^-	$3.295e-01^{**}$	$1.285e-01$	2.564	$2.645e-01^{*}$	$1.340e-01$	1.974
	β_{22}^+	$-3.318e-02$	$1.300e-01$	-0.255	$5.627e-03$	$1.355e-01$	0.042

(continued)

Table 1. (*continued*)

ALLIANZ					BMW		
Model	Coef.	Estimate	Std. Error	t–value	Estimate	Std. Error	t value
		Adjusted R^2:: 0.5418			Adjusted R^2: 0.4955		
CG	β_0	$1.210e-05^{***}$	$2.708e-06$	4.468	$2.164e-05^{***}$	$4.374e-06$	4.947
	β_1^-	$7.664e-01^{***}$	$4.947e-02$	15.494	$6.433e-01^{***}$	$5.219e-02$	12.325
	β_1^+	$2.702e-01^{***}$	$5.398e-02$	5.005	$2.327e-01^{***}$	$5.193e-02$	4.481
	β_5^-	$3.768e-01^{***}$	$7.414e-02$	5.082	$3.978e-01^{***}$	$7.763e-02$	5.125
	β_1^+	$2.117e-01^{*}$	$8.358e-02$	2.533	$3.517e-01^{***}$	$6.995e-02$	5.027
	β_{22}^-	$3.320e-01^{**}$	$1.286e-01$	2.582	$2.613e-01^{*}$	$1.339e-01$	1.951
	β_{22}^+	$-3.538e-02$	$1.301e-01$	-0.272	$6.770e-03$	$1.355e-01$	0.050
	β_{J1}	$-2.939e-01$	$1.470e-01$	-2.000	$-3.129e-01^{*}$	$1.541e-01$	-2.031
		Adjusted R^2:: 0.5417			Adjusted R^2: 0.4962		
HAR-RV-SJ	β_0	$1.164e-05^{***}$	$2.701e-06$	4.311	$1.964e-05^{***}$	$4.269e-06$	4.602
	β_{C1}	$5.184e-01^{***}$	$2.535e-02$	20.445	$4.506e-01^{***}$	$2.551e-02$	17.662
	$\beta_{\Delta1}$	$-2.443e-01^{***}$	$4.476e-02$	-5.459	$-2.189e-01^{***}$	$4.430e-02$	-4.941
	β_5	$3.083e-01^{***}$	$2.848e-02$	10.824	$3.675e-01^{***}$	$2.957e-02$	12.427
	β_{22}	$1.460e-01^{***}$	$2.376e-02$	6.143	$1.393e-01^{***}$	$2.628e-02$	5.300
		Adjusted R^2:: 0.5398			Adjusted R^2: 0.5029		
HAR-RV-CSJ	β_0	$1.252e-05^{***}$	$2.712e-06$	4.616	$2.150e-05^{***}$	$4.387e-06$	4.901
	β_{C1}	$5.126e-01^{***}$	$2.535e-02$	20.222	$4.321e-01^{***}$	$2.591e-02$	16.673
	$\beta_{\Delta1}$	$-2.398e-01^{***}$	$4.457e-02$	-5.381	$-2.114e-01^{***}$	$4.444e-02$	-4.756
	β_{C5}	$3.027e-01^{***}$	$2.936e-02$	10.312	$3.867e-01^{***}$	$2.989e-02$	12.938
	$\beta_{\Delta5}$	$-9.126e-02$	$7.278e-02$	-1.254	$-2.175e-02$	$6.588e-02$	-0.330
	β_{C22}	$1.491e-01^{***}$	$2.419e-02$	6.164	$1.359e-01^{***}$	$2.663e-02$	5.103
	$\beta_{\Delta22}$	$-1.937e-01$	$1.272e-01$	-1.522	$-1.273e-01$	$1.317e-01$	-0.967
		Adjusted R^2:: 0.5404			Adjusted R^2: 0.4949		
HAR-RV-Sjd	β_0	$1.165e-05^{***}$	$2.700e-06$	4.314	$1.960e-05^{***}$	$4.225e-06$	4.638
	$\beta_{\Delta1}^-$	$-2.043e-01^{*}$	$8.746e-02$	-2.336	$-9.717e-02$	$8.792e-02$	-1.105
	$\beta_{\Delta1}^+$	$-2.882e-01^{*}$	$9.035e-02$	-3.190	$-3.530e-01^{***}$	$9.400e-02$	-3.755
	β_{C1}	$5.082e-01^{***}$	$3.111e-02$	16.334	$4.227e-01^{***}$	$3.002e-02$	14.081
	β_5	$3.095e-01^{***}$	$2.856e-02$	10.839	$3.701e-01^{***}$	$2.966e-02$	12.478
	β_{22}	$1.464e-01^{***}$	$2.377e-02$	6.159	$1.406e-01^{***}$	$2.606e-02$	5.397
		Adjusted R^2:: 0.5399			Adjusted R^2: 0.5049		
HAR-RV-CSjd	β_0	$1.251e-05^{***}$	$2.711e-06$	4.615	$2.181e-05^{***}$	$4.306e-06$	5.065
	$\beta_{\Delta1}^-$	$-2.895e-01^{**}$	$8.996e-02$	-3.219	$-8.182e-02$	$8.750e-02$	-0.935
	$\beta_{\Delta1}^+$	$-1.956e-01^{**}$	$8.710e-02$	-2.246	$-3.526e-01^{***}$	$9.435e-02$	-3.737
	β_{C1}	$5.010e-01^{***}$	$3.104e-02$	16.143	$4.024e-01^{***}$	$3.037e-02$	13.250
	$\beta_{\Delta5}^-$	$-1.754e-01$	$1.298e-01$	-1.351	$-1.316e-01$	$1.201e-01$	-1.096
	$\beta_{\Delta5}^+$	$7.065e-03$	$1.512e-01$	0.047	$5.078e-02$	$1.580e-01$	0.321
	β_{C5}	$2.917e-01^{***}$	$3.268e-02$	8.926	$3.980e-01^{***}$	$3.402e-02$	11.698
	$\beta_{\Delta22}^-$	$-2.296e-01$	$2.154e-01$	-1.066	$5.074e-02^{*}$	$2.386e-01$	0.213
	$\beta_{\Delta22}^+$	$-1.404e-01$	$2.960e-01$	-0.474	$-3.334e-01^{\cdot}$	$1.949e-01$	-1.711
	β_{C22}	$1.493e-01^{***}$	$2.657e-02$	5.621	$1.475e-01^{***}$	$2.826e-02$	5.218
		Adjusted R^2:: 0.5405			Adjusted R^2: 0.5025		

Table 2. Out of sample forecasting results for Allianz (top) and BMW (bottom). For each loss function the entries are: the ratio of the average loss for each model to the average loss of the HAR-RV model and the *p*-value of the MCS.

ALLIANZ

Model	QLIKE		MSE		AMSE		MAE		MAD		MSD	
	Loss	p-value	Loss	p-value	Loss	p-value	Loss	p-value	Loss	p-value	Loss	p-value
HAR-RV	1.0000	0.8632	1.0000	0.9593	1.0000	0.9613	1.0000	0.6241	1.0000	0.9597	1.0000	0.3926
HAR-RV-J	0.9957	0.9747	0.9997	0.9593	1.0001	0.9613	1.0003	0.4773	0.9982	0.9597	0.9997	0.7550
HAR-RV-CJ	0.9948	0.9747	0.9998	0.9593	0.9994	0.9613	1.0001	0.4773	0.9971	0.9597	0.9998	0.7550
TS	0.9949	0.9747	0.9981	0.9593	0.9826	0.9613	0.9987	0.3385	0.9894	0.9597	0.9981	0.3926
Pslev	0.9950	0.9747	0.9992	0.9593	0.9880	0.9613	1.0006	0.2762	0.9911	0.9597	0.9992	0.3926
HAR-RS	1.0096	0.7044	0.9873	0.9593	0.9805	0.9613	0.9844	0.8342	0.9848	0.9597	0.9873	0.7550
CG	1.0056	0.7423	0.9874	0.9593	0.9801	0.9613	0.9848	0.6241	0.9826	0.9597	0.9874	0.7550
HAR-RV-SJ	0.9904	1.0000	0.9978	0.9593	0.9827	0.9613	0.9990	0.2762	0.9870	0.9597	0.9978	0.3926
HAR-RV-CSJ	1.0006	0.8963	0.9854	1.0000	0.9794	1.0000	0.9835	1.0000	0.9800	1.0000	0.9854	1.0000
HAR-RV-Sjd	1.0087	0.0552	1.0067	0.2166	1.0173	0.2397	1.0119	0.0401	1.0115	0.0845	1.0067	0.0698
HAR-RV-CSjd	1.0165	0.0552	0.9920	0.3235	1.0182	0.3445	0.9955	0.4773	1.0042	0.3542	0.9920	0.7550

BMW

HAR-RV	1.0000	0.0394	1.0000	0.2297	1.0000	0.2214	1.0000	0.0540	1.0000	0.0414	1.0000	0.0518
HAR-RV-J	0.9967	0.0493	0.9998	0.2443	0.9981	0.2397	0.9996	0.0689	0.9988	0.0466	0.9998	0.0518
HAR-RV-CJ	0.9913	0.1356	0.9991	0.1409	1.0048	0.1358	1.0016	0.0689	1.0024	0.0380	0.9991	0.0697
TS	0.9734	0.1898	0.9873	0.5013	0.9692	0.5051	0.9861	0.3506	0.9668	0.7121	0.9873	0.3704
Pslev	0.9779	0.1898	0.9918	0.8159	0.9694	0.8289	0.9894	0.3506	0.9716	0.7121	0.9918	0.3704
HAR-RS	0.9746	0.1448	0.9825	0.8159	0.9622	0.8289	0.9807	0.8114	0.9618	0.7150	0.9825	0.6742
CG	0.9713	0.1898	0.9813	1.0000	0.9599	1.0000	0.9795	0.9010	0.9603	0.7150	0.9813	0.6742
HAR-RV-SJ	0.9680	0.2497	0.9868	0.5013	0.9679	0.5051	0.9854	0.3506	0.9658	0.7121	0.9868	0.3704
HAR-RV-CSJ	0.9662	0.2497	0.9827	0.8159	0.9654	0.8289	0.9823	0.7890	0.9629	0.7121	0.9827	0.5631
HAR-RV-Sjd	0.9661	0.2497	0.9853	0.8159	0.9654	0.8289	0.9840	0.4992	0.9626	0.7121	0.9853	0.3704
HAR-RV-CSjd	0.9586	1.0000	0.9788	0.8655	0.9611	0.8662	0.9790	1.0000	0.9570	1.0000	0.9788	1,0000

with their sign, the negative ones are insignificant for serie BMW and significant at 5% for series Allianz. Finally, in the HAR-RV-CSjd model, the signed jumps are insignificant or significant at 1%, in contrast the continuous component.

In the following of this section we analyse and compare the forecasting performances of the eleven different specifications (7)–(17) by looking at their ability to predict future values. To this aim, an initial subsample of length R is used to estimate the models, and for each of them the one-step-ahead out-of-sample forecast is produced. Here we choose $R = 2291$, so that the evaluation set has length 500. We assess the statistical significance of differences in the forecasting performance of the models by using the MCS proposed in [5]: it uses an arbitrary loss function whose choice depends on the nature of the candidate models. Here we consider the following six loss functions:

$$QLIKE = (T - R)^{-1} \sum_{t=R+1}^{T} \left[\frac{y_t}{\widehat{y}_t} - \log \frac{y_t}{\widehat{y}_t} - 1 \right], \qquad MSE = \frac{\sum_{t=R+1}^{T} (y_t - \widehat{y}_t)^2}{T - R},$$

$$AMSE = (T - R)^{-1} \sum_{t=R+1}^{T} \left[1 + \frac{(y_t - \widehat{y}_t)^2}{y_t} I_{(\widehat{y}_t < y_t)} \right] (y_t - \widehat{y}_t)^2, \quad MAE = \frac{\sum_{t=R+1}^{T} |y_t - \widehat{y}_t|}{T - R},$$

$$MAD = \frac{\sum_{t=R+1}^{T} (\sqrt{y_t} - \sqrt{\widehat{y}_t})^2}{T - R}, \qquad MSD = \frac{\sum_{t=R+1}^{T} |\sqrt{y_t} - \sqrt{\widehat{y}_t}|}{T - R}.$$

where y_t is the actual RV at day t and $\widehat{y}_t$ is the forecast of RV. Table 2 shows the ratio of the average loss for each model to the average loss of the HAR-RV model and the p-value of the MCS procedure. We use the confidence level of 90%. For Allianz, in the MCS with the QLIKE loss function, the HAR-RV-Sjd and the HAR-RV-CSjd models have worse performance than all the other models. The model which has the smallest average loss function is the HAR-RV-Sj model. The MSE and AMSE loss functions seem not able to discriminte among the proposed models. Moreover the MCS has the same structure for MAE, MAD and MSD where the model HAR-RV-Sjd is always eliminated from the set of superior models and the HAR-CSJ model presents the smallest average loss function.

For BMW series in the MCS with the QLIKE loss function, also in this case the HAR-RV and the HAR-RV-J models have worse forecasting performances. The model which has the smallest average loss function is the HAR-RV-CSjd model. The MSE and AMSE loss functions seem not to be able to discriminte among the proposed models. Moreover, the MCS has the same structure for MAE, MAD and MSD where the HAR-RV, HAR-RV-J and HAR-RV-CJ models are always eliminated from the set of superior models and again the HAR-RV-CSjd with the smallest average loss function.

4   Concluding Remarks

The paper focuses on eleven HAR-type models to describe realized volatility dynamics. To the aim of comparing the out-of-sample performances of the considered models, the Model Confidence Set procedure has been implemented with different specifications of the loss functions. The results give evidence that, for both the considered datasets, the models with better forecasting performances are those introducing the signed jump as esplicative variable. Moreover, as expected, the QLIKE loss function seems to better discriminate among models, while when MSE and AMSE are used the MCS includes almost all the considered models. Finally, MAE, MAD and MSD lead to the same composition of the MCS.

References

1. Corsi, F.: A simple approximate long-memory model of realized volatility. J. Fin. Econom. **7**(2), 174–196 (2009)
2. Giot, P., Lauren, S., Petitjeanad, M.: Trading activity, realized volatility and jumps. J. Empir. Finance **17**(1), 168–175 (2010)
3. Sevi, B.: Forecasting the volatility of crude oil futures using intraday data. Eur. J. Oper. Res. **235**, 643659 (2014)
4. Wang, Y., Ma, F., Wei, Y., Wu, C.: Forecasting realized volatility in a changing world: a dynamic model averaging approach. J. Bank. Finance **64**, 136–149 (2016)
5. Hansen, P.R., Lunde, A., Nason, J.M.: The model confidence set. Econometrica **79**(2), 453–497 (2011)
6. Barndorff-Nielsen, O.E., Shephard, N.: Power and bipower variation with stochastic volatility and jumps. J. Financial Econom. **2**(1), 1–37 (2004)

On the Successive Passage Times of Certain One-Dimensional Diffusions

Mario Abundo$^{(\boxtimes)}$ and Maria Beatrice Scioscia Santoro

Tor Vergata University, Rome, Italy
abundo@mat.uniroma2.it, beatricescioscia@gmail.com

Abstract. We study the distribution of the nth-passage time of a one-dimensional diffusion, obtained by a space or time transformation of Brownian motion, through a constant barrier a. Some explicit examples are reported.

Keywords: First-passage time · Second-passage time · One-dimensional diffusion

Mathematics Subject Classification: 60J60 · 60H05 · 60H10

1 Introduction

This is a continuation of [2], where we studied excursions of Brownian motion with drift, and we found explicitly the distribution of the nth-passage time of Brownian motion (BM) through a linear boundary. Here, we study more in depth the nth-passage time of a one-dimensional diffusion process $X(t)$, which is obtained from BM by a space transformation or a time change. Particularly, focusing on the second passage-time, we find explicit formulae for the density and the Laplace transforms of the second-passage time (for more, see [1]). The successive-passage times of a diffusion $X(t)$ through a boundary $S(t)$ are related to the excursions of $Y(t) := X(t) - S(t)$; indeed, when $Y(t)$ is entirely positive or entirely negative on the time interval (s, u), it is said that it is an excursion of $Y(t)$. Excursions have interesting applications in Biology, Economics, and other applied sciences (see e.g. [2,15]). They are also related to the last passage time of a diffusion through a boundary; actually, last passage times of continuous martingales play an important role in Finance, for instance, in models of default risk (see e.g. [7,9]). In the special case when $X(t)$ is BM, its excursions follow the arcsine law, namely the probability that BM has no zeros in the time interval (s, u) is $\frac{2}{\pi} \arcsin \sqrt{s/u}$ (see e.g. [11]). Really, Salminen's formula for the last passage time of BM through a linear boundary (see [14]) was used in [2] to find the law of the excursions of drifted BM, and the distribution of the nth passage time of BM through a linear boundary; in this article, we obtain analogous results for transformed BM through a constant boundary.

© Springer Nature Switzerland AG 2020
R. Moreno-Díaz et al. (Eds.): EUROCAST 2019, LNCS 12013, pp. 195–203, 2020.
https://doi.org/10.1007/978-3-030-45093-9_24

2 Preliminary Results on the nth-Passage Time of BM

The first-passage time of BM starting from x, through the linear boundary $S(t) = a + bt$, is $\tau_1^B(S|x) = \inf\{t > 0 : x + B_t = a + bt\}$, and will be denoted in this section by $\tau_1(x)$, dropping, for simplicity, the superscript B which refers to BM and the dependence on S; we recall the Bachelier-Levy formula for the distribution of $\tau_1(x)$:

$$P(\tau_1(x) \leq t) = 1 - \Phi((a-x)/\sqrt{t} + b\sqrt{t}) + exp(-2b(a-x))\Phi(b\sqrt{t} - (a-x)/\sqrt{t}), \ t > 0$$

where $\Phi(y) = \int_{-\infty}^{y} \phi(z)dz$, with $\phi(z) = e^{-z^2/2}/\sqrt{2\pi}$, is the cumulative distribution function of the standard Gaussian variable. If $(a - x)b > 0$, then $P(\tau_1(x) < \infty) = e^{-2b(a-x)}$, whereas, if $(a - x)b \leq 0$, $\tau_1(x)$ is a.s finite and it has the Inverse Gaussian density, which is non-defective (see e.g. [6,10]):

$$f_{\tau_1}(t) = f_{\tau_1}(t|x) = \frac{d}{dt}P(\tau_1(x) \leq t) = \frac{|a - x|}{t^{3/2}} \ \phi\left(\frac{a + bt - x}{\sqrt{t}}\right), \ t > 0; \qquad (1)$$

moreover, if $b \neq 0$, the expectation of $\tau_1(x)$ is finite, being $E(\tau_1(x)) = \frac{|a-x|}{|b|}$.

The second-passage time of BM starting from x through $S(t)$, is $\tau_2(x) := \tau_2^B(S|x) = \inf\{t > \tau_1(x) : x + B_t = a + bt\}$, and generally, for $n \geq 2$, $\tau_n(x) = \inf\{t > \tau_{n-1}(x) : x + B_t = a + bt\}$ denotes the nth-passage time of BM through $S(t)$. We assume that $b \leq 0$ and $x \leq S(0) = a$, or $b \geq 0$ and $x \geq a$, so that $P(\tau_1(x) < \infty) = 1$. Then, for fixed $t > 0$, we consider a kind of *conditional last-passage time prior to t* of BM through $S(t)$:

$$\lambda_S^t = \begin{cases} \sup\{0 \leq u \leq t : x + B_u = S(u)\} & \text{if } \tau_1(x) \leq t \\ 0 & \text{if } \tau_1(x) > t. \end{cases} \qquad (2)$$

The following explicit result holds (see [2]):

Theorem 1. *Let be $\psi_t(u) = \frac{d}{du}P(\lambda_S^t \leq u|\tau_1(x) \leq t)$, the conditional density of λ_S^t, under the condition $\tau_1(x) \leq t$. Then, for $0 < u < t$:*

$$\psi_t(u) = \frac{e^{-\frac{b^2}{2}u}}{\pi\sqrt{u(t-u)}}\left[e^{-\frac{b^2}{2}(t-u)} + \frac{b}{2}\sqrt{2\pi(t-u)}\left(2\Phi(b\sqrt{t-u}) - 1\right)\right]. \qquad (3)$$

Notice that ψ_t is independent of a; if $b = 0$, one gets $\psi_t(u) = \frac{1}{\pi\sqrt{u(t-u)}}$, $0 < u < t$, that is, the arc-sine law with support in $(0, t)$. For $x < a$ and $b \leq 0$, we set $T_1(x) = \tau_1(x)$ and $T_2(x) = \tau_2(x) - \tau_1(x)$. It can be proved (see [2]) that $T_2(x)$ is finite with probability one, only if $b = 0$. Moreover:

$$P(\tau_2(x) > \tau_1(x) + t|\tau_1(x) = s) = P(\lambda_S^{s+t} \leq s|\tau_1(x) \leq s + t) = \int_0^s \psi_{s+t}(y)dy$$

and so

$$P(T_2(x) \le t | \tau_1(x) = s) = 1 - \int_0^s \psi_{s+t}(y) dy. \tag{4}$$

Then:

$$P(T_2(x) \le t) = 1 - \int_0^{+\infty} f_{\tau_1}(s) ds \int_0^s \psi_{s+t}(y) dy. \tag{5}$$

By taking the derivative with respect to t, one gets the density of $T_2(x)$:

$$f_{T_2}(t) = \int_0^{+\infty} f_{\tau_1}(s) \left[e^{-b^2(s+t)/2} \frac{\sqrt{s}}{\pi \sqrt{t(s+t)}} \right] ds. \tag{6}$$

Notice that $\int_0^s \psi_{s+t}(y) dy$ is decreasing in t and $f_{T_2}(t) \sim const/\sqrt{t}$, as $t \to 0^+$. Moreover, from (5) it follows that, if $b \ne 0$, the distribution of $T_2(x)$ is defective. In fact (see [2]), $P(T_2(x) = +\infty) = 2\,\mathrm{sgn}(b)E\left(\varPhi(b\sqrt{\tau_1(x)}) - \frac{1}{2}\right) > 0$. On the contrary, if $b = 0$, from the formula above we get $P(T_2(x) = +\infty) = 0$, that is, $T_2(x)$ is a proper random variable, and also $\tau_2(x) = T_2(x) + \tau_1(x)$ is a.s finite; precisely, by calculating the integral in (5) we have:

$$P(T_2(x) \le t) = \int_0^{+\infty} \frac{2}{\pi} \arccos \sqrt{\frac{s}{s+t}} \frac{|a-x|}{\sqrt{2\pi s^{3/2}}} e^{-(a-x)^2/2s} ds. \tag{7}$$

Then, taking the derivative with respect to t, the density of $T_2(x)$ is:

$$f_{T_2}(t) = \int_0^{+\infty} \frac{1}{\pi(s+t)\sqrt{t}} \frac{|a-x|}{\sqrt{2\pi s}} e^{-(a-x)^2/2s} ds. \tag{8}$$

The expectation of $\tau_2(x)$ is obviously infinite for $b \ne 0$, while $E(\tau_1(x))$ is finite. If $b = 0$, $E(\tau_1(x))$ and $E(\tau_2(x))$ are both infinite; for $\tau_1(x)$ this is a well-known fact, for $\tau_2(x)$ it is a consequence of $\tau_2(x) > \tau_1(x)$. By taking the derivative with respect to t in (4), one obtains the density of $T_2(x)$ conditional to $\tau_1(x) = s$, that is:

$$f_{T_2|\tau_1}(t|s) = -\frac{d}{dt} \int_0^s \psi_{s+t}(y) dy = e^{-b^2(s+t)/2} \frac{\sqrt{s}}{\pi(s+t)\sqrt{t}}, \tag{9}$$

and, for $b = 0$:

$$f_{T_2|\tau_1}(t|s) = \frac{\sqrt{s}}{\pi(s+t)\sqrt{t}}. \tag{10}$$

Since $\tau_2(x) = \tau_1(x) + T_2(x)$, by the convolution formula, the density of $\tau_2(x)$ follows:

$$f_{\tau_2}(t) = \int_0^t f_{T_2|\tau_1}(t-s|s) f_{\tau_1}(s) ds = \frac{e^{-b^2 t/2}}{\pi t} \int_0^t \frac{|a-x|}{\sqrt{2\pi s}\sqrt{t-s}} e^{-(a+bs-x)^2/2s} ds. \tag{11}$$

Of course, the distribution of $\tau_2(x)$ is defective for $b \ne 0$, namely $\int_0^{+\infty} f_{\tau_2}(t) dt = 1 - P(\tau_2(x) = +\infty) < 1$, since $P(\tau_2(x) = +\infty) = P(T_2(x) = +\infty) > 0$.

If $b = 0$, we obtain the non-defective density:

$$f_{\tau_2}(t) = \frac{1}{\pi t} \int_0^t \frac{|a - x|}{\sqrt{2\pi s}\sqrt{t - s}} e^{-(a-x)^2/2s} ds, \tag{12}$$

Proposition 1 (see [2]). *Let be* $T_1(x) = \tau_1(x)$, $T_n(x) = \tau_n(x) - \tau_{n-1}(x)$, $n = 2, \ldots$ *Then:*

$$P(T_1(x) \le t) = 2(1 - \Phi(a - x/\sqrt{t})), \tag{13}$$

$$P(T_n(x) \le t) = 1 - \int_0^{+\infty} f_{\tau_{n-1}}(s)ds \int_0^s \psi_{s+t}(y)dy, \ n = 2, \ldots \tag{14}$$

Moreover, the density of $\tau_n(x)$ *is:*

$$f_{\tau_n}(t) = \int_0^t f_{T_n|\tau_{n-1}}(t - s|s)f_{\tau_{n-1}}(s)ds, \tag{15}$$

where $f_{\tau_{n-1}}$ *and* $f_{T_n|\tau_{n-1}}$ *can be calculated inductively, in a similar way, as done for* f_{τ_2} *and* $f_{T_2|\tau_1}$. *If* $b = 0$, $T_1(x)$, $T_2(x)$, ... *are finite with probability one.* □

For the explicit forms of the Laplace transforms (LT) of first and second passage-time of B_t through the boundary $S(t) = a + bt$, see [1].

3   The nth-Passage Time of a Diffusion Conjugated to BM

Let be $X(t)$ a one-dimensional, time-homogeneous diffusion process, driven by the SDE $dX(t) = \mu(X(t))dt + \sigma(X(t))dB_t$, $X(0) = x$, where the coefficients $\mu(\cdot)$ and $\sigma(\cdot)$ satisfy the usual conditions for the existence and uniqueness of the solution (see e.g. [8]). We recall that $X(t)$ is said to be *conjugated to BM* (see [5]), if there exists an increasing function v with $v(0) = 0$, such that:

$$X(t) = v^{-1}\left(B_t + v(x)\right), \ t \ge 0. \tag{16}$$

Examples of diffusions conjugated to BM are the Feller process or Cox-Ingersoll-Ross (CIR) model, and the Wright and Fisher-like process (see [5]). Let $X(t)$ be conjugated to BM via the function v, and let $\tau_1(a|x) = \inf\{t > 0 : X(t) = a\}$ be the FPT of $X(t)$ through the level $a > x$; as easily seen, one has $\tau_1(a|x) = \tau_1^B(v(a)|v(x))$, where $\tau_1^B(\alpha|y)$, denotes the FPT of BM, starting from $y < \alpha$, through the level α. Then, from (1), one gets the density of $\tau_1(a|x)$:

$$f_{\tau_1}(t) = \frac{v(a) - v(x)}{t^{3/2}} \phi\left(\frac{v(x) - v(a)}{\sqrt{t}}\right), \tag{17}$$

implying that $E\left[\tau_1(a|x)\right] = +\infty$. For $x < a$, one has:

$$\tau_n(a|x) = \inf\{t > \tau_{n-1}(a|x) : B_t = v(a) - v(x)\} = \tau_n^B(v(a)|v(x)),$$

where the superscript B refers to BM, namely $\tau_n^B(a|x)$, $n = 2, \ldots$ is the nth-passage time of BM through the level a, when starting from $x < a$. As in the previous section, we set $T_1(x) = \tau_1(a|x) = \tau_1^B(v(a)|v(x))$, $T_2(x) = \tau_2(a|x) - \tau_1(a|x) = \tau_2^B(v(a)|v(x)) - \tau_1^B(v(a)|v(x))$; from (7) $(b = 0)$, we get

$$P\left(T_2(x) \le t\right) = \frac{2}{\pi} \int_0^{+\infty} f_{\tau_1}(s) \arccos \sqrt{\frac{s}{s+t}} \, ds. \qquad (18)$$

Thus, the density of T_2 is $f_{T_2}(t)$

$$= \int_0^{+\infty} f_{\tau_1}(s) \frac{\sqrt{s}}{\pi\sqrt{t(s+t)}} \, ds = \int_0^{+\infty} \frac{v(a) - v(x)}{s^{3/2}} \frac{e^{-\frac{(v(x)-v(a))^2}{2s}}}{\sqrt{2\pi}} \frac{\sqrt{s}}{\pi\sqrt{t(s+t)}} \, ds$$

$$= \frac{v(a) - v(x)}{\pi\sqrt{2\pi}} \int_0^{+\infty} \frac{1}{(s+t)\sqrt{t}} \frac{1}{s} e^{-\frac{(v(x)-v(a))^2}{2s}} \, ds. \qquad (19)$$

The expectation of $\tau_2(a|x)$ is infinite, because $E\left[T_2(x)\right] = +\infty$. From (9) with $b = 0$, we get

$$f_{T_2|\tau_1}(t|s) = -\frac{d}{dt} \int_0^s \psi_{s+t}(y) \, dy = \frac{\sqrt{s}}{\pi(s+t)\sqrt{t}}. \qquad (20)$$

Moreover, from (12) $(b = 0)$:

$$f_{\tau_2}(t) = \int_0^t \frac{v(a) - v(x)}{\pi t s \sqrt{t-s}\sqrt{2\pi}} \cdot e^{-\frac{(v(x)-v(a))^2}{2s}} \, ds. \qquad (21)$$

Setting $T_n(x) = \tau_n(a|x) - \tau_{n-1}(a|x)$, and using Proposition 1, we get

$$P\left(T_1(x) \le t\right) = 2\left(1 - \Phi\left(\frac{v(a) - v(x)}{\sqrt{t}}\right)\right), \qquad (22)$$

$$P\left(T_n(x) \le t\right) = 1 - \int_0^{+\infty} f_{\tau_{n-1}}(s) \, ds \int_0^s \psi_{s+t}(y) \, dy, \quad n = 2, \ldots. \qquad (23)$$

The density of $\tau_n(a|x)$ is:

$$f_{\tau_n}(t) = \int_0^t f_{T_n|\tau_{n-1}}(t - s|s) \cdot f_{\tau_{n-1}}(s) \, ds, \qquad (24)$$

where $f_{\tau_{n-1}}$ and $f_{T_n|\tau_{n-1}}$ can be calculated inductively, in a similar way, as done for f_{τ_2} and $f_{T_2|\tau_1}$. In conclusion, $T_1(x), T_2(x), \ldots$ are finite with probability one.

4   The nth-Passage Time of Time-Changed BM

The previous arguments can be applied also to time-changed BM, namely $X(t) = x + B(\rho(t))$, where $\rho(t) \ge 0$ is an increasing, differentiable function of $t > 0$, with $\rho(0) = 0$. Such kind of diffusion process $X(t)$ is a special case of Gauss-Markov process (see [3,4]). Let us consider the constant barrier $S = a$, and $x < a$; then, the FPT of $X(t)$ through a is $\tau_1(a|x) = \rho^{-1}(\tau_1^B(a|x))$, where $\tau_1^B(a|x)$ denotes the first-passage time of BM, starting from x, through the barrier a. Thus, the density of $\tau_1(a|x)$ is:

$$f_{\tau_1}(t) = f_{\tau_1^B}(\rho(t))\rho'(t) = \frac{|a-x|}{\rho(t)^{3/2}} \phi\left(\frac{a-x}{\sqrt{\rho(t)}}\right)\rho'(t). \tag{25}$$

The expectation is:

$$E\left[\tau_1(a|x)\right] = \int_0^{+\infty} t \, f_{\tau_1}(t) \, dt = \int_0^{+\infty} t \frac{|a-x| \cdot \rho'(t)}{\rho(t)^{3/2}} \frac{e^{-\frac{(x-a)^2}{2\rho(t)}}}{\sqrt{2\pi}} \, dt \tag{26}$$

Suppose that $\rho(t) \sim ct^\alpha$, as $t \to +\infty$; then, as easily seen, for $\alpha > 2$ the integral converges, namely $E\left[\tau_1(a|x)\right] < +\infty$, unlike the case of BM, for which the expectation of FPT is infinite. One has $\tau_n(a|x) = \rho^{-1}(\tau_n^B(a|x))$ with $\tau_n^B(a|x) = \inf\{s > \tau_{n-1}^B(a|x) : x + B_s = a\}$; $T_1(x) = \tau_1(a|x) = \rho^{-1}(\tau_1^B(a|x))$, $T_2(x) = \tau_2(a|x) - \tau_1(a|x) = \rho^{-1}(\tau_2^B(a|x)) - \rho^{-1}(\tau_1^B(a|x))$. The density of $\tau_2(a|x)$ is $f_{\tau_2}(t) = \frac{d}{dt}P\left(\tau_2^B(a|x) \le \rho(t)\right) = f_{\tau_2^B}(\rho(t)) \cdot \rho'(t)$

$$= \frac{\rho'(t)}{\pi\rho(t)} \int_0^{\rho(t)} \frac{|a-x|}{\sqrt{2\pi s}\sqrt{\rho(t)-s}} e^{-\frac{(a-x)^2}{2s}} \, ds. \tag{27}$$

The density of $\tau_n(a|x)$ is $f_{\tau_n}(t) = \frac{d}{dt}P\left(\tau_n^B(a|x) \le \rho(t)\right)$

$$= f_{\tau_n^B}(\rho(t))\rho'(t) = \int_0^{\rho(t)} f_{T_n^B|\tau_{n-1,B}}(\rho(t)-s|s)f_{\tau_{n-1}^B}(s) \, ds \cdot \rho'(t), \tag{28}$$

where $f_{T_n^B|\tau_{n-1}^B}$ and $f_{\tau_{n-1}^B}$ can be calculated inductively.

5   Some Examples

(i) $B(t^3/3)$ **and integrated BM.** Let be $X(t) = \int_0^t B_s \, ds$ the integrated BM; it has many important application (see e.g. [3,4]), and is a Gaussian process with mean zero and covariance function $cov(X(t), X(s)) = s^2(t/2 - s/6)$, $s \le t$ (see e.g. [13]). Thus, its variance is $Var(X(t)) = t^3/3$, and $X(t)$ has the same distribution as $B(\frac{t^3}{3})$. Motivated from this, we study the successive-passage time of $B(\frac{t^3}{3})$ through a constant barrier a. Since the starting point is zero, we consider $\tau_1(a|0) = \inf\{t > 0 : X(t) = a\} = \inf\{t > 0 : B(t^3/3) = a\}$, $\tau_2(a|0) = \inf\{t > \tau_1(a|0) : B(t^3/3) = a\}$, and $T_1(x) = \tau_1(a|0)$, $T_2(x) = \tau_2(a|0) - \tau_1(a|0)$. Then, by using (25), we get the density of $\tau_1(a|0)$:

$$f_{\tau_1}(t) = \frac{|a-x|}{(t^3/3)^{3/2}} \cdot \phi\left(\frac{a-x}{\sqrt{t^3/3}}\right) t^2 = \frac{3\sqrt{3}\,|a-x|}{t^{5/2}} \frac{e^{-\frac{3(a-x)^2}{2t^3}}}{\sqrt{2\pi}}. \tag{29}$$

In the Fig. 1 (panel a) we report the FPT density of $B(\frac{t^3}{3})$ obtained from (29), for various values of the barrier: $a = 1$, $a = 2$, $a = 3$. From (26) we get that

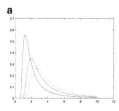 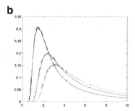

Fig. 1. Comparison between the shapes of the FPT density of $B(t^3/3)$, for different values of the barrier a; from top to the bottom, with respect to the peak of the curve: $a = 1, a = 2, a = 3$ (panel (a)); density of $\tau_2(a|x)$ in the case of $B(t^3/3)$, for various values of the parameter a; from top to the bottom: $a = 1, a = 2, a = 3$ (panel (b)).

$E\left[\tau_1(a|0)\right] = \frac{3\sqrt{3}}{\sqrt{2\pi}}\frac{|a-x|}{\int_0^{+\infty}}\frac{e^{-\frac{3(a-x)^2}{2s}}}{t^{3/2}}\,dt$. Notice that, unlike the case of BM, it is finite (in fact $\rho(t) = t^3/3 \sim ct^\alpha$ with $\alpha = 3 > 2$, see the observation after Eq. (26)). By using (27), one gets the density of $\tau_2(a|x)$:

$$f_{\tau_2}(t) = \frac{3}{\pi t}\int_0^{\frac{t^3}{3}}\frac{|a-x|}{\sqrt{2\pi}s\sqrt{\frac{t^3}{3}-s}}e^{-\frac{(a-x)^2}{2s}}\,ds. \tag{30}$$

The shape of the density of $\tau_2(a|x)$, for different values of the parameter a, is shown in the Fig. 1 (panel b). The Laplace transform of $\tau_2(a|x)$ is

$$\tilde{\psi}(\lambda) = \int_0^{+\infty}e^{-\lambda t}f_{\tau_2}(t)dt = \frac{3|a-x|}{\pi\sqrt{2\pi}}\int_0^{+\infty}\frac{e^{-\lambda t}}{t}[\int_0^{\frac{t^3}{3}}\frac{e^{-\frac{(a-x)^2}{2s}}}{s\sqrt{\frac{t^3}{3}-s}}ds]dt, \ \lambda > 0.$$

$$E(\tau_2(a|0)) = -\frac{d}{d\lambda}\tilde{\psi}(\lambda)\Big|_{\lambda=0} = \frac{3|a|}{\pi\sqrt{2\pi}}\int_0^{+\infty}\left[\int_0^{\frac{t^3}{3}}\frac{e^{-\frac{a^2}{2s}}}{s\sqrt{\frac{t^3}{3}-s}}ds\right]dt.$$

(ii) The Ornstein-Uhlenbeck (OU) process. It is the solution of $dX(t) = -\mu X(t)dt + \sigma dB(t)$, $X(0) = x$, with $\mu, \sigma > 0$. Explicitly, $X(t) = e^{-\mu t}\left(x + \int_0^t \sigma e^{\mu s}\,dB(s)\right)$. Moreover, (see e.g. [5]) $X(t)$ can be written as $e^{-\mu t}(x + B(\rho(t)))$, where $\rho(t) = \frac{\sigma^2}{2\mu}(1 - e^{-2\mu t})$, namely in terms of time-changed BM. Therefore, $X(t)$ has normal distribution with mean $xe^{-\mu t}$ and variance $\frac{\sigma^2}{2\mu}(1 - e^{-2\mu t})$. We choose the time-varying barrier $S(t) = ae^{-\mu t}$, with $x < a$, and we reduce to the passage-times of BM through the constant barrier a. In fact, $\tau_1(S|x) = \inf\{t > 0 : X(t) = S(t)\} = \inf\{t > 0 : x + B(\rho(t)) = a\}$ and $\tau_1(S|x) = \rho^{-1}\left(\tau_1^B(a|x)\right)$, where $\tau_1^B(a|x)$ is the FPT of BM, through the barrier a. Moreover, $\tau_2(S|x) = \rho^{-1}\left(\tau_2^B(a|x)\right),\ldots,\tau_n(S|x) = \rho^{-1}\left(\tau_n^B(a|x)\right)$, where $\tau_n^B(a|x)$ are the successive-passage times of BM through a. Thus, by using (25), we get the density of $\tau_1(S|x)$:

$$f_{\tau_1}(t) = \frac{2\mu\sqrt{2\mu}|a-x|}{\sigma\left(e^{2\mu t}-1\right)^{3/2}}\frac{1}{\sqrt{2\pi}}e^{-\frac{\mu(a-x)^2}{\sigma^2(e^{2\mu t}-1)}+2\mu t}. \tag{31}$$

In the Fig. 2, panel (a), we report the density of $\tau_1(S|x)$, for $x = 0$, $\mu = 2$, $\sigma = 0.2$, and various values of a. The expectation of $\tau_1(a|x)$ is $E[\tau_1(S|x)] = \frac{|a-x| \cdot 2\mu\sqrt{\mu}}{\sigma\sqrt{\pi}} \cdot \int_0^{+\infty} te^{-\frac{(x-a)^2 \cdot \mu}{\sigma^2 \cdot (e^{2\mu t}-1)}+2\mu t}\, dt$. By (27), we obtain the density of τ_2;

$$f_{\tau_2}(t) = \frac{2\mu|a-x|e^{2\mu t}}{\pi\sqrt{2\pi} \cdot (e^{2\mu t}-1)} \cdot \int_0^{\frac{\sigma^2}{2\mu}(e^{2\mu t}-1)} \frac{e^{-\frac{(a-x)^2}{2s}}}{s\sqrt{\frac{\sigma^2}{2\mu}(e^{2\mu t}-1)-s}}\, ds. \tag{32}$$

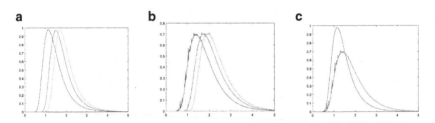

Fig. 2. Panel (a): FPT density of OU through the boundary $S(t) = ae^{-\mu t}$, for $x = 0$, $\mu = 2$, $\sigma = 0.2$ and from top to the bottom: $a = 1, 2, 3$. Panel (b): density of $\tau_2(S|x)$ for $x = 0$, $\mu = 2$, $\sigma = 0.2$ and from top to the bottom: $a = 1, 2, 3$. Panel (c): comparison between the density (32) of $\tau_2(S|x)$ (lower peak) and the density (31) of $\tau_1(S|x)$ (upper peak), for $x = 0$, $\mu = 2$, $\sigma = 0.2$ and $a = 1$.

Its shape is shown in the Fig. 2, panel (b), for $x = 0$, $\sigma = 0.2$, $\mu = 2$ and various values of the parameter a. The LT of $\tau_2(S|x)$ and its expectation cannot be explicitly found, because of the complexity of the integrals involved in the calculation; so they have been estimated by Monte Carlo simulation (see [1]).

References

1. Abundo, M., Scioscia Santoro, M.B.: On the successive passage times of certain one-dimensional diffusions. Preprint 2018 arXiv:1803.09910 (2018)
2. Abundo, M.: On the excursions of drifted Brownian motion and the successive passage times of Brownian motion. Physica A **457**, 176–182 (2016). https://doi.org/10.1016/j.physa.2016.03.052
3. Abundo, M.: On the representation of an integrated Gauss-Markov process. Sci. Math. Jpn. **77**, 719–723 (2013). Online e-2013
4. Abundo, M.: On the first-passage time of an integrated Gauss-Markov process. Sci. Math. Jpn. **28**, 1–14 (2015). Online e-2015
5. Abundo, M.: An inverse first-passage problem for one-dimensional diffusions with random starting point. Stat. Probab. Lett. **82**(1), 7–14 (2012)
6. Abundo, M.: Some results about boundary crossing for Brownian motion. Ricerche di Matematica **50**(2), 283–301 (2001)
7. Elliot, R.J., Jeanblanc, M., Yor, M.: On models of default risk. Math. Finance **10**, 179–196 (2000)

8. Ikeda, N., Watanabe, S.: Stochastic Differential Equations and Diffusion Processes. North-Holland Publishing Company, Amsterdam (1981)
9. Jeanblanc, M., Yor, M., Chesney, M.: Mathematical Methods for Financial Markets. Springer, London (2009). https://doi.org/10.1007/978-1-84628-737-4
10. Karatzas, I., Shreve, S.: Brownian Motion and Stochastic Calculus. Springer, New York (1998)
11. Klebaner, F.C.: Introduction to Stochastic Calculus with Applications, 2nd edn. Imperial College Press, London (2005)
12. Revuz, D., Yor, M.: Continous Martingales and Brownian Motion. Springer, Heidelberg (1991). https://doi.org/10.1007/978-3-662-06400-9
13. Ross, S.M.: Introduction to Probability Models, 10th edn. Academic Press, Elsevier, Burlington (2010)
14. Salminen, P.: On the first hitting time and the last exit time for a Brownian motion to/from a moving boundary. Adv. Appl. Probab. **20**(2), 411–426 (1988)
15. Schumpeter, J.A.: Business Cycles: A Theoretical Historical and Statistical Analysis of the Capitalist Process. McGraw-Hill, New York (1939)

Diffusion Processes for Weibull-Based Models

Antonio Barrera[1]([✉]), Patricia Román-Román[2], and Francisco Torres-Ruiz[2]

[1] Departamento de Análisis Matemático, Estadística e Investigación Operativa y
Matemática Aplicada, Facultad de Ciencias, Universidad de Málaga,
Bulevar Louis Pasteur 31, Campus de Teatinos, 29010 Málaga, Spain
antonio.barrera@uma.es
[2] Departamento de Estadística e Investigación Operativa,
Instituto de Matemáticas (IEMath-GR), Facultad de Ciencias s/n,
Universidad de Granada, Campus de Fuentenueva, 18071 Granada, Spain
{proman,fdeasis}@ugr.es

Abstract. A stochastic diffusion process based on a generalization of
the Weibull curve, depending on a parametric function and initial values, is considered. One particular case is the hyperbolastic curve of type
III, successfully applied in growth dynamics. We also proposed a basic
methodology to address the issue related with the estimation of the
parameters of the generic function from a discrete sampling of observations.

Keywords: Diffusion process · Weibull curve · Hyperbolastic model

1 Introduction

Description and modelling of biological dynamic phenomena such as growth is
well done by means of differential equations. In the particular, growth curves
lead to interesting models which are applied in research fields such as biology,
medicine, ecology or finance.

Nevertheless, these deterministic models do not concern about influence from
unknown sources, as well as the inherent random nature of other factors. In order
to obtain more accurate and realistic models, all of these elements must be taken
into account.

Stochastic diffusion processes have been used for years to construct models
able to describe deterministic dynamics influenced by random, external factors.
By controlling certain characteristics of the process, they can be designed and
adapted to the behaviour of data (see, for example, [1,4]).

There exist diverse approaches to build stochastic models to describe dynamical behaviour, for example by adding a random term into a deterministic model.

This work was supported in part by the Ministerio de Economía y Competitividad,
Spain, under Grant MTM2017-85568-P.

© Springer Nature Switzerland AG 2020
R. Moreno-Díaz et al. (Eds.): EUROCAST 2019, LNCS 12013, pp. 204–210, 2020.
https://doi.org/10.1007/978-3-030-45093-9_25

Other approach consists on define the model in a way such that its trend follows a deterministic curve (examples in [2,4]).

Apart from issues related with these aspects, the number of growth curves and the complexity of the models are increasing, leading to difficulties to apply them in an appropriate context. Many of these special models are indeed particular cases of classical growth curves. Weibull curve and models based on it are of particular interest due to its capacity to adapt and describe growth behaviour. One of these modifications is the hyperboslatic curve of type III [7]. This curve, as well as the other two types of hyperbolastic, have been successfully applied in different fields, showing great performance in the description of biological growth processes such as cell growth, among others (see [3,5,6] and references therein).

In this work we address the problem related with the complexity of growth models, by considering a generalization of the Weibull curve which leads to a stochastic model able to reproduce Weibull-based growth behaviour. The hyperbolastic model of type III could be viewed as a particular case of the general model. As a result of this, we proposed a methodology to obtain an estimation of the intrinsic parameters of the unknown function which generalizes the model.

2 Generalized Weibull Model

A generalization of the classical Weibull model can be approached by solving the following ordinary differential equation,

$$\frac{dx(t)}{dt} = (\alpha - x(t)) \, g(t) \tag{1}$$

with initial condition $x(t_0) = x_0 \in \mathbb{R}$ at times $t \geq t_0$, and being $g(t)$ a continuous function such that $\int^t g(u)du \to \infty$ when $t \to \infty$.

The solution of (1) is done by

$$x(t) = \alpha - \frac{\alpha - x_0}{\varepsilon(t_0)} \, \varepsilon(t),$$

where $\varepsilon(t) := \exp\left(-\int^t g(u)du\right)$. We note that the carrying capacity is the parameter $\alpha = \lim_{t \to \infty} x(t)$. Nevertheless, in several applications, carrying capacity must be dependent on initial value x_0. In order to do this, the curve can be modified as follows,

$$x(t) = x_0 \frac{\eta - \varepsilon(t)}{\eta - \varepsilon(t_0)} \tag{2}$$

where η is another parameter depending on initial values x_0 and t_0. The rest of the parameters are encoded in function g and hence ε. So g can be defined as a parametric function with unbounded integral. Note that with this formulation, the carrying capacity (upper bound) is written as $\alpha = \eta \, x_0 \, (\eta - \varepsilon(t_0))^{-1}$.

In particular, for $g(t) = \beta\gamma t^{\gamma-1}$, curve (2) becomes classical Weibull curve with growth parameters β and γ.

2.1 Hyperbolastic Model of Type III

If we set $g(t) = \beta\gamma t^{\gamma-1} + \theta\left(1 + \theta^2 t^2\right)^{-\frac{1}{2}}$, resulting curve (2) is known as the hyperbolastic curve of type III. Introduced in 2005 [7], the family of hyperbolastic curves have shown remarkable results describing and predicting behaviour of growth phenomena (see, for example, [3,5–7] and references therein). In particular, hyperbolastic curve of type III outperforms classical curves such as logistic, Gompertz or Richards.

The curve of type III depends on two parameters β and γ related with growth and a third parameter θ as a distance from the classical Weibull model. Indeed, when θ vanishes, the hyperbolastic of type III becomes the Weibull curve. This additional term gives the curve an exceptional flexibility due to the mobility of the inflection point.

A model for the hypebolastic curve of type I, which is a modification of the logistic model, was proposed in [2], addressing the problem of inference by means of metaheuristic algorithms. In the case of type III, the generalization of the Weibull curve allows us to define the model in a more generic fashion.

3 Stochastic Model

In order to build a stochastic version of Eq. (1), we consider the non-homogeneous stochastic process $X(t)$ defined on $\mathbb{R}^+$, verifying the stochastic differential equation

$$dX(t) = \frac{\varepsilon(t)}{\eta - \varepsilon(t)}\, g(t)\, X(t)dt + \sigma X(t)\, dW(t) \tag{3}$$

with $X(t_0) = x_0$ a.s., σ a diffusion parameter (representing random fluctuations) and $W(t)$ the standard Wiener process. The solution of (3) is the lognormal diffusion process with exogenous factors, characterized by drift and diffusion, respectively,

$$A_1(x,t) = h(t)\, x, \qquad A_2(x) = \sigma^2 x^2,$$

where $h(t) := \frac{\varepsilon(t)}{\eta - \varepsilon(t)}\, g(t)$. Expression of the drift is obtained from (2) by

$$\frac{d}{dt}x(t) = -\frac{x_0}{\eta - \varepsilon(t_0)}\frac{d}{dt}\varepsilon(t) = \frac{x_0}{\eta - \varepsilon(t_0)}\varepsilon(t)g(t) = x(t)\frac{\varepsilon(t)}{\eta - \varepsilon(t)}g(t).$$

Thus, mean function of diffusion solution of (3) is done by

$$E(X(t)) = X(t_0)\, e^{\int_{t_0}^t h(u)du} = x_0\, \frac{\eta - \varepsilon(t)}{\eta - \varepsilon(t_0)} \quad a.s.$$

which is the generalized Weibull curve. Hence, the diffusion process $X(t)$ evolves following the trend imposed by the curve.

A remarkable feature of this kind of processes is the availability of explicit expressions for several characteristics. For example by solving the Kolmogorov equation (noting ∂_x the operator for partial derivative with respect to x)

$$\partial_s f(x,t|y,s) + \frac{y\,\varepsilon(t)g(t)}{\eta - \varepsilon(t)}\,\partial_y f(x,t|y,s) + \frac{y^2\sigma^2}{2}\,\partial_y^2 f(x,t|y,s) = 0,$$

the transition probability density function $f(x,t|y,s) := \partial_x P(X(t) \leq x | X(s) = y)$ for $s < t$ can be obtained. In particular, the conditional distribution is also lognormal,

$$X(t)|X(s) = y \sim \Lambda\left(\log y + \log\frac{\eta - \varepsilon(t)}{\eta - \varepsilon(s)} - \frac{\sigma^2}{2}(t-s), \sigma^2(t-s)\right).$$

In Fig. 1 diverse behaviour of functions g and h, as well as original curve and simulated paths of the hyperbolastic diffusion process of type III, can be observed.

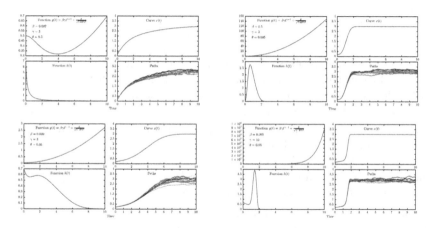

Fig. 1. Functions g, x, h and simulated paths, for different parameters, of a hyperbolastic diffusion of type III.

3.1   Inference with Unknown g

For practical purposes, function g must be known. In several cases, the parametric expression of g can be proposed in the context of the study, focusing on the meaning of the parameters and the mathematical relations between them.

In order to address this issue, we propose a methodology to estimate the parameters of a known expression of g by means of a transformation of the original observations.

Procedure is based on fitting a suggested function to sampling coming from the primitive of g, denoted by G, and defining a diffusion as in previous section.

Indeed, solving Eq. (2) for $\varepsilon(t)$ leads to

$$\varepsilon(t) = \eta \left(1 - \frac{1}{\alpha} x(t) \right),$$

so,

$$G(t) = - \log \eta - \log \left(1 - \frac{1}{\alpha} x(t) \right). \tag{4}$$

Let us now consider a discrete sampling $\{x_{i,j}\}_{i,j}$ for $i = 1, \ldots, d$ and $j = 1, \ldots, n$ of d sample paths observed at same time instants t_j.

At first glance, for a fixed path i, we obtain from (4),

$$\int_{t_{j-1}}^{t_j} g(u)du = \log \frac{\alpha - x_{i,j-1}}{\alpha - x_{i,j}},$$

which can be used to obtain characteristics of the function g, compute hitting times of the process or address other inference issues.

Nevertheless, a sampling for G cannot be obtained due to unknown parameter η. But in order to define a final diffusion process is enough to consider a modified sampling

$$\left\{ -\log \left(1 - \frac{1}{\alpha} x_{\cdot j} \right) \right\}_j, \tag{5}$$

where $x_{\cdot j}$ is the mean of observations at time j along the paths.

Indeed, by considering a suggested parametric function Γ fitting data coming from (4) the expression $\Gamma(t) = \Gamma_0(t) - \log \eta$ holds for all t, where Γ_0 is another function fitting set (5). The drift of final diffusion depends on $\frac{d}{dt}\Gamma(t)$ and not on $\Gamma(t)$; on the other hand, $\frac{d}{dt}\Gamma(t) = \frac{d}{dt}\Gamma_0(t)$ for all t. Hence, for our purposes, function Γ_0 can be consider instead of Γ.

Finally, thanks to the drift of the diffusion process, parameter η (generally unknown) can be ignored in order to obtain a discrete sampling of G from $\{x_{ij}\}$ by means of the following procedure (see Fig. 2):

1. To estimate the upper bound $\hat{\alpha}$ from data.
2. To estimate parameter of random fluctuations $\hat{\sigma}$ by, for example, maximum likelihood.
3. To modify original observations by means of the transformation

$$\rho(x) := -\log \left(1 - \frac{1}{\hat{\alpha}} x \right).$$

4. To fit a suggested differentiable function Γ_0 to modified data from 3.
5. To define a lognormal stochastic process with drift

$$A_1(x,t) = x \left(e^{\Gamma_0(t)} - 1 \right)^{-1} \frac{d}{dt}\Gamma_0(t)$$

and diffusion $A_2(x) = \hat{\sigma} x$.

Difficulties in step 4 may arise depending on the number of parameters and the complexity of the suggested function. In that case, numerical analysis methods or even metaheuristic algorithms could be used.

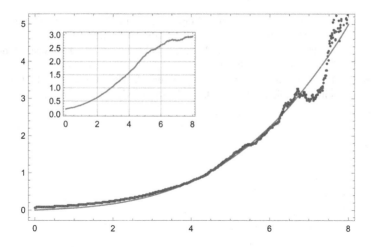

Fig. 2. Function G (red) and modified data (blue) from simulated path (inset plot) of a hyperbolastic diffusion process of type III. (Color figure online)

4   Conclusions

In this work a new stochastic model based on a generalization of the Weibull curve is proposed. The use of a generic function g with just a few properties leads to expressions for different characteristics of the process, allowing us to obtain particular results directly for every function g.

One particular case is the hyperbolastic curve of type III. In this context, the diffusion can describe random sigmoidal dynamics with more flexibility than other models based on classical curves.

In a practical way, the problem related with the lack of a known function g is addressed. We propose a method based on fitting a suggested parametric function to modified data in order to avoid unknown suitable parameter of the model.

References

1. Albano, G., Giorno, V., Román-Román, S., Román-Román, P., Torres-Ruiz, F.: Estimating and determining the effect of a therapy on tumor dynamics by means of a modified Gompertz diffusion process. J. Theor. Biol. **364**, 206–219 (2015)
2. Barrera, A., Román-Román, P., Torres-Ruiz, F.: A hyperbolastic type-I diffusion process: parameter estimation by means of the firefly algorithm. Biosystems **163**, 11–22 (2018)
3. Eby, W., Tabatabai, M., Bursac, Z.: Hyperbolastic modeling of tumor growth with a combined treatment of iodoacetate and dimethylsulphoxide. BMC Cancer **10**, 509 (2010)
4. Román-Román, P., Torres-Ruiz, F.: Modeling logistic growth by a new diffusion process: application to biological systems. Biosystems **110**, 9–21 (2012)

5. Tabatabai, M., Bursac, Z., Eby, W., Singh, K.P.: Mathematical modeling of stem cell proliferation. Med. Biol. Eng. Comput. **49**(3), 253–262 (2011)
6. Tabatabai, M., Eby, W., Singh, K.P.: Hyperbolastic modeling of wound healing. Math. Comput. Model. **53**, 5–6 (2011)
7. Tabatabai, M., Williams, D., Bursac, Z.: Hyperbolastic growth models: theory and application. Theor. Biol. Med. Modell. **2**, 14 (2005)

On the Integration of Fractional Neuronal Dynamics Driven by Correlated Processes

Enrica Pirozzi[(✉)] [ID]

Dipartimento di Matematica e Applicazioni, Università di Napoli FEDERICO II,
Via Cintia, Monte S.Angelo, 80126 Napoli, Italy
enrica.pirozzi@unina.it

Abstract. The stochastic Leaky Integrate-and-Fire (LIF) model is revisited adopting a fractional derivative instead of the classical one and a correlated input in place of the usual white noise. The aim is to include in the neuronal model some physiological evidences such as correlated inputs, codified input currents and different time-scales. Fractional integrals of Gauss-Markov processes are considered to investigate the proposed model. Two specific examples are given. Simulations of paths and histograms of first passage times are provided for a specific case.

Keywords: Fractional LIF model · Ornstein-Uhlenbeck process · Simulation

1 Introduction

Many recent contributions have been given to design stochastic neuronal models specialized to include in the classical Leaky Integrate-and-Fire (LIF) model some phenomenological evidences, such as: correlated inputs, adaptation phenomena, different time-scales, etc. (see, for instance, [1–4]). The classical stochastic LIF model is based on the following stochastic differential equation (SDE), $\forall t > 0$:

$$dV(t) = -\frac{g_L}{C_m}[V(t) - V_L]dt + \frac{\sigma}{C_m}dW, \qquad V(0) = V_0. \qquad (1)$$

The stochastic process $V(t)$ represents the voltage of the neuronal membrane, whereas the other parameters and functions are: C_m the membrane capacitance, g_L the leak conductance, V_L a specific value of potential, $dW(t)$ the noise (with $W(t)$ a standard Brownian motion (BM)) and σ its intensity. A spike is emitted when $V(t)$ attains a threshold value V_{th}. Unfortunately, this model has some limitations (see, for instance, [5]). It cannot include a correlated input like colored noise, but only a delta-correlated white noise. It is not suitable to represent any adaptation phenomena: indeed, relating the firing adaptation to memory effects, the LIF solution process is a Markov process, hence it has no memory. Moreover,

Partially supported by GNCS.

R. Moreno-Díaz et al. (Eds.): EUROCAST 2019, LNCS 12013, pp. 211–219, 2020.
https://doi.org/10.1007/978-3-030-45093-9_26

different time-scales cannot be adopted for different confluent dynamics, such as those of the membrane, ionic channels, synaptic inputs and noise. In [6–9], we tried to specialize the LIF model, and its SDE, in such a way some of the above phenomenological evidences can be described and reliably predicted.

Specifically, inspired by results of [3] and [10], a LIF model with a colored noise was considered in [6] and for it an approximating time-non-homogeneous Gauss-Markov (GM) process was proposed. The purpose was to obtain some approximations of the first passage time (FPT) through a constant level (the firing threshold). Again in [6], a fractional Langevin equation, proposed for neuronal models [10,11], was considered for simulations and comparisons. Then, in [7], further investigations were carried out about firing rates.

In [8] a detailed analysis of a LIF model with correlated inputs was made, using both mathematical tools and simulation approaches. A preeminent role was played by the theory of integrated GM processes [12]. Indeed, the neuronal voltage $V(t)$ was obtained as the integral over time of an Ornstein-Uhlenbeck (OU) process modeling a correlated input.

With the aim to generalize the study of those processes, in [9] some specific results were obtained for fractional integrals of GM processes. Then, an embryonic neuronal model was proposed in order to highlight the possible application of those processes in neuronal modeling.

Our twofold interest, for this kind of processes and for the neuronal modeling, leads us to propose here a fractional SDE whose solution is the fractional integral of a correlated OU process. This approach has the aim to employ jointly the correlated input, as done in [6] and [8] for the integer (non fractional) case, and the fractional integrals [9] in a LIF-type model, as done in [6] and [7] for a fractional model with non correlated inputs. Here, the discussion will be regarding the joint effect of these two approaches, already considered in previous papers, but separately.

Specifically, the idea is to consider a more general model based on the following two coupled differential equations:

$$\mathcal{D}^\alpha V(t) = F(t, V, V_L, \eta), \qquad\qquad V(0) = V_0, \qquad\qquad (2)$$

$$d\eta(t) = G(t, \eta, I)dt + \gamma dW(t), \quad \eta(0) = \eta_0, \qquad\qquad (3)$$

where $\mathcal{D}^\alpha$ is the fractional Caputo derivative [13] with $\alpha \in (0,1)$, $\eta(t)$ is a GM process and it represents a stochastic correlated input. I stands for an input current. Here, in order to apply the theory of the integrated GM processes in neuronal models, we adopt the following forms for the functions F and G and parameter γ:

$$F(t, V, V_L, \eta) = f(t, V, V_L) + \frac{\eta(t)}{C_m}, \quad G(t, \eta, I) = -\frac{\eta(t) - I}{\tau}, \quad \gamma = \frac{\sigma}{\tau}. \qquad (4)$$

In this way, the membrane potential $V(t)$ will be a process that involves the fractional Caputo integral of $\alpha-$order of the correlated input $\eta(t)$.

The fractional order of the derivative in (2) can allow to describe dynamics of the membrane potential $V(t)$ on a time-scale different from that of the input $\eta(t)$; furthermore, the input process $\eta(t)$ drives the dynamics of $V(t)$ with a correlation on the time.

In the next section, we first consider the model (2)–(3) for a specific choice of F, G and γ. We remark that, due the shortness of this communication, we do not specify all appropriate hypotheses for differential problems ensuring the consistency of our results (see, for instance, [14]) neither the suitable probability space (see, for instance, [15]), we only give some mathematical details and provide simulations of the considered processes. In the last section, we consider the model (2)–(3) of LIF-type addressing a preliminary investigation.

2   A First Model

Firstly, we investigate the model of [9], that is obtained from (2)–(3), with $f(t, V, V_L) \equiv \frac{g_L}{C_m} V_L$ and G and γ as in (4), i.e.

$$\mathcal{D}^\alpha V(t) = \frac{g_L}{C_m} V_L + \frac{\eta(t)}{C_m}, \qquad V(0) = V_0$$

$$\tag{5}$$

$$d\eta(t) = -\frac{\eta(t) - I}{\tau} dt + \frac{\sigma}{\tau} dW(t), \quad \eta(0) = \eta_0.$$

Here, I represents a synaptic or an injected current that arrives, codified by the correlated input $\eta(t)$, to the neuron. As physiological justification, we can say that the above dynamics can happen when the firing activity of a neuron is affected by the complex interactions with many others neurons in a networks: in this case, correlated inputs, codified currents and different time-scales have to be considered. Indeed, the neuronal activity (in this case without leakage) is modeled by the process $V(t)$ as a fractional integral of the whole evolution of the input process $\eta(t)$ from an initial time, but on a (more or less) finer time scale that can be regulated by choosing the fractional order α of integration suitably adherent to the neuro-physiological evidences.

2.1   The $\eta(t)$ Process

The stochastic process $\eta(t)$ in (5) is a GM process: specifically, it is an OU process with the following representation in terms of a time-transformed BM process W:

$$\eta(t) = m_\eta(t) + h_2(t) W(r(t)) \tag{6}$$

with mean $m_\eta(t)$, covariance $c(s, t) = h_1(s) h_2(t)$, and $W(t)$ a standard BM process in $r(t) = \frac{h_1(t)}{h_2(t)}$ increasing function.

We remark that the initial values V_0 and η_0 in (5) can be specified like constants or random variables, making the model suitable for the description of endogenous stimuli or exogenous stimuli, respectively [8]. Here, for fixed constant initial values, we have

$$\eta(t) = I + e^{-t/\tau}[\eta_0 - I + W(r(t))], \tag{7}$$

where $W(t)$ is a standard BM and $r(t) = \frac{\tau}{2}\left(e^{2t/\tau} - 1\right)$ (with $r(0) = 0$). In particular, $\eta(t)$ is the non-stationary OU process with mean:

$$m_\eta(t) = I + e^{-t/\tau}(\eta_0 - I) \tag{8}$$

and covariance

$$c(s,t) = h_1(s)h_2(t) = \frac{\sigma^2}{2\tau}\left(e^{-(t-s)/\tau} - e^{-(s+t)/\tau}\right), \ 0 \le s \le t, \tag{9}$$

with

$$h_1(t) = \frac{\sigma}{2}\left(e^{t/\tau} - e^{-t/\tau}\right), \ h_2(t) = \frac{\sigma}{\tau}e^{-t/\tau}. \tag{10}$$

2.2 The $V(t)$ Process

The process $V(t)$ involves the fractional Riemann-Liouville (RL) integrals of the GM process $\eta(t)$. For sake of readability, we give some essentials of fractional calculus. Let be $\alpha \in (0,1)$; if $f(t)$ is a real-valued differentiable function on $\mathbb{R}$, we recall that the *Caputo fractional derivative* of f of order α is defined as follows [13]

$$\mathcal{D}^\alpha f(t) = \frac{1}{\Gamma(1-\alpha)} \int_0^t \frac{f'(s)}{(t-s)^\alpha} \, ds, \tag{11}$$

where f' denotes the ordinary derivative of f. If f is a continuous function, its *fractional RL integral* of order α is defined as the following:

$$\mathcal{I}^\alpha(f)(t) = \frac{1}{\Gamma(\alpha)} \int_0^t (t-s)^{\alpha-1} f(s)ds, \tag{12}$$

where Γ is the Gamma Euler function, i.e. $\Gamma(z) = \int_0^{+\infty} t^{z-1}e^{-t}dt$, $z > 0$.

Referring to the model (5), assuming that $V(0) = 0$, the RL fractional integral $\mathcal{I}^\alpha$ can be used as the left-inverse of the Caputo derivative $\mathcal{D}^\alpha$ [17], in such a way we obtain

$$V(t) = \mathcal{I}^\alpha(\mathcal{D}^\alpha V)(t) = \mathcal{I}^\alpha\left(\frac{g_L}{C_m}V_L\right)(t) + \frac{1}{C_m}\mathcal{I}^\alpha(\eta)(t). \tag{13}$$

Now, we focus our attention on $\mathcal{I}^\alpha(\eta)(t)$. Here, we can exploit some results about the (path-wise) fractional Riemann-Liouville integral of GM processes [9]. Indeed, recalling also (6), the fractional RL integral of $\eta(t)$ is:

$$
\begin{aligned}
\mathcal{I}^\alpha(\eta)(t) &= \frac{1}{\Gamma(\alpha)} \int_0^t \frac{\eta(s)}{(t-s)^{1-\alpha}} ds \\
&= \frac{1}{\Gamma(\alpha)} \int_0^t \frac{m_\eta(s) + h_2(s)W(r(s))}{(t-s)^{1-\alpha}} ds \\
&= \mathcal{I}^\alpha(m_\eta)(t) + \mathcal{I}^\alpha(h_2 W_r)(t),
\end{aligned}
\tag{14}
$$

where W_r stands for $W(r(t))$. Furthermore, we have that $\mathcal{I}^\alpha(\eta)(t)$ has normal distribution with mean $\mathcal{I}^\alpha(m_\eta)(t)$ and covariance

$$
Cov_\eta^\alpha(u,t) = cov\left[\mathcal{I}^\alpha(h_2 W_r)(u)\mathcal{I}^\alpha(h_2 W_r)(t)\right].
$$

Specifically, from [9], for $0 \le u < t$, we have $Cov_\eta^\alpha(u,t) = \frac{1}{\Gamma^2(\alpha)}\left[\widetilde{I}_1 + \widetilde{I}_2 + \widetilde{I}_3\right]$ with

$$
\begin{aligned}
\widetilde{I}_1 &= \frac{\sigma^2}{\tau} \int_0^u ds(u-s)^{\alpha-1} e^{-s/\tau} \int_0^s dv(t-v)^{\alpha-1} \sinh(v/\tau), \\
\widetilde{I}_2 &= \frac{\sigma^2}{\tau} \int_0^u ds(u-s)^{\alpha-1} \sinh(s/\tau) \int_s^u dv(t-v)^{\alpha-1} e^{-v/\tau}, \\
\widetilde{I}_3 &= \frac{\sigma^2}{\tau} \left(\int_0^u ds(u-s)^{\alpha-1} \sinh(s/\tau)\right) \left(\int_u^t dv(t-v)^{\alpha-1} e^{-v/\tau}\right).
\end{aligned}
\tag{15}
$$

Finally, the $V(t)$ process has mean:

$$
\begin{aligned}
m_V(t) &= \mathcal{I}^\alpha\left(\frac{g_L}{C_m} V_L\right)(t) + \frac{1}{C_m}\mathcal{I}^\alpha(m_\eta)(t) \\
&= \frac{g_L V_L}{C_m} \frac{t^\alpha}{\Gamma(\alpha+1)} + \frac{1}{C_m}\mathcal{I}^\alpha(I + e^{-t/\tau}(\eta_0 - I))(t) \\
&= \left(\frac{g_L V_L + I}{C_m}\right) \frac{t^\alpha}{\Gamma(\alpha+1)} + \frac{1}{C_m}(\eta_0 - I)t^\alpha E_{1,1+\alpha}(-t/\tau),
\end{aligned}
\tag{16}
$$

where $E_{\alpha,\beta}(z) = \sum_{k=0}^\infty \frac{z^k}{\Gamma(\alpha k+\beta)}$ is the Mittag-Leffler function [17], and covariance

$$
Cov_V^\alpha(u,t) = \frac{1}{C_m^2} Cov_\eta^\alpha(u,t).
$$

2.3   Some Simulations Results

We now focus on the simulation of paths of the process $V(t)$ for specific cases. In particular, due to the key rule of the process $\mathcal{I}^\alpha(\eta)(t)$ in the dynamics of $V(t)$, here, we perform simulations of such a process when the $V(t) = \mathcal{I}^\alpha(\eta)(t)$. This occurs for a specific setting of the parameters, i.e. for $V_L = \eta_0 = 0$, $\sigma = \tau = C_m = 1$, $I = 0$, (disregarding the measure units). The simulation algorithm of sample paths exploits the expression of the covariance specified in the previous section. At first, a numerical evaluation of the values of the covariance matrix on a (time) square grid is performed; then, the successive steps of each paths are evaluated by means of sequences of pseudo-random Gaussian numbers. The codes for simulations are prepared as R scripts [16]. As in [9], we follow these steps: we evaluate the elements of $N \times N$ covariance matrix $C(t_i, t_j)$ at times t_i, $i = 1, \ldots, N$, of an equi-spaced temporal grid; we apply the Cholesky decomposition algorithm to the covariance matrix C to obtain a lower triangular matrix $L(i, j)$ such that $C = LL^T$; we generate a N-dimensional array $\mathbf{z}$ of standard pseudo-Gaussian numbers; we construct the sequence of simulated values of the correlated fractional integrated process as the array $\mathbf{v} = L\mathbf{z}$. Finally, the array $\mathbf{v}$ provides the simulated path $(V(t_1), \ldots, V(t_N))$, which components have the assigned covariance.

In Fig. 1 some simulated sample paths of $V(t)$ are shown. The paths show an increasing roughness for decreasing values of α. By means of 10^4 simulated sample paths of $V(t)$ until the first time of crossing of the threshold value V_{th}, and recording these FPTs, we can construct their histograms as approximations of the FPT probability density. These results can be used for a statistical prediction of the first firing time of the neuron. In Fig. 2, histograms of simulated FPTs are shown for different values of α and thresholds V_{th}. The investigations, by means of the simulation approach, suggest that for decreasing values of α the tail of the histograms becomes more and more heavy (see Fig. 2, right). This evidence can be related with the long memory of the process $V(t)$, that is Gaussian but non Markovian. As a consequence in the application to the neuronal models, we have that much more probability mass is distributed in correspondence of long times for firing.

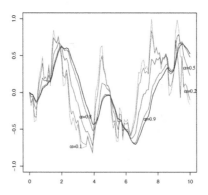

Fig. 1. Simulated sample paths of $V(t)$ with $V_L = \eta_0 = 0$, $\sigma = \tau = C_m = 1$, $I = 0$ and some indicated values of α.

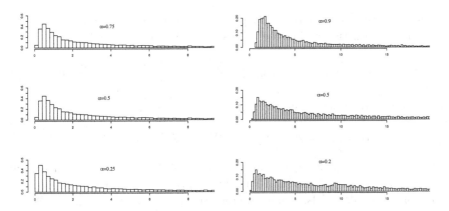

Fig. 2. Histograms of FPT of $V(t)$ through the threshold $V_{th} = 0.5$ (left) and $V_{th} = 1$ (right) and some values of α. The other parameters are the same of Fig. 1.

3   A Fractional LIF-type Model with a Correlated Input

If in the model (2)–(3), and in (4), we set $f(t, V, V_L) = -\frac{g_L}{C_m}[V(t) - V_L]$, we obtain

$$\mathcal{D}^\alpha V(t) = -\frac{g_L}{C_m}[V(t) - V_L] + \frac{\eta(t)}{C_m}, \quad V(0) = V_0$$

$$(17)$$

$$d\eta(t) = -\frac{\eta(t) - I}{\tau} dt + \frac{\sigma}{\tau} dW(t), \qquad \eta(0) = \eta_0$$

a fractional LIF-type model with $\eta(t)$ correlated input. In these setting, the stochastic process $\eta(t)$ is the same of the previous sections, while the process $V(t)$ stands for a neuronal membrane potential that now shows a time decay towards the state V_L (the resting potential). For $V_0 = 0$, we can also consider the integral version of the first equation of (17):

$$V(t) = \mathcal{I}^\alpha \left(-\frac{g_L}{C_m}[V(t) - V_L] \right)(t) + \frac{1}{C_m} \mathcal{I}^\alpha(\eta)(t). \tag{18}$$

Referring to [13], [17] and [18] for deterministic counterpart of this problem, finally, we can consider the process $V(t)$ with mean

$$m_V(t) = V_0 t^{\alpha-1} E_{\alpha,\alpha} \left(-\frac{g_L t^\alpha}{C_m} \right)$$

$$+ \int_0^t (t - s)^{\alpha-1} E_{\alpha,\alpha} \left(-\frac{g_L(t - s)^\alpha}{C_m} \right) \left[\frac{g_L V_L}{C_m} + \frac{m_\eta(s)}{C_m} \right] ds. \tag{19}$$

In this case, the form of the $V(t)$ process involves a fractional integral of a Mittag-Leffler function and of the process $\eta(t)$. In particular, setting:

$$\mathcal{J}^{(E_\alpha)}(\eta)(t) = \int_0^t (t-s)^{\alpha-1} E_{\alpha,\alpha}\left(-\frac{g_L(t-s)^\alpha}{C_m}\right)\left[\frac{g_L V_L}{C_m} + \frac{\eta(s)}{C_m}\right] ds, \quad (20)$$

and recalling (6), for the covariance of $V(t)$ we can write

$$Cov_V^\alpha(u,t) = cov\left[\mathcal{J}^{(E_\alpha)}(h_2 W_r)(u)\mathcal{J}^{(E_\alpha)}(h_2 W_r)(t)\right].$$

Specific algorithms for the simulation of this dynamics can be realized and relative investigations about the FPTs through the level V_{th} will be really interesting for neuronal models. Finally, we remark that the model (17) gives rise to a process $V(t)$ with memory, involving a fractional integral, and, in addition, the process integrates a stochastic input with correlation. Further theoretical and numerical investigations are mandatory and they will be carried out in future works.

References

1. Buonocore, A., Caputo, L., Pirozzi, E., Carfora, M.F.: A leaky integrate-and-fire model with adaptation for the generation of a spike train. Math. Biosci. Eng. **13**(3), 483–493 (2016)
2. Kim, H., Shinomoto, S.: Estimating nonstationary inputs from a single spike train based on a neuron model with adaptation. Math. Bios. Eng. **11**, 49–62 (2014)
3. Sakai, Y., Funahashi, S., Shinomoto, S.: Temporally correlated inputs to leaky integrate-and-fire models can reproduce spiking statistics of cortical neurons. Neural Netw. **12**, 1181–1190 (1999)
4. Kobayashi, R., Kitano, K.: Impact of slow K^+ currents on spike generation can be described by an adaptive threshold model. J. Comput. Neurosci. **40**, 347–362 (2016)
5. Shinomoto, S., Sakai, Y., Funahashi, S.: The Ornstein-Uhlenbeck process does not reproduce spiking statistics of cortical neurons. Neural Comput. **11**, 935–951 (1997)
6. Pirozzi, E.: Colored noise and a stochastic fractional model for correlated inputs and adaptation in neuronal firing. Biol. Cybern. **112**(1–2), 25–39 (2018)
7. Ascione, G., Pirozzi, E.: On fractional stochastic modeling of neuronal activity including memory effects. In: Moreno-Díaz, R., Pichler, F., Quesada-Arencibia, A. (eds.) EUROCAST 2017. LNCS, vol. 10672, pp. 3–11. Springer, Cham (2018). https://doi.org/10.1007/978-3-319-74727-9_1
8. Ascione, G., Pirozzi, E.: On a stochastic neuronal model integrating correlated inputs. Math. Biosci. Eng. **16**(5), 5206–5225 (2019)
9. Abundo, M., Pirozzi, E.: On the Fractional Riemann-Liouville Integral of Gauss-Markov processes and applications. arXiv preprint arXiv:1905.08167
10. Bazzani, A., Bassi, G., Turchetti, G.: Diffusion and memory effects for stochastic processes and fractional Langevin equations. Phys. A Stat. Mech. Appl. **324**, 530–550 (2003)
11. Teka, W., Marinov, T.M., Santamaria, F.: Neuronal spike timing adaptation described with a fractional leaky integrate-and-fire model. PLoS Comput. Biol. **10**, e1003526 (2014)

12. Abundo, M., Pirozzi, E.: Integrated stationary Ornstein-Uhlenbeck process, and double integral processes. Phys. A: Stat. Mech. Appl. **494**, 265–275 (2018). https://doi.org/10.1016/j.physa.2017.12.043
13. Podlubny, I.: Fractional Differential Equations. Academic Press, New York (1999)
14. Samko, S.G., Kilbas, A.A., Marichev, O.I.: Fractional Integrals and Derivatives: Theory and Applications. Gordon and Breach Science Publishers, London (1993)
15. Baldi, P.: Stochastic Calculus, Universitext. Springer, Cham (2017). https://doi.org/10.1007/978-3-319-62226-2
16. Haugh, M.: Generating Random Variables and Stochastic Processes, Monte Carlo Simulation. Columbia University, New York (2017)
17. Garrappa, R., Kaslik, E., Popolizio, M.: Evaluation of fractional integrals and derivatives of elementary functions: overview and tutorial. Mathematics **7**, 407 (2019). https://doi.org/10.3390/math7050407
18. Kazem, S.: Exact solution of some linear fractional differential equation by Laplace transform. Int. J. Nonlinear Sci. **16**(1), 3–11 (2013)

Simulation of an α-Stable Time-Changed SIR Model

Giacomo Ascione[(✉)]

Dipartimento di Matematica e Applicazioni "Renato Caccioppoli",
Università degli Studi di Napoli Federico II, 80126 Napoli, Italy
giacomo.ascione@unina.it

Abstract. Starting from the stochastic SIR model, we give a time-changed SIR model obtained via an inverse α-stable subordinator. In particular, we study the distribution of its inter-jump times and we use it to describe a simulation algorithm for the model. Finally, such algorithm is tested and used to describe graphically some properties of the model.

Keywords: Epidemics model · Subordinator · Mittag-Leffler function

1 Introduction

The SIR model was introduced by Kermack and McKendrick in 1927 [14] and since then it has been one of the most important mathematical model in epidemics. Its stochastic counterpart has been introduced lately by Bartlett, in 1949 [7], and it is based on the same compartmental scheme of the original SIR model, this time described by using a Continuous Time Markov Chain (CTMC). Such kind of models have been widely studied and generalized through time.

A particular generalization of the deterministic model is given by the fractional order SIR model, which overcomes the classical one in describing some particular infection dynamics (see [1,12]). At the same time, for the stochastic case, the semi-Markov property can be preferred to the Markov one to describe, for instance, the infection dynamics of the AIDS (see [15] and reference therein).

Here we propose a model which can be seen as a link between the fractional order deterministic SIR model and the stochastic semi-Markov one, focusing on simulation procedures for such process. In Sect. 2 we give some basics on the classical stochastic SIR model. In Sect. 3 we construct our time-changed SIR model starting from the classical one. In particular, we exploit the distribution of the inter-jump times and we use such property to describe a simulation algorithm. In Sect. 4 we show some simulation results while in Sect. 5 we give some concluding remarks.

2 The Stochastic SIR Model

Let us consider a classical *general* stochastic SIR model (see for instance [11]) on a filtered probability space $(\Omega, \mathcal{F}, \mathcal{F}_t, \mathbb{P})$. It is described by three processes

© Springer Nature Switzerland AG 2020
R. Moreno-Díaz et al. (Eds.): EUROCAST 2019, LNCS 12013, pp. 220–227, 2020.
https://doi.org/10.1007/978-3-030-45093-9_27

$S^1(t)$, $I^1(t)$ and $R^1(t)$ (for $t \geq 0$) that represent respectively the number of susceptible, infective and removed individuals of the population. Let us suppose that the population is **closed**, that is to say $S^1(t) + I^1(t) + R^1(t) = N$, where $N \in \mathbb{N}$ is a constant. This first assumption allows us to take in consideration just the pair $(S^1(t), I^1(t))$ since the third process $R^1(t)$ can be obtained from the other two processes. Moreover, let us suppose that the pair $(S^1(t), I^1(t))$ is a bivariate (cádlág) CTMC with (finite) state space $\mathcal{S} = \{(s, i) \in \mathbb{N}^2 : s + i \leq N\}$ and with infinitesimal transition probabilities given by

$$\mathbb{P}\left[(S^1(t + \delta t), I^1(t + \delta t)) = (s - 1, i + 1)\,|\,(S^1(t), I^1(t)) = (s, i)\right] = \beta s i \delta t + o(\delta t)$$
$$\mathbb{P}\left[(S^1(t + \delta t), I^1(t + \delta t)) = (s, i - 1)\,|\,(S^1(t), I^1(t)) = (s, i)\right] = \gamma i \delta t + o(\delta t)$$
$$\mathbb{P}\left[(S^1(t + \delta t), I^1(t + \delta t)) = (s, i)\,|\,(S^1(t), I^1(t)) = (s, i)\right] = 1 - (\beta s + \gamma)i \delta t + o(\delta t)$$

where β and γ are respectively the pairwise infection and the removal parameters. In particular, since the removed individuals do not take part in the infection dynamics, it is not restrictive to consider $R(0) = 0$ almost surely.

2.1 Asymptotic Behaviour

It is interesting to observe that the states $(s, 0) \in \mathcal{S}$ are absorbent states. In particular, let us define $T^1 = \inf\{t > 0 : I^1(t) = 0\}$, that is the duration of the outbreak. Hence, it is easy to see that $T^1 < +\infty$ almost surely and then, for any $\omega \in \Omega$, $S^1(t, \omega)$ is definitely constant. We can thus define the random variable $S_\infty = \lim_{t \to +\infty} S^1(t)$ and ask for the probability distribution of S_∞. This quantity characterizes *how much the infection has spread*. Indeed, let us fix an initial number of susceptibles S_0 and infectives I_0, and let us define the random variable $I_{tot} = S_0 - S_\infty$, that is called the **ultimate size of the epidemics**, since it counts exactly the total of individuals that have been infected since the start of the outbreak. Another interesting random variable is **the intensity of the epidemics** defined as $I_{rel} = \dfrac{I_{tot}}{S_0}$ that represents the proportion of susceptible individuals that have been infected. Finally, let us denote with π the distribution of I_{rel}, that is to say

$$\pi : \xi \in (0, 1) \mapsto \mathbb{P}\left[I_{rel} \leq \xi\,|\,(S(0), I(0)) = (S_0, I_0)\right].$$

We say that a major outbreak occurs (for the initial value (S_0, I_0) and the couple of parameters (β, γ)) if there exists a $\xi \in (0, 1)$ such that $\pi(\xi) < 1$, that is to say that at least a susceptible individual has been infected. This quantity is strictly linked to the **basic reproductive number** $\mathcal{R}_0$ defined as $\mathcal{R}_0 = \dfrac{S_0 \beta}{\gamma}$, that represents the mean number of susceptible individuals that a single infective individual can infect during its mean lifetime. The strict connection between this value and the function π has been exploited by Whittle in [22]. In particular, we recall the following consequence of Whittle's threshold theorem, known as $\mathcal{R}_0$-dogma (see for instance [20] and references therein).

Proposition 1. *A major outbreak occurs if and only if $\mathcal{R}_0 > 1$.*

2.2   The Jump Chain and the Inter-jump Times

Now let us introduce some other important quantities. The following construction is common for any CTMC (see, for instance, [2]), hence we will not give the details. Let us denote for $n \geq 1$ with (S_n, I_n) the jump-chain (i. e. the embedded Markov chain) of $(S^1(t), I^1(t))$, A_n^1 the jump times and $T_n^1 = A_n^1 - A_{n-1}^1$ (where $A_0^1 = 0$) the inter-jump times (or **holding times**). As it is well known, these random variables are conditionally exponentially distributed, that is to say

$$\mathbb{P}[T_n^1 \leq t | (S_n, I_n) = (s, i)] = 1 - e^{-r(s,i)t}, \ \forall t \geq 0,$$

where the rate function is given by $r(s, i) = (\beta s + \gamma)i$. Let us observe that the distribution of S_∞ (and then the function π) depends only on the jump chain (S_n, I_n). Finally, one can use the properties of the jump chain to determine the mean outbreak duration $\mathbb{E}[\mathcal{T}^1]$, which is shown to be finite (see [11]).

2.3   Simulation of the Model

The inter-jump times $(T_n^1)_{n \in \mathbb{N}}$ and the jump chain $(S_n, I_n)_{n \in \mathbb{N}}$ can be used then to simulate the model. Indeed, to simulate a CTMC, a well-known algorithm is *Gillespie's algorithm* (see [13]). In our case, it adapts as

- Initialize the process choosing $S^1(0)$ and $I^1(0)$;
- Suppose we have simulated the process $(S^1(t), I^1(t))$ up to a jump time A_n^1. Then we have explicitly (S_n, I_n);
- We can simulate T_{n+1}^1 as an exponentially distributed random variable with parameter $r(S_n, I_n)$ to obtain $A_{n+1}^1 = T_{n+1}^1 + A_n^1$ (this can be done by inversion of the distribution function method, see [6]);
- To choose what event will happen in A_{n+1}^1, define $p_n = \frac{\gamma I_n}{r(S_n, I_n)}$ and simulate a uniform random variable U in $(0, 1)$;
- If $U < p_n$, then A_{n+1}^1 is a removal time and we set $S_{n+1} = S_n$ and $I_{n+1} = I_n - 1$;
- If $U \geq p_n$, then A_{n+1}^1 is an infection time and we set $S_{n+1} = S_n - 1$ and $I_{n+1} = I_n + 1$;
- For any $t \in (A_n^1, A_{n+1}^1)$ we set $S^1(t) = S_n$ and $I^1(t) = I_n$;
- Stop as $I_{n+1} = 0$.

3   The α-Stable Time-Changed SIR Model

Now let fix $\alpha \in (0, 1)$ and let us consider an α-stable subordinator $\{\sigma_\alpha(t), t \geq 0\}$ (for details, see [8]) which is independent from $\{(S^1(t), I^1(t)), t \geq 0\}$ and its inverse process

$$L_\alpha(t) = \inf\{y > 0 : \sigma_\alpha(y) \geq t\}$$

called inverse α-stable subordinator. An almost complete review of the properties of such process is given in [17]. We define our α-stable Time-Changed SIR model as $(S^\alpha(t), I^\alpha(t)) = (S^1(L_\alpha(t)), I^1(L_\alpha(t)))$. The first important thing we *gain* after this time change is the semi-Markov property, which is a typical property of time-changed Markov process (see [18]). As we stated in the introduction, semi-Markov property is quite an important property for such kind of models, since it is much more realistic than the assumption that the process is Markov. Moreover, defining the jump chain in the same way as we did for the classical model, since we only considered a random time-rescaling, it has the same distribution as the classical jump chain (and then we will still denote it by (S_n, I_n)). An important consequence of such observation is that, since the distribution of S_∞ and the function π depend only on the jump chain, Whittle's threshold theorem still holds for such processes.

Let us also denote with T^α the outbreak duration. Then, since T^1 is the exit time of $(S^1(t), I^1(t))$ from an open set and $\mathbb{E}[T^1] < +\infty$, we have that the function $t \mapsto \mathbb{P}(T^\alpha > t) \simeq Ct^{-\alpha}$ for $t \to +\infty$ (see [5]), hence its decay is slower then the one of $\mathbb{P}(T^1 > t)$.

3.1 The Inter-jump Times

Now our aim is to determine the distribution of the inter-jump times. To do this, let us recall the definition of the Mittag-Leffler function E_α for some $\alpha \in (0,1)$, i. e. for $t \in \mathbb{R}$

$$E_\alpha(t) = \sum_{k=0}^{+\infty} \frac{t^k}{\Gamma(\alpha k + 1)}.$$

In particular we say that a random variable T^α is **Mittag-Leffler distributed** of parameter $\lambda > 0$ and order $\alpha \in (0,1)$ if $\mathbb{P}[T^\alpha \le t] = 1 - E_\alpha(-\lambda t^\alpha)$, $\forall t \ge 0$. We can then prove the following Proposition.

Proposition 2. *Let us denote with* $(T_n^\alpha)_{n \in \mathbb{N}}$ *the inter-jump times of* $(S^\alpha(t), I^\alpha(t))$. *Then the* T_n^α *are conditionally Mittag-Leffler distributed, i. e.*

$$\mathbb{P}[T_n^\alpha \le t \,|\, (S_n, I_n) = (s,i)] = 1 - E_\alpha(-r(s,i)t^\alpha), \ \forall t \ge 0$$

where the rate function is given by $r(s,i) = (\beta s + \gamma)i$.

Proof. First of all, let us observe that from the definition of inverse α-stable subordinator we have that

$$S^\alpha(t) = S^1(y), \ \sigma_\alpha(y-) \le t < \sigma_\alpha(y)$$

hence $T_n^\alpha = \sigma_\alpha(T_n^1-)$. Moreover, σ_α does not have any fixed discontinuity (i.e. for any $t \ge 0$ $\sigma_\alpha(t-) = \sigma_\alpha(t)$ almost surely) hence, by using a conditioning argument and the fact that σ_α and T_n^1 are independent, it is easy to show that $\sigma_\alpha(T_n^1-) \overset{d}{=} \sigma_\alpha(T_n^1)$ (where with $\overset{d}{=}$ we denote the equality in distribution). Now, since by definition $L_\alpha(\sigma_\alpha(t)) = t$, we have that

$$\mathbb{P}\left[T_n^\alpha > t \,|(S_n, I_n) = (s, i)\right] = \mathbb{P}\left[\sigma_\alpha(T_n^1) > t \,|(S_n, I_n) = (s, i)\right]$$
$$= \mathbb{P}\left[T_n^1 > L_\alpha(t) \,|(S_n, I_n) = (s, i)\right].$$

Since T_n^1 and $L_\alpha(t)$ are independent, T_n^1 is conditionally exponentially distributed and L_α is independent from (S_n, I_n), we have

$$\mathbb{P}\left[T_n^\alpha > t \,|(S_n, I_n) = (s, i)\right] = \mathbb{P}\left[T_n^1 > L_\alpha(t) \,|(S_n, I_n) = (s, i)\right]$$
$$= \int_0^{+\infty} \mathbb{P}(y > L_\alpha(t))e^{-r(s,i)y}dy = E_\alpha(-r(s,i)t^\alpha)$$

where the last equality follows from the Laplace transform of the distribution of the inverse α-stable subordinator (see [9]).

3.2   Simulation of a Mittag-Leffler Random Variable

Now that we have shown that the inter-jump times T_n^α are conditionally Mittag-Leffler distributed, to recovery Gillespie's algorithm we need to show how to simulate Mittag-Leffler distributed random variables. The Proposition we are going to recall follows straightforward from the scaling properties of the α-stable subordinator and the Laplace transform of the distribution of the inverse α-stable subordinator.

Proposition 3. *Let T^1 be an exponential random variable of parameter $\lambda > 0$, T^α a Mittag-Leffler random variable of parameter $\lambda > 0$ and fractional order $\alpha \in (0, 1)$ and $\sigma_\alpha(t)$ an α-stable subordinator independent from T^1. Then*

$$T^\alpha \stackrel{d}{=} (T^1)^{\frac{1}{\alpha}}\sigma_\alpha(1). \tag{1}$$

The proof is given, for instance, in [4].

From this proposition we know how to simulate a Mittag-Leffler random variable T^α of parameter $\lambda > 0$ and fractional order $\alpha \in (0, 1)$:

- Simulate an exponential random variable T^1 of parameter $\lambda > 0$;
- Simulate a one-sided α-stable random variable $\sigma_\alpha(1)$ (this can be done by a generalization of the Box-Muller algorithm with a suitable choice of parameters, see [16]);
- Compute $T^\alpha = (T^1)^{\frac{1}{\alpha}}\sigma_\alpha(1)$.

3.3   Simulation of the α-Stable Time-Changed SIR Model

Now we can completely recovery Gillespie's algorithm (as done in [4,10]) to simulate an α-stable Time-Changed SIR Model:

- Initialize the process choosing $S^\alpha(0)$ and $I^\alpha(0)$;
- Suppose we have simulated the process $(S^\alpha(t), I^\alpha(t))$ up to a jump time A_n^α. Then we have explicitly (S_n, I_n);

- We can simulate T_{n+1}^α as a Mittag-Leffler distributed random variable with parameter $r(S_n, I_n)$ and fractional order α (we know how to do it from the previous subsection) to obtain $A_{n+1}^\alpha = T_{n+1}^\alpha + A_n^\alpha$;
- To choose what event will happen in A_{n+1}^α, define $p_n = \frac{\gamma I_n}{r(S_n, I_n)}$ and simulate a uniform random variable U in $(0, 1)$;
- If $U < p_n$, then A_{n+1}^α is a removal time and we set $S_{n+1} = S_n$ and $I_{n+1} = I_n - 1$;
- If $U \geq p_n$, then A_{n+1}^α is an infection time and we set $S_{n+1} = S_n - 1$ and $I_{n+1} = I_n + 1$;
- For any $t \in (A_n^\alpha, A_{n+1}^\alpha)$ we set $S^\alpha(t) = S_n$ and $I^\alpha(t) = I_n$;
- Stop as $I_{n+1} = 0$.

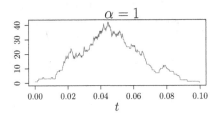 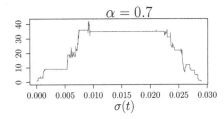

Fig. 1. On the left: a sample path of $I^1(t)$ with $I_0 = 1$, $S_0 = 200$, $\beta = 1$, $\gamma = 110$ discretized with a time interval of $\Delta t = 0.0001$. On the right: the respective time-changed sample path by means of an α-stable inverse subordinator with $\alpha = 0.7$.

4 Simulation Results

We are now able to simulate both the classical stochastic SIR model than the α-stable Time-Changed one. Simulation of such models have been performed by using R [19]. The scripts we used to simulate such processes can be found in [3]. To simulate one-sided α-stable random variables, we will make use of the R package `stabledist` [23]. In particular, by using such package, we are able to simulate α-stable subordinators. Thus, following the lines given in [16, Example 5.21], to show the effect of the time-change, we first simulate separately a classical stochastic SIR model (in particular the process $I^1(t)$) and an α-stable subordinator $\sigma_\alpha(t)$, then we plot the curves $(t, I^1(t))$ and $(\sigma_\alpha(t), I^1(t))$. To attach these (and the following) images, they have been converted into TikZ code by using the package `tikzDevice` [21]. We can see such plots in Fig. 1. One must pay attention to the fact that on the right we have the curve $(\sigma_\alpha(t), I^1(t))$ and not the plot of $I^\alpha(t)$, that is, however, identical in shape to such curve. Here we can see how the process is delayed by the plateaus of the inverse α-stable subordinator $L_\alpha(t)$. In Fig. 2 we show the plots of the function $\pi(\xi)$ first for the classical SIR model and then for the α-stable time-changed one, this time simulated by using the modified Gillespie's algorithm we described before. We can see that the plots are almost identical: this is due to the fact that the function

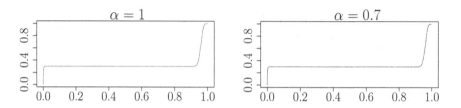

Fig. 2. On the left: plot of the function $\pi(\xi)$ for a classical stochastic SIR model with $I_0 = 1$, $S_0 = 500$, $\beta = 1$, $\gamma = 150$ for 10000 trajectories. On the right: plot of the function $\pi(\xi)$ for an α-stable time-changed SIR model with $\alpha = 0.7$ and the same parameters.

π depends only on the jump chain, which is identically distributed independently from the choice of $\alpha \in (0, 1)$. In [3], we provide also the scripts to generate some histograms for the distribution of S_∞ and the plots of the function $\mathbb{P}(\mathcal{T}^1 > t)$ and $\mathbb{P}(\mathcal{T}^\alpha > t)$.

5   Conclusions

We have introduced a semi-Markov SIR model which is obtained from the classical one by means of a time-change via an inverse α-stable subordinator. Thus, we have given the distribution of its inter-jump times and we used such information to describe a simulation algorithm for such model. Being a semi-Markov process, our model exhibits a form of memory, as it is requested for such kind of processes to be more realistic (see for instance [15]). However, our model cannot be used to obtain a different outcome for the ultimate size of the infection, since it preserves the distribution of the jump chain, but it alters the transient behaviour and the duration of the outbreak. Indeed, there is a link between our model and the fractional order deterministic SIR model (see for instance [1,12]), which will be exploited in a work that is actually in preparation. However, here we have given a generalization of Gillespie's algorithm, adapted to our modified SIR model, that can be used to simulate this model, in order to numerically verify some of its properties.

References

1. Al-Sulami, H., El-Shahed, M., Nieto, J.J., Shammakh, W.: On fractional order dengue epidemic model. Math. Probl. Eng. **2014**, 1–6 (2014)
2. Anderson, W.J.: Continuous-Time Markov Chains: An Applications-Oriented Approach. Springer, New York (2012). https://doi.org/10.1007/978-1-4612-3038-0
3. Ascione, G.: STCSIR, May 2019. https://doi.org/10.5281/zenodo.3229269
4. Ascione, G., Leonenko, N., Pirozzi, E.: Fractional Erlang queues. arXiv preprint arXiv:1812.10773 (2018)

5. Ascione, G., Pirozzi, E., Toaldo, B.: On the exit time from open sets of some semi-Markov processes. arXiv preprint arXiv:1709.06333 (2017)
6. Asmussen, S., Glynn, P.W.: Stochastic Simulation: Algorithms and Analysis, vol. 57. Springer, New York (2007). https://doi.org/10.1007/978-0-387-69033-9
7. Bartlett, M.: Some evolutionary stochastic processes. J. Roy. Stat. Soc. Ser. B (Methodological) **11**(2), 211–229 (1949)
8. Bertoin, J.: Lévy Processes, vol. 121. Cambridge University Press, Cambridge (1996)
9. Bingham, N.: Limit theorems for occupation times of Markov processes. Probab. Theory Relat. Fields **17**(1), 1–22 (1971)
10. Cahoy, D.O., Polito, F., Phoha, V.: Transient behavior of fractional queues and related processes. Methodol. Comput. Appl. Probab. **17**(3), 739–759 (2015)
11. Daley, D.J., Gani, J.: Epidemic Modelling: An Introduction, vol. 15. Cambridge University Press, Cambridge (2001)
12. Diethelm, K.: A fractional calculus based model for the simulation of an outbreak of dengue fever. Nonlinear Dyn. **71**(4), 613–619 (2013)
13. Gillespie, D.T.: Exact stochastic simulation of coupled chemical reactions. J. Phys. Chem. **81**(25), 2340–2361 (1977)
14. Kermack, W.O., McKendrick, A.G.: A contribution to the mathematical theory of epidemics. Proc. Roy. Soc. London Ser. A Contain. Papers Math. Phys. Charact. **115**(772), 700–721 (1927)
15. Lefèvre, C., Simon, M.: SIR epidemics with stages of infection. Adv. Appl. Probab. **48**(3), 768–791 (2016)
16. Meerschaert, M.M., Sikorskii, A.: Stochastic Models for Fractional Calculus, vol. 43. Walter de Gruyter, Berlin (2011)
17. Meerschaert, M.M., Straka, P.: Inverse stable subordinators. Math. Model. Nat. Phenom. **8**(2), 1–16 (2013)
18. Meerschaert, M.M., Straka, P., et al.: Semi-Markov approach to continuous time random walk limit processes. Ann. Probab. **42**(4), 1699–1723 (2014)
19. R Core Team: R: A Language and Environment for Statistical Computing. R Foundation for Statistical Computing, Vienna, Austria (2016). https://www.R-project. org/
20. Reluga, T.C., Medlock, J., Perelson, A.S.: Backward bifurcations and multiple equilibria in epidemic models with structured immunity. J. Theor. Biol. **252**(1), 155–165 (2008)
21. Sharpsteen, C., Bracken, C.: tikzDevice: R Graphics Output in LaTeX Format, R package version 0.11 (2018). https://CRAN.R-project.org/package=tikzDevice
22. Whittle, P.: The outcome of a stochastic epidemic—a note on Bailey's paper. Biometrika **42**(1–2), 116–122 (1955)
23. Wuertz, D., Maechler, M., Rmetrics core team members.: stabledist: Stable Distribution Functions (2016). https://CRAN.R-project.org/package=stabledist, R package version 0.7-1

Some Results on a Growth Model Governed by a Fractional Differential Equation

Antonio Di Crescenzo$^{(\boxtimes)}$ and Alessandra Meoli

Dipartimento di Matematica, Università di Salerno, Fisciano, Italy
{adicrescenzo,ameoli}@unisa.it

Abstract. We define a deterministic growth model which generalizes both the Gompertz and the Korf law in a fractional way. We provide lower bounds for the solution of the corresponding initial value problem and discuss how the introduction of "memory effects" affects the shape of such functions. We also compute maximum and inflection points.

Keywords: Population growth · Caputo fractional derivative · Inflection points

1 Introduction

The stability of an ecosystem strongly depends on the size of species populations living in that community. Thus, the study of how the population sizes of species change over time and space is of fundamental importance and poses many exciting challenges for mathematicians and statisticians. The exponential function is a useful, but naive, model of population dynamics, because changes in limiting factors are not taken into account. Over the years more complicated models have been developed that mainly involve regulatory effects (see, for instance, [4,16,17,19]). A general model for population growth can be described by

$$\frac{dN(t)}{dt} = \xi(t)N(t), \qquad t > 0, \tag{1}$$

where $N(t)$ represents the population size and $\xi(t)$ is a time-dependent growth rate. In the present paper we describe the response of the population to different environmental characteristics by also letting the instantaneous rate of change depend on the past state. Specifically, in Eq. (1) we replace the ordinary time derivative with a derivative of fractional order, thus introducing "memory effects". The modelling of fractional order systems has gained greater and greater significance and attention over recent decades and it is now well established in the literature that the dynamics of some biological or physical real-world phenomena can be better explained by means of fractional tools (see, for instance, [1]). A reference book in this field has been provided by Baleanu et al. [2].

Paper partially supported by MIUR - PRIN 2017, project "Stochastic Models for Complex Systems". The authors are members of the INdAM Research group GNCS.

The paper is organized as follows. In Sect. 2 we specialize the relative growth rate in (1). This leads to a growth model inspired both by Gompertz and Korf laws, and review its main features. In Sect. 3 we consider the time-fractional counterpart to Eq. (1). Then we give some lower bounds of the solution and solve it numerically, providing suitable estimates for the inflection point.

2   Background on Some Growth Models

If in Eq. (1) the relative growth rate $\xi(t)$ decays exponentially or in power-law form, i.e. if

$$\xi(t) = \xi_G(t) := \alpha e^{-\beta t} \quad \text{or} \quad \xi(t) = \xi_K(t) := \alpha t^{-(\beta+1)}, \tag{2}$$

then we obtain the differential equation governing respectively the Gompertz growth and the Korf growth. The parameters $\alpha, \beta > 0$ in (2) are the growth and decay rates respectively. One can easily prove that, if $y > 0$ and $t > 0$, the solution of (1) becomes

$$N_G(t) = y \exp\left\{ \frac{\alpha}{\beta} \left(1 - e^{-\beta t}\right) \right\}, \qquad N_G(0) = y,$$

$$N_K(t) = y \exp\left\{ \frac{\alpha}{\beta} \left(1 - t^{-\beta}\right) \right\}, \qquad N_K(0) = 0.$$

The Gompertz model is one of the most frequently used sigmoid curves fitted to growth data. There is now a large body of literature on the applications of the Gompertz model to describe bacterial growth curves [19], the prediction of experimental tumor growth [3], plant growth [12], animal growth [11], just to name a few. The Korf growth function has found specific applications in forest science; see, for example [15] and references therein. However, since growth processes are influenced by a variety of factors, there is an increasing need for suitable equations. Di Crescenzo and Spina [8] recently contributed to the issue raised by proposing a new growth model which has the same carrying capacity as the Gompertz and the Korf laws, but captures different growth dynamics, under an alternative time-dependent growth rate, given by

$$\xi(t) = \alpha(1+t)^{-(\beta+1)}, \qquad t > 0.$$

The corresponding growth model is, for $\alpha, \beta > 0$, and $y > 0$,

$$N(t) = y \exp\left\{ \frac{\alpha}{\beta} \left[1 - (1+t)^{-\beta}\right] \right\}, \qquad N(0) = y.$$

Hereafter we summarize the main features of the three growth curves.

- They have initial values and derivatives

$$N_K(0) = 0, \quad N(0) = N_G(0) = y > 0,$$

$$N_K'(0) = 0, \quad N'(0) = N_G'(0) = \alpha y > 0.$$

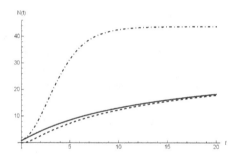

Fig. 1. The proposed curve (full), Korf curve (dashed), and Gompertz curve (dot-dashed), for $y = 0.8$, $\alpha = 2$, $\beta = 0.5$, respectively, with $C \simeq 43.68$.

- They are increasing, with $N_K(t) < N(t) < N_G(t)$, $t > 0$.
- They are bounded by the same carrying capacity:

$$\lim_{t \to +\infty} N_G(t) = \lim_{t \to +\infty} N_K(t) = \lim_{t \to +\infty} N(t) = C \equiv y e^{\alpha/\beta}.$$

- The three curves tend to the carrying capacity C according to different rules, since

$$\frac{\beta}{\alpha} \left| \log \frac{N_G(t)}{C} \right| = e^{-\beta t}, \quad \frac{\beta}{\alpha} \left| \log \frac{N_K(t)}{C} \right| = t^{-\beta}, \quad \frac{\beta}{\alpha} \left| \log \frac{N(t)}{C} \right| = (1+t)^{-\beta}.$$

For fixed choices of the parameters y, α, β, the new model describes an intermediate growth between the (lower) Korf and (upper) Gompertz curves. Furthermore, for small values of t the curve $N(t)$ behaves similarly to $N_G(t)$, whereas for large times it grows similarly to $N_K(t)$, since it tends to the carrying capacity polynomially fast. In brief, such a model is motivated by the need of describing evolutionary dynamics characterized by non-zero initial values and approximately linear initial slope, and tending to a carrying capacity from below through a long-term power-law growth (Fig. 1).

Let us now focus on the inflection point of the growth model. Clearly, this is of high interest in population growth since for sigmoidal curves such point expresses the instant when the growth rate is maximum. When the inflection points exist, we have $N_K(t_K) = N(t_N) < N_G(t_G)$, so that the new model and the Korf model evaluated at the inflection points have identical population size, which is smaller than that of the Gompertz model, whereas $N_K(t) < N(t) < N_G(t)$ for all $t > 0$.

3   Fractional Extension of a Growth Model

Applications of fractional calculus to growth models recently have gained considerable attention and recognition since fractional operators change the solutions we usually obtain in classical systems. Indeed, they can be used to describe the variation of a population in which the instantaneous rate of change depends on the past state, thus introducing memory effects. It is widely acknowledged that memory effects are a translation of the environment resistance. A fractional logistic equation has been proposed by Varalta et al. [18], whilst a modified fractional logistic equation has been discussed in D'Ovidio et al. [6] (see also references therein). Generalizations of the Gompertz law via fractional calculus approach have been considered by Bolton et al. [5] and, more recently, by Frunzo et al. [9]. To the best of our knowledge, no fractional Korf growth model has been provided so far.

To extend Eq. (1), we consider the model for population growth described by

$$\begin{cases} {}^C D_{0+}^\nu N(t) = \xi(t) N(t), & t > 0, \\ N(0) = n_0 \end{cases} \tag{3}$$

where

$$\xi(t) = \alpha(1+t)^{-(\beta+1)}, \qquad t > 0,$$

with ${}^C D_{0+}^\nu$ the Caputo fractional derivative of order $\nu \in (0,1)$:

$$ {}^C D_{0+}^\nu u(t) = \frac{1}{\Gamma(1-\nu)} \int_0^t (t-s)^{-\nu} \frac{d}{ds} u(s) \, ds. $$

The latter involves the convolution between a power law function and the derivative of the considered function. It is generally preferred to deal with the Caputo derivative. Indeed, when a Cauchy problem is expressed by means of the Caputo fractional differentiation operator, standard initial conditions in terms of integer-order derivatives are involved. They have clear physical meaning as initial position, initial velocity etc. On the contrary, when a Cauchy problem is defined by means of other fractional differentiation operators, e.g. the Riemann-Liouville one, fractional initial conditions are required. They have no clear physical interpretation, and this makes solutions of such initial value problems useless [13].

As shown in [14], the solution of Eq. (3) in an interval $[0, T]$ can be expressed formally as

$$ N(t) = n_0 e^{\mathcal{I}_{0+}^\nu \xi(t)}, \qquad 0 < t < T, $$

where $\mathcal{I}_{0+}^\nu$ is the Riemann-Liouville fractional integral of order $\nu \in (0,1)$

$$ \mathcal{I}_{0+}^\nu \xi(t) = \frac{1}{\Gamma(\nu)} \int_0^t (t-s)^{\nu-1} \xi(s) \, ds, \tag{4} $$

and

$$ e^{\mathcal{I}_{0+}^\nu \xi(t)} \equiv 1 + \mathcal{I}_{0+}^\nu \xi(t) + \frac{1}{2!} (\mathcal{I}_{0+}^\nu \xi(t))^2 + \ldots + \frac{1}{n!} (\mathcal{I}_{0+}^\nu \xi(t))^n + \ldots, $$

where

$$ \frac{(\mathcal{I}_{0+}^\nu \xi(t))^n}{n!} = \underbrace{\mathcal{I}_{0+}^\nu \left(\xi(t) \, \mathcal{I}_{0+}^\nu \left(\xi(t) \, \mathcal{I}_{0+}^\nu \left(\ldots \left(\xi(t) \, \mathcal{I}_{0+}^\nu \xi(t) \right) \right) \right) \right)}_{n \text{ times}}. $$

When $\nu = \beta$, for the solution

$$ N(t) = n_0 \left[1 + \mathcal{I}_{0+}^\beta \xi(t) + \frac{1}{2!} (\mathcal{I}_{0+}^\beta \xi(t))^2 + \ldots + \frac{1}{n!} (\mathcal{I}_{0+}^\beta \xi(t))^n + \ldots \right] \tag{5} $$

we have

$$\mathcal{I}_{0+}^{\beta}\xi(t) = \frac{\alpha t^{\beta}}{\Gamma(\beta+1)}\frac{1}{1+t},$$

$$\frac{(\mathcal{I}_{0+}^{\beta}\xi(t))^2}{2!} = \frac{\alpha^2 t^{2\beta}}{\Gamma(2\beta+1)} {}_2F_1(\beta+2, \beta+1; 2\beta+1; -t),$$

and

$$\frac{(\mathcal{I}_{0+}^{\beta}\xi(t))^3}{3!} = \frac{\alpha^3 t^{3\beta}}{\Gamma(3\beta+1)}\frac{F_3\left(\beta+1, \beta+2, \beta, \beta+1; 3\beta+1; \frac{t}{t+1}, -t\right)}{(1+t)^{\beta+1}},$$

where, for $0 < \operatorname{Re} b < \operatorname{Re} c$ and $|\arg(1-z)| < \pi$,

$$_2F_1(a, b; c; z) = \frac{\Gamma(c)}{\Gamma(b)\Gamma(c-b)}\int_0^1 t^{b-1}(1-t)^{c-b-1}(1-zt)\,dt$$

is (the Euler integral representation of) the Gauss hypergeometric function and, for $0 < \operatorname{Re} b < \operatorname{Re} c$,

$$F_3(a, a', b, b'; c; w, z) = \frac{\Gamma(c)}{\Gamma(b)\Gamma(c-b)}\int_0^1 u^{c-b-1}(1-u)^{b-1}(1-w+wu)^{-a}$$
$$\times\, {}_2F_1(a', b'; c-b; zu)\,du$$

is (the integral representation of) the Appell function F_3. The above results are essential to define a sequence of increasing lower bounds for $N(t)$ as k grows:

$$N_k(t) := \sum_{i=1}^{k}\frac{1}{k!}(\mathcal{I}_{0+}^{\beta}\xi(t))^k, \quad k \in \mathbb{N}. \tag{6}$$

Some plots of the solution (5) and of the related lower bounds (6) are shown in Fig. 2 for different choices of the parameter ν. The initial value problem (3) has been solved by means of computational tools. Specifically, we employed some implicit fractional linear multistep methods of the second order (Flmm2 MAT-LAB package, see Garrappa [10]). In the first three cases of Fig. 2 the population grows fast and then decreases in size until it reaches the carrying capacity. Such behaviour becomes more abrupt as $\nu \to 0^+$, that is to say as the influence of the past on the future becomes more and more significant. Instead, for $\nu = 0.95$ the population will continue to grow until it reaches the carrying capacity.

Hereafter we derive numerically maximum and inflection points for the solution of (3). Specifically, maximum and inflection points are depicted in Fig. 3 for two choices of the fractional parameter ν, whilst in Fig. 4 the dependence of the number of inflection points on the parameter α, when ν is fixed, is highlighted (See also Tables 1 and 2).

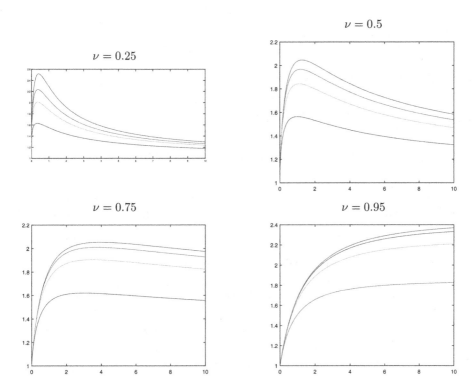

Fig. 2. Plots of the lower bounds in (6) and of the solution in (5) for $n_0 = 1$, $\alpha = 1$ and $\beta = \nu$. From top to bottom, the cases $k = 1$, $k = 2$, $k = 3$ and the exact solution are considered.

Concluding Remarks

A fractional population growth model inspired by the Gompertz and the Korf laws has been proposed. Due to the mathematical difficulties involved in the analysis of inhomogeneous fractional differential equations, we have solved numerically the initial value problem of interest and we have determined analytically some lower bounds of the exact solution in a special case. The introduction of memory effects in the governing fractional differential equation is of special relevance since the corresponding solution exhibits non-monotonic behaviour, as it is shown in Figs. 2 and 3. There are many ways in which this work can be extended, for example we can consider new modifications based on different implementations of the fractionalization. Moreover, we can apply different fractional approaches to logistic growth (cf. Di Crescenzo and Paraggio [7]) and perform comparisons with the proposed model.

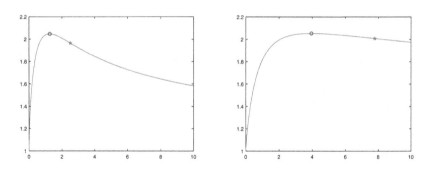

Fig. 3. The proposed curve for $n_0 = 1$, $\alpha = 1$ and $\nu = 0.5$ (left), $\nu = 0.75$ (right).

Table 1. Maximum and inflection points

ν	Maximum point	Maximum value	Inflection point	Population at the inflection point
0.5	1.2561	2.0470	2.5161	1.9646
0.75	3.9502	2.0531	7.791	2.0079

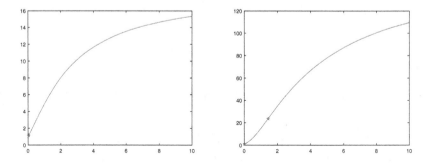

Fig. 4. The proposed curve for $\nu = 0.95$, $n_0 = 1$ and $\alpha = 3$ (left), $\alpha = 5$ (right).

Table 2. Inflection points when $\nu = 0.95$

α	Inflection points	Population at the inflection points
3	0.0359	1.1339
5	0.0117, 1.4390	1.0768, 23.6594

References

1. Almeida, R., Bastos, N.R., Monteiro, M.T.T.: Modeling some real phenomena by fractional differential equations. Math. Methods Appl. Sci. **39**(16), 4846–4855 (2016)
2. Baleanu, D., Diethelm, K., Scalas, E., Trujillo, J.J.: Fractional Calculus. World Scientific Publishing Company, Singapore (2016)
3. Benzekry, S., et al.: Classical mathematical models for description and prediction of experimental tumor growth. PLOS Comput. Biol. **10**(8), e1003800 (2014)
4. Bhowmick, A.R., Bhattacharya, S.: A new growth curve model for biological growth: some inferential studies on the growth of Cirrhinus mrigala. Math. Biosci. **254**, 28–41 (2014)
5. Bolton, L., Cloot, A.H., Schoombie, S.W., Slabbert, J.P.: A proposed fractional-order Gompertz model and its application to tumour growth data. Math. Med. Biol. **32**(2), 187–207 (2014)
6. D'Ovidio, M., Loreti, P., Ahrabi, S.S.: Modified fractional logistic equation. Physica A **505**, 818–824 (2018)
7. Di Crescenzo, A., Paraggio, P.: Logistic growth described by birth-death and diffusion processes. Mathematics **7**(6), 489 (2019)
8. Di Crescenzo, A., Spina, S.: Analysis of a growth model inspired by Gompertz and Korf laws, and an analogous birth-death process. Math. Biosci. **282**, 121–134 (2016)
9. Frunzo, L., Garra, R., Giusti, A., Luongo, V.: Modelling biological systems with an improved fractional Gompertz law. Commun. Nonlinear Sci. Numer. Simul. **74**, 260–267 (2019)
10. Garrappa, R.: Trapezoidal methods for fractional differential equations: theoretical and computational aspects. Math. Comput. Simulat. **110**, 96–112 (2015)
11. Hafiz, M.A.W., Hifzan, M.R., Bahtiar, I.A.J., Ariff, O.M.: Describing growth pattern of Brakmas cows using non-linear regression models. Mal. J. Anim. Sci. **18**(2), 37–45 (2015)
12. Paine, C.E.T., et al.: How to fit nonlinear plant growth models and calculate growth rates: an update for ecologists. Methods Ecol. Evol. **3**(2), 245–256 (2012)
13. Podlubny, I.: Fractional differential equations: an introduction to fractional derivatives, fractional differential equations, to methods of their solution and some of their applications, vol. 198. Elsevier (1998)
14. Sambandham, B., Aghalaya, V.: Basic results for sequential Caputo fractional differential equations. Mathematics **3**(1), 76–91 (2015)
15. Sedmk, R., Scheer, L.: Modelling of tree diameter growth using growth functions parameterised by least squares and Bayesian methods. J. For. Sci. **58**(6), 245–252 (2012)
16. Talkington, A., Durrett, R.: Estimating tumor growth rates in vivo. Bull. Math. Biol. **77**(10), 1934–1954 (2015). https://doi.org/10.1007/s11538-015-0110-8
17. Tsoularis, A., Wallace, J.: Analysis of logistic growth models. Math. Biosci. **179**(1), 21–55 (2002)
18. Varalta, N., Gomes, A.V., Camargo, R.F.: A prelude to the fractional calculus applied to tumor dynamic. TEMA Tendências em Matemática Aplicada e Computacional **15**(2), 211–221 (2014)
19. Zwietering, M.H., Jongenburger, I., Rombouts, F.M.: Modeling of the bacterial growth curve. Appl. Environ. Microbiol. **56**(6), 1875–1881 (1990)

Theory and Applications of
Metaheuristic Algorithms

Surrogate-Assisted Multi-Objective Parameter Optimization for Production Planning Systems

Johannes Karder[1,3]([✉]), Andreas Beham[1,3], Andreas Peirleitner[2],
and Klaus Altendorfer[2]

[1] Heuristic and Evolutionary Algorithms Laboratory, University of Applied Sciences
Upper Austria, 4232 Hagenberg, Austria
{johannes.karder,andreas.beham}@fh-hagenberg.at
[2] Department of Operations Management, University of Applied Sciences
Upper Austria, 4400 Steyr, Austria
{andreas.peirleitner,klaus.altendorfer}@fh-steyr.at
[3] Institute for Formal Models and Verification, Johannes Kepler University,
4040 Linz, Austria

Abstract. Efficient global optimization is, even after over two decades of research, still considered as one of the best approaches to surrogate-assisted optimization. In this paper, material requirements planning parameters are optimized and two different versions of EGO, implemented as optimization networks in HeuristicLab, are applied and compared. The first version resembles a more standardized version of EGO, where all steps of the algorithm, i.e. expensive evaluation, model building and optimizing expected improvement, are executed synchronously in sequential order. The second version differs in two aspects: (i) instead of a single objective, two objectives are optimized and (ii) all steps of the algorithm are executed asynchronously. The latter leads to faster algorithm execution, since model building and solution evaluations can be done in parallel and do not block each other. Comparisons are done in terms of achieved solution quality and consumed runtime. The results show that the multi-objective, asynchronous optimization network can compete with the single-objective, synchronous version and outperforms the latter in terms of runtime.

1 Introduction

To improve the overall performance of real-world systems, researchers transform these systems into simulation models with a certain level of abstraction. These models can then be used to gain behavioral insights into the system by conducting analysis experiments and to subsequently optimize these systems by means of *simulation-based optimization*. Since executing such simulations can be computationally expensive and conventional optimization approaches require an effort of hundreds of thousands of evaluations to traverse the search space in a reasonable way, the idea of *surrogate-assisted optimization* has emerged.

© Springer Nature Switzerland AG 2020
R. Moreno-Díaz et al. (Eds.): EUROCAST 2019, LNCS 12013, pp. 239–246, 2020.
https://doi.org/10.1007/978-3-030-45093-9_29

This technique requires a surrogate model to replace the runtime expensive simulation, which then approximates the simulation's outputs and can be evaluated in shorter time frames. Surrogate models can be created by various (supervised) machine learning algorithms, including simpler ones such as *linear regression* or more sophisticated ones such as *Gaussian processes* [8]. By learning correlations between inputs and outputs of the underlying expensive model, these models can then be used to make predictions of output values for unobserved input values. Knowledge of the expensive model's internals are not used for model building. Hence, learning from in- and outputs only is considered a black-box approach. A state-of-the-art algorithm for expensive black-box optimization problems has been proposed by Jones et al. [5] in 1998. The principle of *Efficient Global Optimization* (EGO) is to iteratively learn a Gaussian process (also known as *kriging*) model and use it to explore and exploit the search space based on the *Expected Improvement* (EI) of new data points.

Within this paper, *Material Requirements Planning* (MRP) [4] parameters of a simplified *flow shop* system [1] are optimized by an *Optimization Network* (ON) [6]. Two objectives are used to evaluate the quality of MRP parameters: (i) inventory costs and (ii) tardiness costs, both are output by the simulated production system. A multi-objective optimization approach based on EGO and ONs is implemented. In previous publications, metaheuristic optimization networks have been successfully applied to different kinds of interrelated problems, e.g. to solve the *location routing problem* [3] or conduct *integrated machine learning* [7]. By using ONs, multiple algorithms can be combined to solve optimization problems in a cooperative fashion. Finally, we compare our approach to a more standardized single-objective, synchronous implementation of EGO as ON, where both objectives are summed up to form a single objective that accounts for the total costs that arise within the system. We analyze both approaches and compare them in terms of runtime and achieved solution quality. The central research question to be answered is: "How well does a multi-objective, asynchronous optimization network perform against a more traditional, single-objective, synchronous EGO approach in the context of MRP parameter optimization?"

The rest of this paper is structured as follows. Section 2 describes the simulated flow shop in detail. The basic concepts of EGO are reviewed in Sect. 3. Section 4 gives a quick introduction to optimization networks in general and explains the two implemented MRP parameter optimization networks. The conducted experiments and achieved results are presented in Sect. 5. Finally, Sect. 6 concludes the paper and gives an outlook on future work.

2 Simulated Flow Shop

In the implemented flow shop, a total of 16 materials are simulated (see also [1]). Those materials are manufactured and have low-level codes (LLC) 0–2, while raw materials are assumed to be always available and have LLC 3. The system consists of a total of 6 machines and on each machine 2–4 materials are produced.

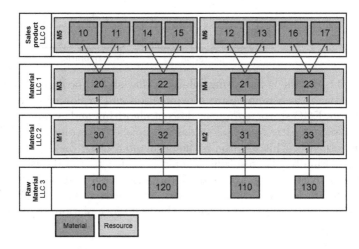

Fig. 1. The schema of the simulated flow shop.

To configure the production system, the following 3 MRP parameters have to be defined for each materials beforehand: (i) lot-size, (ii) planned lead time and (iii) safety stock. Therefore, a total of 48 input parameters must be given to the simulation and can be optimized. The simulation outputs a number of result values, including the production system's utilization, service level and costs. More specifically, inventory and tardiness costs have been chosen as quality values used to evaluate MRP parameters. Both costs, when summed up, make up the total costs within the system, and have to be minimized. Figure 1 depicts the flow shop with all materials, machines and lines.

3   Efficient Global Optimization

Efficient Global Optimization (EGO) consists of two concepts: (i) adaptive sampling and (ii) *Expected Improvement* (EI). EGO starts off by generating and evaluating n points (i.e. solutions) of the problem to be optimized using the true (expensive) objective function, therefore gathering n samples (i.e. n inputs and corresponding outputs). Using these samples, a dataset is created. This dataset is then used to build a surrogate model that represents the actual evaluation function. EGO uses *kriging* (i.e. a *Gaussian process*) to build a surrogate model. This model cannot only be used to predict outputs for unobserved inputs, but also provides uncertainty for each prediction. This uncertainty can be used to calculate the EI for each input $\mathbf{x}$, as shown in Eq. 1, where $f_{\min}$ denotes the current best objective value observed so far (i.e. in the dataset), $\hat{y}$ denotes the model's predicted output for $\mathbf{x}$, s is the model's uncertainty of prediction $\hat{y}$, and Φ and ϕ denote the standard normal density and distribution functions, respectively.

$$E[I(\mathbf{x})] = \underbrace{(f_{\min} - \hat{y})\Phi\left(\frac{f_{\min} - \hat{y}}{s}\right)}_{\text{exploitation}} + \underbrace{s\phi\left(\frac{f_{\min} - \hat{y}}{s}\right)}_{\text{exploration}} \qquad (1)$$

EI is used to find new promising solutions that should be expensively evaluated and balances the search between exploration and exploitation. In its standard definition, EGO starts a gradient descent algorithm and evaluates solutions using EI. Therefore, after the gradient descent finishes, the solution with highest EI found is returned. This solution is then evaluated using the expensive objective function. Finally, the resulting sample is added to the dataset, the best solution found so far is updated and the kriging model is rebuilt. EGO stops after its termination criterion has been met, e.g. a certain number of iterations have been conducted. Algorithm 1 shows the EGO formulation in pseudo-code.

$model \leftarrow NULL$
$points \leftarrow$ Sample()   ▷ create and evaluate initial collection of points

$bestPoint, bestQuality \leftarrow$ FindBest($points$)

while termination criterion not met **do**
    $model \leftarrow$ BuildModel($points$)   ▷ build surrogate model
    $point \leftarrow$ OptimizeEI($model, points$)   ▷ find (new) point with highest EI
    $quality \leftarrow$ EvaluateExpensively($point$)   ▷ evaluate actual quality

    $points$.Add($point, quality$)   ▷ add point to collection of points

    if IsBetter($quality, bestQuality$) **then**   ▷ update best found so far
        $bestPoint \leftarrow point$
        $bestQuality \leftarrow quality$
    end if
end while

Algorithm 1. Pseudo-code describing the EGO workflow.

4   Optimization Networks

Optimization Networks (ONs) are a relatively young concept implemented in HeuristicLab[1], a paradigm-independent and extensible environment for heuristic optimization. Using ONs, it is possible to combine multiple algorithms and let them work together in a cooperative fashion. The main entities of optimization networks are nodes, which can communicate with each other. Each node of an optimization network fulfills a certain task, e.g. it can execute algorithms, serve problem parameters, or conduct other tasks defined by the user. In this paper, two different ONs have been implemented.

---

[1] https://dev.heuristiclab.com.

4.1 Single-Objective, Synchronous Optimization Network

This single-objective (SO) network resembles an algorithm comparable to the default EGO algorithm. The network is started using the control node. The whole communication in the network happens via an orchestrator node. The orchestrator retrieves algorithm and problem parameters used by other nodes from the parameters provider and forwards them accordingly. Initially, the orchestrator creates n samples (i.e. solutions) and sends them to the evaluator. The evaluator executes the production simulation for each incoming sample and returns the simulated total costs. With these results, the orchestrator builds an initial dataset. This dataset then gets sent to the total costs model builder. This algorithm node executes a Gaussian process regression algorithm, which returns a Gaussian process model. This model is sent back to the orchestrator and forwarded to the last algorithm node, the optimizer. The optimizer uses the model to find a new sample (i.e. solution) with high expected improvement. After the optimizer finishes, the algorithm node sends the best found solution (w.r.t. its expected improvement) to the orchestrator, who then again forwards this single solution to the evaluator for simulation. The inputs, together with the simulated output make up a new row to be added to the dataset. A new kriging model is built, a new sample with high expected improvement is found and the simulation will be executed again. The ON repeats these steps until its termination criterion, i.e. a maximum number of expensive evaluations, has been reached. Figure 2 depicts the workflow within this ON.

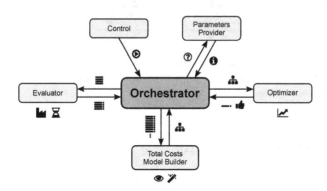

Fig. 2. The workflow of the single-objective, synchronous optimization network. Model building, optimization of EI and expensive evaluations block each other and are executed one at a time.

4.2 Multi-Objective, Asynchronous Optimization Network

The multi-objective (MO) network conceptually works similar to the aforementioned single-objective, synchronous network, however, the workflow is now asynchronous. Two model builders are now executed. The first model builder creates surrogate models that predict inventory costs, the second model builder is used to generate surrogate models for tardiness costs. Both model builders run in parallel. As soon as the first models have been built, the optimizer uses both to find new samples with high expected improvement. A multi-objective algorithm is used here, which returns a set of Pareto-optimal solutions in the end. All solutions are then sent to the evaluator for simulation. Once the optimizer finishes, it is restarted immediately to find new points with high expected improvement. The model builders are also immediately restarted each time after they finish. Figure 3 depicts the workflow with subtle differences compared to Fig. 2. It is noteworthy that the evaluated multi-objective optimization network uses only a single evaluator node, two model builders and one optimizer, but due to its asynchronous nature, it would also be possible to extend the implementation and include multiple evaluators, model builders and optimizers. This is indicated in Fig. 3 by dashed nodes.

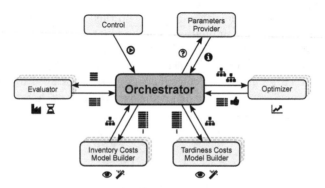

Fig. 3. The workflow of the multi-objective, asynchronous optimization network. Model building, optimization of EI and expensive evaluations do not block each other and are executed at the same time.

5   Experiments and Results

The experiments have been set up as follows. Single- and multi-objective variants of the genetic algorithm have been used to optimize the expected improvement of points. In the single-objective case, a default genetic algorithm is used. In the multi-objective case, an NSGA-II [2] is executed. Both networks use Heuristic-Lab's implementation of Gaussian processes to build surrogate models. Initially, a total of 10 samples are generated randomly and simulated. The simulation is executed with 10 replications for each sample and 10 runs have been conducted with both networks, whereby the execution of a single run is stopped once 1,000 expensive evaluations (i.e. $1,000 * 10 = 10,000$ simulations) have been conducted.

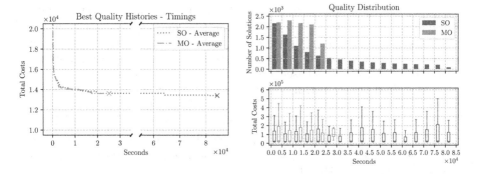

Fig. 4. The best quality histories (10 runs' average) and quality distribution for all conducted runs.

Figure 4 visualizes the obtained results. To be able to compare the single- and multi-objective results, the objectives in the MO case have been summed to calculate the respective total costs. Both networks achieve rather similar quality values after their evaluation limit (i.e. 1,000 evaluations) has been reached. On average, the SO network yields slightly better results. However, due to its asynchronous nature, the MO network is able to evaluate more solutions in a shorter time frame and thus yields more good solutions in a shorter amount of time. It can also be observed that the SO network is not able to achieve much to any progress in solution quality after some time. The bigger the dataset becomes, the longer the model building takes. On the one hand, this explains the longer runtime of the SO network, because here, model building is a blocking operation. On the other hand, this means that in the MO network, where model building is a non-blocking operation that runs asynchronously, less frequent updates of the surrogate models occur. While a new model is being built, the network still uses the old model and tries to find new solutions with high expected improvement and expensively evaluates those. New knowledge is only incorporated once the model is updated. This might explain why it performs a bit worse than its SO counterpart, where the model is updated after each single gain of knowledge (i.e. expensive evaluation).

6   Conclusion and Outlook

The EGO "framework" has been used to optimize MRP parameters for a simplified flow shop system. Two optimization networks have been implemented, one single-objective version that represents a sequential, more standardized workflow of EGO, and a multi-objective version that asynchronously builds surrogate models, tries to find new promising solutions with high expected improvement and conducts expensive evaluations. The achieved results show that the single-objective version yields slightly better results, however, the runtime required for 1,000 evaluations is much higher and there is not much progress in terms of solution quality in later search periods. The multi-objective network yields more high-quality solutions in a shorter time and finishes faster, but all in all yields slightly worse solutions when compared to the single-objective ON. Further work on this matter can be done by conducting simulation-based optimization in order to generate a baseline for comparison and to determine how well the implemented networks truly perform. Additionally, one should also explore how to prioritize expensive evaluations and model building to incorporate new information, since spending expensive evaluations for solutions whose EI has been calculated with outdated surrogate models is wasteful. With such prioritization, a trade-off between achieved solution quality and consumed runtime could probably be maintained.

Acknowledgments. The work described in this paper was done within the *Produktion der Zukunft* Project *Integrated Methods for Robust Production Planning and Control* (SIMGENOPT2, #858642), funded by the Austrian Research Promotion Agency (FFG).

References

1. Altendorfer, K., Felberbauer, T., Jodlbauer, H.: Effects of forecast errors on optimal utilisation in aggregate production planning with stochastic customer demand. Int. J. Prod. Res. **54**(12), 3718–3735 (2016)
2. Deb, K., Pratap, A., Agarwal, S., Meyarivan, T.: A fast and elitist multiobjective genetic algorithm: NSGA-II. IEEE Trans. Evol. Comput. **6**(2), 182–197 (2002)
3. Hauder, V.A., Karder, J., Beham, A., Wagner, S., Affenzeller, M.: A general solution approach for the location routing problem. In: Moreno-Díaz, R., Pichler, F., Quesada-Arencibia, A. (eds.) EUROCAST 2017. LNCS, vol. 10671, pp. 257–265. Springer, Cham (2018). https://doi.org/10.1007/978-3-319-74718-7_31
4. Hopp, W.J., Spearman, M.L.: Factory Physics. Waveland Press, Long Grove (2011)
5. Jones, D.R., Schonlau, M., Welch, W.J.: Efficient global optimization of expensive black-box functions. J. Global Optim. **13**(4), 455–492 (1998)
6. Karder, J., Wagner, S., Beham, A., Kommenda, M., Affenzeller, M.: Towards the design and implementation of optimization networks in HeuristicLab. In: Proceedings of the Genetic and Evolutionary Computation Conference Companion, GECCO 2017, pp. 1209–1214. ACM (2017)
7. Kommenda, M., et al.: Optimization networks for integrated machine learning. In: Moreno-Díaz, R., Pichler, F., Quesada-Arencibia, A. (eds.) EUROCAST 2017. LNCS, vol. 10671, pp. 392–399. Springer, Cham (2018). https://doi.org/10.1007/978-3-319-74718-7_47
8. Rasmussen, C.E.: Gaussian processes in machine learning. In: Bousquet, O., von Luxburg, U., Rätsch, G. (eds.) ML 2003. LNCS (LNAI), vol. 3176, pp. 63–71. Springer, Heidelberg (2004). https://doi.org/10.1007/978-3-540-28650-9_4

Surrogate-Assisted Fitness Landscape Analysis for Computationally Expensive Optimization

Bernhard Werth[1,2(✉)], Erik Pitzer[1], and Michael Affenzeller[1,2]

[1] Heuristic and Evolutionary Algorithms Laboratory,
University of Applied Sciences Upper Austria, Hagenberg, Austria
bernhard.werth@fh-hagenberg.at
[2] Johannes Kepler University, Altenberger Straße 68, 4040 Linz, Austria

Europäische Union Investitionen in Wachstum & Beschäftigung, Österreich.

Exploratory fitness landscape analysis (FLA) is a category of techniques that try to capture knowledge about a black-box optimization problem. This is achieved by assigning features to a certain problem instance utilizing only information obtained by evaluating the black-box. This knowledge can be used to obtain new domain knowledge but more often the intended use is to automatically find an appropriate heuristic optimization algorithm [9]. FLA-based algorithm selection and parametrization hinges on the idea, that, while no optimization algorithm can be the optimal choice for all black-box problems, algorithms are expected to work similarly well on problems with similar statistical characteristics [8,15].

For some applications of heuristic optimization the objectives or constraint functions are not provided in a closed mathematical form. Rather, highly precise simulation models are used to assign objective values to candidate solutions. Examples for application areas of simulation-based optimization are production, energy distribution, fluid dynamics and traffic systems [5,12,17]. In such situations a single fitness evaluation can easily require minutes or hours of execution time. This restricts the use of many conventional heuristic algorithms like Genetic Algorithms [18], that are usually expected to converge after millions of function evaluations [13]. An alternative approach to solving computationally expensive problems is the use of surrogate models.

Surrogate-assisted optimization algorithms use general purpose regression techniques in order to create computationally cheaper approximations of the expensive optimization problem.

The work described in this paper was done within the project "Connected Vehicles" which is funded by the European Fund for Regional Development (EFRE; further information on IWB/EFRE is available at www.efre.gv.at) and the country of Upper Austria.

R. Moreno-Díaz et al. (Eds.): EUROCAST 2019, LNCS 12013, pp. 247–254, 2020.
https://doi.org/10.1007/978-3-030-45093-9_30

- *Gaussian process models* are often called *Kriging* models and are especially attractive for surrogate models as they can be made to fit every point in the training data set exactly and provide a measure of the models' own uncertainty. In the *efficient global optimization* (EGO) scheme [3] this uncertainty is used to sample new data points that are promising in quality and contribute substantially to the next model.
- *Support vector machines* have been used in conjunction with surrogate-assisted algorithms both as regression models [2] or as ranking models, identifying the most promising candidate solution of a population [6].
- A severe drawback of both support vector machines and Gaussian processes is that the time to build them scales, limiting the applicability of these models to scenarios where only a few sample points are used to build the model. *Random forests* on the other hand are fast to build, can approximate complex landscapes [1] and provide a form of confidence measure as the agreement of the different trees. A distinct drawback of random forests as surrogate models is that at any given point the derivative is $\mathbf{0}$, which can hinder heuristic algorithms that fail to obtain a meaningful search direction.
- *Polynomial regression models* are often used as baseline models in comparisons as they are cheap and easy to build. Actual uses for polynomial models are as parts of ensemble model structures comprising multiple surrogate models [19]. Another application scenario was presented in [4], where polynomial representations were used to approximate very high dimensional data.

Extending a heuristic algorithm with surrogate-assistance can drastically increase the number of parameters for these algorithms. This as well as the increased execution time make the automatic selection of algorithms, models and parameters for expensive optimization even more important than for conventional problems. However, for such an automated selection system to be employed successfully, a way to extract FLA features from computationally expensive problems needs to be found. The central research question of this paper is, how much a set of FLA features deviates when taken from the surrogate models rather than the actual functions.

A standard set of fitness landscape features are *walk-based* FLA measures, that are calculated by employing a *random walk*. The random walks used in this work are series of evaluations $\{y_0, y_1, \ldots, y_m\} = \{f(x_0), f(x_1), \ldots, f(x_m)\}$ with x_0 being sampled uniformly random from the search space and $x_{i+1} = x_i + \delta$, where δ is sampled uniformly random from the surface of a sphere with radius r.

- The *autocorrelation* function of a random walk $\rho(\delta) = \frac{E(y_t \cdot y_{t+\delta}) - E(y_t)E(y_{t-\delta})}{\text{Var}(y_t)}$ of a random walk is one of the oldest FLA measures [16]. Most prominently $\rho(1)$ is used as a singular value, which will be used going forward and called autocorrelation for brevity.
- The *information content* [14] of a random walk is calculated by first transforming the series $\boldsymbol{y} = \{y_0, y_1, \ldots, y_m\}$ to $\boldsymbol{\phi} = \{p_0, p_1, \ldots, p_{m-1}\}$ where $p_i = 1$ if $y_{i+1} - y_i < -\epsilon$, $p_i = -1$ if $y_{i+1} - y_i > \epsilon$ and $p_i = 0$ else. Then, for all combinations $s \in S = \{(0,1), (0,-1), (1,0), (-1,0), (1,-1), (-1,1)\}$ their

occurrence in the transformed series is counted as c_s and the information content is calculated as $\sum_{s \in S} \left(\frac{c_s}{(m-1)} * -\log_3 \left(\frac{c_s}{(m-1)} \right) \right)$. Going forward ϵ is chosen as 0.1.

- The *density basin information* [14] is calculated similarly to the information content with the only difference being that the set of symbols is complementary $S = \{(0,0), (1,1), (-1,-1)\}$
- The *partial information content* [14] is obtained by removing all 0's from ϕ to create ϕ' and calculating the ratio of lengths of the altered and original series $\frac{\phi'}{\phi}$.
- The *information stability* [14] is the highest fitness difference between neighbours in y.

In the following experiment, the above FLA measures from both the actual fitness function and a surrogate thereof are compared. Table 1 lists the experiment parameters. As optimization problems four academic test functions were selected. For analysing the deviations introduced by the surrogate, choosing well researched test problems is more sensible than using full simulations. This way, the expected FLA results are known beforehand and it has the added benefit of allowing for better interpretation of the FLA results. All selected test functions have a configurable degree of dimensionality (number of input variables). As surrogate-assisted optimization is mostly used with lower-dimensional problems the dimensionality n of the test functions is varied from 1 to 10. The length m of a random walk as well as the size between random steps r can have significant impact on the experienced FLA features and are therefore varied pseudo-logarithmically. Four different types of models are compared: a Gaussian process model, a random forest, a support vector machine and, as a baseline, a linear regression model. The number of fitness evaluations is scaled linearly with the problem dimensionality.

Table 1. Experiment parameters

Name	Range
FLA measures	autocorrelation, information content, density basin information, partial information content, information stability
Test function	Ackley, Griewank, Rastrigin, Rosenbrock (see [10] for definitions)
Problem dimension n	$\{1, 2, 3, 4, 5, 6, 7, 8, 9, 10\}$
Walk length m	$\{100, 200, 500, 1000, 2000, 5000\}$
Step size r	$\{0.01\%, 0.1\%, 1\%, 5\%, 10\%, 50\%\}$
Sample size s	$\{1 \cdot n, 2 \cdot n, 3 \cdot n, 5 \cdot n, 10 \cdot n, 20 \cdot n\}$
Model type	linear regression, Gaussian process, support vector machine, random forest

Lastly, in preliminary results it has emerged, that the way bounding box constraints are handled has significant influence on the obtained FLA measures. Strategies for dealing with bounding box violations are resampling the step in the random walk, enforcing the bound by setting the position of the next step at the bounds themselves. Additionally, while the above mentioned test functions provide bounding box constraints for input variables, they actually can be calculated for any vector of real numbers, therefore the bounding box constraints could just be ignored. Figure 1 shows an example of a two dimensional random walk that has exceeded its bounds and the three possible strategies that could be employed to continue the walk. For the experiments in this work we decided to employ a resampling strategy. Figure 2 displays the median autocorrelation values measured from the Rastrigin function and a Gaussian process model over both FLA inherent parameters step size and walk length. While both surfaces have roughly the same shape, the characteristic "dip", that is caused by the many equally sized local optima of the Rastrigin function, is reflected as high variance in the right surface. In the following analysis the effects of walk length and step size are compensated for by assigning every combination of test function, dimensionality, model type and sample size a surface of FLA values rather than a singular value.

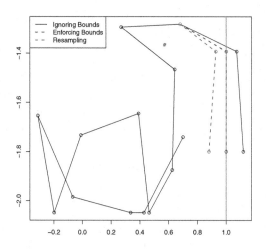

Fig. 1. Example of bounding strategies for a random walk with random step sizes

In order to evaluate how much the FLA results obtained from surrogate models differ from surface to surface, the root mean squared difference between surfaces is chosen as a distance function. Utilizing t-distributed Stochastic Neighbour Embedding (t-SNE) [7] the results of the fitness landscape analysis can be visualized in a two-dimensional space. Figure 3 shows such an embedding of the autocorrelation surfaces from the actual test functions. The embedded points form easily distinguishable clusters that correlate to the individual test func-

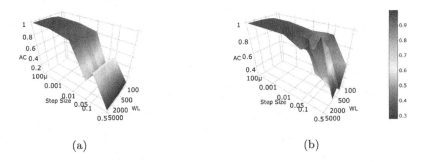

(a)   (b)

Fig. 2. Autocorrelation (AC) of the Rastrigin test function [11] measured with different walk lengths (WL) and step sizes on the actual function (a) and on a Gaussian process model (b) created from 100 sample points.

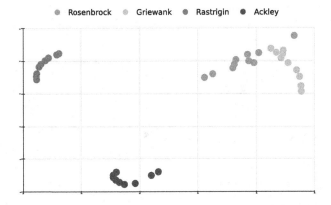

Fig. 3. t-SNE projection of the autocorrelation surfaces obtained from the actual test functions

tions. The clusters relating to the Rosenbrock and Griewank functions are, while still separable, closer to each other than to other clusters.

For surrogate-based fitness landscape analysis as described above to provide meaningful information and prediction features, similar clusters corresponding to the underlying test functions should also emerge when embedding the FLA surfaces taken from surrogate models. Figure 4 presents the results of such embeddings for each selected FLA measure. Immediately, several very particular structures can be identified. Excluding information stability, every embedding contains a very pronounced ring structure, a long narrow dense cluster, a loose cloud spanning a large area of the embedded space and strongly distorted rectangle. As can be seen when colouring the same embedded points by the underlying fitness function (column (a)) and by model type (column (b)), these structures do not correlate to the fitness functions they were taken from, but very clearly to the type of surrogate model used. The ring structure contains all FLA surfaces obtained from linear surrogate models. The distorted rectangle consists almost solely of results obtained from random forest models. The strong

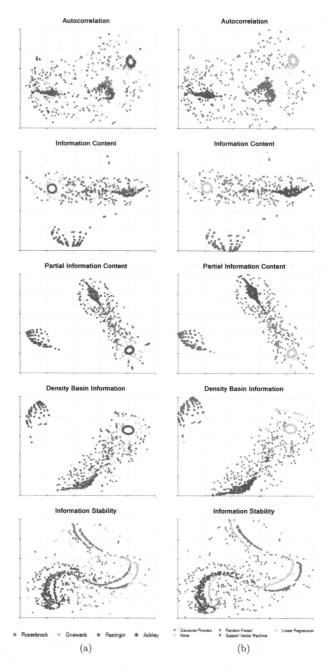

Fig. 4. Embedding of the FLA-feature surfaces. Points are coloured by test function in column (a) and by model type in column (b).

difference between the random forest cluster and the others for the FLA features information content, partial information content and density basin information is likely a result of the fact that random forest surrogates consist of many patches of flat surfaces and changes in predicted fitness will appear less often in a random walk, than with other types of surrogates. The dense and loose clouds of points correspond clearly to the Gaussian process and support vector machine model types respectively.

The embedding for information stability displays a drastically different clustering and when comparing both colourings, it can be seen, that both test functions and model types correlate to several line-like clusters. It is important to note, that the information content value is not invariant to scaling of the objective values and the main reason why clusters corresponding to the test functions emerge is due to the very different scaling in objectives for the different fitness functions (e.g.: $y \in [0, 23]$ for Ackley and $y \in [0, 35146]$ for Rosenbrock).

A number of conclusions can be drawn from these results. First, approximating fitness landscape measurements by applying the landscape analysis to the surrogate model rather than to the expensive actual fitness function is not trivial. At least for the selected measures, that are mainly designed to measure bumpiness and ruggedness, the FLA results are far more indicative of the type of surrogate model than the original landscape. On the other hand, it might be more beneficial to tune ruggedness-sensitive parameters to the surrogate landscape rather than the original one, as in typical surrogate-assisted scenarios the surrogate model is queried far more often than the expensive function. Other FLA measures that aim at capturing more global properties, however, might be considerably easier to approximate. Future research will not only have to establish which measures can effectively be obtained from surrogate models, but also to what degree an optimization algorithm should be tuned towards the actual or the surrogate objective function. Lastly, there might exist a number of more sophisticated ways to obtain surrogate-based FLA results. For example, instead of using sample points that are independent from the random walk, selecting sample points in a fashion so that the predictions are maximally accurate for the walk might immediately improve the approximation.

References

1. Fleck, P., et al.: Box-type boom design using surrogate modeling: introducing an industrial optimization benchmark. In: Andrés-Pérez, E., González, L.M., Periaux, J., Gauger, N., Quagliarella, D., Giannakoglou, K. (eds.) Evolutionary and Deterministic Methods for Design Optimization and Control With Applications to Industrial and Societal Problems. CMAS, vol. 49, pp. 355–370. Springer, Cham (2019). https://doi.org/10.1007/978-3-319-89890-2_23
2. Jin, Y.: Surrogate-assisted evolutionary computation: recent advances and future challenges. Swarm Evol. Comput. 1(2), 61–70 (2011)
3. Jones, D.R., Schonlau, M., Welch, W.J.: Efficient global optimization of expensive black-box functions. J. Global Optim. 13(4), 455–492 (1998). https://doi.org/10.1023/A:1008306431147

4. Kubicek, M., Minisci, E., Cisternino, M.: High dimensional sensitivity analysis using surrogate modeling and high dimensional model representation. Int. J. Uncertain. Quantif. **5**(5), 393–414 (2015)

5. Li, Z., Shahidehpour, M., Bahramirad, S., Khodaei, A.: Optimizing traffic signal settings in smart cities. IEEE Trans. Smart Grid **8**(5), 2382–2393 (2017)

6. Loshchilov, I., Schoenauer, M., Sebag, M.: Self-adaptive surrogate-assisted covariance matrix adaptation evolution strategy. In: Proceedings of the 14th Annual Conference on Genetic and Evolutionary Computation, pp. 321–328. ACM (2012)

7. van der Maaten, L., Hinton, G.: Visualizing data using t-SNE. J. Mach. Learn. Res. **9**(Nov), 2579–2605 (2008)

8. Malan, K.M., Engelbrecht, A.P.: Fitness landscape analysis for metaheuristic performance prediction. In: Richter, H., Engelbrecht, A. (eds.) Recent Advances in the Theory and Application of Fitness Landscapes. ECC, vol. 6, pp. 103–132. Springer, Heidelberg (2014). https://doi.org/10.1007/978-3-642-41888-4_4

9. Mersmann, O., Bischl, B., Trautmann, H., Preuss, M., Weihs, C., Rudolph, G.: Exploratory landscape analysis. In: Proceedings of the 13th Annual Conference on Genetic and Evolutionary Computation, pp. 829–836. ACM (2011)

10. Molga, M., Smutnicki, C.: Test functions for optimization needs, p. 101 (2005)

11. Potter, M.A., De Jong, K.A.: A cooperative coevolutionary approach to function optimization. In: Davidor, Y., Schwefel, H.-P., Männer, R. (eds.) PPSN 1994. LNCS, vol. 866, pp. 249–257. Springer, Heidelberg (1994). https://doi.org/10.1007/3-540-58484-6_269

12. Saffari, M., de Gracia, A., Fernández, C., Cabeza, L.F.: Simulation-based optimization of PCM melting temperature to improve the energy performance in buildings. Appl. Energy **202**, 420–434 (2017)

13. Tang, K., et al.: Benchmark functions for the CEC 2008 special session and competition on large scale global optimization. Nature Inspired Computation and Applications Laboratory, USTC, China, 24 (2007)

14. Vassilev, V.K., Fogarty, T.C., Miller, J.F.: Information characteristics and the structure of landscapes. Evol. Comput. **8**(1), 31–60 (2000)

15. Watson, J.-P.: An introduction to fitness landscape analysis and cost models for local search. In: Gendreau, M., Potvin, J.Y. (eds.) Handbook of Metaheuristics. ISOR, vol. 146, pp. 599–623. Springer, Boston (2010). https://doi.org/10.1007/978-1-4419-1665-5_20

16. Weinberger, E.: Correlated and uncorrelated fitness landscapes and how to tell the difference. Biol. Cybern. **63**(5), 325–336 (1990)

17. Werth, B., Pitzer, E., Ostermayer, G., Michael, A.: Surrogate-assisted high-dimensional optimization on microscopic traffic simulators. In: Proceedings of the 30th European Modeling and Simulation Symposium EMSS 2018, pp. 46–52, (2018)

18. Wright, A.H.: Genetic algorithms for real parameter optimization. In: Foundations of genetic algorithms, vol. 1, pp. 205–218. Elsevier (1991)

19. Zhou, Z., Ong, Y.S., Nguyen, M.H., Lim, D.: A study on polynomial regression and Gaussian process global surrogate model in hierarchical surrogate-assisted evolutionary algorithm. In: 2005 IEEE Congress on Evolutionary Computation, vol. 3, pp. 2832–2839. IEEE (2005)

VNS and PBIG as Optimization Cores in a Cooperative Optimization Approach for Distributing Service Points

Thomas Jatschka[1]([✉]), Tobias Rodemann[2], and Günther R. Raidl[1]

[1] Institute of Logic and Computation, TU Wien, Vienna, Austria
{tjatschk,raidl}@ac.tuwien.ac.at
[2] Honda Research Institute Europe, Offenbach, Germany
tobias.rodemann@honda-ri.de

Abstract. We present a cooperative optimization approach for distributing service points in a geographical area with the example of setting up charging stations for electric vehicles. Instead of estimating customer demands upfront, customers are incorporated directly into the optimization process. The method iteratively generates solution candidates that are presented to customers for evaluation. In order to reduce the number of solutions presented to the customers, a surrogate objective function is trained by the customers' feedback. This surrogate function is then used by an optimization core for generating new improved solutions. In this paper we investigate two different metaheuristics, a variable neighborhood search (VNS) and a population based iterated greedy algorithm (PBIG) as core of the optimization. The metaheuristics are compared in experiments using artificial benchmark scenarios with idealized simulated user behavior.

Keywords: Cooperative optimization · Facility location problem · Metaheuristics

1 Introduction

Usually, locations for setting up charging stations for electric vehicles are optimized by some algorithm w.r.t. previously estimated customer demand. However, estimating the demand of a customer upfront is challenging, as this task is usually based on various uncertain data and uncertain assumptions about the behavior of customers. Among other things, customer demand may be fulfilled in alternative ways that cannot all be predicted in advance. Sometimes, customers might not even completely be aware themselves at which conditions their demands can be fulfilled best until they can see an actual system configuration. At this point, however, it is usually too late to make any further impactful changes.

Thomas Jatschka acknowledges the financial support from Honda Research Institute Europe.

R. Moreno-Díaz et al. (Eds.): EUROCAST 2019, LNCS 12013, pp. 255–262, 2020.
https://doi.org/10.1007/978-3-030-45093-9_31

Therefore, we propose in [4] a *Cooperative Optimization Approach (COA)* for solving this problem by incorporating potential customers into a combined data acquisition and optimization process. The method iteratively generates solutions that are presented to the customers for evaluation. Based on the customer's feedback a surrogate objective function is trained and used by an optimization core to generate new, improved solutions. This process is iterated on a large scale with many potential customers and several rounds until a satisfactory solution is reached.

In this contribution, we focus in particular on the optimization part of our approach. A proper configuration of the used optimization core is vital for the cooperative approach to work. We investigate a variable neighborhood search (VNS) and a population based iterated greedy algorithm (PBIG) for this purpose. First, however, we formally introduce the problem to be solved – the Service Point Distribution Problem (SPDP).

2 The Service Point Distribution Problem

While this paper considers the distribution of charging stations for electric vehicles as particular example, we define the underlying problem we consider – the Service Point Distribution Problem (SPDP)– in a more general way to be independent of the actual application scenario. In the SPDP we are given a set of locations V at which service points may be built and a set of users U for which the service points are built. The fixed costs for setting up a service point at location $v \in V$ are $c_v \geq 0$, and this service point's maintenance over a defined time period is supposed to induce variable costs $z_v \geq 0$. The total construction costs must not exceed a maximum budget $B > 0$. Erected service stations may satisfy customer demand, and for each unit of satisfied customer demand a prize $p > 0$ is earned.

A solution to the SPDP is given by a binary incidence vector $x = (x_v)_{v \in V}$, where $x_v = 1$ indicates that a service point is to be set up at location v.

The objective is to find a feasible solution that maximizes the prizes earned for satisfied customer demands reduced by the variable costs for maintaining the service points

$$f(x) = p \cdot \sum_{u \in U} \sum_{v \in V} d(u, v, x) - \sum_{v \in V} z_v x_v, \tag{1}$$

where $d(u, v, x)$ specifies the demand of user $u \in U$ fulfilled at location $v \in V$ in solution x. The problem is incompletely specified in the sense that we do not know how to calculate $d(u, v, x)$. The function can only be evaluated by presenting the solution x to user u and collecting the user's feedback. Clearly, the number of candidate solutions that are evaluated in this interactive way must be kept low, as we cannot confront each user with hundreds of evaluation requests. Hence, we additionally make use of a surrogate function $\tilde{d}(u, v, x)$ trained by obtained user feedback, which is actually used by the optimization core for evaluating solutions.

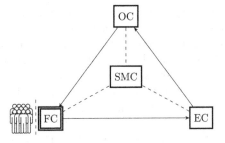

Fig. 1. The cycle of the COA framework. Users evaluate candidate solutions provided by the FC. The feedback is used to train a surrogate function in the EC which is used by the OC to find new optimized solutions.

3   Related Work

The SPDP is a variant of the *uncapacitated Facility Location Problem* (uFLP) [3]. The uFLP generally deals with selecting a subset from a set of potential facility sites in order to serve a set of demand points w.r.t. some optimization goal subject to a set of constraints.

With the rise of electric vehicles, the problem of finding optimal locations for charging stations has gained increased attention recently. There already exists a large number of studies concerning this topic. Moreover, these studies all have to address the problem of how to determine the demands of potential customers. Chen et al. [2] derive the customer demand for charging electric vehicles by using parking demand gained from a travel survey. In [5] charging stations for an on-demand bus system are located using taxi probe data of Tokyo.

For a survey on interactive optimization algorithms see [9]. Continuous user interactions will eventually result in user exhaustion [7], negatively influencing the reliability of the obtained feedback. A way to unburden the users is to use a surrogate-based approach. Surrogate models are typically used as a proxy of functions which are either unknown or extremely time consuming to compute [6].

4   Cooperative Optimization Approach (COA)

In this section, we summarize the COA as proposed in [4], which consists of: an *evaluation component (EC)*, an *optimization core (OC)*, a *feedback component (FC)*, and a *solution management component (SMC)*, see also Fig. 1.

The SMC stores and manages solutions and all associated data such as user feedback or surrogate objective values. All framework components can access the SMC.

The FC provides the interface to the users and is responsible for deriving an individual set of solutions for each user that is then presented to the user for evaluation. Solutions are derived from the so far best found solutions stored in the SMC with the purpose of identifying new relevant locations for a user and determining the relationship between a user's relevant locations. Each user gives feedback to the proposed solutions by stating how much of the user's demand would actually be satisfied at which locations, i.e., a user $u \in U$ returns the values $d(u, v, x)$ for all $v \in V$ w.r.t. a solution x. Weaker preferences of locations are hereby expressed by smaller values.

Once the feedback is obtained from the users, the EC builds a surrogate function $\tilde{d}(u, v, x)$ based on machine learning (ML) models $g_{u,v}$ for each pair $(u, v) \in U \times V_u$, where V_u is the set of so far identified relevant locations of user $u \in U$, i.e., the set of locations for which a user has expressed a positive demand at least once. Each $g_{u,v}$ gets as input a binary incidence vector $x = (x_w)_{w \in V_u, w \neq v}$, with $x_w = 1$ indicating that a service point exists at location w. Initially, each $g_{u,v}$ starts out as a linear regression model. However, once the mean squared error (MSE) of $\tilde{d}(u, v, x)$ exceeds some threshold $\tau = 0.075$, the model is upgraded to a perceptron. The model can be further upgraded to a neural network with a single hidden layer and initially two neurons in the hidden layer. Afterwards, whenever the MSE of $\tilde{d}(u, v, x)$ exceeds τ, an additional neuron is added to the hidden layer of the neural network until a maximum number of neurons is reached.

In the last step of a COA iteration, new, improved solutions are generated in the OC using the surrogate function for evaluating solutions. The OC is implemented as a black-box optimization model and returns one or multiple close-to-optimal solutions w.r.t. $\tilde{d}$. In the next section we investigate two different methaheuristics serving as core optimization of COA.

4.1 VNS and PBIG as Core Optimization of COA

A proper choice of optimization algorithm and corresponding configuration is vital for the COA to work. The optimization algorithm should not only provide close-to-optimal solutions but should also not need too many candidate solution evaluations as they are rather time-expensive and the OC needs to be repeatedly performed. For this purpose we consider two different metaheuristics – a VNS and a PBIG – as OC.

The VNS follows the classical scheme from [10]. An initial solution is generated via the randomized construction heuristic that considers all locations in random order and sets up a station at a location as long as the budget is not exceeded.

Our local search uses an exchange & complete neighborhood following a first improvement strategy. In the first step of the neighborhood, a location in the solution is replaced by an unused location, i.e., a location at which no service point exists. As this exchange might decrease the current budget of the solution, we afterwards add further unused locations in a random order to the solution as long as the budget allows it. The k-th shaking removes k randomly selected locations from the solution and then iteratively adds unused locations in a uniform random order until no more locations can be added. Note that the VNS considers only feasible solutions.

For the general principles of the PBIG we refer to [1]. An initial population of solutions is generated via the same randomized construction used for generating the initial solution of the VNS. Then, in each major iteration a new solution is derived from each solution in the current population by applying a destroy & recreate operation. The best solutions from the joint set of original and newly derived solutions are accepted as new population for the next iteration. Our destroy & recreate operation first removes a number of selected locations from the solution and then again iteratively adds unused locations in a uniform random order, such that the solution stays feasible and no more stations can be added. We reuse the exchange & complete neighborhood as well as the shakings operators of the VNS as destroy & recreate operations of the PBIG. Hence, the PBIG also considers only feasible solutions. Note that the PBIG returns its final population while the VNS only returns a single solution as result of the optimization.

5 Experimental Evaluation

As previously mentioned, the COA is tested in a proof-of-concept manner on artificial benchmark scenarios using an idealized simulation of all user interaction.

The primary parameters for our benchmark scenarios are the number of potential locations for service stations n and the number of users m, and we consider here the combinations $n = 50, 60, \ldots, 100$ with $m = 50$ and $n = 50$ with $m = 50, 60, \ldots, 100$. The n locations correspond to points in the Euclidean plane with coordinates chosen uniformly at random from a grid with height and width $\lceil 10\sqrt{n} \rceil$. The fixed costs c_v as well as the variable costs z_v for setting up a service station at each location $v \in V$ are uniformly chosen at random from $\{50, \ldots, 100\}$. The budget is chosen in such a way that about 10% of the service points can be set up on average.

Each user has a set of *use case locations* distributed in the same grid as the potential service point locations V. The number of use case locations is determined by a shifted Poisson distribution with offset one and expected value three. The use case locations themselves are determined with a set of randomly chosen attraction points, i.e., the closer a location is to an attraction point, the more likely the location is to be chosen as a use case location. Each use case location is associated with a maximal demand that can be partially fulfilled by service points within a maximal walking distance. The satisfied demand is calculated by a distance decay function. It is assumed that a user always chooses the service point closest to a use case location, i.e., the service point that can satisfy most of the use case location's demand.

For each combination of n and m 30 independent scenarios were created. They are available at www.ac.tuwien.ac.at/research/problem-instances/#spdp. The instances were also specifically designed with the ability in mind to calculate proven optimal solutions to which we will compare the solutions of our framework.

<table>
<tr><td colspan="2"></td><td colspan="4">Table 1. VNS vs. PBIG.</td></tr>
</table>

Table 1. VNS vs. PBIG.

n	m	VNS %-gap	$\sigma_{\%\text{-gap}}$	n_{it}^{best}	t[s]	PBIG %-gap	$\sigma_{\%\text{-gap}}$	n_{it}^{best}	t[s]
50	50	0.26	0.48	29	3	0.28	1.22	696	8
50	60	0.20	0.61	35	4	0.31	1.36	656	9
50	70	0.00	0.02	32	4	0.10	0.42	634	8
50	80	0.31	0.74	31	4	0.09	0.28	631	9
50	90	0.14	0.42	32	4	0.37	1.01	648	11
50	100	0.37	1.03	33	4	0.01	0.07	706	15
60	50	0.25	0.64	43	4	0.07	0.34	862	9
70	50	0.34	0.62	54	5	0.28	0.57	1113	17
80	50	0.43	0.54	56	6	0.24	0.51	1389	23
90	50	0.30	0.42	64	7	0.19	0.60	1628	30
100	50	0.37	0.47	65	9	0.42	0.82	1754	39

Table 2. COA[VNS] vs. COA[PBIG].

n	m	COA[VNS] %-gap	$\sigma_{\%\text{-gap}}$	$\overline{n}_{it}$	t[s]	COA[PBIG] %-gap	$\sigma_{\%\text{-gap}}$	$\overline{n}_{it}$	t[s]
50	50	0.28	0.70	11	2259	0.20	0.45	10	2662
50	60	0.73	1.27	9	2343	0.06	0.18	9	2643
50	70	0.14	0.37	10	3107	0.09	0.44	10	3764
50	80	0.19	0.36	10	3588	0.04	0.15	10	3919
50	90	0.42	0.68	10	3596	0.19	0.71	9	4516
50	100	0.12	0.26	10	4391	0.02	0.08	10	4995
60	50	0.48	0.72	11	2460	0.05	0.11	10	2944
70	50	0.46	0.66	11	2533	0.13	0.43	10	3658
80	50	0.22	0.58	12	2864	0.11	0.28	11	4810
90	50	0.37	0.52	11	2910	0.26	0.65	11	5435
100	50	0.49	1.02	12	3460	0.07	0.15	11	7197

5.1 Computational Results

The OC was implemented in C++, compiled with GNU G++ 5.5.0, while the remaining components of the framework were realized in Python 3.7. All test runs have been executed on an Intel Xeon E5-2640 v4 with 2.40 GHz machine.

Initially, we determined the parameter configurations of standalone variants of the VNS and the PBIG, in which it is naively assumed that users can evaluate all intermediate solutions. The parameters have been determined with irace [8] on a separate set of benchmark instances and have been tuned with the goal to minimize the number of iterations it takes the metaheuristics to generate near optimal solutions, i.e., solutions with an optimality gap of 0.5%. For the VNS we obtained to best use shaking neighborhoods $k \in \{1, 2\}$, for PBIG $k \in \{1, \ldots, 10\}$ with a population size of 100. The termination criteria of the metaheuristics have been chosen according to the number of iterations that were necessary on average to find the close-to-optimal solutions, which was 60 iterations without improvement for the VNS, and 300 iterations (or three generations) without improvement for PBIG. Table 1 shows a comparison of the standalone variants of the metaheuristics using above parameter configurations. The table shows for each instance group with n locations and m users the average optimality gap ($\overline{\%\text{-gap}}$) and their corresponding standard deviation ($\sigma_{\%\text{-gap}}$) of the metaheuristics. Moreover, the table also shows the iteration in which the best solution has been found on average ($\overline{n}_{it}^{best}$) and the median of the total computation times ($t[s]$).

PBIG produces slightly better optimality gaps but also needs significantly more time than the VNS. Moreover, it takes PBIG much more iterations to find the best solution as opposed to the VNS.

Next, we tested COA in conjunction with the VNS (denoted as COA[VNS]) and the PBIG (denoted as COA[PBIG]), respectively, as OC. Further tests have shown that COA[PBIG] yields slightly better results when using the so far best found solutions stored in the SMC as initial population. In case there are not enough solutions available, the remaining solutions are generated by the randomized construction heuristic. The other parameters remain unchanged. The COA

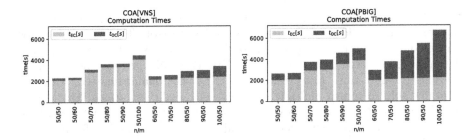

Fig. 2. Median computation times of the COA components for each instance set

terminated after five iterations without improvement or after two hours. The comparison can be seen in Table 2. The table shows again the average optimality gaps ($\overline{\%\text{-gap}}$) and their corresponding standard deviations ($\sigma_{\%\text{-gap}}$). Moreover, the table also shows the average number of COA iterations ($\overline{n_{it}}$) and the median of the total computation times ($t[s]$). While the optimality gaps in Table 1 are somewhat comparable between the VNS and the PBIG, COA[PBIG] clearly outperforms COA[VNS] w.r.t. the optimality gaps. This difference can be explained by the number of solutions returned by the VNS and the PBIG. While the OC of COA[VNS] only returns one solution in every COA iteration, the OC of COA[PBIG] returns 100 solutions in every iteration. A higher number of solutions in the SMC results in more diversified solutions not only in the FC but also in the EC. Hence, the accuracy of the surrogate function increases w.r.t. larger areas of the search space, whereas for less diversified training data the accuracy of the surrogate function usually only increases in a small part of the search space. Moreover, the high number of shakings in the PBIG additionally increases the diversity of the solutions returned by the OC. Note however that COA[PBIG] does not scale so well w.r.t. computation times, especially for an increasing number of locations. The reason for this is the generous termination criterion of the PBIG in comparison to the VNS, since it takes the PBIG much more time to find a near optimal solution.

Finally, Fig. 2 shows the computation times of the individual components of the COA framework. The total computation time is primarily split between the EC and the OC while the FC has barely any impact. Moreover, the figure also shows that the computation time of the EC mainly depends on the number of users, while the computation time of the OC primarily scales with the number of service point locations. Finally, Fig. 2 also shows large differences in computation times between COA[VNS] and COA[PBIG].

6 Conclusion and Future Work

We considered a Cooperative Optimization Approach (COA) for distributing service points within a geographical area in mobility applications under incomplete information. Instead of estimating user demands upfront, our method directly incorporates potential customers in the optimization process. In this contribution we specifically focused on the optimization part of COA. We could show that for our instances a variable neighborhood search (VNS) as well as a population based iterated greedy algorithm (PBIG) reliably generate near optimal solutions in short time. However, within COA, not quality matters: PBIG is naturally able to return multiple different high quality solutions that can all be further exploited. While COA[PBIG] has the edge over COA[VNS] w.r.t. to optimality gaps, COA[PBIG] does not scale as well as COA[VNS] for instances with a large number of service point locations.

In future work, we aim at adapting COA to also work for larger instances in a reasonable of time. Alternative models will be considered for the surrogate function, and we plan to change the optimization core (OC) from a black-box optimization to a white-box or at least a gray box model.

References

1. Bouamama, S., Blum, C., Boukerram, A.: A population-based iterated greedy algorithm for the minimum weight vertex cover problem. Appl. Soft Comput. **12**(6), 1632–1639 (2012)
2. Chen, T., Kockelman, K.M., Khan, M.: The electric vehicle charging station location problem: a parking-based assignment method for Seattle. In: 92nd Annual Meeting of the Transportation Research Board in Washington DC (2013)
3. Cornuéjols, G., Nemhauser, G.L., Wolsey, L.A.: The uncapacitated facility location problem. In: Mirchandani, P.B., Francis, R.L. (eds.) Discrete Location Theory, pp. 119–171. Wiley, New York (1990)
4. Liefooghe, A., Paquete, L. (eds.): EvoCOP 2019. LNCS, vol. 11452. Springer, Cham (2019). https://doi.org/10.1007/978-3-030-16711-0
5. Kameda, H., Mukai, N.: Optimization of charging station placement by using taxi probe data for on-demand electrical bus system. In: König, A., Dengel, A., Hinkelmann, K., Kise, K., Howlett, R.J., Jain, L.C. (eds.) KES 2011. LNCS (LNAI), vol. 6883, pp. 606–615. Springer, Heidelberg (2011). https://doi.org/10.1007/978-3-642-23854-3_64
6. Koziel, S., Ciaurri, D.E., Leifsson, L.: Surrogate-based methods. In: Koziel, S., Yang, X.S. (eds.) Computational Optimization, Methods and Algorithms. SCI, vol. 356, pp. 33–59. Springer, Heidelberg (2011). https://doi.org/10.1007/978-3-642-20859-1_3
7. Llorà, X., Sastry, K., Goldberg, D.E., Gupta, A., Lakshmi, L.: Combating user fatigue in iGAs: Partial ordering, support vector machines, and synthetic fitness. In: Proceedings of the 7th Annual Conference on Genetic and Evolutionary Computation, GECCO 2005, pp. 1363–1370. ACM, New York (2005)
8. López-Ibáñez, M., Dubois-Lacoste, J., Pérez Cáceres, L., Birattari, M., Stützle, T.: The irace package: iterated racing for automatic algorithm configuration. Oper. Res. Perspect. **3**, 43–58 (2016)
9. Meignan, D., Knust, S., Frayret, J.M., Pesant, G., Gaud, N.: A review and taxonomy of interactive optimization methods in operations research. ACM Trans. Interact. Intell. Syst. **5**(3), 17:1–17:43 (2015)
10. Mladenović, N., Hansen, P.: Variable neighborhood search. Comput. Oper. Res. **24**(11), 1097–1100 (1997)

Concept for a Technical Infrastructure for Management of Predictive Models in Industrial Applications

Florian Bachinger$^{(\boxtimes)}$ and Gabriel Kronberger

Josef Ressel Center for Symbolic Regression, Heuristic and Evolutionary Algorithms Laboratory, University of Applied Sciences Upper Austria, Hagenberg, Austria
florian.bachinger@fh-hagenberg.at

Abstract. With the increasing number of created and deployed prediction models and the complexity of machine learning workflows we require so called model management systems to support data scientists in their tasks. In this work we describe our technological concept for such a model management system. This concept includes versioned storage of data, support for different machine learning algorithms, fine tuning of models, subsequent deployment of models and monitoring of model performance after deployment. We describe this concept with a close focus on model lifecycle requirements stemming from our industry application cases, but generalize key features that are relevant for all applications of machine learning.

Keywords: Model management · Machine learning workflow · Model lifecycle · Software architecture concept

1 Motivation

In recent years, applications of machine learning (ML) algorithms grew significantly, leading to an increasing number of created and deployed predictive models. The iterative and experimental nature of ML workflows further increases the number of models created for one particular ML use case. We require so called model management systems to support the full ML workflow and to cover the whole lifecycle of a predictive model. Such a model management system should improve collaboration between data scientist, ensure the replicability of ML pipelines and therefore increase trust in the created predictive models. To highlight the need for model management systems we follow a typical machine learning process and identify shortcomings or problems occurring in practice, that could be mitigated by a model management system.

1.1 Model Management in the Machine Learning Workflow

A typical ML workflow, as described by the CRISP-DM data mining guide [1] and illustrated in Fig. 1, is a highly iterative process. *Business Understanding*

© Springer Nature Switzerland AG 2020
R. Moreno-Díaz et al. (Eds.): EUROCAST 2019, LNCS 12013, pp. 263–270, 2020.
https://doi.org/10.1007/978-3-030-45093-9_32

and *Data Understanding* is gained through assessment of the particular ML use case and the initial gathering and analysis of data. Meticulous *Data Preparation*, cleaning of data and feature engineering, is an important prerequisite for good modeling results, as selection and data transformation are critical for successful application of ML methods. In the subsequent *Modeling* task, different ML frameworks and algorithms are applied. Therein, it is necessary to test a variety of algorithm configurations. It is common that the data preparation step and modeling step are repeated and fiddled with, until satisfying results are achieved. Analysis of applied data preparation techniques and their affect on model quality can provide insights on the physical system and improve future modeling tasks in this domain. Similarly, comparison of all *Evaluation* results can provide additional insights about suitable algorithm configurations for similar ML problems.

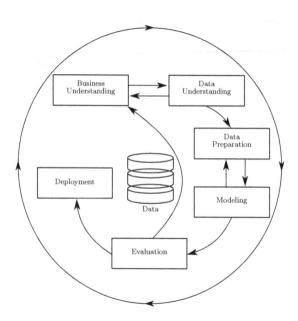

Fig. 1. Visualization of the typical machine learning workflow as described in the CRISP-DM 1.0 Step-by-step data mining guide [1]

Finally, when a suitable model is discovered, it is deployed to the target system. Typical ML workflows often end with this step. However, because of changes to the system's environment, i.e. concept drift, the model's predictive accuracy can deteriorate over time. Information about the several ML workflow steps that led to the deployed model might be forgotten, lost, or scattered around different files or knowledge systems. Model management systems should aid by tracking the complete ML workflow and saving every intermediate artefact in order to ensure replicability.

1.2   Managing the Model Lifecycle - Industrial Applications

Predictive models are increasingly deployed to so-called edge computing devices, which are installed close to the physical systems, e.g. for controlling production machines in industry plants. In such scenarios, predictive models usually need to be tuned for each particular installation and environment, resulting in many different versions of one model. Additionally, models need to be updated or re-tuned to adapt to slowly changing systems or environmental conditions, i.e. concept drift. Once a tuned model is ready for deployment, the model needs to be validated to ensure the functional safety of the plant. The heterogenous landscape of ML frameworks, their different versions and software environments, further increases difficulty of deployment. We argue that model management system need to ML should borrow well established concepts from software development, i.e. continuous integration, continuous delivery, to cope with fast model iterations, and to cover the whole model lifecycle.

In a subsequent phase, the deployed model's prediction performance, during production use, needs to be monitored to detect concept drift or problems in the physical system. Continuous data feedback from the physical system back to the model management system provides additional data for training, and future model validation.

2   Related Work

Kumar et al. [5] have fairly recently published a survey on research on data management for machine learning. Their survey covers different systems, techniques and open challenges in this areas. Each surveyed project is categorized into one of three data centric categorizations:

ML in Data Systems cover projects that combine ML frameworks with existing data systems. Projects like Vertica [7], Atlas [9] or Glade [2] integrate ML functionality into existing DBMS system by providing user-defined aggregates that allow the user to start ML algorithms in an SQL like syntax, and provide models as user-defined functions.

DB-Inspired ML Systems describe projects that apply DB proven concepts to ML workloads. Most projects apply these techniques in order to speedup or improve ML workloads. This includes techniques like asynchronous execution, query rewrites and operator selection based on data clusters, or application of indices or compression. ML.NET Machine Learning as described by Interlandi et al. [4] introduces the so called DataView abstraction which adapts the idea of views, row cursors or columnar processing to improve learning performance.

ML Lifecycle Systems go beyond simply improving performance or quality of existing ML algorithms. These systems assist the data-scientist in different phases of the ML workflow. In their survey, Kumar et al. [5] further detail the area of *ML Lifecycle Systems* and introduce so called *Model Selection and Management* systems. These systems assist not one but many phases of the ML lifecycle.

One representative of *Model Selection and Management* systems is described by Vartak et al. [8]. Their so called ModelDB, is a model management system for the spark.ml and scikit-learn machine learning frameworks. ModelDB provides instrumented, wrapped APIs replacing the standard Python calls to spark.ml or scikit-learn. The wrapped method calls the ML framework functionality and sends parameters or metadata of the modelling process to the ModelDB-Server. This separation allows ModelDB to be ML framework agnostic, given that the ModelDB API is implemented. Their system provides a graphical user interface (GUI) that compares metrics of different model versions and visualizes the ML pipeline which lead to each model as a graph. However, ModelDB only stores the pipeline comprised of ML instructions that yielded the model. ModelDB does not store the model itself, training data, or metadata about the ML framework version. External changes to the data, for example, are not recognized by the system and could hamper replicability.

Another representative, ProvDB, as described by Miao et al. [6] uses a version control system (e.g. git) to store the data and script files and model files created during the ML lifecycle in a versioned manner. Therein, scripts can be used for either preprocessing or to call ML framework functionalities. Git itself only recognizes changes to the files and therefore treats changes to data, script or model files equally. In order to store semantic connections between e.g. data versions and their respective preprocessing scripts or the connection between data, the ML script and the resulting model, ProvDB uses a graph database (e.g. Neo4j) on top. Provenance of files and metadata about the ML workflow, stored in the graph database, can be recorded through the ProvDB command line or manually defined through the file importer tool or the ProvDB GUI. This design allows ProvDB to support data and model storage of any ML framework given that these artefacts can be stored as files and are committed to git. However, automatic parsing and logging of instructions to the ML framework only works in the ProvDB command line environment. Inside the ProvDB environment, calls to the ML framework and their parameters are first parsed and then forwarded which requires ML frameworks that provide a command line interface.

Similar to ProvDB we aim to store all artefacts created during the ML life-cycle and allow the definition of semantic connections between these artefacts. Our approach differs from ProvDB as we plan to use a relational database for the data persistence and aim to develop a tighter integration to the actual ML framework, comparable to the API approach of ModelDB. Moreover, we aim to support tuning, automated validation and deployment of ML models and provide functionality to monitor performance of deployed models.

In their conclusion Kumar et al. [5] identify the area of *"Seamless Feature Engineering and Model Selection"*, systems that support end-to-end ML work-flows, as important open areas in the field of data management for machine learning. They highlight the need for fully integrated systems that support the machine learning lifecycle, even if it only covers a single ML system/framework.

3   Design and Architecture

In the following section we describe our concept of a model management system. This system is designed to be ML framework agnostic. Integration into the open source ML framework HeuristicLab[1], which is being developed and maintained by our research group, will serve as a prototypical template implementation. Figure 2 serves as illustration of the data-flows and individual components of the model management system described in the following sections. The described system can be used either locally by a single user or as a centralized instance to enables collaboration of different users. The model management system is described by the following key features:

- Centralized and versioned data storage for all artefacts of the ML framework.
- Definition of semantic connections between the different artefacts.
- Storage API for ML framework integration.
- Automatic evaluation of models on semantically connected snapshots.
- Bundling of models for deployment and subsequent monitoring.

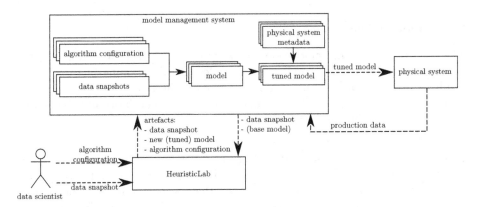

Fig. 2. Visualization of interaction between the envisioned model management system, a machine learning framework (e.g. HeuristicLab) and an external physical target system.

3.1   Data Management

The accuracy of prediction models, achievable by different ML algorithms, depends on the quality of the (training) data. Errors in data recording, or wrong assumptions made during business- and data-understanding phase affect data preparation and are therefore carried over to the modeling phase and will subsequently result in bad models. Though, bad models are not solely caused by poor data quality, as a model can become biased if certain information, contained in

---

[1] https://dev.heuristiclab.com.

the dataset, was then not present in the training portion of the ML algorithm. These problems are especially hard to combat or debug if the connection between a specific model version and its training data was not properly documented or if the information is scattered around different knowledge bases and therefore hard to connect and retrieve. A model management system should therefore provide an integrated, versioned data-storage.

ML frameworks and supported preprocessing tools need to be able to query specific version of data, i.e. snapshots, from the database. Preprocessing tools also need to be able to store modified data as new snapshots. Additionally, data scientists should be able define semantical relations between snapshots. This connections can be used to mark compatible datasets that stem from similar physical systems or are a more recent data recording of the same system, as also discussed in Sect. 3.4. In such cases a model management system could automatically evaluate a model's prediction accuracy on compatible snapshots. The same semantic connections can be used to connect base datasets with the specific datasets from physical systems for model tuning. When a new, better model on the base dataset is created the model management system can automatically tune it to all connected specific datasets.

3.2   Model Management

The section model management loosely encompasses all tasks and system components related to the ML model, this includes the ML training phase and the resulting predictive model, fine tuning the model to fit system specific data sets, evaluation of models (on training sets or physical system data) and the subsequent deployment of validated models.

Model Creation or Model Training. In order to conveniently support the ML workflow, a model management system should impose no usability overhead. In case of the *Modeling* phase this means that necessary instrumentation of ML framework methods, to capture ML artefacts, should not affect usage or require changes in existing pipelines/scripts. Therefore, method signatures of the ML framework must stay the same. Functionally, the model management system has to be able to capture all ML framework artefacts, configurations and metadata necessary to fully reproduce the training step. In our concept for a model management system we intend to provide an easy to use API for capturing artefacts that can be integrated by any ML framework. The resulting knowledge base of tried and tested ML algorithm parameters for a variety of ML problems can serve as a suggestion for suitable configurations for future experiments, or meta-heuristic optimization for problem domains.

Model Evaluation. Besides ensuring replicability of the ML workflow, a model management system should also aid in the evaluation of models. In practice we require evaluation of a model's prediction accuracy not only on the test section of data, but also on "older" data snapshots, to evaluate whether a new model actually has achieved equal or better predictive quality than it's predecessor.

Similarly, functional safety of the prediction models in their production environments can be ensured by automated validation of models on past production data or on simulations of the physical system. This model evaluation process can be seen as the analogy of unit tests in the continuous development process. Facilitation of the semantic connections between datasets provides the necessary information these evaluation steps.

Model Tuning. Predictive models often need to be tuned in order to describe a specific target system. Model tuning refers to the task of using ML algorithms to adapt an existing predictive model, or model structure, to fit to a specific previously unknown environment. Model tuning can be used to fit an existing model to its changed environment after a concept drift was detected. Likewise, tuning can improve or speedup the ML workflow by using an existing, proven model as a starting point to describe another representative of a similar physical system. If a model type and ML framework support tuning, the model management system can trigger this tuning process and subsequent evaluation automatically. This process reassembles automated software build processes. If concept drift is detected or automated model tuning is enabled for a physical system, the model management system can take action autonomously.

3.3   Model Deployment

The model management system should aid in the deployment of prediction models. This means providing the model bundled with all libraries necessary for execution of the model. The heterogenous landscape of ML frameworks and their different versions and software environments can cause compatibility issues on the target system. Crankshaw et al. [3] describe a system called Clipper, that solves this problem by deploying the model and its libraries bundled inside Docker images. This technique could also aid in distribution of the model, as the Docker ecosystem includes image management applications, that host image versions and provide deployment mechanisms. By applying this technique, the model management system can ensure executability and solve delivery of models to the target system.

3.4   Data Feedback

Our concept for a model management system includes a data gathering component to capture feedback in the form of production data from the edge device. Monitoring of the model's prediction accuracy during deployment and evaluation of the model on the production data enables the model management system to detect concept drift and to tune and subsequently re-deploy tuned models. The software necessary for monitoring and data gathering can be deployed to the edge device by addition to the bundle created during model deployment.

4   Summary

In this work we described our technological concept for a model management system. We describe the different features and components that are necessary to fully capture the machine learning lifecycle to ensure replicability of modeling results. Application of predictive models in industrial scenarios provides additional challenges regarding validation, monitoring, tuning and deployment of models that are addressed by the model management system. We argue that advances in model management are necessary to facilitate the transition of machine learning from an expert domain into a widely adopted technology.

Acknowledgement. The financial support by the Christian Doppler Research Association, the Austrian Federal Ministry for Digital and Economic Affairs and the National Foundation for Research, Technology and Development is gratefully acknowledged.

References

1. Chapman, P., et al.: CRISP-DM 1.0 step-by-step data mining guide (2000)
2. Cheng, Y., Qin, C., Rusu, F.: GLADE: big data analytics made easy. In: Proceedings of the 2012 ACM SIGMOD International Conference on Management of Data, SIGMOD 2012, pp. 697–700. ACM, New York (2012)
3. Crankshaw, D., Wang, X., Zhou, G., Franklin, M.J., Gonzalez, J.E., Stoica, I.: Clipper: a low-latency online prediction serving system. In: 14th USENIX Symposium on Networked Systems Design and Implementation (NSDI 17), pp. 613–627 (2017)
4. Interlandi, M., Matusevych, S., Amizadeh, S., Zahirazami, S., Weimer, M.: Machine learning at microsoft with ML.NET. In: NIPS 2018 Workshop MLOSS (2018)
5. Kumar, A., Boehm, M., Yang, J.: Data management in machine learning. In: Proceedings of the 2017 ACM International Conference on Management of Data, pp. 1717–1722. ACM Press (2017)
6. Miao, H., Chavan, A., Deshpande, A.: ProvDB: lifecycle management of collaborative analysis workflows. In: Proceedings of the 2nd Workshop on Human-In-the-Loop Data Analytics, HILDA 2017, pp. 7:1–7:6. ACM, New York (2017)
7. Prasad, S., et al.: Large-scale predictive analytics in vertica: fast data transfer, distributed model creation, and in-database prediction. In: Proceedings of the 2015 ACM SIGMOD International Conference on Management of Data, SIGMOD 2015, pp. 1657–1668. ACM, New York (2015)
8. Vartak, M., et al.: ModelDB: a system for machine learning model management. In: Proceedings of the Workshop on Human-In-the-Loop Data Analytics, HILDA 2016, pp. 14:1–14:3. ACM, New York (2016)
9. Wang, H., Zaniolo, C., Luo, C.R.: ATLAS: a small but complete SQL extension for data mining and data streams. In: Proceedings 2003 VLDB Conference, pp. 1113–1116. Elsevier (2003)

Guiding Autonomous Vehicles Past Obstacles – Theory and Practice

Bin Hu[1]([⊠]), Clovis Seragiotto[1], Matthias Prandtstetter[1], Benjamin Binder[2], and Markus Bader[3]

[1] Center for Mobility Systems – Dynamic Transportation Systems, AIT Austrian Institute of Technology, Giefinggasse 2, 1210 Vienna, Austria
{bin.hu,clovis.seragiotto,matthias.prandtstetter}@ait.ac.at
[2] DS AUTOMOTION GmbH, Lunzerstrae 60, 4030 Linz, Austria
b.binder@ds-automotion.com
[3] Institute of Computer Engineering – Automation Systems Group, Vienna University of Technology, Treitlstrasse 1-3, 1040 Vienna, Austria
markus.bader@tuwien.ac.at

Abstract. Driverless systems are currently used in a variety of applications such as self-driving cars, automated transportation systems at harbors, factories and hospitals. However, only a few applications are actually designed to work in a shared environment with humans because such a system would require an autonomous system instead of automated vehicles. In the project *Autonomous Fleet*, we consider a system-of-system problem where the coordination and navigation of a fleet of autonomous vehicles for dynamic environments with people is required. On the one hand, the *coordination system* considers the whole fleet, i.e., the position and goal of each vehicle. Its duty is to propose a routing solution for all vehicles by setting checkpoints. This poses similarities with the traffic control system for vehicle drivers on roads - it affects the driving behavior but cannot control it. On the other hand, each vehicle has its own *navigation system* which is responsible for how to drive between checkpoints and avoid collisions with (dynamic) obstacles and is therefore autonomous. The focus of this work is on the coordination system with algorithms for collision-free routing for a fleet of autonomous vehicles.

Keywords: Vehicle automation · Shortest path · Dynamic programing

1 Introduction

Automated guided vehicles (AGVs) are used for several decades in a number of different industrial applications where they allow for a robust and efficient operation. These automated vehicles typically follow fixed predefined lanes and stop movement in case of unexpected obstacles blocking their way.

This work is partially funded by the Austrian Research Promotion Agency (FFG) under grant number 855409 ("AutonomousFleet").

R. Moreno-Díaz et al. (Eds.): EUROCAST 2019, LNCS 12013, pp. 271–278, 2020.
https://doi.org/10.1007/978-3-030-45093-9_33

This class of automated vehicles are therefore not suitable for dynamic environments with people. Thus, the great potential of autonomous vehicles for transport tasks while driving among people in public spaces, such as shopping centers, hospitals or public infrastructures, cannot be exploited. For a fully functioning system with multiple AGVs, the coordination system must be able to recognize mutual obstructions or blocking in time to initiate adequate countermeasures. Moreover, the holistic view of the system is of vital importance for the acceptance especially in the environment of people. Processes have to be plausible and people need to be able to contribute their knowledge to the current system. In Fig. 1 we show two example scenarios that are typical for the use of AGVs: areas with long corridors (e.g., hospitals) and open space with multiple rooms (e.g., shopping centers).

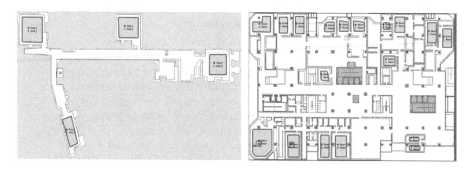

Fig. 1. Example scenario 1 with long corridors (left) and example scenario 2 with open space and multiple rooms (right).

In the research project "Autonomous Fleet" we explore the autonomy of vehicles as a system-of-system problem, which has not been studied so far: the coordination and management of a significant number of autonomous vehicles for dynamic environments with people. Here, the coordination system of an autonomous vehicle poses similarities with the traffic control system for vehicle drivers on roads – it may only affect the driving behavior but cannot control it. In addition, other challenges arise since environments with pedestrians have greater degrees of freedom concerning the choice of lanes.

The envisioned framework consists of three main components:

- Multilevel knowledge base: It stores and processes historical data and real-time data of vehicles, their sensors, people, and the physical environment.
- Fleet coordination system: It computes for each vehicle a route in form of checkpoints and aims for a global system optimum.
- On-board navigation system: It is responsible for the navigation between the checkpoints of the fleet coordination system and performs small-scale corrections, collision avoidance, etc. for each individual vehicle.

The emphasis of this paper is on the fleet coordination system where we implemented and tested algorithms for route planning and deadlock prevention. There are numerous path-finding algorithms in the literature on this topic, e.g., in the area of warehouse logistics [8], deadlock-free routing on network-on-chip architectures [6] or deadlock-free scheduling in manufacturing [4]. From a theoretical point of view, these problems are equal when represented as a graph problem.

This work is based on the knowledge of [1–3] where path-finding algorithms were implemented for single and multiple vehicles settings, controller mechanisms were presented, and in overall the topic of AGV as support for logistic operations addressed.

2 Collision-Free Route Planning in Theory

We consider the fleet coordination system also as a graph problem with nodes representing points where the AGVs can navigate along. The edges represent the direct connections between them. Each vehicle starts at a node and has to reach a target node. In our problem we assume a homogenous fleet of AGVs, so the travel time on each edge is fixed. There are multiple variations of how the objective can be defined – in our case we consider the minimization of the total amount of time required for all vehicles to reach their targets. The solution is a sequence of nodes for each vehicle, which are basically the checkpoints used for the onboard navigation unit.

Figure 2 shows two example graphs. The first one is a test-graph with four nodes (A, B, C, D) and two vehicles (R0 and R1). The target is to swap the positions of both vehicles, which requires a small amount of evasion. The second example is a grid graph generated from example 2 in Fig. 1. The open area was processed into 2 m × 2 m cells and polished at the borders.

Fig. 2. A small test-graph (left) and a grid graph based on example scenario 2 (right).

2.1 Heuristic Approach

Based on the concepts of [1] we implemented following two algorithms and extended thereby the previous work [3]. First, we use a *heuristic algorithm* from Adriaan *et al.* [7] that computes shortest paths sequentially for one vehicle at a time.

It is based on the A* shortest path algorithm, extended by a so-called free time window graph and works as follows. For each considered vehicle and each iteration in the path-finding process, the partial path is extended with the next node so that the value of $f = g + h$ is minimal, where g is the actual cost of the partial path, and h is a heuristic estimate of reaching the target node. This part resembles the classical A* algorithm. To ensure that there is no collision and deadlock for all vehicles, we have to guarantee that each extension of the partial path is possible with respect to a so-called *free time window* (FTW). The FTW is defined on a resource component – in our case an edge – and states that the time window is long enough for a for a vehicle to enter the resource, traverse it, and leave it. Initially each edge has a single FTW across the whole time horizon and each time a vehicle traverses it, the FTW splits up into two FTWs – one before the vehicle enters the edge and one after the vehicle leaves the edge. For a detailed description of this algorithm, we refer to [7]. Note that the original algorithm also considers the capacity of a resource, but in our case we assume that each edge can only be traversed by one vehicle at a time.

The computed path for each vehicle is optimal, but since the vehicles are considered sequentially, the whole solution is heuristic, or the algorithm might not be able to find a feasible solution at all (e.g., due to a bad order). Therefore we extended the heuristic by an iterative process where we change the order in which the vehicles are considered. Each time the heuristic fails to find a path for a vehicle, we move the vehicle to the first position in the order and recompute the paths. The idea is that in this case we assume the problematic vehicle to be more critical in the whole process and should be considered earlier.

In general, the vehicle order can be regarded as a combinatorial optimization problem itself, since it does not only impact the possibility of finding a feasible solution, but also the solution quality. However, we omitted to apply a meta-heuristic approach on it, because we were able to find a feasible solution within 2 tries at most. In addition, the environment is dynamic due to interactions with pedestrians. Hence re-calculations are necessary and it is not reasonable to sacrifice too much computation time for the sake of optimality.

2.2   Exact Approach

The second algorithm is a state-based dynamic programming approach from Prandtstetter *et al.* [5]. The essential difference to the heuristic approach is that each step considers the system state for all vehicles. To be more precise, each state encodes a pointer to its previous state, a time stamp, and the positions of all vehicles. A vehicle can further be:

- active: it has started the journey
- inactive: it did not start yet
- finished: it finished the journey at target node

The algorithm starts with the state where all vehicles are inactive at the starting positions. In each iteration we generate for a given state a full list of feasible follower states, i.e., where at least one vehicle makes a movement along an edge. Among the follower states, the inferior ones are discarded (if the vehicle positions equals a stat that has been considered before and the time stamp is worse). In the basic version, all remaining follower states are examined at some point. This algorithm is able to find an optimal solution, but it is obvious that the search space grows exponentially with the number of vehicles and the size of the graph. For a detailed description, we refer to [5].

It is possible to accelerate the algorithm by optimizing the order in which the follower states are examined (e.g., most promising ones first) or by restricting the set of follower states (e.g., forbid certain actions like vehicles moving back-and-forth). Latter is risky though since the optimality may be affected when certain states cannot be reached anymore. In our experiments, this exact algorithm still had a terrible scalability, even with some very restrictive rules for expanding follower states. At the end, this approach is only applicable for small graphs, e.g., there are small examples that are explicitly designed to be tricky for the heuristic approach. In real-world scenarios such as grid graphs generated from Fig. 1, the heuristic algorithm is completely sufficient and in every respect superior.

3   Collision-Free Route Planning in Practice

The real world is much more complex than what is usually considered by classical route planning algorithms. The latter assume a perfect world where each vehicle moves in a fully predictable way. Typically neglected aspects are:

- orientations of the vehicles and time for making turns,
- lack of accuracy caused by sensors and/or motion control units, and
- unpredicted obstacles such as people getting in the way.

The straightforward output of the route planning algorithms are route plans for each vehicle, i.e., sequences of instructions where and when to move next. In our first implementation we did not think about orientations and the necessity that vehicles will spend time with turning. As a result, the vehicles were not able to move according to the calculated time plan.

The second issue is usually neglected in most algorithmic approaches, but appears when deploying real vehicles. If the grid is too fine grained, a vehicle could occupy multiple nodes at once and collide or block each other. Possible ways to handle this issue is to avoid grid graphs with too short edges and/or introduce a buffer which restricts the vehicles to move too close in time and space. In our case the graphs have a edge length of at least 2 m and the problem was not present.

The problem with unpredictability is a large one, since classical route planning algorithms generate solutions which consist of a sequence of instructions for each vehicle. These instructions follow a strict chronology according to a time-table, which is fault-prone in practice. In our approach based on [3], we transform the time-table based solutions into *dependency based solutions* via post-processing. The idea is instead of telling the vehicles when to do what, we state for each instruction the *preconditions* that have to be met, e.g., vehicle x may start with instruction a once vehicle y completed instruction b. With this we guarantee that the sequence of instructions is applied in the planned order without the necessity to synchronize all vehicles globally using a time-table.

We visualize this procedure in Fig. 3. In the T-shaped example graph from Fig. 2, the two vehicles R0 and R1 want to swap their positions. With the routing planning algorithms (heuristic or exact), we obtain a solution with instructions based on a time-table. To transform the time-table based instructions to dependency based instructions, we examine the critical steps for each vehicle and observe that

– step 1 of vehicle R0 may start once vehicle R1 finished step 2, and
– step 3 of vehicle R1 may start once vehicle R0 finished step 2.

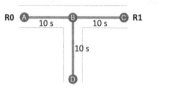

time	R0	R1
0	A	C
10	A	B
20	B	D
30	C	B
40	C	A

step	R0	R1
1	A→B after R1:2	C→B
2	B→C	B→D
3		D→B after R0:2
4		B→A

Fig. 3. Example for the transformation procedure: small instance from Fig. 2 (left). Routing solution based on time-table (center). Routing solution based on dependencies (right).

4   Experimental Results

We tested our algorithm with up to 120 vehicles on a relatively sparse random graph with 3500 nodes and 3800 edges. This graph is based on a grid graph with 70×50 cells, but much sparser. The reason is that a full grid graph (which would have 6880 edges) is too easy since deadlocks are very unlikely to occur.

Table 1 shows the results where the columns represent the number of vehicles, number of tries (i.e., if the heuristic approach fails to find a solution, it changes the vehicle order and restarts, see Sect. 2.1), total CPU time (on a Intel Core i7-4600U PC running at 2.7 GHz and 16 GB main memory), and the success rate of finding a feasible solution. The success rate is based on 100 random instances per line. We observe that until 100 vehicles the heuristic approach is always able to solve the instance. For some instances with 110 and 120 vehicles, a restart with different vehicle order was necessary to obtain a feasible solution. At the moment, the size of the graph and the number of vehicles are reasonably large for applications in the real world, indicating that the approach is feasible.

Table 1. Results of the heuristic approach on random instances.

Num vehicles	Num tries	Time [s]	Success rate
10	1	12	100%
20	1	27	100%
30	1	85	100%
40	1	88	100%
50	1	121	100%
60	1	180	100%
70	1	240	100%
80	1	313	100%
90	1	439	100%
100	1	534	100%
110	1	676	97%
120	1	836	99%
110	2	774	100%
120	2	930	100%

5   Conclusions and Future Work

In this paper, we discussed approaches for implementing the coordination system as a part of a framework for fleet of automated guided vehicles (AGVs). While the exact approach is able to find optimal results, it is only applicable for smallest problem instances. The heuristic approach scales well with reasonably large graphs and number of vehicles. Therefore it is the algorithm of choice for real-world environments where unpredictable events – especially interactions with pedestrians – occur and dynamic adaptions are necessary. We showed some disparities between the theory of route planning for AGVs and practice, as well as how we addressed them in our approach.

Within the project, there is a demonstration with multiple AGVs at premises of the Vienna University of Technology planned. The area corresponds to example scenario 2 in Fig. 1, or a subsection of it. With these real data, further adjustments in all components of the framework will be carried out. For the fleet coordination system, we will see, for example, how well the planned routes will correspond with those performed in reality, and how often we need to re-plan dynamically.

References

1. Bader, M., et al.: Balancing centralized control with vehicle autonomy in AGV systems. In: Proceedings 11th International Conference on Autonomic and Autonomous Systems (ICAS), vol. 11, pp. 37–43, May 2015

2. Bader, M., Todoran, G., Beck, F., Binder, B., Buchegger, K.: TransportBuddy: autonomous navigation in human accessible spaces. In: Proceedings of 7th Transport Research Arena TRA 2018, Vienna, Austria, April 2018 (2018)
3. Binder, B.: Spatio-temporal prioritized planning. Master's thesis, Vienna University of Technology (2018)
4. Lei, H., Xing, K., Han, L., Xiong, F., Ge, Z.: Deadlock-free scheduling for flexible manufacturing systems using petri nets and heuristic search. Comput. Ind. Eng. **72**, 297–305 (2014)
5. Prandtstetter, M., Seragiotto, C.: Towards system-aware routes. In: Moreno-Díaz, R., Pichler, F., Quesada-Arencibia, A. (eds.) EUROCAST 2017. LNCS, vol. 10671, pp. 291–298. Springer, Cham (2018). https://doi.org/10.1007/978-3-319-74718-7_35
6. Schafer, M.K.F., Hollstein, T., Zimmer, H., Glesner, M.: Deadlock-free routing and component placement for irregular mesh-based networks-on-chip. In: IEEE/ACM International Conference on Computer-Aided Design, pp. 238–245 (2005)
7. ter Mors, A.W., Witteveen, C., Zutt, J., Kuipers, F.A.: Context-aware route planning. In: Dix, J., Witteveen, C. (eds.) MATES 2010. LNCS (LNAI), vol. 6251, pp. 138–149. Springer, Heidelberg (2010). https://doi.org/10.1007/978-3-642-16178-0_14
8. Zhang, Z., Guo, Q., Chen, J., Yuan, P.: Collision-free route planning for multiple AGVs in automated warehouse based on collision classification. IEEE Access **6**, 26022–26035 (2018)

Casual Employee Scheduling with Constraint Programming and Metaheuristics

Nikolaus Frohner$^{(\boxtimes)}$, Stephan Teuschl, and Günther R. Raidl

Institute of Logic and Computation, TU Wien, Vienna, Austria
{nfrohner,raidl}@ac.tuwien.ac.at, e0608934@student.tuwien.ac.at

Abstract. We consider an employee scheduling problem where many casual employees have to be assigned to shifts defined by the requirement of different work locations. For a given planning horizon, locations specify these requirements by stating the number of employees needed at specific times. Employees place offers for shifts at locations they are willing to serve. The goal is to find an assignment of employees to the locations' shifts that satisfies certain hard constraints and minimizes an objective function defined as weighted sum of soft constraint violations. The soft constraints consider ideal numbers of employees assigned to shifts, distribution fairness, and preferences of the employees. The specific problem originates in a real-world application at an Austrian association. In this paper, we propose a Constraint Programming (CP) model which we implemented using MiniZinc and tested with different backend solvers. As the application of this exact approach is feasible only for small to medium sized instances, we further consider a hybrid CP/metaheuristic approach where we create an initial feasible solution using a CP solver and then further optimize by means of an ant colony optimization and a variable neighborhood descent. This allows us to create high-quality solutions which are finally tuned by a manual planner.

Keywords: Employee scheduling · Constraint programming · Ant Colony Optimization · Multi-objective optimization · Variable neighborhood descent

1 Introduction

We consider an employee scheduling problem that arises as a real-world problem in an Austrian association. It deals with assigning employees to shifts at work locations within a given planning horizon so that certain hard constraints are fulfilled and the violation of soft constraints regarding demand satisfaction of the locations, fairness, and preferences of the employees is minimized. The problem falls into the broad class of personnel scheduling [1] with strong ties to the nurse rostering problem [2] but has some distinguishing features. One is the substantial fluctuation of employees and the high variance of their availabilities over different planning horizons and within each; therefore employees are

© Springer Nature Switzerland AG 2020
R. Moreno-Díaz et al. (Eds.): EUROCAST 2019, LNCS 12013, pp. 279–287, 2020.
https://doi.org/10.1007/978-3-030-45093-9_34

coined "casual". Another specialty is that employees specify individual maximum numbers of shifts they desire to work. The variance of the ratio of actual shifts assigned divided by the desired shifts should be minimized to balance the fulfillment of the employees' desires. Likewise, the fulfillment ratio of the locations' requirements shall be balanced as well. Fig. 1a shows the availabilities of employees to serve shifts on given days over a month compared with the requirements by the locations. This exemplary month starts with a weekend where a lack of employees is evident, whereas on other days there is substantial overcapacity. A hard constraint is that employees cannot be assigned every day they are available since they offer a desired number of shifts for a month which also acts as a hard upper limit. The corresponding distribution can be seen in Fig. 1b. Typically, in our application there is always a shortage of workers which makes it highly desirable to distribute this shortage evenly over the shifts.

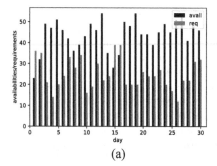
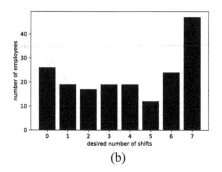

(a)   (b)

Fig. 1. (a) Availabilities of employees to serve a shift vs. requirements of locations over an exemplary month. (b) Distribution of maximum number of shifts employees offer to serve in the considered month.

In the following Sect. 2, we will formally define the optimization problem including its hard and soft constraints. This formulation gives rise to an exact approach by means of Constraint Programming (CP) which we will describe in Sect. 3. In Sect. 4, we introduce a hybrid algorithm that makes use of this CP-model, Ant Colony Optimization, and Variable Neighborhood Descent to be able to tackle large problem instances. In Sect. 5 we conduct a computational study for both approaches on artificially generated data and real data provided by an Austrian association, after which we conclude in Sect. 6.

2   Problem Formulation

Given locations L and a planning horizon D consisting of days $d \in D$, the tuples $(d, l) =: s \subset D \times L$ constitute shifts. Each shift s has a requirement $R_s \in \mathbb{Z}_+$ of employees that should ideally serve it. Each employee $w \in W$ chooses certain shifts $S_w \subset S$ which he could potentially serve and a desired number of shifts N_w, which also acts as upper limit of the shifts w will be assigned to. The locations are comprised of houses H that have shifts on a more regular basis and events E whose shifts are more sparsely distributed over the planning horizon but possibly with higher requirement peaks. Furthermore, there are two special locations: *standby*, denoted by b, which is used in case someone becomes sick, and *floating*, denoted by f, for employees that are assigned dynamically to a house on the very day of the shift. Only employees whose numbers of shifts N_w are above a given threshold are eligible for standby and floating shifts. If an employee selects a shifts (d, l) where $l \in H$, then all the other houses are selected automatically for that day as well, so that there is enough flexibility for the planners. However, the employees provide a nonempty preference list of houses $H_w \subset H$. Events on the other hand can be selected separately.

The goal is to find an assignment of employees to shifts, which we denote by the sets of shifts $A_w \subset S_w$ each worker $w \in W$ is assigned to, that satisfies a number of hard constraints and minimizes violations of a number of soft constraints. The most relevant hard constraints are that employees have to be assigned at least once, at most in accordance to their desired/maximum number of shifts, and on each day at most once; each shift has a time duration and the total duration for each worker must stay within legal bounds. Furthermore, the standby and floating shifts must be fully covered, otherwise their purpose of being a backup would be defeated.

We are thus facing a multi-objective optimization problem where soft constraint violations are modeled as different objectives with different priorities. Since shifts provide service to paying customers, the fulfillment of the corresponding requirements is by far the most important objective. Given an assignment A, each shift has a relative shortage $u_s = 1 - |\{A_w \in A \mid s \cap A_w \neq \emptyset\}|/R_s$ that should be kept small and balanced over all shifts, which we implement by minimizing the mean squared error of the vector $\mathbf{u} = (u_s)_{s \in S}$ with respect to the desired optimum $\mathbf{u}^* = \mathbf{0}$. Next priority is to distribute the shifts over employees as fair as possible, taking their numbers of desired shifts into account. To achieve this, we minimize the variance of the sum of the assigned shift durations divided by the number of desired shifts multiplied by the maximum shift duration over the employees. The shift duration is denoted as Δ_s. The fractions of floating shift hours over the assigned hours should also be distributed evenly among all workers. Last but not least, we aim at keeping the ratios workers are assigned to non-preferred houses small and balanced over all workers, for which again minimizing the corresponding mean squared error is deemed suitable. Putting everything together yields the vector-valued objective function $\mathbf{f} \colon \mathcal{A} \to [0, 1]^4$:

$$\mathbf{f}(A) := (g_u(A), g_f(A), g_{ff}(A), g_{np}(A))^T \tag{1}$$

$$g_u(A) := \frac{1}{|S|} \sum_{s \in S} \left(1 - \frac{|\{A_w | s \cap A_w \neq \emptyset\}|}{R_s} \right)^2 \tag{2}$$

$$g_f(A) := \mathrm{Var} \left[\frac{\sum_{s \in A_w} \Delta_s}{N_w \max_{s \in S} \Delta_s} \right], g_{ff}(A) := \mathrm{Var} \left[\frac{\sum_{s \in A_w \wedge l(s)=f} \Delta_s}{\sum_{s \in A_w} \Delta_s} \right] \tag{3}$$

$$g_{np}(A) := \frac{1}{|W|} \sum_{w \in W} \left(\frac{|\{s \in A_w | l(s) \in H \setminus H_w\}|}{|A_w|} \right)^2 \tag{4}$$

3   Exact Solution Approach

We model the problem as a constraint program using MiniZinc [5]. We consider different formulations. The first one is based on a set formulation and the second one on binary decision variables. In the set formulation, the main decision variable is a two-dimensional array of size $|L| \cdot |D|$ containing sets of integers representing an assignment $A_{(l,d)} \subset W$, $\forall(d,l) \in S$ of specific employees to shifts. Table 1 shows a simplified assignment example of three employees, being assigned to three houses with different requirements over three days where every employee is assigned the desired number of shifts.

Table 1. Small example of a assignment with three houses, three days and three employees.

w	N_w	H_w	$R_{(l,d)}$	0	1	2	$A_{(l,d)}$	0	1	2
0	3	$\{2\}$	0	1	1	1	0	$\{0\}$	$\{2\}$	$\{1\}$
1	2	$\{1,2,3\}$	1	0	1	1	1	$\{\}$	$\{1\}$	$\{0\}$
2	1	$\{0\}$	2	2	1	0	2	$\{\}$	$\{0\}$	$\{\}$

For the minimization of the soft constraint violations we choose the weighted sum approach, where the vector-valued objective function is condensed to a real-valued function:

$$f(A) := \lambda_u \cdot g_u(A) + \lambda_f \cdot g_f(A) + \lambda_{ff} \cdot g_{ff}(A) + \lambda_{np} \cdot g_{np}(A) \tag{5}$$

We designed the soft constraint violations to yield values between zero and one, therefore no special re-scaling is necessary. To put more weight on the unassigned shifts objective and keep the others equally weighted, we use the weights $\lambda_u = 10$ and $\lambda_f = \lambda_{ff} = \lambda_{np} = 1$. Since we make use of floating point variables, we solve the MiniZinc models by two different, float-capable solver backends, namely Gecode and JaCoP.

In the approach where we greedily extend a small initial solution as described in the next section, an alternative CP model is used to only satisfy the hard constraints. It is based on $|L| \cdot |D| \cdot |W|$ binary decision variables, stating whether an employee w is assigned to a shift (d, l). It does not consider the soft constraints and is thus only used for pure hard constraint satisfaction. This model can then also be solved by the solvers Chuffed and Gurobi that as of the time of writing neither support sets nor floats in combination with MiniZinc.

4   Hybrid Approach

First tests with MiniZinc using real-world instances indicate that our problem instances are too big to be solved to optimality within reasonable time. This gives rise to a hybrid approach, where we use CP to create an initial solution that satisfies the hard constraints and which is then fed into a metaheuristic for further improvement. We propose two different combinations of CP and metaheuristics, where both our CP-models, the one with soft constraint optimization and the one with hard-constraints only, come into play:

1. The MiniZinc optimization model, which also considers the soft constraints, is solved until a first feasible solution is obtained in order to obtain a rather "complete" solution from CP, which is then passed to a variable neighborhood descent (VND) for possible further improvement.
2. The MiniZinc hard constraint satisfaction model is used to create a small feasible solution. This solution typically has a high objective value and can be substantially improved by assigning further employees. This is done by a successively applying an Ant Colony Optimization (ACO) [3,4] with a min/max pheromone model [6]. Each ant starts from the initial solution and iteratively extends it according to the ACO's usual probabilistic principles in combination with certain greedy criteria. The so far best solution is used for pheromone update. After we hit a time limit, we take the best solution and try to improve it further by the variable neighborhood descent (VND), either up to local optimality or until an overall time limit is hit.

In the first variant, the CP solver is used in optimization mode to create a rather complete (many assignments), initial solution, whereas in the second variant the binary decision variable model is used in satisfaction mode to create a rather small (few assignments) solution. The latter is realized by using the value choice heuristic *indomain_min* which prefers to set decision variables to zero.

For the VND, we use neighborhood structures induced by the following operations, in the given order: *MoveEmployeeInDay*: For a given employee w and a day d, change the location of his assignment, *AssignEmployee*: Assign a shift with shortage to an employee, *ReassignShift*: Unassign the shift from an employee and assign it to a different employee, *ReassignEmployee*: Unassign an employee from a shift and assign the employee to a different shift, and *SwapShifts*: Swap shifts of two employees. All these neighborhoods are searched in a first improvement fashion.

Each operator has impact on a different set of soft constraints. For example, swapping the assignment of two employees keeps the distribution fairness among locations unchanged but may improve the preference satisfaction of the employees.

For the ACO we use a min/max pheromone update system as described in [6], which bounds the pheromone values to lie within the interval $[\tau^{\min}, \tau^{\max}]$. The pheromone matrix elements $\tau_{s,w}$ encode a bias for assigning shift s to employee w. Given an ant's current assignment A and $S_u(A) \ni s$ be the shifts with shortage and employees $W_s \ni w_s$ that are available for this shift and fulfill all the hard constraints, then we add the assignment $s \leftrightarrow w_s$ as extension to A resulting in A' with a probability depending on this bias and an attractiveness depending on the decrease in the objective value $\Delta f(A, A') = f(A) - f(A')$:

$$p_{s,w_s} \sim \tau_{s,w_s}^{\alpha} \cdot (1 + \Delta f(A, A'))^{\beta}$$

We start with a high objective value since many shifts are unassigned and are rewarded for assigning those. This is done until no more extensions are possible. $\Delta f(A, A')$ might also be slightly negative (increase in objective value) due to the fairness constraints, therefore we shift this difference by one. After each iteration when all ants have constructed their solutions, pheromone evaporation is performed, controlled by the parameter ρ, and the so-far-best solution A_{bs} is used to increase the respective pheromones by $\Delta\tau = \frac{1}{f(A_{\mathrm{bs}})}$.

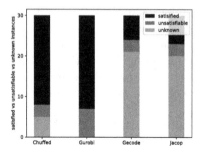 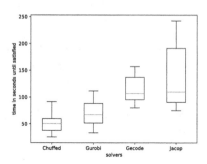

Fig. 2. Comparisons of CP solvers for 30 artificial test instances with a one hour time limit. Left: Feasibility statistics, where we see the stronger performance of Chuffed and Gurobi. Right: Boxplots for the times until a feasible solution was found for satisfied instances.

5   Computational Study

We created 30 random artificial instances for a CP-solver benchmark and test the whole exact and hybrid approach on four real-world instances. For the artificial instances, we sampled the numbers of houses from $\{6, \ldots, 8\}$, the numbers of events from $\{2, \ldots, 11\}$, and the number of employees from $\{170, \ldots, 249\}$. We considered a planning horizon of 30 days and created shift requirements following

Table 2. Comparisons of exact and hybrid algorithms on four different real-world instances with resulting weighted objective values after a time limit of one hour.

Instance	$|W|$	$\sum_s R_s$	$|D|$	t_f [s]	Exact	CSP	VND	ACO	ACO+VND
Sep 2018	184	928	30	123	1.290	7.206	0.418	0.530	**0.407**
Oct 2018	253	1079	31	238	–	6.901	–	0.553	**0.501**
Feb 2019	172	685	28	56	0.243	6.634	**0.093**	0.239	0.141
Apr 2019	170	947	30	124	2.547	7.218	1.018	0.849	**0.716**

different load patterns (peaky, steady, weekend-only, etc.) and randomly sampled employees' desired shifts and availabilities following observations from the real-world instances. 23 of the artificial instances are satisfiable, seven not; all of the real-world instances are satisfiable.

All tests were conducted on an Intel Xeon E5-2640 processor with 2.40 GHz in single-threaded mode and a memory limit of 32 GB. We used Python 3.6 for the implementation of the ACO, Java 11 for the implementation of the VND, and MiniZinc 2.2.3 with the backends Chuffed 0.10.3, Gecode 6.1.0, Gurobi 8.1.0, and JaCoP 4.6.

In Fig. 2 we see a comparison of the four different CP solvers we tested on the artificial instances. Chuffed and Gurobi using the binary formulation gave superior results in terms of number of instances they could satisfy or prove unsatisfiable within a CPU time limit of one hour, and the CPU time needed to satisfy an instance. We chose Gurobi as basis for the hybrid algorithm with the ACO, since it could satisfy every satisfiable instance within a couple of minutes. In the other variant, where we start from a complete solution provided by the CP-solver, we use JaCoP which performed better than Gecode in our experiments.

In Table 2, the main results of our real-world instances are described. Every algorithm is given a time limit of one hour. We compare the exact approach using our constraint optimization model in the set formulation with JaCoP as backend solver with the hybrid variants. In ACO and ACO+VND we start from a small basic feasible solution provided by Gurobi, extended it by an ACO and possibly further improve it with the VND. Values t_f denote the times needed until a feasible solution was provided, the rest of the time is then used either by ACO or shared evenly among ACO and VND. In the other variant, we take the first feasible solution from JaCoP in optimization mode which is rather complete and feed it directly into the VND. We conducted the tests with six ants per iteration and ACO parameters $\alpha = \beta = 1$ and $\tau^{\min} = 1.0, \tau^{\max} = 10.0$, and $\rho = 0.9$. For the Feb 2019 instance, the CP+VND only approach gave the best objective value after one hour, for the others, CP+ACO+VND gave better results. For the Oct 2019 instance JaCoP could not find a feasible solution within the allowed time limit.

In Table 3, we take for each real-world instance the best solution and show the shift coverage compared to a theoretical upper bound and the unweighted soft constraints. In the February instance there is high availability of workers which allows for a high shift coverage, in the September and October instance, we get close to the theoretical upper bounds. Since the VND always hits the time limit, further improvements are expected to be possible; a corresponding analysis of converged solutions will be provided in the master thesis of Stephan Teuschl [7].

Table 3. Shift coverage and unweighted soft constraint objective values for best solutions of real-world instances, where u is the number of unassigned shifts, c the shift coverage, and u_c the upper bound of the shift coverage, calculated by the sum of the number of shifts of the employees divided by the total requirements.

Instance	Best f	c	u_c	u	g_u	g_f	g_{ff}	g_{np}
Sep 2018	0.407	0.873	0.888	117	0.153^2	0.243^2	0.172^2	0.292^2
Oct 2018	0.501	0.868	0.918	142	0.174^2	0.319^2	0.130^2	0.281^2
Feb 2019	0.093	0.988	1.168	8	0.022^2	0.262^2	0.092^2	0.107^2
Apr 2019	0.716	0.767	0.817	221	0.227^2	0.303^2	0.161^2	0.291^2

6   Conclusion

We introduced a casual employee scheduling problem arising in an Austrian association, where employees have to be assigned to shifts to satisfy demands at locations for a given planning horizon where a number of hard constraints has to be met and violations of fairness and preference soft constraints shall be minimized. We created two different MiniZinc constraint programming models for this problem. The first model is used only for satisfying the hard constraints by assigning a small number of employees to shifts, which is then the basis for further extensions by an ant colony optimization algorithm together with a variable neighborhood descent. We compared four different backend solvers on 30 artificial benchmark instances for our problem, and Chuffed and Gurobi turned out to be the best choices. In the second model, the float-capable backend solvers JaCoP and Gecode are used to solve the optimization model considering the soft constraints up to the first feasible solution, which is then passed to the VND for further improvement. We compared the exact and the hybrid algorithms on four real-world instances with a CPU time limit of one hour and found that the hybrid approaches provided superior results. Further research is to be conducted to make use and measure the impact of different initial solutions, different orderings of neighborhoods in the VND, and parameter tuning of the ACO. An in-depth study of the presented casual employee scheduling problem and its solution methods will be given in the master thesis of Stephan Teuschl [7].

References

1. Van den Bergh, J., Beliën, J., De Bruecker, P., Demeulemeester, E., De Boeck, L.: Personnel scheduling: a literature review. Eur. J. Oper. Res. **226**(3), 367–385 (2013)
2. Burke, E.K., De Causmaecker, P., Berghe, G.V., Van Landeghem, H.: The state of the art of nurse rostering. J. Sched. **7**(6), 441–499 (2004)
3. Dorigo, M., Stützle, T.: Ant colony optimization: overview and recent advances. In: Gendreau, M., Potvin, J.Y. (eds.) Handbook of Metaheuristics, pp. 311–351. Springer, Boston (2019). https://doi.org/10.1007/978-1-4419-1665-5_8
4. Gutjahr, W.J., Rauner, M.S.: An ACO algorithm for a dynamic regional nurse-scheduling problem in austria. Comput. Oper. Res. **34**(3), 642–666 (2007)
5. Nethercote, N., Stuckey, P.J., Becket, R., Brand, S., Duck, G.J., Tack, G.: MiniZinc: towards a standard CP modelling language. In: Bessière, C. (ed.) CP 2007. LNCS, vol. 4741, pp. 529–543. Springer, Heidelberg (2007). https://doi.org/10.1007/978-3-540-74970-7_38
6. Stützle, T., Hoos, H.H.: MAX-MIN ant system. Future Gener. Comput. Syst. **16**(8), 889–914 (2000)
7. Teuschl, S.: Casual employee scheduling with constraint programming and metaheuristics. Master's thesis, TU Wien, Austria (2020, to appear)

White Box vs. Black Box Modeling: On the Performance of Deep Learning, Random Forests, and Symbolic Regression in Solving Regression Problems

Michael Affenzeller[1,2], Bogdan Burlacu[1], Viktoria Dorfer[1], Sebastian Dorl[1,2], Gerhard Halmerbauer[3], Tilman Königswieser[4], Michael Kommenda[1], Julia Vetter[1], and Stephan Winkler[1,2(✉)]

[1] School for Informatics, Communications and Media, University of Applied Sciences Upper Austria, Hagenberg, Austria
{michael.affenzeller,bogdan.burlacu,sebastian.dorl,michael.kommenda,
julia.vetter,stephan.winkler}@fh-hagenberg.at
[2] Department of Computer Science, Johannes Kepler University, Linz, Austria
[3] School for Management, University of Applied Sciences Upper Austria, Steyr, Austria
gerhard.halmerbauer@fh-steyr.at
[4] Salzkammergut-Klinikum, Oberösterreichische Gesundheitsholding GmbH, Gmunden, Austria
tilman.koenigswieser@ooeg.at

Abstract. Black box machine learning techniques are methods that produce models which are functions of the inputs and produce outputs, where the internal functioning of the model is either hidden or too complicated to be analyzed. White box modeling, on the contrary, produces models whose structure is not hidden, but can be analyzed in detail. In this paper we analyze the performance of several modern black box as well as white box machine learning methods. We use them for solving several regression and classification problems, namely a set of benchmark problems of the PBML test suite, a medical data set, and a proteomics data set. Test results show that there is no method that is clearly better than the others on the benchmark data sets, on the medical data set symbolic regression is able to find the best classifiers, and on the proteomics data set the black box modeling methods clearly find better prediction models.

1 Introduction: Machine Learning

In general, machine learning (ML) is understood as that branch of computer science that is dedicated to the development of methods for learning knowledge and models from given data.

The work described in this paper was supported by the Josef Ressel Center *SymReg* as well as *TIMED*, FH OÖ's Center of Excellence for Technical Innovation in Medicine.

© Springer Nature Switzerland AG 2020
R. Moreno-Díaz et al. (Eds.): EUROCAST 2019, LNCS 12013, pp. 288–295, 2020.
https://doi.org/10.1007/978-3-030-45093-9_35

In fact, machine learning is an essential step within the overall system identification process [8]. The system, which is investigated, might be of technical nature, a biological system, a finance system, or any other system - the essential point is that there are variables/features, which can be measured (here denoted as $x_1, x_2, \ldots, x_n$).

Machine learning algorithms are then used to identify models on the basis of data collections; these models are then often used to predict the investigated systems' behavior, to identify relevant relationships between variables, and in general to gain further insight into the investigated systems. Obviously, the collection of data is a very important step, as we will only be able to understand the system and its internal functionality correctly if its characteristic behavior is actually represented in the data.

There are two main classes of machine learning scenarios, namely supervised learning and unsupervised learning:

- In *supervised learning* the modeling algorithm is given a set of samples with input features as well as target/output values - i.e., the data are "labeled". The goal is to find a model (or a set of models) that is (are) able to map the input values to the designated target variable(s). Assuming we use the data matrix x described above, for each target variable x_t the goal is to find a model that allows us to define x_t as a function of all available/allowed variables: $x_t = f(x_1, x_2, \ldots, x_n)$.
- In *unsupervised learning* the modeling algorithm is given a set of samples that are "unlabeled", as there are no target variables, and the task is to model the structure of the data. The goal here is to identify clusters of similar samples or to describe the distribution of the features' values via density estimation.

2   Black-Box vs. White Box Modeling Methods

In this research we analyze the performance of several modern black box as well as white box methods for solving supervised learning problems. Black box ML techniques are methods that produce models which are functions of the inputs and produce outputs, where the internal functioning of the model is either hidden or too complicated to be analyzed. White box modeling, on the contrary, produces models whose structure is not hidden, but can be analyzed in detail.

2.1   Black-Box Modeling

For this study we apply the following **black-box modeling methods**:

Random Forests: (RFs, [3]) are ensembles of decision trees, each created on a set of randomly chosen samples and features from the available training data basis. The best known algorithm for inducing random forests combines bagging and random feature selection [3]:

- For each tree in the forest, a certain number of input variables is used to determine the decision at a node of the tree.
- A certain number of samples is randomly drawn from the training data base; the rest of the samples is used as internal validation set for estimating the model's prediction error (out-of-bag error).

RFs are a very popular machine learning method as they are known to be one of the most accurate learning algorithms available [14], robust against overfitting, and widely considered a very efficient machine learning method.

Artificial Neural Networks: (ANNs, [4], Fig. 1) are networks of simple processing elements, which are called neurons and are in most cases structured in layers. Neurons are connected to neurons of other layers. The first layer consists of input neurons, which receive the input variables x, and the last layer of output neurons, which define the network's output y. For each neuron the behavior is clearly defined, in most cases as a transfer function ϕ of the weighted sum of the connected neurons' values. For neuron n the output is defined as

$$y(n) = \phi(\sum_{i=1}^{k}(w_i \cdot y(m_i))) \qquad (1)$$

where y denotes the output of a node and m the k nodes in the previous layer that are connected to the node n. The transfer function ϕ can, for example, be implemented as a sigmoid function or another nonlinear activation function.

During the training of the network, the most important task is to optimize the connections between the nodes, i.e., their weights; this is often done using methods such as back propagation [12].

ANNs have been used in computer science for several decades, in recent years they have massively gained popularity due to development of deep learning algorithms [5,13]. ANNs are, e.g., successfully applied in image analysis, nonlinear system identification, pattern recognition, text mining, speech recognition, game engines, and medical and financial applications.

These methods have in common that a basic model structure is assumed and then parameters are optimized during the training phase. Additionally, the structure of the model does not give information about the structure of the analyzed system; the models are used as *black box models*.

2.2   White-Box Modeling

For **white box modeling** we use *symbolic regression*, a method for inducing mathematical expressions on data. The key feature of this technique is that the object of search is a symbolic description of a model, whose structure is not predefined. This is in sharp contrast with other methods of nonlinear regression, where a specific model is assumed and the parameters/coefficients are optimized

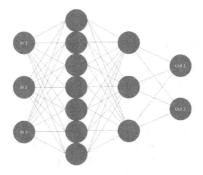

Fig. 1. An exemplary artificial network with three input nodes, 10 hidden nodes in two layers, and two output neurons.

during the training process. Using a set of basic functions we apply *genetic programming* (GP, [1]) as search technique that evolves symbolic regression models as combinations of basic functions through an evolutionary process. As the complexity of these models is limited, the process is forced to include only relevant variables in the models. We use standard GP as well as GP with offspring selection [1] (Fig. 2).

The functions set described in [16] (including arithmetic as well as logical ones) was used for building composite function expressions.

Applying offspring selection has the effect that new individuals are compared to their parents; in the strict version, children are passed on to the next generation only if their quality is better than the quality of both parents. Figure 2 shows GP with OS as used in the research discussed here.

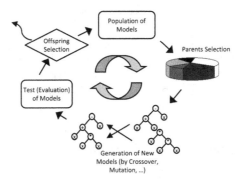

Fig. 2. Genetic programming with offspring selection [16].

3   Test Studies

3.1   Solving Regression Problems of the PMLB Benchmark Data Set Suite

First we used black box as well as white box modeling methods for solving regression problems included in PMLB, a benchmark suite for machine learning evaluation and comparison [9]. In the empirical part of this paper we analyze the performance of all here investigated modeling methods; their performance is here measured by means of the fit (R^2) on test data as defined for the data sets.

RFs were trained using the AlgLib implementation available in HeuristicLab [15], a framework for prototyping and analyzing optimization techniques in which evolutionary algorithms as well as numerous machine learning algorithms and analysis functions are available. ANNs were trained using the Python packages sklearn [11] and pytorch [10]. For white box modeling we used GP with strict OS as implemented in HeuristicLab [7].

All methods were executed with dozens of varying parameter settings, such as population sizes, number of trees, number of hidden layers and nodes, etc. In Table 1 we summarize the test results of the best models returned by the training algorithms. These results have been identified using the following settings:

- RF: number of trees: 50, m (ratio of features used in each tree): 0.5, r (ratio of training set used for constructing each tree): 0.3
- ANN: number of layers, nodes per layers, type of activation function, and droput values were determined for each data set individually using a grid search on the training data
- GP: population size: 100, strict offspring selection, gender specific parent selection (random & roulette), maximum selection pressure: 100, maximum model size: 50

Table 1. Average quality of best regression models for PMLB benchmark data sets.

Data set	Average quality of regression models (R^2)		
	RF	ANN	GP
1027_ESL	0.843	0.833	0.801
1028_SWD	0.402	0.443	0.403
1029_LEV	0.556	0.478	0.572
1030_ERA	0.296	0.254	0.302
228_elusage	0.781	0.591	0.718
542_pollution	0.251	0.248	0.555
584_fri_c4_500_25	0.708	0.274	0.975
617_fri_c3_500_5	0.888	0.841	0.980
695_chatfield_4	0.809	0.707	0.680

Further results achieved for these data sets using HeuristicLab can be found in [6].

3.2   Identification of Classifiers for Medical Data

As second test scenario we used data provided by Oberösterreichische Gesund-heitsholding. It contains data from 7747 patients, for which information about their diseases and treatments are known. In total, there are 120 variables that are available; there is one binary target variable, which is to be modeled in relation to the other available variables.

Again, we used RFs, ANNs and GP using the same varying settings as described above and here report the performance of the best models produced by the training algorithms. In Table 2 we summary the test results of the best models. These results have been identified using the following settings:

- RF: number of trees: 100, maximum tree size: 10
- ANN: 3 hidden layers (each containing 64 rectified linear units), output node sigmoid, training algorithm: back propagation, 500 epochs
- GP: population size: 100, strict offspring selection, gender specific parent selection (random & roulette), maximum selection pressure: 100, maximum model size: 100

Table 2. Quality of best patient classification models for a medical data set.

Modeling method	Average quality of prediction models (accuracy)
OS-GP	87.02%
RF	81.06%
ANN	78.64%

3.3   Prediction of Peptide Fragment Intensities in Tandem Mass Spectrometry

Over the last years, mass spectrometry based proteomics has emerged to a pow-erful and widely used technique in the analysis of biological samples [2]. Tandem mass spectrometers are nowadays used to identify fragments of peptides, which are fragments of proteins, in biological samples; the so obtained mass spectra contain peaks as mass-to-charge ratios and respective ion intensities of peptide fragments.

In this study we used the mouse HCD peptide library from the National Institute of Standards and Technology (NIST) containing 216,677 mass spectra and their peptide sequences. For all these spectra our goal is to predict the relative intensities of singly charged y fragment ions. In total, there are 89 input features mainly containing the information about the presence of amino acids left and right of the fragmentation site and the charge state, all of them being Boolean or integer.

For this data set we see that artificial neural nets and random forests were by far more able to produce good prediction models. Again we executed 5-fold cross validation test series with symbolic regression (OS-GP), random forests and artificial neural networks with varying settings as described in the previous section. The best results are summarized in Table 3. These results have been identified using the following settings:

- RF: number of trees: 500, m (ratio of features used in each tree): 0.5, r (ratio of training set used for constructing each tree): 0.5
- ANN: 1 hidden layer containing 16 rectified linear units, 2 hidden layers containing sigmoid units (8 and 4, resp.), sigmoid output node, training algorithm: back propagation, 100 epochs
- GP: population size: 100, strict offspring selection, gender specific parent selection (random & roulette), maximum selection pressure: 100, maximum model size: 200

Table 3. Quality of best prediction models for y fragment ion intensities.

Modeling Method	Average quality of prediction models (R^2)
OS-GP	0.187
RF	0.606
ANN	0.690

4 Discussion and Conclusion

As expected, these test results do not show that any of the applied methods is better than the others, no matter which data set is analyzed. On some of the PBML data sets RFs perform best, on some GP, and on others ANNs. For the medical data set, this is not the case as it seems that GP is able to find better models than the other methods. Interestingly, the models identified using GP only use 11.7 (out of 120) variables, which highlights GP's variable selection capabilities; obviously there is only a small number of variables that carry the really relevant information in this particular case. On the third analyzed data set including proteomics data, black box models are clearly better than those found by symbolic regression. This is not surprising, as we here do not need feature selection (all amino acids are relevant) but rather bigger and more complex models that are able to incorporate information of a high number of variables.

References

1. Affenzeller, M., Winkler, S., Wagner, S., Beham, A.: Genetic Algorithms and Genetic Programming - Modern Concepts and Practical Applications, Numerical Insights, vol. 6. Chapman & Hall, CRC Press (2009)

2. Angel, T.E., et al.: Mass spectrometry-based proteomics: existing capabilities and future directions. Chem. Soc. Rev. **41**(10), 3912–3928 (2012). https://doi.org/10.1039/c2cs15331a
3. Breiman, L.: Random forests. Mach. Learn. **45**(1), 5–32 (2001)
4. Duda, R.O., Hart, P.E., Stork, D.G.: Pattern Classification, 2nd edn. Wiley Interscience, New York (2000)
5. Goodfellow, I., Aaron Courville, Y.B.: Deep Learning (Adaptive Computation and Machine Learning). MIT Press, Cambridge (2016)
6. Kommenda, M., Burlacu, B., Kronberger, G., Affenzeller, M.: Parameter identification for symbolic regression using nonlinear least squares. Genet. Program Evolvable Mach. (2019, in revision)
7. Kommenda, M., Kronberger, G., Wagner, S., Winkler, S., Affenzeller, M.: On the architecture and implementation of tree-based genetic programming in Heuristiclab. In: Proceedings of the 14th Annual Conference Companion on Genetic and Evolutionary Computation, GECCO 2012, pp. 101–108. ACM, New York (2012). https://doi.org/10.1145/2330784.2330801
8. Ljung, L. (ed.): System Identification: Theory for the User, 2nd edn. Prentice Hall PTR, Upper Saddle River (1999)
9. Olson, R.S., Cava, W.L., Orzechowski, P., Urbanowicz, R.J., Moore, J.H.: PMLB: a large benchmark suite for machine learning evaluation and comparison. CoRR abs/1703.00512 (2017). http://arxiv.org/abs/1703.00512
10. Paszke, A., et al.: Automatic differentiation in PyTorch. In: NIPS Autodiff Workshop (2017)
11. Pedregosa, F., et al.: Scikit-learn: machine learning in Python. J. Mach. Learn. Res. **12**, 2825–2830 (2011)
12. Rumelhart, D.E., Hinton, G.E., Williams, R.J.: Learning representations by back-propagating errors. Nature **323**, 533–536 (1986)
13. Schmidhuber, J.: Deep learning in neural networks. Neural Netw. **61**(C), 85–117 (2015). https://doi.org/10.1016/j.neunet.2014.09.003
14. Segal, M.R.: Machine Learning Benchmarks and Random Forest Regression. Center for Bioinformatics & Molecular Biostatistics (2004)
15. Wagner, S., et al.: Architecture and design of the Heuristiclab optimization environment. In: Advanced Methods and Applications in Computational Intelligence, Topics in Intelligent Engineering and Informatics, vol. 6, pp. 197–261 (2013)
16. Winkler, S.M.: Evolutionary System Identification: Modern Concepts and Practical Applications. Schriften der Johannes Kepler Universität Linz, Universitätsverlag Rudolf Trauner (2009)

Concept Drift Detection with Variable Interaction Networks

Jan Zenisek[1,2(✉)], Gabriel Kronberger[1], Josef Wolfartsberger[1], Norbert Wild[1], and Michael Affenzeller[1,2]

[1] Center of Excellence for Smart Production,
University of Applied Sciences Upper Austria, Campus Hagenberg Softwarepark 11,
4232 Hagenberg, Austria
jan.zenisek@fh-hagenberg.at

[2] Institute for Formal Models and Verification, Johannes Kepler University Linz,
Altenberger Straße 69, 4040 Linz, Austria

Abstract. The current development of today's production industry towards seamless sensor-based monitoring is paving the way for concepts such as Predictive Maintenance. By this means, the condition of plants and products in future production lines will be continuously analyzed with the objective to predict any kind of breakdown and trigger preventing actions proactively. Such ambitious predictions are commonly performed with support of machine learning algorithms. In this work, we utilize these algorithms to model complex systems, such as production plants, by focussing on their variable interactions. The core of this contribution is a sliding window based algorithm, designed to detect changes of the identified interactions, which might indicate beginning malfunctions in the context of a monitored production plant. Besides a detailed description of the algorithm, we present results from experiments with a synthetic dynamical system, simulating stable and drifting system behavior.

Keywords: Machine learning · Predictive Maintenance · Concept drift detection · Structure learning · Regression

1 Motivation

The increasing amount of data recorded by today's automatized and sensor-equipped production plants is an essential impetus for current developments in the industrial area. In order to face the challenge of actually making use of the recordings, machine learning algorithms can be employed to create models and hence, fully utilize the available data. In reference to a real-world production system, the relationships between multiple variables representing inputs (cf. configured process- and recipe parameters), internal states (cf. measured time series from condition monitoring) and dependent outputs (cf. measured product quality indicators) have to be covered by such models to identify a system

© Springer Nature Switzerland AG 2020
R. Moreno-Díaz et al. (Eds.): EUROCAST 2019, LNCS 12013, pp. 296–303, 2020.
https://doi.org/10.1007/978-3-030-45093-9_36

comprehensively. One modeling approach for the analysis of complex systems are variable interaction networks [8] – directed graphs representing system variables as nodes and their impact on others as weighted edges. Primarily, variable interaction networks have been employed to gain a better understanding of the interdependencies within a modeled system [4]. In this work however, we utilize them to detect changing system behavior – so-called concept drifts [3] – online. Therefore, identified system relationships are tracked over time and analyzed for changes, which might give an indication for beginning malfunctions when applied at production plants.

The objective of this approach is closely related to the currently intensively investigated topic Predictive Maintenance [7], which is concerned with forecasting the remaining useful lifetime of a production system based on its current condition and scheduling specific preventing actions proactively. However, data which enables such predictions is quite difficult to gather, starting by consolidating data from various sources, up to carrying out a large number of run-to-failure experiments [9] under continuous assessment of the actual system condition. Tracking of changing system relationships however, is applicable also for less strictly controlled environments and allows a closer look into a system's dynamics.

In Sect. 2, we describe an algorithm to model variable interaction networks and present a sliding window based evaluation method, performing concept drift detection. Further on, a test problem is introduced in Sect. 3, which we use to generate synthetic data streams and validate and discuss our approach in Sect. 4. Finally, we give a brief summary and an outlook for possible future extensions in Sect. 5.

2   Variable Interaction Networks for Drift Detection

The developed two-phase concept drift detection approach may be categorized as supervised learning based detector [5]. While the aim of the first phase is to develop a comprehensive model, which describes an initially stable system, during the second phase this network is used to detect structural interaction changes on a continuous stream of new data from the respective system. All parts of the approach have been implemented and tested using the open source framework HeuristicLab[1].

2.1   Network Modeling

As a first step, we define a set of variables within the system of interest. For each of them a regression model is trained using the other variables as inputs. For this task various machine learning algorithms may be employed, including multivariate linear regression, random forests or symbolic regression. Subsequently, we determine the relevance of each input variable for the respective target within a model: The impact of a variable is calculated based on the increasing regression

---

[1] https://dev.heuristiclab.com/trac.fcgi/ticket/2288.

error of the developed model when re-evaluating it on a data set, for which the values of the variable have been randomly shuffled [2]. By this means, the information value of the respective variable is removed from the data set, without changing its distribution. Eventually, the calculated value is normalized to the range 0..1. Subsequently, the directed, weighted graph for the variable interaction network can be constructed by creating a node for each participating variable and creating weighted edges from input to target nodes, by using the calculated impacts.

Several post-processing measures are advisable in order to prune less important nodes and edges and hence, determine more robust network structures: Regression models (and the derived variable impacts) with an estimation accuracy below a problem dependent threshold should not be taken into account right from the start, as they might not identify the system correctly. Further on, variable impacts below a user-defined threshold may be pruned to sparsify the networks, without loosing much information.

Moreover, we developed a routine, which uses the previously described impact computation as base, but assembles *acyclic* graphs in order to support identifying the correct variable interaction *direction*, as summarized in Listing 1.1. Instead of creating edges for any computed impact, the routine adds edges step-wise, alternating with removal of the weakest links to break up cycles.

Listing 1.1. Creating an acyclic variable interaction network.

```
1   foreach variable // inputs and targets
2    create a node
3
4   foreach node: create edge for highest incoming impact
5   while any edges added // after loop: final acyclic graph
6    find shortest cycles
7    while any cycles found
8      delete weakest link // based on impact
9      find shortest cycles
10   foreach node: create edge for next highest incoming impact
```

2.2   Network Evaluation

Within the second phase, as depicted in Fig. 1, we consider a stream of new, unseen data, which is partition-wise analyzed (A). The described calculation of variable impacts based on the previously built models, as well as the successive creation of networks is constantly repeated (B.1), while a window slides over this data stream. For identifying drifts in the underlying system, we compute the similarity of the initially built network – representing a stable state – and the updated networks, as part of the sliding window evaluation (B.2). Presumably, changing system behavior affects internal variable dependencies to some extent, which hence, should be reflected by the freshly created networks. We apply the Spearman's rank correlation coefficient and the normalized discounted cumulative gain (NDCG), as proposed in [6], to compare the network structures. The Spearman's rank correlation considers only deviations in ranks, such that

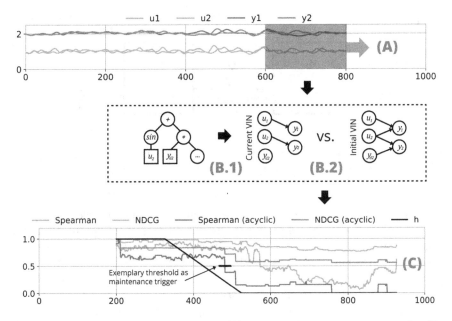

Fig. 1. Sliding window based evaluation (A) of variable interaction networks (B.1, B.2) for concept drift detection (C). The decreasing network similarity scores and the also decreasing drift indicator h, represent the intended functioning of the proposed algorithm.

top-ranked variables are treated equally to lower ranked variables. In contrast, the NDCG puts more weight on top-ranked variables by using an exponential weighting scheme.

A system may be declared *drifting* if the similarity score during the evaluation drops below a threshold. If the actual drift state is known, as e.g. for synthetic data sets, the correlation between the drift value and the computed similarity might give a good indication of how well the drift detection performed (C).

3   Test Problem: Clogging Communicating Vessels

In order to test the proposed concept drift detection algorithm, we designed a synthetic problem based on the system of communicating vessels as illustrated in Fig. 2. The vessels – y_1 and y_2 represent their current fill state – are continuously filled with fluid from two inlets. The flow rates of these inlets – u_1 and u_2 – are independent, dynamic and defined by stationary, auto-regressive models with normally distributed terms. The outlet flow rate for each vessel depends on the current fill state and hence, helps to preserve their stationarity. The communication channel between the vessels transports fluid into the vessel with the currently lower fill state and is described by the flow rate y_3. The dynamics of the system are defined by a system of differential Eqs. (1), (2) and (3). For this

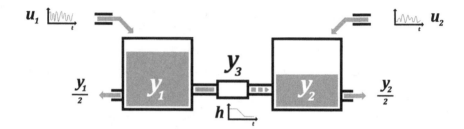

Fig. 2. Vessel fill states y_1 and y_2 with inlets (u_1, u_2), outlets ($y_1/2$, $y_2/2$) and the gradually clogging (cf. h) communication channel with flow rate y_3.

particular example we designed a channel that may gradually clog over time, thus, eventually resulting in a malfunctioning of the vessel communication, controlled by the parameter h (3). This clogging channel represents the maintenance problem, which is aimed to be found by the proposed detection algorithm.

$$\dot{y}_1 = u_1 + y_3 - \frac{y_1}{2} \tag{1}$$

$$\dot{y}_2 = u_2 - y_3 - \frac{y_2}{2} \tag{2}$$

$$\dot{y}_3 = -(y_1 - y_2) - hy_3 \tag{3}$$

Based on the system definition we compiled a set of 10 training instances representing stable system states (i.e. h remains constant) and 10 evaluation instances with drifting behavior (i.e. h slowly decreases), each consisting of 1000 data points. The variables allowed for model training and evaluation, as both inputs and targets, are u_1, u_2, y_1 and y_2. Further on, the first numerical derivative (as defined by the Eqs. (1) and (2)) and the second numerical derivative for each vessel fill state are provided as additional input variables. The current flow between the vessels, represented by y_3 and the clogging factor h however, remain unknown to the regression models. This limitation is inspired by real-world problems, in which availability and quality of data are not always fully ensured, either for technical, monetary or security related reasons. It is the essential motivation and goal of the proposed drift detection algorithm to estimate and monitor the changing variable interactions, when this cannot be observed directly.

4   Experiments and Results

The training of regression models, representing the variable interaction networks' foundation, was performed with multi-variate linear regression (LR), random forest (RF) and symbolic regression (SR). To tune the random forest and the symbolic regression algorithm, we performed a parameter grid search for reasonable configurations:

– Random Forest: R: 0.5, M: 0.2, 100 trees
– Symbolic Regression: Offspring Selection Genetic Algorithm (OSGA) [1], population size: 100, generations: 1000, selection pressure: 100, proportional and random selection, mutation rate: 25%, crossover rate: 100%, unary functions $(\sin, \exp, \log)$, binary functions $(+, -, \times, \div)$, max. tree length: 25 nodes

The modeling results, aggregated for all 10 training instances, are summarized in Table 1. Linear and symbolic regression both achieved almost perfect fits on the training as well as the test partition. The random forest models however, tend to overfit, no matter the tested algorithm parameters.

Table 1. Model estimation qualities R^2 and Normalized Mean Squared Error (NMSE) for the selected targets on training/test partition (cf. split at 66%/34%).

	Linear Regression (LR)		Random Forest (RF)		Symbolic Regression (SR)	
	R^2	NMSE	R^2	NMSE	R^2	NMSE
u_1	0.99/0.99	0.00/0.00	0.96/0.76	0.06/0.30	0.97/0.97	0.02/0.03
u_2	0.99/0.99	0.00/0.00	0.95/0.70	0.06/0.34	0.97/0.97	0.02/0.03
y_1	0.99/0.99	0.00/0.00	0.95/0.67	0.06/0.41	0.93/0.92	0.06/0.09
y_2	0.99/0.99	0.00/0.00	0.95/0.71	0.06/0.36	0.94/0.92	0.05/0.09

Based on the regression models, the initial variable interaction networks, representing stable system behavior, have been computed. Further on, we defined a threshold for the minimum NMSE of 0.2 a model has to achieve to be considered for the network creation. For the random forest model, the threshold has been set higher, to an NMSE of 0.5, because in the first modeling step we observed that the predictive quality of RF is lower compared to SR and LR. Furthermore, we set a minimum variable impact threshold of 0.1 to prune less important edges from the final networks. After the modeling phase (cf. Sect. 2.1), one cyclic and one acyclic network version for each algorithm and each of the 10 training sets has been created.

The second phase (cf. Sect. 2.2) has been performed using the same configurations for the creation of networks during the sliding window evaluation. The results of the drift detection method for each regression modeling algorithm, with varying sliding window size and aggregated for all 10 drift data sets are depicted in Fig. 3. The bar chart illustrates the computed correlation between the network similarity and the synthetic drift, as described in Fig. 1.

According to the computed correlation scores, the linear and the symbolic regression models detect the synthetically introduced drifts quite well with a correlation of roughly 0.7. Although the performance of the random forest models clearly lags behind, one can observe that the drifts are still detected to some extent. In conclusion, the detection algorithm is agnostic to the used regression models, however, accurate models with the ability to generalize (i.e. not to overfit) are necessary.

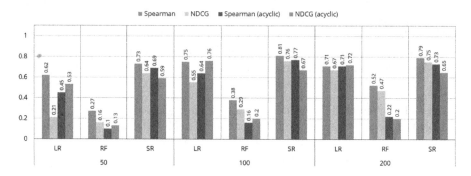

Fig. 3. Correlation of the decreasing network similarity scores and the known, also decreasing drift indicator h, representing the gradually clogging communication channel.

One key factor of the detection algorithm is the sliding window size, which has to be tuned for any problem. With a large window size more stable network structures can be identified, which are valid for a longer period while moving over the data stream. This results in a smoother curve shape of the similarity score, however decreases the reaction speed to underlying trends and hence, should be limited to a reasonable level. In this example, sizes between 100 and 200 showed similar good results. Furthermore, the acyclic networks achieved smoother similarity score curve shapes (cf. Fig. 1), than networks with cycles. Although, there is no advantage according to the computed detection quality by using these networks, it is easier to define a threshold as a minimum similarity score, when the values do not vary too much within a certain period, which is a clear benefit of the acyclic networks.

5   Conclusion and Outlook

In this work we presented a machine learning based approach for identifying changing relationships of dynamical systems, such as industrial production plants. We show how variable interaction networks are developed and utilized to evaluate a continuous data stream and identify deviations from the original behavior, which eventually might enable triggering maintenance actions proactively. We implemented the algorithm using the open source framework HeuristicLab and tested the approach on a synthetic problem successfully.

As a promising next step to enhance the described approach, we consider to investigate how a closer integration of modeling and evaluation phase might lead to a more accurate calculation of variable impacts and hence, more robust networks. A repeated or open-ended training of regression models – which represent the foundation of the variable interaction networks – on the continuously updated data stream might provide valuable information concerning the current impact of variables. Proceeding from drift detection, investigating the dependency changes closely, might eventually enable tracking a system change back to its beginnings. Especially considering the potential value for domain experts, such a root-cause analysis would be a powerful component for future production systems.

Acknowledgments. The work described in this paper was done within the project "Smart Factory Lab" which is funded by the European Fund for Regional Development (EFRE) and the country of Upper Austria as part of the program "Investing in Growth and Jobs 2014–2020".

Europäische Union Investitionen in Wachstum & Beschäftigung. Österreich.

Gabriel Kronberger gratefully acknowledges the financial support by the Austrian Federal Ministry for Digital and Economic Affairs and the National Foundation for Research, Technology and Development within the Josef Ressel Centre for Symbolic Regression.

References

1. Affenzeller, M., Winkler, S., Wagner, S., Beham, A.: Genetic Algorithms and Genetic Programming: Modern Concepts and Practical Applications. CRC Press, New York (2009)
2. Breiman, L.: Random forests. Mach. Learn. **45**(1), 5–32 (2001)
3. Gama, J., Žliobaitė, I., Bifet, A., Pechenizkiy, M., Bouchachia, A.: A survey on concept drift adaptation. ACM Comput. Surv. (CSUR) **46**(4), 44 (2014)
4. Kommenda, M., Kronberger, G., Feilmayr, C., Affenzeller, M.: Data mining using unguided symbolic regression on a blast furnace dataset. In: Di Chio, C., et al. (eds.) EvoApplications 2011. LNCS, vol. 6624, pp. 274–283. Springer, Heidelberg (2011). https://doi.org/10.1007/978-3-642-20525-5_28
5. Krawczyk, B., Minku, L.L., Gama, J., Stefanowski, J., Woźniak, M.: Ensemble learning for data stream analysis: a survey. Inf. Fusion **37**, 132–156 (2017)
6. Kronberger, G., Burlacu, B., Kommenda, M., Winkler, S., Affenzeller, M.: Measures for the evaluation and comparison of graphical model structures. In: Moreno-Díaz, R., Pichler, F., Quesada-Arencibia, A. (eds.) EUROCAST 2017. LNCS, vol. 10671, pp. 283–290. Springer, Cham (2018). https://doi.org/10.1007/978-3-319-74718-7_34
7. Lee, J., Kao, H.A., Yang, S.: Service innovation and smart analytics for industry 4.0 and big data environment. Procedia CIRP **16**, 3–8 (2014)
8. Rao, R., Lakshminarayanan, S.: Variable interaction network based variable selection for multivariate calibration. Analytica Chimica Acta **599**(1), 24–35 (2007)
9. Saxena, A., Goebel, K., Simon, D., Eklund, N.: Damage propagation modeling for aircraft engine run-to-failure simulation. In: International Conference on Prognostics and Health Management 2008, pp. 1–9. IEEE (2008)

Visualization of Solution Spaces
for the Needs of Metaheuristics

Czesław Smutnicki[✉]

Faculty of Electronics, Wrocław University of Science and Technology,
Wroclaw, Poland
czeslaw.smutnicki@pwr.edu.pl

Abstract. We present several technologies recommended for making
2-dimensional (2D) or 3-dimensional (3D) representations of selected
discrete solution spaces as well as features of the solution algorithms
occurring in combinatorial optimization (CO) tasks. We provide some
results of theoretical as well as experimental investigations of so called
landscape of the space with reference to some exemplary hard CO problems. Theoretical analysis starts from various measures of the distance
between solutions represented by typical combinatorial objects, namely
permutations, composition of permutations, set partition and so forth.
Then, we propose some mapping of n-dimensional space of permutations
into 2- or 3 dimensional Euclidean space to extract factors responsible
for hardness of the problem and to illustrate behavior of approximation
algorithms.

1 Introduction

Sequential, parallel and distributed metaheuristic algorithms (MA) currently
become the dominant solution tool used and recommended for very hard combinatorial optimization (CO) problems, [7]. This is a direct consequence of some
numerical problems recognized for CO, let we mention at least: NP-hardness,
curse of dimensionality, multiple local extremes, uneven distribution of extremes,
exponential growth of the number of extremes and deception extremes. The list of
known metaheuristics now contains now 40+ various technologies in the sequential version and 80+ including parallel variants, [8]. Simultaneously, the success
or failure of a metaheuristic algorithm depends on skillful composition of some
structural elements and a number of tuning parameters. All crucial elements are
usually tailored through experimental research. Visualization of the search trajectories and search space allow us to understand the behavior of the algorithm,
to select its best configuration and employ structure of the space in order to
guide the search into the most promising areas of the solution space. Unstable
and capricious behavior of approximate methods, already on standard benchmarks, incline scientists to define and identify factors of the instance hardness,
which allow one to differentiate hard cases from easy ones. "No free lunch" theorem fully justifies rich variety of metaheuristics, [11]. A lot of recent papers refer
to trajectories and landscape of the solution space to analyze some phenomenon

© Springer Nature Switzerland AG 2020
R. Moreno-Díaz et al. (Eds.): EUROCAST 2019, LNCS 12013, pp. 304–311, 2020.
https://doi.org/10.1007/978-3-030-45093-9_37

in the CO area, [1,10] or in the process of developing new algorithms [4,5], for various CO problems, especially scheduling problems from the project [1].

In this paper we present fundamentals and propose some new ideas for making 2D and 3D representations of selected discrete solution spaces and features of the solution algorithms occurring in CO. Results extend and develop essentially our previous findings from [3]. Theoretical analysis starts from various measures of the distance between solutions represented by typical combinatorial objects, namely permutations, composition of permutations, set partition and so on. These metrics based on Cayley, Ulam, Kendall and Hamming distances between permutations. Then, we discuss methods of transforming n-dimensional space into 2- or 3 dimensional space, [6] to extract factors responsible for hardness of the problem (e.g. roughness of the landscape), and to illustrate behavior of a sequential algorithm (e.g. searching trajectory, goal search paths, scatter search, relinking paths, wandering around, sampling, diversification, intensification) and parallel algorithms (e.g. distribution of populations, cooperation).

2   Distance Measures

The visualization can be perceived as a mapping from the solution space, usually having the combinatorial character, onto the flat plane or a surface (spherical model is recommended to this aim). We expect that the mapping will be computationally inexpensive and additionally preserves the distance from the solution space, i.e. transforms close solutions onto close images on the plane, whereas far solutions - onto far images on the plane. It may be hard since we reduce significantly dimension of the space, so mapping needn't preserve the distance. We need at least two types of distance measures. The former refers to the solution space of the considered discrete optimization task, usually represented by an combinatorial object or their composition. This subject we discuss in detail in the next section and partially in here. The latter refers to 2D ($n = 2$) or 3D ($n = 3$) Euclidean space, that is collected from the literature and discussed in the sequel.

For the given two points $x = (x_1, \ldots, x_n)$ and $y = (y_1, \ldots, y_n)$, $x, y \in R^n$, the *Minkowski distance* $L_p(x, y)$ is defined as

$$L_p(x, y) = (\sum_{i=1}^{n} |x_i - y_i|^p)^{\frac{1}{p}}. \tag{1}$$

Setting in (1) $p = 1$ we obtain the *Manhattan distance*

$$L_1(x, y) = \sum_{i=1}^{n} |x_i - y_i| \tag{2}$$

---

[1] Paper is supported by funds of National Science Centre, grant OPUS no. DEC 2017/25/B/ST7/02181.

whereas for $p = 2$ from (1) we get the *Euclidean distance*

$$L_2(x, y) = \sqrt{\sum_{i=1}^{n}(x_i - y_i)^2}. \tag{3}$$

By applying the limit we get the *Chebyshev distance*

$$L_\infty(x, y) = \lim_{p \to \infty} L_p(x, y) = \max_{1 \leq i \leq n} |x_i - y_i|. \tag{4}$$

Let us note that formulae (2)–(4) can be applied also for discrete x, y. For example, assuming that x, y represent permutations, then from (2) we obtain *footrule measure*, whereas taking square of (3) – the *Spearman's rank correlation*.

3   Distance Between Combinatorial Objects

Most metaheuristics for CO problems operate on solutions represented by a combinatorial object. Various objects have been used depending on the problem type. Therefore, adequate distance measure should be defined for each considered combinatorial object with some of them are common. Many problems, e.g. Traveling Salesman Problem (TSP), scheduling problems and Quadratic Assignment Problem (QAP), uses permutation on the set $N = \{1, 2, \ldots, n\}$ to represent a solution. There are at least two fundamental types of metaheuristic approaches: (LS) local search with single search trajectory or multiple parallel independent search trajectories, (PS) population-based approach. Distance topic appeared in both cases but in the different context and are used for different aims. Specific measures are recommended for the local search metaheuristics (e.g. simulated annealing, simulated jumping, tabu search, random search, greedy random search, variable neighborhood search, etc.) where the given solution is modified slightly step-by-step using so-called moves. At least three types of moves have been commonly distinguished in the literature: adjacent swap (A), non-adjacent swap (S) and insert (I). These type of moves correspond directly to various technologies of generating permutations. Measures between permutations α, β denoted as $D^Z(\alpha, \beta)$ ($Z \in \{A, S, I\}$) have been already partially studied in the literature, see e.g. the review in [2]. Their selected properties, fundamental for our applications, have been collected in Table 1 (α^{-1} is the inverse permutation to α). Table 1 refers to auxiliary results known in the literature measures: Kendall's tau, Cayley's and Ulam's. The last one can be calculated using so-called Floyd's game.

Another group of measures considers solutions represented by a string of characters. Strings are suitable for some scheduling, location, distribution or collocation problems, including solutions represented by binary strings. Since any sequence of integers can be converted to a binary string, then the possible application is wide. Levenshtein distance is dedicated to evaluate the difference between two strings of characters. The measure includes elementary actions (insertion, deletion, substitution) required to pass from the one string to

Table 1. Distance between permutations.

Move type	Adjacent swap	Swap	Insert
Denotation	$D^A(\alpha, \beta)$	$D^S(\alpha, \beta)$	$D^I(\alpha, \beta)$
Source	Kendall's tau	Cayley's	Ulam's, Floyd's game
Algorithm	Number of inversions in $\alpha^{-1} \circ \beta$	n minus the number of cycles in $\alpha^{-1} \circ \beta$	n minus the length of the maximal increasing subsequence in $\alpha^{-1} \circ \beta$
Mean	$\frac{n(n-1)}{4}$	$n - \sum_{i=1}^{n} \frac{1}{i}$	$n - O(\sqrt{n})$
Maximum	$\frac{n(n-1)}{2}$	$n - 1$	$n - 1$
Variance	$\frac{n(n-1)(2n+5)}{72}$	$\sum_{i=1}^{n}(\frac{1}{i} - \frac{1}{i^2})$	$\Theta(n^{\frac{1}{3}})$
Complexity	$O(n^2)$	$O(n)$	$O(n \log n)$

the other. It can be calculated using dynamic programing in the time $O(nm)$, where n and m are string lengths. The extension of the Levenshtein measure which allows the additional elementary action to "swap two adjacent characters in the string" is called Damerau-Levenshtein measure. The last measure can be perceived as a generalization of the Hamming distance defined for two strings of equal length. Hamming measure value finds the number of corresponding positions in strings which are different. It can be calculated in the time $O(n)$.

For complex combinatorial objects, one can define measure as a combination of a few components that is desirable to be homogenous. For example, in our previous paper [9] there was defined the measure for the job-shop scheduling problem having a solution composed of m actually dependent permutations α_l, β_l of various length, $l = 1, 2, \ldots, m$. Assuming that they are independent, one can naturally define a measure as follows

$$D^*(\alpha, \beta) = \sum_{l=1}^{m} D_l(\alpha_l, \beta_l), \tag{5}$$

where $D_l(\alpha_l, \beta_l)$ is one measure shown in Table 1 for component permutations α_l, β_l. Using features enumerated in Table 1 we can find basic properties of the measure (5), namely

$$D^*_{l,max} = D_{max} = n(n-1)/2 \quad \beta_l = a_l^{-1}, \quad D^*_{max} = m \cdot D_{max}, \tag{6}$$

$$AV(D^*) = D^*_{max} \cdot \frac{1}{2}, \quad VAR(D^*) = D^*_{max} \cdot \frac{2n+5}{36} \tag{7}$$

Theoretical distribution of the number of solutions located on the distance equal k namely $T(k) = card\{\beta : D(\alpha, \beta) = k\}$ for any α is also shown in [9].

Maximum value D_{max} of each chosen measure defines the *space diameter*. Because D_{max} varies, it is useful to normalize the distance

$$D'(\alpha, \beta) = 100\% \cdot \frac{D(\alpha, \beta)}{D_{max}}. \tag{8}$$

Similarly, we can the normalize goal function with respect to a reference value.

4 Mappings

As we already mentioned, the visualization can be perceived as mapping from the solution space onto the flat plane or a surface (e.g. sphere). Studying the literature we find the similar topic, which are the mappings from R^n to R^2 for the needs of exploratory pattern analysis as seen in the survey in [6]. They can be applied, after a modification in our case. This subject needs further extensive research. Skipping consciously the review of these methods, we only mention a few important features: (a) methods (linear/nonlinear mappings), (b) complexity of mapping calculation, (c) usage of the increased history (incremental calculation) and (d) non-standard approaches. Among linear mappings we refer to: principal components, generalized declustering, last squares, projection pursuit). Among nonlinear mappings we refer to: Sammon's, triangular, distance from two means, k-nearest neighbor.

Visualization can be applied for: (1) graphical representation of the solution space, (2) observing features of the solution space, (3) local extremes distribution, (4) detection of valleys (attractors), (5) trajectory visualization and examination and (6) guiding the search. Because tracing the search trajectory has a dynamic unpredictable character, self-adapting method of visualization is desirable. We outline our proposal called *reference points* mapping T to transform permutations from the set P^n into points in Euclidean R^2 2-D space, namely

$$T : P^n \to R^2. \tag{9}$$

The method consists of two phases. In the first phase we transform r points $\sigma_1, \ldots, \sigma_r$ from P^n into r points $e_1, \ldots, e_r$ in R^2 called images of the mapping. Images remain unchanged next and create the reference set. In the second phase, we transform step-by-step each individual point τ into its image t in R^2. Distance in P^n is represented by a measure $D(..)$ and distance in R^2 by a measure $d(..)$. Images are set by solving the following optimization task

$$\sum_{i=1}^{r} \sum_{j=1}^{r} \|d(e_i, e_j) - D(\sigma_i, \sigma_j)\| \to \min_{e_1, \ldots, e_r} \tag{10}$$

where $\|..\|$ denotes a certain norm, e.g. $(..)^2$ or $|..|$. Mapping for any individual permutation τ is to assign point $t \in R^2$ so that

$$\sum_{i=1}^{r} \|d(e_i, t) - D(\sigma_i, \tau)\| \to \min_{t} \tag{11}$$

Optimization task (10) is a nonlinear case with $2r$ continuous variables, whereas (11) has 2 variables and is also nonlinear. So one can think that this approach is too expensive for space visualization. We will show a more efficient technology. We set $r = 3$ and for $\sigma_1, \sigma_2, \sigma_3 \in P^n$ find images $e_1, e_2, e_3 \in R^2$ so that

$$d(e_1, e_2) = D(\sigma_1, \sigma_2), \quad d(e_1, e_3) = D(\sigma_1, \sigma_3), \quad d(e_3, e_2) = D(\sigma_3, \sigma_2). \tag{12}$$

Triangle inequality in P^n ensures suitable diversity of images. One can hope that we do not select linear points. For each individual point $\tau \in P^n$ we create the image t using a geometric construction based on distances $D(\tau, \sigma_1)$, $D(\tau, \sigma_2)$, $D(\tau, \sigma_3)$ and orthogonal projection on the plane, see Figs. 1 and 2.

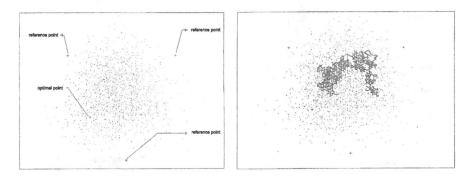

Fig. 1. Reference points (left) and images. Random search trajectory (right).

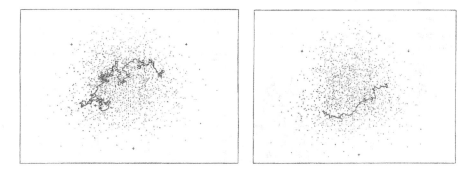

Fig. 2. Simulated annealing trajectory (left). Tabu search trajectory (right).

Another mapping has been proposed to calculate on-time images on the plane, with r reference points $\sigma_1, \ldots, \sigma_r$, see (13). Illustrations of the search trajectories for a few commonly known metaheuristics are shown in Fig. 3.

$$\begin{bmatrix} x \\ y \end{bmatrix} = \sum_{i=1}^{r} D(\tau, \sigma_i) \begin{bmatrix} \sin(\frac{(i-1)\pi}{r}) \\ \cos(\frac{(i-1)\pi}{r}) \end{bmatrix} \tag{13}$$

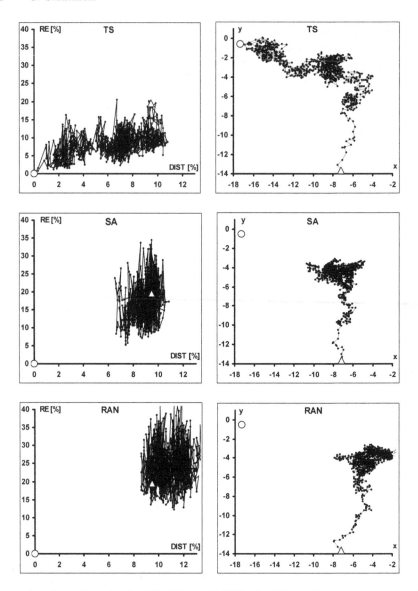

Fig. 3. Search trajectories for TS, SA and RAND algorithms, from start △ to goal ○. *RE* is the normalized goal function value, DIST is a measure based on swap moves.

5   Conclusion

The following conclusions can be drawn: (A) we still need a convenient model of the space (plane, sphere, ellipsoid, ...), thus the further research is needed in this area, (B) evaluate/approximate/verify/validate experimentally the shape model of the space, (C) sample space to detect its features, center, big valley

(BV), roughness, (D) best algorithm should go quickly to BV and search BV long time intensively, (E) evaluate quality of the neighborhood by random wandering around and (F) select combinatorial representation of the problem so that the space size is as small as possible.

References

1. Mattfeld, D.C., Bierwirth, C., Kopfer, H.: A search space analysis of the job shop scheduling problem. Ann. Oper. Res. **86**, 441–453 (1999)
2. Knuth, D.E.: The Art of Computer Programming. Addison Wesley, Longman (1977)
3. Nowicki, E., Smutnicki, C.: 2D and 3D representations of solution spaces for CO problems. In: Bubak, M., van Albada, G.D., Sloot, P.M.A., Dongarra, J. (eds.) ICCS 2004. LNCS, vol. 3037, pp. 483–490. Springer, Heidelberg (2004). https://doi.org/10.1007/978-3-540-24687-9_61
4. Nowicki, E., Smutnicki, C.: An advanced tabu search algorithm for the job shop problem. J. Sched. **8**(2), 145–159 (2005)
5. Nowicki, E., Smutnicki, C.: Some aspects of scatter search in the flow-shop problem. Eur. J. Oper. Res. **169**(2), 654–666 (2006)
6. Siedlecki, W., Siedlecka, K., Sklansky, J.: An overview of mapping techniques for exploratory pattern analysis. Pattern Recognit. **21**(5), 411–429 (1988)
7. Smutnicki, C.: Optimization technologies for hard problems. In: Fodor, J., Klempous, R., Araujo, C.P.S. (eds.) Recent Advances in Intelligent Engineering Systems, pp. 79–104. Springer, Heidelberg (2012). https://doi.org/10.1007/978-3-642-23229-9_4
8. Smutnicki, C., Bożejko, W.: Parallel and distributed metaheuristics. In: Moreno-Díaz, R., Pichler, F., Quesada-Arencibia, A. (eds.) EUROCAST 2015. LNCS, vol. 9520, pp. 72–79. Springer, Cham (2015). https://doi.org/10.1007/978-3-319-27340-2_10
9. Smutnicki, C., Bożejko, W.: Tabu search and solution space analyses. The job shop case. In: Moreno-Díaz, R., Pichler, F., Quesada-Arencibia, A. (eds.) EUROCAST 2017. LNCS, vol. 10671, pp. 383–391. Springer, Cham (2018). https://doi.org/10.1007/978-3-319-74718-7_46
10. Watson, J.P., Howe, A.E., Whitley, L.D.: Deconstructing Nowicki and Smutnicki's i-TSAB tabu search algorithm for the job-shop scheduling problem. Comput. Oper. Res. **33**(9), 2623–2644 (2006)
11. Wolpert, D.H., Macready, W.G.: No free lunch theorems for optimization. IEEE Trans. Evol. Comput. **1**(1), 67–82 (1997)

Preprocessing and Modeling of Radial Fan Data for Health State Prediction

Florian Holzinger[(✉)] and Michael Kommenda

Heuristic and Evolutionary Algorithms Laboratory,
University of Applied Sciences Upper Austria, Hagenberg, Austria
{florian.holzinger,michael.kommenda}@fh-hagenberg.at

Abstract. Monitoring critical components of systems is a crucial step towards failure safety. Affordable sensors are available and the industry is in the process of introducing and extending monitoring solutions to improve product quality. Often, no expertise of how much data is required for a certain task (e.g. monitoring) exists. Especially in vital machinery, a trend to exaggerated sensors may be noticed, both in quality and in quantity. This often results in an excessive generation of data, which should be transferred, processed and stored nonetheless. In a previous case study, several sensors have been mounted on a healthy radial fan, which was later artificially damaged. The gathered data was used for modeling (and therefore monitoring) a healthy state. The models were evaluated on a dataset created by using a faulty impeller. This paper focuses on the reduction of this data through downsampling and binning. Different models are created with linear regression and random forest regression and the resulting difference in quality is discussed.

Keywords: Radial fan · Sampling · Binning · Machine learning

1 Introduction

With the latest advance in electronics and computer science, many new technologies and trends have emerged. Sensors are becoming cheaper and more powerful, enabling the seamless integration of smartphones and wearables in our lives. Many people augment their life by carrying multiple sensors around, providing performance indications such as amount of steps taken today or average heart rate. The industry is changing in the same way, pushing several new trends such as Industry 4.0 and the Internet of Things. Machinery is being monitored, relevant data is collected, stored, and analysed. These trends induced a slow paradigm shift from preventive to predictive maintenance [5], with the goal of predicting the current health state and the remaining useful lifetime of machinery. Imperative for any modeling approach or goal is the acquisition of relevant data. Radial fans [3] are possible candidates for a predictive maintenance approach as they are often a crucial part of factories and an unforeseen outage may have a serious, negative impact. However, hardly any data from real world industrial radial fans is publicly available.

© Springer Nature Switzerland AG 2020
R. Moreno-Díaz et al. (Eds.): EUROCAST 2019, LNCS 12013, pp. 312–318, 2020.
https://doi.org/10.1007/978-3-030-45093-9_38

Therefore, instead of pursuing the ambitious goal of calculating the remaining useful lifetime, this paper focuses on the necessary preprocessing steps and their influence on modeling the current health state of a specific radial fan. This radial fan is the key element of an ongoing project, exploring the applicability of predictive maintenance on industrial radial fans [4, 6]. For this purpose, a test setup has been prepared and several different, relevant sensors are mounted on a radial fan. As the most common failure or sign of wear of radial fans is the abrasion and caking of an impeller, two different impellers were provided. One of them is new and flawless, and the other one artificially pre-damaged to simulate long-term abrasive stress. Several test runs have been carried out and data has been collected for each of these impellers. One of the current flaws of the setup is the huge volume of data generated, which roughly corresponds to about 1 GB/h for one radial fan. This amount of data is acceptable for offline analysis but could pose a difficulty for online processing, be it for either monitoring or prediction purposes.

Two major concerns of a predictive maintenance approach are the prediction quality and the costs, which are necessary to implement such a solution. As higher priced sensors tend to have a higher possible sampling rate, a reduction of sampling rate may render higher priced sensors superfluous and reduces the total cost of the system. Reducing the sampling rate is equivalent to a reduction of the total amount of data and ultimately a loss of information. The data loss can be mitigated by introducing additional features with common statistical key figures such as mean, variance or range while simultaneously reducing the total amount of data. There will always be a trade-off between the amount of data available and quality, but the decrease of quality can be attenuated by extracting beneficial features and reducing the sampling rate nonetheless. To achieve this goal, two different strategies are examined. First, a simple downsampling is applied and the resulting change in quality is shown. Second, the data is binned into bins of certain time intervals and several features are calculated, representing the bins. These calculated features are used in the modeling approach.

For measuring the quality of the models created for the downsampling and binning approach, two different modeling strategies are used. These are linear regression (LR) [2] and random forest regression (RF) [1]. The rotational speed was chosen as the target variable, as it can be manually adjusted. Therefore, additional test runs can be carried out, which can be used for later validation.

2 Setup

The setup for this paper consists of a radial fan with two differently prepared impellers. Each of them has a diameter of 625 mm, weights about 38 kg and has 12 impeller blades. They show different signs of wear, the first one being in perfect condition, the second one was artificially damaged to imitate worn blades (abrasion at the edges), simulating long-term abrasive stress. The radial fan is powered by a 37 kW two-pole electric motor. The engine shaft is connected to the engine by a coupling. An electric frequency converter allows the adjustment of the rotational speed up to 2960 rpm and a control flap mimics pressure loss of the system. This setup is commonly found in industrial applications and therefore a suitable candidate for experimentation.

For the data-acquisition, four gyroscopes (BMX055) provide acceleration and rotation signals in all three dimensions. They were mounted on different positions on the bearings and the case. An additional sensor (GS100102) measures the rotational speed. The gyroscope signals are sampled with 1000 Hz, the rotational speed with 100 Hz. The setup can be extended with additional sensors to measure more properties (vibration, temperature, pressure, humidity, etc.). As for the downsampling and binning approach this paper focuses on, properties such as temperature and humidity with a significant, inherent inertia are not taken into account.

A test routine was defined, enabling reproducible test runs and the generation of comparable data between different configurations (summer/winter, different impeller, etc.). To gather the data from different possible runtime configurations, the rotational speed was incremented stepwise from 0 rpm to 2960 rpm in 370 rpm-steps, each with a recording period of 15 min, resulting in a total observation period of about 2 h. Figure 1 illustrates such a test run. Because the rotational speed is measured, slight noise is present in the raw data, which also increases with high rotational speed. This whole procedure was executed twice, differing in the mounted impeller and with otherwise, as far as possible, near identical parameters (same season, installation site, engine, control-flap position, etc.).

3 Methodology

To summarize, two approaches to reduce the amount of data are examined. The first one is linear downsampling, the second one is binning the data into bins of certain time intervals and calculating features, representing the bins. Therefore, we chose six different configurations for the downsampling approach, using either 50%, 25%, 10%, 1%, 0.1%, or 0.01% of the available, raw data. For the binning approach seven configurations were defined, using bin sizes of 100, 500, 1000, 2500, 5000, 10000 and 50000, containing consecutive samples of the raw data. Each of these bins is represented by key figures from descriptive statistics, namely the mean, standard deviation, range, median and kurtosis. The available data, generated with the previously described setup was preprocessed (downsampled and binned, a total of 13 different configurations) and further split into two different training and test partitions for each of the runs. Firstly, the complete data was shuffled and the first 67% where used for training, the later 33% for test. Secondly, the training partition consists of the first, third, fifth and seventh rpm-step and the test partition was defined with the remaining three, as one of them was removed due to faulty data. The ascends in between each rpm-step were removed from both datasets. Also, considering the different sampling rates of the rotational speed and gyroscope sensors, the rotational speed was padded with the last measured value to remove missing values from the datasets. These datasets are later referred to as the shuffled dataset and the partitioned

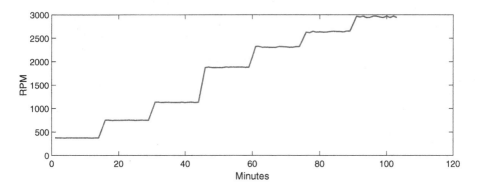

Fig. 1. Visualisation of a test run. The rotational speed was increased from 0 to 2960 rpm in 370 rpm-steps.

dataset. The shuffled dataset provides a general estimation of the change in quality with different downsampling rates and bin sizes, whereas the partitioned dataset suits for examining the ability of the chosen algorithms to generalize with the given configuration. The experiments were conducted with HeuristicLab, an open source framework for heuristic and evolutionary algorithms [7].

4   Results

As a baseline for the raw data, the NMSE of the shuffled dataset, using RF is 0.1398 (Test) and 0.0701 (Train) and the NMSE of the partitioned dataset is 0.4953 (Test) and 0.0419 (Train). For the LR, the corresponding values are 0.9774, 0.9773, 1.0586 and 0.9739. The results of the previously defined setup and methodology for the downsampled datasets (both shuffled and partitioned) are illustrated in Fig. 2. In comparison with the results of the LR using raw models, only a marginal decrease in quality occurred. The more the data is reduced, the bigger the divergence of the NMSE between training and test gets. For example, the NMSE for the LR on the shuffled dataset using 0.01% of the data is 1.0312 (Test) and 0.9048 (Train) and with the partitioned dataset 1.2617 and 0.8589 respectively. The RF behaves similar on both partitions (e.g. 0.6792 (Test) and 0.1661 (Train) on the partitioned dataset using 0.01% of the data). By considering the different character of the partitioned dataset, challenging the ability of the chosen algorithms to interpolate, the difference can be explained. For the given data, linear downsampling is a feasible way to reduce the total amount of data with little negative impact on the NMSE for both LR and RF with the best results being achieved by the RF.

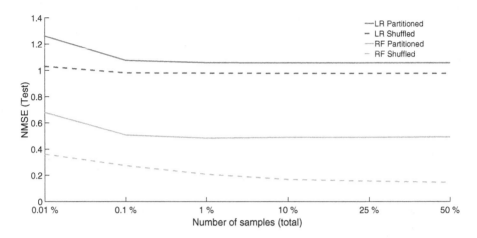

Fig. 2. Results of downsampling for LR and RF on the partitioned and shuffled dataset.

Figure 3 illustrates the results of the binning approach for the shuffled dataset, and Fig. 4 for the partitioned dataset. The algorithms are additionally configured to use either just the mean, the mean and the standard deviation, or all of the previously defined key figures. In contrast to the downsampling approach, the NMSE improves with increasing bin size and yields better results. This seems reasonable, considering an existent noise, which is reduced by calculating the mean of a bin. By adding more features, the results improve when using the shuffled dataset and the RF creates better results than the LR again. Just as with the downsampling approach, the partitioned dataset poses the bigger challenge for both algorithms. The best configuration was achieved by using LR with just the mean, as the RF apparently tends to overfit on the datasets,

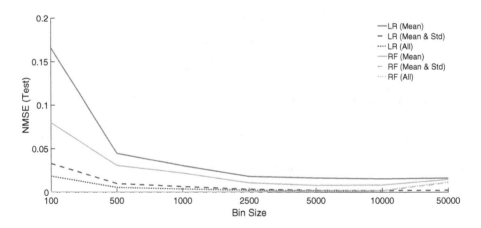

Fig. 3. Results of binning for LR and RF on the shuffled dataset.

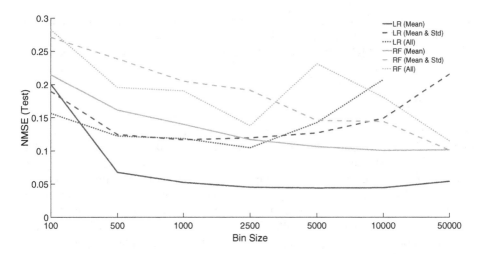

Fig. 4. Results of binning for LR and RF on the partitioned dataset.

especially if more calculated features are added (NMSE of 0.1379 (Test) and 0.0003 (Train) for RF on the partitioned dataset with a bin size of 2500, using all available features, and 0.0004 and 0.0002 for the shuffled dataset, respectively).

As the goal is to reduce the amount of data while keeping an eye on the decline of model quality, which is possible according to the results, a differentiation between a healthy and damaged state (in our case discriminated by the impeller) must still be possible. Therefore, the models generated by the RF were evaluated on the corresponding dataset created with the faulty impeller. The results of such an evaluation, using RF with a bin size of 5000 and utilizing all generated bin representatives on the shuffled dataset, are shown in Fig. 5. The NMSE (calculated for the whole dataset) worsened from 0.0014 (healthy) to 0.0992 (damaged) and a visual distinction is possible. A similar behaviour

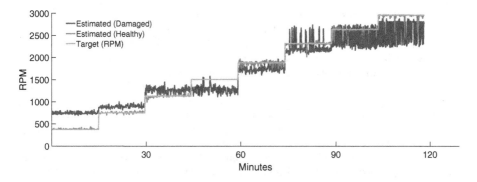

Fig. 5. Evaluation of a random forest model on the dataset with the healthy and damaged impeller.

can be observed for the other configurations as well, especially noteworthy for the LR using only the mean, which yielded the best results on the partitioned dataset (change of NMSE from 0.0146 (healthy) to 3.8796 (damaged), calculated for the whole dataset).

5 Conclusion

We analysed the impact of downsampling and binning on the quality of models generated with LR and RF to approximate the rotational speed of a radial fan with a healthy impeller, using data from four gyroscopes. The reduction of data achieved by using the downsampling approach had little influence on the model quality (NMSE) of both algorithms with the RF having the better model quality, up to a point where only 0.01% of the available data was used and a decline started to manifest. Similar behaviour could be observed by using the binning approach, even having a positive impact on model quality. The ability of the LR and RF to create generalizable models suffered from too many features, representing a bin. Using only the mean as a bin representation yielded the best results in this respect. Evaluating a selected model on an additional dataset generated with a damaged impeller illustrated that the so generated models can still be used to differ between a healthy and damaged state. By representing bins with a mean, the data is implicitly smoothed and possible noise is reduced. Binning may be used to reduce the data while simultaneously improving the model quality if the data has certain characteristics such as a stationary trend and existent noise.

Acknowledgments. The work described in this paper was done within the project #862018 "Predictive Maintenance für Industrie-Radialventilatoren" funded by the Austrian Research Promotion Agency (FFG) and the Government of Upper Austria.

References

1. Breiman, L.: Random forests. Mach. Learn. **45**(1), 5–32 (2001)
2. Draper, N.R., Smith, H.: Applied Regression Analysis, vol. 326. Wiley, Hoboken (2014)
3. Goetzler, W., Guernsey, M., Chung, G.: Pump and fan technology characterization and R&D assessment. Technical Report, October 2015
4. Holzinger, F., Kommenda, M., Strumpf, E., Langer, J., Zenisek, J., Affenzeller, M.: Sensor-based modeling of radial fans. In: The 30th European Modeling and Simulation Symposium-EMSS, Budapest, Hungary (2018)
5. Mobley, R.K.: An Introduction to Predictive Maintenance. Elsevier, Amsterdam (2002)
6. Strumpf, E., Holzinger, F., Eibensteiner, F., Langer, J.: A cost optimized data acquisition system for predictive maintenance. In: 6. Tagung Innovation Messtechnik (2019)
7. Wagner, S., et al.: Architecture and design of the heuristiclab optimization environment. In: Klempous, R., Nikodem, J., Jacak, W., Chaczko, Z. (eds.) Advanced Methods and Applications in Computational Intelligence. TIEI, vol. 6, pp. 197–261. Springer, Heidelberg (2014). https://doi.org/10.1007/978-3-319-01436-4_10

On Modeling the Dynamic Thermal Behavior of Electrical Machines Using Genetic Programming and Artificial Neural Networks

Alexandru-Ciprian Zăvoianu[1]($\boxtimes$), Martin Kitzberger[2], Gerd Bramerdorfer[2], and Susanne Saminger-Platz[3]

[1] School of Computing Science and Digital Media, Robert Gordon University, Aberdeen, Scotland, UK
c.zavoianu@rgu.ac.uk
[2] Institute for Electrical Drives and Power Electronics, Johannes Kepler University, Linz, Austria
[3] Department of Knowledge-based Mathematical Systems, Johannes Kepler University, Linz, Austria

Abstract. We describe initial attempts to model the dynamic thermal behavior of electrical machines by evaluating the ability of linear and non-linear (regression) modeling techniques to replicate the performance of simulations carried out using a lumped parameter thermal network (LPTN) and two different test scenarios. Our focus falls on creating highly accurate simple models that are well-suited for the real-time computational demands of an envisioned symbiotic interaction paradigm. Preliminary results are quite encouraging and highlight the very positive impact of integrating synthetic features based on exponential moving averages.

Keywords: Data driven modeling · Time series · Genetic programming · Dynamic thermal behavior · Electrical machines · Lumped parameter thermal networks

1 Introduction and Motivation

Given the multitude of challenges imposed by the present trend of automation and data exchange in manufacturing processes (i.e., Industry 4.0), it is feasible that in the near future ever more electrical machines will not only be optimized according to multiple criteria [4,12], but will also need to operate inside a symbiotic interaction paradigm. This shift from the current state in which electrical machines (as well as other process components) act as slaves of a master control system will require electrical drives and actuators to actively exchange data with their superior system and operating environment and to incorporate some decision-making autonomy that can ensure good long term operational characteristics.

R. Moreno-Díaz et al. (Eds.): EUROCAST 2019, LNCS 12013, pp. 319–326, 2020.
https://doi.org/10.1007/978-3-030-45093-9_39

In this newly envisioned operational context, locally-embedded processors like micro-controllers would require: (i) information regarding the current state of the electrical machine, (ii) information regarding environment conditions, (iii) the ability to predict power, maintenance and lifetime reserves for upcoming load conditions. In case of the latter, the usage of highly accurate wear and load models – underpinned by linear and especially non-linear regression techniques – has yielded very promising results [3].

The ability to model the dynamic (real-time) thermal behavior is very important, as the strong interaction between electromagnetic and thermal aspects is known to affect the operational performance of electrical machines: losses are dependent upon temperature and vice versa. According to [13], traditional approaches for performing an accurate thermal analysis of electrical machines fall within two broad categories:

A. *Numerical methods* – which group techniques based on finite element analysis or computational fluid dynamics. These approaches are quite general with regard to device geometry but very computationally-demanding.
B. *Analytical modeling* – which is usually instance specific and requires highly specialized domain expertise during model development, but produces models that are very fast to evaluate and could thus be suitable for real-time approaches.

The state-of-the-art approach in analytical modeling aims to construct and correctly parameterize complex *lumped parameter thermal networks* (LPTNs) [2] that accurately model the main heat-transfer paths of an electrical machine by aggregating several non-linear loss sources (i.e., copper and iron losses).

2   Research Focus and Approach

The aim of the present work is to explore if standard data-driven modeling techniques can be applied to predict the dynamic thermal behavior of electrical machines. The main focus falls on obtaining low-complexity models as these:

– can meet the real-time computational demands of micro-controllers;
– can be interpreted more easily by domain experts (i.e., electrical engineers), thus facilitating both domain acceptance and the future development of mixed thermal models.

While the final modeling scenario will consist of time series that aggregate sensor-based temperature measurements collected from electrical drive prototypes, in this incipient study we focus on modeling data streams obtained using simulation programs with integrated circuit emphasis (SPICE) [14]. Thus, using the expert-designed LPTN from Fig. 1, we simulate two realistic electrical machine operational scenarios.

Although simplistic, this approach enables an initial comparative analysis of the ability of data-driven modeling approaches to replicate LPTN-based results and allows for a better grounding of expectations regarding future sensor-based performance. Furthermore, by simulating the data, we have:

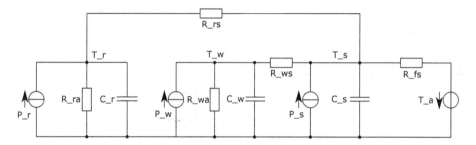

Fig. 1. Basic expert-designed LPTN used for generating data via LTspice©.

- full control of the input stream – i.e., the time-dependent values of P_w, P_s and P_r, the three power sources from Fig. 1 ;
- cost-effective instant access to the output stream – i.e., (LPTN-estimated) temperatures in the three key parts of the electrical motor we are interested in monitoring: the winding (T_w), the stator lamination (T_s) and the rotor lamination (T_r).

Given the characteristics and demands of the envisioned application scenarios (time series data, unknown output values during operation, wish for large prediction horizons, aim for simple models, real-time modeling), we have chosen to model each temperature output at time t – e.g., $T\_w(t)$ – using both the three power inputs at time t – i.e., $P\_w(t)$, $P\_s(t)$, $P\_r(t)$ – and three new *synthetic features* obtained by computing *exponential moving averages* (EMAs) for each input. For a given (power) input $\mathbf{x}$, at each generation t, we have that:

$$EMA_{\mathbf{x}}(t) = \begin{cases} \mathbf{x}(t), & \text{if } t = 1 \\ 0.01 \cdot \mathbf{x}(t) + 0.99 \cdot EMA_{\mathbf{x}}(t-1), & \text{if } t > 1 \end{cases}$$

Using the six inputs, we attempted to model each temperature output using three regression techniques: linear regression [9], genetic programming [10] and (shallow) artificial neural networks [7].

3 Experimental Setup

The Simulation Scenarios
Using the LPTN from Fig. 1, two operational scenarios were simulated using the LTspice© software. The first simulation lasts for 2000 s and concerns a constant load as all three power inputs are simultaneously ramped up to predefined thresholds. The second simulation lasts for 5000 s and concerns a more complicated variable load scenario. The input and output time-wise variations for both scenarios are plotted in Fig. 2.

Data Sets and Experimental Setup
While simulation results were initially sampled equidistantly at every 100 ms, the final data sets used for modeling were obtained by keeping only every 10th

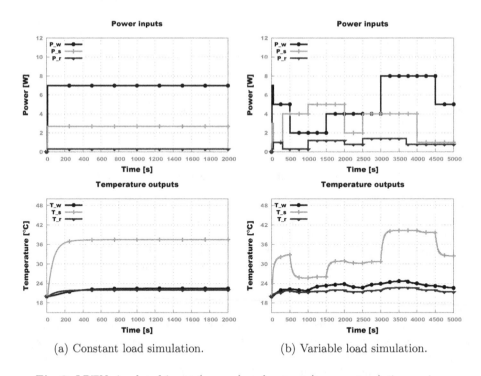

(a) Constant load simulation.   (b) Variable load simulation.

Fig. 2. LPTN-simulated input (powers) and output (temperature) time series.

sample (subsampling). This resulted in a data set with 2000 samples for the first operational scenario and a data set of 5000 samples for the second one. After computing the associated EMA features for each sample, the data sets were randomly shuffled. Afterwards, during the modeling process, 50% of the samples were used for training and 50% for testing.

The plots in Fig. 3 show the distribution of training and test samples across the two data sets and are extremely insightful regarding the importance of the synthetic EMA features, as they highlight the contrast between the very low variance of the original power input features (top subplots) and the diverse range of the temperature outputs (bottom subplots).

The linear regression and artificial neural networks (ANN) implementations we tested with are the ones provided by the WEKA machine learning platform [5]. In case of the linear modeling, we used a rather standard parameterization: M5 attribute selection [11], elimination of collinear attributes, a value of 10^{-8} for the ridge parameter.

For the neural networks, a limited but systematic series of tests using 10-fold cross validation yielded that, across both scenarios and all three outputs, one can obtain a simple but highly accurate model with a single hidden layer that contains 6 hidden units (one for each input) by using the following training parameters: a learning rate of 0.2, a momentum of 0.2 and at most 1000 epochs

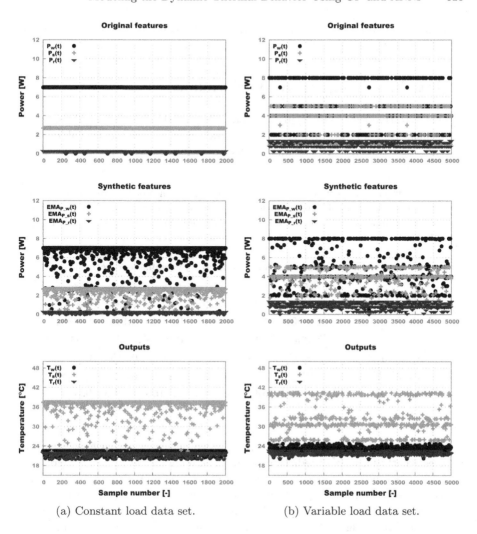

Fig. 3. Shuffled data sets used for modeling the dynamic thermal behavior.

(i.e., training iterations). As a measure aimed at preventing overfitting, ANN training was stopped before performing all the 1000 iterations whenever the approximation error over a validation set containing 20% of the initial training samples constantly worsened for 20 consecutive iterations.

The genetic programming (GP) implementation we employed is the one provided by the HeuristicLab framework. More specifically, we opted for an offspring selection flavour of genetic programming (OSGP) that also features constant optimization [8], as this has shown the ability to produce high-quality models in several other application domains [6]. During the run, we evolved 20 generations of 500 individuals each. the maximum selection pressure was set to 50%

(i.e., early stopping mechanism) and the best model was chosen based on a validation set containing 20% of the training samples.

4   Results - Comparative Performance

The comparative results of trying to model the two dynamic thermal scenarios using linear and non-linear regression techniques are presented in Table 1 and they indicate that both OSGP and ANNs can deliver very competitive results.

Table 1. Comparative test data performance of regression modeling techniques measured via the coefficient of determination (R^2) and the mean absolute error (MAE). Across both scenarios, the best performance in each temperature - error indicator triple is highlighted.

Technique-Indicator	Constant load scenario			Variable load scenario		
	T_w	T_s	T_r	T_w	T_s	T_r
LinReg–R^2	0.9622	0.9980	0.8297	0.9954	0.9978	0.9487
ANN–R^2	**1.0000**	0.9990	0.9978	**0.9998**	**0.9996**	**0.9992**
OSGP–R^2	**1.0000**	**0.9997**	**0.9999**	**0.9998**	0.9985	0.9921
LinReg–MAE	0.2394	0.0108	0.1806	0.2194	0.0156	0.1496
ANN–MAE	0.0117	0.0083	0.0246	0.0489	**0.0075**	**0.0280**
OSGP–MAE	**0.0041**	**0.0044**	**0.0049**	**0.0455**	0.0125	0.0695

It is noteworthy that complementing the two data sets with EMA synthetic features greatly improves prediction accuracy and enables even linear regression models to generally deliver competitive results. Without EMAs, on the simpler constant load scenario, the best regression models can only achieve (test) R^2 values smaller than 0.55 across all three outputs.

To give some rough insight related to comparative model training times, we mention that constructing the linear regression and ANN models took at most a few seconds when using a single computing core of a standard laptop computer while the OSGP runs took between 20 and 50 min when using all four cores.

As there were no attempts to actively limit the size of the evolved GP regression models during the evolutionary run, the resulting solution sizes are bordering the predefined maximal model sizes (i.e., tree depth = 12, tree length = 50). However, these solutions can be simplified by up to 40% without any significant loss of accuracy and OSGP has been shown to produce far more parsimonious solutions (without sacrificing accuracy) when combined with active runtime bloat control techniques such as iterated tournament pruning and dynamic depth limits [15].

5   Conclusions and Future Work

The obtained results show that data-driven techniques can be applied to model (near-perfectly) the dynamic thermal behavior of electrical machines at the level currently obtained by (basic) LPTNs and LTspice© [1]. Furthermore, the high accuracy and quite simple structure of the non-linear ANN and OSGP models makes them highly suitable for the envisioned real-time applications. Results indicate a very slight accuracy edge for GP and an overall model simplicity and training time advantage for ANNs.

In the future, we aim to build upon these initial findings and use the data-driven models to construct prediction intervals based on estimated load conditions for different horizons. Furthermore, as sensor data from electrical machine prototypes becomes available, we also plan to replicate the modeling experiment on the real-world data sets. Last but not least, the long term goal is to explore identified synergies and to help the development of hybrid models that combine data-driven and LPTN-based thermal modeling strategies.

Acknowledgments. This work has been supported by the COMET-K2 "Center for Symbiotic Mechatronics" of the Linz Center of Mechatronics (LCM) funded by the Austrian federal government and the federal state of Upper Austria.

References

1. https://www.analog.com/en/design-center/design-tools-and-calculators/ltspice-simulator.html , LTspice. Accessed 30 May 2019
2. Boglietti, A., Cavagnino, A., Staton, D., Shanel, M., Mueller, M., Mejuto, C.: Evolution and modern approaches for thermal analysis of electrical machines. IEEE Trans. Ind. Electron. **56**(3), 871–882 (2009)
3. Bramerdorfer, G., et al.: Using FE calculations and data-based system identification techniques to model the nonlinear behavior of PMSMs. IEEE Trans. Ind. Electron. **61**(11), 6454–6462 (2014)
4. Bramerdorfer, G., Zăvoianu, A.C.: Surrogate-based multi-objective optimization of electrical machine designs facilitating tolerance analysis. IEEE Trans. Mag. **53**(8), 1–11 (2017)
5. Eibe, F., Hall, M., Witten, I.: The WEKA Workbench. Practical Machine Learning Tools and Techniques. Morgan Kaufmann, Online Appendix for Data Mining (2016)
6. Fleck, P., et al.: Box-type boom design using surrogate modeling: introducing an industrial optimization benchmark. In: Andrés-Pérez, E., González, L.M., Periaux, J., Gauger, N., Quagliarella, D., Giannakoglou, K. (eds.) Evolutionary and Deterministic Methods for Design Optimization and Control With Applications to Industrial and Societal Problems. CMAS, vol. 49, pp. 355–370. Springer, Cham (2019). https://doi.org/10.1007/978-3-319-89890-2_23
7. Haykin, S.: Neural Networks: A Comprehensive Foundation, 2nd edn. Prentice Hall Inc., Upper Saddle River (1999)
8. Kommenda, M., Kronberger, G., Winkler, S., Affenzeller, M., Wagner, S.: Effects of constant optimization by nonlinear least squares minimization in symbolic regression. In: Proceedings of the 15th Annual Conference Companion on Genetic and Evolutionary Computation, pp. 1121–1128. ACM (2013)

9. Marquardt, D.W., Snee, R.D.: Ridge regression in practice. Am. Stat. **29**(1), 3–20 (1975)
10. Poli, R., Langdon, W.B., McPhee, N.F., Koza, J.R.: A field guide to genetic programming. Lulu. com (2008)
11. Quinlan, J.R., et al.: Learning with continuous classes. In: 5th Australian Joint Conference on Artificial Intelligence, vol. 92, pp. 343–348. World Scientific (1992)
12. Silber, S., Koppelstätter, W., Weidenholzer, G., Segon, G., Bramerdorfer, G.: Reducing development time of electric machines with SyMSpace. In: 2018 8th International Electric Drives Production Conference (EDPC), pp. 1–5. IEEE (2018)
13. Staton, D., Pickering, S., Lampard, D.: Recent advancement in the thermal design of electric motors. In: Proceedings of SMMA 2001 Fall Technology Conference - Emerging Technologies for Electric Motion Industry, Durham, North Carolina, USA, 3–5 October 2001, pp. 1–11 (2001)
14. Wang, T.Y., Chen, C.C.P.: SPICE-compatible thermal simulation with lumped circuit modeling for thermal reliability analysis based on modeling order reduction. In: Proceedings of the 2004 5th International Symposium on Quality Electronic Design, pp. 357–362. IEEE (2004)
15. Zăvoianu, A.-C., Kronberger, G., Kommenda, M., Zaharie, D., Affenzeller, M.: Improving the parsimony of regression models for an enhanced genetic programming process. In: Moreno-Díaz, R., Pichler, F., Quesada-Arencibia, A. (eds.) EUROCAST 2011. LNCS, vol. 6927, pp. 264–271. Springer, Heidelberg (2012). https://doi.org/10.1007/978-3-642-27549-4_34

Solving a Flexible Resource-Constrained Project Scheduling Problem Under Consideration of Activity Priorities

Viktoria A. Hauder[1,2(✉)], Andreas Beham[1,3], Sebastian Raggl[1], and Michael Affenzeller[1,3]

[1] Heuristic and Evolutionary Algorithms Laboratory, School of Informatics, Communications and Media, University of Applied Sciences Upper Austria, Hagenberg, Austria
{viktoria.hauder,andreas.beham,sebastian.raggl, michael.affenzeller}@fh-hagenberg.at
[2] Institute for Production and Logistics Management, Johannes Kepler University, Linz, Austria
[3] Institute for Formal Models and Verification, Johannes Kepler University Linz, Linz, Austria

Abstract. In the context of real-world optimization problems in the area of production and logistics, multiple objectives have to be considered very often. Precisely such a situation is also regarded in this work. For a resource-constrained project scheduling problem with activity selection and time flexibility, a new bi-objective extension is developed. Motivated by a steel industry production case, each of two already existing objective functions, makespan minimization and time balance maximization, is deployed together with a newly developed objective, the so-called activity priority maximization. To solve the resulting two bi-objective optimization problems and provide all existing trade-off solutions, the ϵ-constraint method is used. A constraint programming model is presented and solved with the CP Optimizer of IBM ILOG CPLEX and the results are compared concerning solution quality and runtime, showing the competitiveness of the developed model.

Keywords: Bi-objective optimization · ϵ-constraint method · Multi-project scheduling

1 Introduction

In many real-world decision situations, contradictory objectives have to be considered and human operators have to determine how to weight and evaluate

The work described in this paper was done within the project Logistics Optimization in the Steel Industry (LOISI) #855325 within the funding program Smart Mobility 2015, organized by the Austrian Research Promotion Agency (FFG) and funded by the Governments of Styria and Upper Austria.

R. Moreno-Díaz et al. (Eds.): EUROCAST 2019, LNCS 12013, pp. 327–334, 2020.
https://doi.org/10.1007/978-3-030-45093-9_40

opposing goals. However, the evaluation of hardly compatible targets is very difficult in many cases precisely due to their contradictory nature. Well-known examples are the conflicting objectives of quality, cost and time, also referred to as "the magic triangle" [12], as illustrated in Fig. 1.

Fig. 1. The magic triangle (inspired by Weckenmann et al. [12]).

Considering two objectives in an optimization context, bi-objective optimization (BOO) avoids the problem of weighting or evaluating opposing goals prior to the optimization process. It produces all trade-off solutions, i.e. it gives a decision maker the possibility of comparing all existing compromise solutions and decide based on this comparison [4]. Exactly such a bi-objective situation is tackled for a real-world scheduling problem motivated by a steel industry case in this work. For an already existing NP-hard single-objective resource-constrained project scheduling problem (RCPSP), alternative activities have to be selected and start and end times have to be assigned to all selected activities [7]. Since the steel producer plans to determine priorities for all alternative activities, this additional goal has to be considered, resulting in a new bi-objective optimization problem.

Different exact and heuristic methods have already been developed for academic and real-world BOO problems [3,4]. However, to the best of the authors' knowledge, no works on activity selection priorities for the above described scheduling case are available yet. Therefore, we extend the already existing single-objective problem in a way that it fits the requirements of a popular and well-working BOO algorithm, the so-called ϵ-constraint method [5]. For solving this new problem, we propose an extended bi-objective version of the related existing single-objective constraint programming (CP) model [7] and solve it with the CP Optimizer of IBM ILOG CPLEX. We consider two objectives of the single-objective version, makespan minimization and time balance maximization [7], and compare each of them with the newly introduced priority maximization objective regarding optimality and runtime of our CP results.

The paper is organized as follows. In Sect. 2 we review the related literature concerning BOO and the RCPSP. Section 3 contains the presentation of the new bi-objective RCPSP and the developments necessary for the application of the ϵ-constraint method. Our optimization results are then presented in Sect. 4. Finally, in Sect. 5 we conclude our work and give directions for further research.

2   Literature Review

In many real-world decision situations, multiple contradictory objectives have to be regarded. In an optimization context, one possible way of evaluating the influence of opposing goals is BOO. With BOO it is possible to generate all existing trade-off solutions, i.e. all solutions where one objective cannot be improved without worsening another one, leading to the generation of all Pareto-optimal solutions (Pareto set). Thus, decision makers do not have to weight or evaluate contradictory goals prior to the optimization. They have the opportunity to compare all compromise solutions, evaluate their respective effects and decide based on this possible comparison [3, 4, 9]. Figure 2 gives an example of Pareto-optimal optimization solutions (represented by black dots) and solutions which are not on the Pareto front (indicated by grey dots) for a BOO problem.

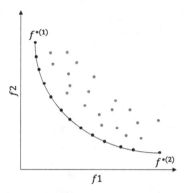

Fig. 2. Exemplarily Pareto set for two objectives $f1$ and $f2$.

Besides many other exact and heuristic BOO methods, such as the balanced box method [2] or the non-dominated sorting genetic algorithm II (NSGA II) [3], one well-researched method is the ϵ-constraint method [5]. It enumerates all non-dominated points by optimizing one objective ($f1$) as a single-objective problem and using the other one ($f2$) as a constraint. After every optimization run, $f2$ is updated by the reduction of a constant δ that has to be defined prior to the optimization. It is noted that this necessary definition, i.e. discretization, is also the greatest drawback of the ϵ-constraint method [9]. However, in the case of our work, the second objective naturally consists of discrete values (see Sect. 3), leading to the exclusion of this drawback.

The BOO problem tackled in this work is a special form of the NP-hard RCPSP [6, 7]. For real-world applications, additional requirements have to be taken into account in many cases, such as the consideration of time flexibility [1], multiple projects [10] or alternative activities [11]. These three components are also integrated into the RCPSP regarded as the starting point of this work. For the production process of a steel producer, various alternative production paths exist for every lot, i.e. alternative activities have to be selected. Start and end times have to be determined for every activity and multiple steel lots have to be produced, leading to a multi-project environment for this real-world RCPSP. Amongst others, the objectives of makespan minimization and time balance maximization, i.e. balanced activity processing time lengths, are tackled. Mixed integer programming and CP models have been presented for this single-objective flexible RCPSP [7]. The steel producer now plans to assign priorities for the production routes and requests new schedules with the highest possible priorities. Thus, we introduce the new objective of priority maximization and a related CP model including the ϵ-constraint method in the following Sect. 3.

3   Priority Maximization for a Flexible RCPSP

We now present a constraint programming model for our new BOO problem that is based on the CP model for the single-objective flexible RCPSP [7]. Section 3.1 includes the necessary formal definitions and notations. Section 3.2 contains the CP model with the newly developed objectives and the related ϵ-constraint method for the two bi-objective versions of the RCPSP.

3.1   Definitions and Notations

In the following, we give the definitions and notations that have to be regarded for our RCPSP. The set $\mathfrak{J}$ indicates all activities $j \in \{0, .., n + 1\}$ of the project and the set $\mathfrak{A}$ illustrates the adjacencies between activities (i, j). The subset $\mathfrak{L} \subseteq \mathfrak{J}$ represents the multi-project case with one sink node per lot. The subset $\mathfrak{M} \subseteq \mathfrak{J}$ together with the parameters S_j and E_j represent the necessary artificial nodes that enable the selection possibilities for alternative activities and the related possible alternative start and end activities per selection possibility. T gives the overall time horizon and $\mathfrak{R}$ the set of all renewable resources $r \in \{1, .., R\}$. The parameters a_j and b_j indicate the minimum and maximum allowed processing time for every activity and D_j represents the delivery time for every sink node $j \in \mathfrak{L}$. The resource availability and the demand are given by C_r and c_{jr}. The function q_r represents the decision on the cumulative resource usage and the interval decision variable w_j decides on the optional selection and the length of an activity j. For a more detailed explanation of all variables and functions of the CP Optimizer, we refer to the work of Laborie et al. [8].

The new set $\mathfrak{P} \subseteq \mathfrak{J}$ represents all nodes consisting of a priority value necessary for the new priority objective and the parameter P_j indicates the priority value of each activity $j \in \mathfrak{P}$. The priority values are defined in a range of 1 to n; the lowest value of 1 indicates the highest (=most preferred) priority, the highest value of n indicates the lowest priority. With the values ϵ and ϵ', the lower and upper bound for the priority objective function are defined. They are calculated by summarizing the minimum necessary (ϵ) and maximum possible (ϵ') priority values for all activities. Together with the lexicographic optimization that is carried out for both bi-objective optimization problems, a discrete value

δ is necessary for the reduction of ϵ' in every optimization run [9]. Due to the determined integer values for the priorities, the discrete value $\delta = 1$ is defined.

3.2 Introduction of Priority Maximization

With the following CP model that is based on the model for the single-objective flexible RCPSP [7], two versions of the new bi-objective RCPSP are presented: The newly developed priority objective function is first combined with the makespan objective and second with the time balance objective. For both bi-objective models, the ϵ-constraint method is introduced.

$$f_1 = \sum_{i \in \mathfrak{P}} \left(\texttt{presenceOf}\left(w_i\right) \right) \cdot p_i \tag{1a}$$

$$f_2 = \texttt{endOf}\left(w_{n+1}\right) \tag{1b}$$

$$f_3 = \left(\max_{i \in \mathfrak{J} \setminus \mathfrak{M}} \left(\texttt{lengthOf}(w_i) - a_i \right) - \min_{i \in \mathfrak{J} \setminus \mathfrak{M}} \left(\texttt{lengthOf}(w_i) - a_i \right) \right) \tag{1c}$$

$$\texttt{lex } \min(f_2, f_1) \tag{2a}$$

$$\texttt{lex } \min(f_3, f_1) \tag{2b}$$

subject to

$$\texttt{startOf}\left(w_0\right) = 1, \tag{3}$$

$$\texttt{presenceOf}\left(w_0\right) = 1, \tag{4}$$

$$\texttt{presenceOf}\left(w_i\right) = 1, \quad \forall\, i \in \mathfrak{L}, \tag{5}$$

$$\texttt{presenceOf}\left(w_{n+1}\right) = 1, \tag{6}$$

$$\texttt{lengthOf}\left(w_i\right) \geq a_i \quad \forall\, i \in \mathfrak{J}, \tag{7}$$

$$\texttt{lengthOf}\left(w_i\right) \leq b_i \quad \forall\, i \in \mathfrak{J}, \tag{8}$$

$$\texttt{endOf}\left(w_i\right) \geq D_i \quad \forall\, i \in \mathfrak{L}, \tag{9}$$

$$\texttt{endOf}\left(w_i\right) \leq T \quad \forall\, i \in \mathfrak{L}, \tag{10}$$

$$\texttt{endAtStart}\left(w_i, w_j\right); \quad \forall\, (i,j) \in \mathfrak{A}, \tag{11}$$

$$\texttt{presenceOf}\left(w_i\right) = \texttt{presenceOf}\left(w_j\right); \quad \forall\, (i,j) \in \mathfrak{A}, \tag{12}$$

$$\texttt{alternative}\left(w_i, \{w_a \in S_i\}\right); \quad \forall\, i \in \mathfrak{M}, \tag{13}$$

$$\texttt{endAtStart}\left(w_i, w_a\right); \quad \forall\, i \in \mathfrak{M}, a \in E_i, \tag{14}$$

$$\texttt{endBeforeStart}\left(w_i, w_{n+1}\right); \quad \forall\, i, j \in \mathfrak{L}, \tag{15}$$

$$q_r \leq C_r \quad \forall\, r \in \mathfrak{R}, \tag{16}$$

$$\epsilon \leq f_1 \leq \epsilon'. \tag{17}$$

Objective functions (1a)–(1c) give an overview of the three goals considered for the BOO. The newly developed priority objective is introduced in (1a). Functions (1b)–(1c) present the objectives makespan and time balance. With (2a), the lexicographic optimization of the makespan and the priorities is introduced. Together with constraints (3)–(17), the objective (2a) leads to the first bi-objective RCPSP. With (2b), the lexicographic optimization of the time balance and the priorities is presented. Together with restrictions (3)–(17), the objective (2b) results in the second bi-objective RCPSP of this work. Restrictions (3) and (4) determine the start of the project. With conditions (5)–(6) it is guaranteed that every lot is produced and that the production process has to be finished. Restrictions (7)–(8) define that the processing time of every activity has to correspond to its minimum and maximum allowed durations. Every lot has to be part of the production process at least until its delivery time, which is determined by restrictions (9). The overall project horizon is defined by constraints (10). Conditions (11)–(12) ensure that idle times are not allowed within the production process of one lot and that all precedence relations are met. Constraints (13)–(15) specify the alternative production routes of every lot and allow idle times only between the production of different lots. With (16), capacity restrictions are satisfied. In connection with the newly introduced functions (2a) and (2b), the new conditions (17) introduce the ϵ-constraint method.

4   Computational Results

The CP model is implemented in OPL and the CPLEX 12.8.0 CP Optimizer is used to solve it. All tests are run on a virtual machine Intel(R) Xeon(R) CPU E5-2660 v4, 2.00 GHz with 28 logical processors, Microsoft Windows 10 Education. The maximum allowed runtime (T) is one hour. The generated benchmark instances are inspired by the description of the regarded single-objective RCPSP [7] and extended in terms of priority assignments. For every alternative activity, a priority value (1 to n) is randomly assigned. We have three different groups (=lot sizes) with 10, 15 and 20 lots, each consisting of five instances. In Table 1, the optimization results are presented. The first column gives the lot size and the second the (randomly generated) number of activities considered per instance. In columns three to four, the solutions for the bi-objective RCPSP with objective function (2a) out of Sect. 3 is presented and in columns five to six, the second considered bi-objective RCPSP with the objective function (2b) out of Sect. 3 is presented. For both considered objectives, the runtime is given in seconds and the number of non-dominated points found is presented.

The results show a very similar picture for both optimization problems. All instances of lot size 10 and three out of five instances of lot size 15 are solved to optimality. For lot size 20, it cannot be proved for any instance that the number of found non-dominated points complies with the complete Pareto set. Concerning the runtime, all optimization runs need below one minute for lot size 10. For lot size 15, the runtime varies strongly from only seconds to one hour. For lot size 20, the maximum runtime is always exploited. In Fig. 3, the results for lot size 10 are additionally illustrated in a diagram, showing the whole Pareto front for all instances. The overlapping non-dominated points indicate an equal solution for different instances. The legend gives the lot size and the number of activities, e.g. "l10_160" means lot size 10 with 160 activities.

Table 1. Computational results for the new bi-objective optimization problem.

Lots	Activities	Makespan / Priorities		Time balance / Priorities	
		Runtime	#Non.-dom.	Runtime	#Non.-dom.
	160	3.21	4	2.40	6
	163	12.42	9	3.79	6
10	158	15.45	9	8.15	7
	163	16.57	11	16.11	8
	167	34.75	9	5.92	6
	232	135.66	6	40.82	5
	259	T	11	T	9
15	238	269.09	12	205.64	12
	252	698.41	20	613.77	13
	247	T	10	T	11
	311	T	5	T	7
	340	T	4	T	6
20	319	T	3	T	5
	340	T	8	T	4
	306	T	7	T	9

Runtime in seconds; T...time limit reached; #Non.-dom. ... amount of non-dominated points found; bold letters indicate that the whole Pareto set is found.

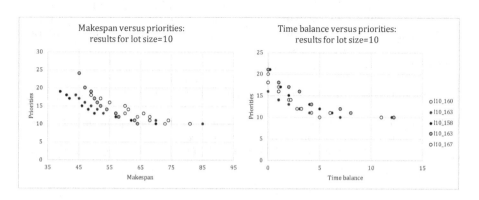

Fig. 3. Pareto fronts of lot size 10.

5   Conclusion and Outlook

In this work, we have developed two new bi-objective versions of an RCPSP motivated by a steel industry case. We have extended a related existing CP model by developing new objectives and applying the ϵ-constraint method for solving the new RCPSP. For both bi-objective cases, the generated results show the competitiveness of our model. Small instances are solved to optimality and for bigger ones, also the complete Pareto set or a part of it is found within the allowed runtime. For future work, additional multi-objective algorithms should be deployed in order to have comparison possibilities concerning solution quality and runtime and to solve even larger instances to optimality.

References

1. Artigues, C.: On the strength of time-indexed formulations for the resource-constrained project scheduling problem. Oper. Res. Lett. **45**(2), 154–159 (2017)
2. Boland, N., Charkhgard, H., Savelsbergh, M.: A criterion space search algorithm for biobjective integer programming: the balanced box method. INFORMS J. Comput. **27**(4), 735–754 (2015)
3. Deb, K.: Multi-objective optimization. In: Burke, E.K., Kendall, G. (eds.) Search Methodologies, pp. 403–449. Springer, New York (2014). https://doi.org/10.1007/978-1-4614-6940-7_15
4. Ehrgott, M.: Multicriteria Optimization, vol. 491. Springer, Heidelberg (2005). https://doi.org/10.1007/3-540-27659-9
5. Haimes, Y.Y., Lasdon, L.S., Wismer, D.A.: On a bicriterion formation of the problems of integrated system identification and system optimization. IEEE Trans. Syst. Man Cybern. **3**, 296–297 (1971)
6. Hartmann, S., Briskorn, D.: A survey of variants and extensions of the resource-constrained project scheduling problem. Eur. J. Oper. Res. **207**(1), 1–14 (2010)
7. Hauder, V.A., Beham, A., Raggl, S., Parragh, S.N., Affenzeller, M.: On constraint programming for a new flexible project scheduling problem with resource constraints. arXiv preprint arXiv:1902.09244 (2019)
8. Laborie, P., Rogerie, J., Shaw, P., Vilím, P.: IBM ILOG CP optimizer for scheduling. Constraints **23**(2), 210–250 (2018). https://doi.org/10.1007/s10601-018-9281-x
9. Laumanns, M., Thiele, L., Zitzler, E.: An efficient, adaptive parameter variation scheme for metaheuristics based on the epsilon-constraint method. Eur. J. Oper. Res. **169**(3), 932–942 (2006)
10. Lova, A., Maroto, C., Tormos, P.: A multicriteria heuristic method to improve resource allocation in multiproject scheduling. Eur. J. Oper. Res. **127**(2), 408–424 (2000)
11. Tao, S., Dong, Z.S.: Scheduling resource-constrained project problem with alternative activity chains. Comput. Ind. Eng. **114**, 288–296 (2017)
12. Weckenmann, A., Akkasoglu, G., Werner, T.: Quality management-history and trends. TQM J. **27**(3), 281–293 (2015)

A Simulated Annealing-Based Approach for Aid Distribution in Post-disaster Scenarios

Alan Dávila de León[1]([⊠]), Eduardo Lalla-Ruiz[2], Belén Melián-Batista[1], and J. Marcos Moreno-Vega[1]

[1] Departamento de Ingeniería Informática y de Sistemas, Universidad de La Laguna, San Cristóbal de La Laguna, Spain
{adavilal,mbmelian,jmmoreno}@ull.es
[2] Department of Industrial Engineering and Business Information Systems, University of Twente, Enschede, The Netherlands
e.a.lalla@utwente.nl

Abstract. Logistics operations have a direct impact on the effectiveness of the humanitarian relief operations and the survival of the population, supplying all demands in a short period of time using the available limited resources. This work addresses the Emergency k-Location Routing Problem (EkLRP) where humanitarian and relief aid has to be distributed from medical infrastructure to the affected people by routing emergency-aimed vehicles minimizing the time required to provide the humanitarian aid. This work proposes a Simulated Annealing with temperature reset in order to promote diversification as well as for escaping from local optima. The numerical experiments indicate that the metaheuristic approach proposed to solve the EkLRP reports high-quality solutions in reasonable computational times.

Keywords: Humanitarian relief · Metaheuristics · Simulated annealing

1 Introduction

Catastrophes and disasters are undesirable events creating potential losses and impacting societies [9]. They are often categorized as natural, i.e., hurricanes, tsunamis, etc., or as man-made such as those caused by socio-political conflicts, terrorist attacks, among others. A study from the Center for Research on the Epidemiology of Disasters (CRED[1]) between the years 1994 and 2013 indicated that 6873 natural disasters were reported around the world. During that period, each year, an average of 68000 deaths and 218 million people were affected. In addition, there were economic losses valued at 147 billion dollars each year. On the other hand, due to conflicts or wars, the average number of refugees

---

[1] http://www.cred.be/.

© Springer Nature Switzerland AG 2020
R. Moreno-Díaz et al. (Eds.): EUROCAST 2019, LNCS 12013, pp. 335–343, 2020.
https://doi.org/10.1007/978-3-030-45093-9_41

per year has been 13 million and 20 million internal displacements. However, according to a report from the International Federation of Red Cross Societies (IFRC)[2] conducted after a flood in 2014 in Afghanistan [8], the planning of the distribution of relief resources was one of the most difficult tasks. In that study, it is pointed out that, despite the complexity involved in emergency logistics, many of those processes and planning are still carried out manually, even though they have proven to be inefficient and inadequate [13]. Finally, an improvement in the response mechanisms is required, specifically in the planning, coordination, and delivery of aid resources since the few existing systems are not sufficiently flexible and dynamic for emergency situations where they aim to use them.

The aid distribution networks managed by institutions and organizations seek to mitigate the damage and suffering caused to the population through the distribution of aid in the form of medicines, food, generators of electricity, medical services, etc. The supply chain is planned to take into account that there are limited resources and that each type of demand has a degree of urgency and a window of service time. On the other hand, emergency scenarios are highly dynamic, which forces the operators in charge to make quick decisions under great pressure. Therefore, having supporting tools and fast solution algorithms to help managers during the decision-making process as well as while designing an aid distribution network is crucial for saving lives and alleviating suffering [1,2,7,12].

In this work, we address a humanitarian supply chain planning problem in catastrophe scenarios where aid distribution facilities have to be located and aid have to be distributed from them to the affected people by routing emergency-aimed vehicles. In this context, the catastrophe is considered in a wide area where a part of the population requires humanitarian help distributed by means of vehicles departing from the located depots (i.e., aid distribution centers). The objective of this problem is to provide aid to the people in need of help as soon as possible (i.e. cumulative objective function). Moreover, given the context where this problem arises, solving and planning time is a crucial factor, hence a metaheuristic algorithm based on Simulated Annealing is proposed. To contextualize its contribution and performance, the proposed approach is compared with other well-known metaheuristic approaches.

The remainder of this paper is organized as follows. The Emergency k-Localization Routing Problem is described in Sect. 2. The description of the Simulated Annealing-based approach is presented in Sect. 3. The computational results obtained by our proposal are discussed in Sect. 4. Finally, Sect. 5 presents the conclusions and several lines for further research.

2   Emergency k-Localization Routing Problem

This section is devoted to introducing the Emergency k-Location Routing Problem (EkLRP). Its goal is to provide humanitarian assistance to people after a disaster situation. In this environment we have to tackle the following decisions:

---

[2] http://www.ifrc.org/.

1. Setting up a well-defined set of medical infrastructures on a ravaged area, where each infrastructure has a fleet of medical vehicles to deliver humanitarian aid and relief.
2. Determining the route that each medical vehicle has to follow in order to provide assistance to the victims.

We are given a set of victims geographically dispersed on a ravaged area and denoted as $N = \{1, 2, \ldots, n\}$. Each victim $c \in N$ requires a strictly positive time to be assisted, $d_c > 0$, and a certain quantity of humanitarian aid, $q_c > 0$.

On the other hand, we are also given a set of locations on the ravaged area denoted as $L = \{1, 2, \ldots, m\}$ in which a medical infrastructure could be set up. Setting up a medical infrastructure at location $l \in L$ involves a positive time, $t_l > 0$, stemming from moving the infrastructure to its destination and deploy it. The first goal is to select a subset of locations, denoted as $L' \subseteq L$ with $|L'| = k$. k is a parameter of the problem which determines the number of locations to select and whose value is selected by the user.

The EkLRP can be modelled by means of a complete graph $G = (H, A)$ with $n + m$ vertices, split into two sets, $H = N \cup L$. Each vertex $v \in H$ is located in a given location (x_v, y_v). Arcs $a_{ij} \in A, \forall i, j \in H$ represent the possibility of moving between the nodes i and j with positive travel time $t_{ij} > 0$. It is worth mentioning that we consider that the travel times are asymmetric for each pair of vertices.

Each medical infrastructure, $l \in L'$, has a fleet of heterogeneous medical vehicles denoted as $V_l = \{1, 2, \ldots, v_l\}$, where each vehicle $v \in V_l$ has a positive capacity, $Q_v > 0$, to carry humanitarian aid. The vehicles are used to assist the victims in such a way that the waiting time of the victims is as short as possible. Therefore, we pursue to determine a set of routes for each selected location, denoted as R_l. Each route r is used by a vehicle $v_r \in V_l$, starts from a medical infrastructure $\sigma_0^r \in L'$, visits a sequence of n_r victims, $\sigma_1^r, \sigma_2^r, \ldots, \sigma_{n_r}^r \in N$, and returns to the same medical infrastructure, that is, $\sigma_{n_r+1}^r = \sigma_0^r$. Additionally, the duration time of the route r is denoted as

$$d(r) = t_{\sigma_0^r \sigma_1^r} + \sum_{i=0}^{n_r} (t_{\sigma_i^r \sigma_{i+1}^r} + d_{\sigma_i^r}) \tag{1}$$

Where:

- Set of selected locations, L'.
- Set of routes of the medical vehicles in the infrastructure $l \in L'$, R_l.
- Number of victims visited by route r, n_r.
- Waiting time of a victim σ_i^r served by a medical vehicle along the route r, $w(\sigma_i^r)$.

The waiting time of a victim σ_i^r served by a medical vehicle along the route r is computed as follows:

$$w(\sigma_i^r) = \begin{cases} t_{\sigma_i^r, r(0)} + d_{\sigma_i^r} & \text{if } i = 0 \\ w(\sigma_{i-1}^r) + t_{\sigma_{i-1}^r \sigma_i^r} + d_{\sigma_i^r} & \text{if } i > 0 \end{cases} \tag{2}$$

Where:

– Waiting time of a victim σ_i^r along the route r, $w(\sigma_i^r)$.
– Travel time required to move between i and j, t_{ij}.
– Time required to set up a medical infrastructure $l \in L'$, t_l.
– Time to assist victim $d_{\sigma_i^r}$.
– σ_i^r i-th vertex visited by the vehicle associated to the route r.

The optimization objective of the EkLRP is to minimize the time required to provide humanitarian aid to all the victims. This time is composed of the time required to set up the k medical infrastructures and deliver the aid to the victims:

$$f(s) = minimize \sum_{l \in L'} \sum_{r \in R_l} \sum_{i=1}^{n_r} w(\sigma_i^r) \tag{3}$$

3   Metaheuristic Approach

Since EkLRP is an $\mathcal{NP}$-hard problem an efficient heuristic algorithm is required in order to provide high-quality solutions within small computational times. In order to achieve these goals a metaheuristic approach based on Simulated Annealing (SA, [3]) is proposed.

3.1   Constructive Heuristic

In order to properly start the SA-based approach, a starting solution is required. In this sense, we propose the Greedy Randomized Algorithm (GRA, [10]) that splits the EkLRP into two interconnected subproblems.

– High-level Problem: Determining the subset of locations to set up medical infrastructures aimed at serving all the victims.
– Low-level Problem: Determining the route that each medical vehicle has to follow in order to provide assistance to the victims.

GRA is used in the generation of initial solutions and in the repair process within Simulated Annealing. Thus, it is used along with a restricted list of candidates (RLC) that contains a subset of the best k candidates to add to the solution constructed so far. By means of this method, we are able to generate a fast initial and feasible solution. Firstly, we use an elitist selection using the average distance of all victims to the medical infrastructure. It tries to insert every victim between each pair of nodes in the routes of the solution. All feasible and possible positions in which non-assigned victims can be placed are included in a restricted candidate list (RLC). Finally, it selects one movement between the best candidates in the RLC with a roulette wheel selection. A greedy function evaluates the impact on the objective function value of selecting the candidate. This fitness level is used to associate a probability of selection with each candidate. This process is repeated until all victim has been visited.

3.2   Simulated Annealing

The version of Simulated Annealing used in this work incorporates the feature of reheating [5,6], that is, once the temperature has decreased until a certain point, it is increased or reset. That feature promotes diversification, and with the SA search procedure permits alternating between diversification and intensification along with search. The number of times that is allowed to reheat the solution is provided by the user. On the other hand, each time that the search is reheated the starting temperature is reduced to avoid randomizing completely the search. To do so, the starting temperature is equalized to the provided temperature by the user, divided by the reheating phase where we are located. For example, if the temperature provided by the user is 1000 and we have 3 reheating phases, the starting temperatures would be: 1000, 500 and 333.33.

In Algorithm 1, the pseudocode of the SA approach using reheating (SA-R) is provided. This procedure receives as input an initial solution, s, generated by the GRA. At each iteration, the destroy method is performed, remove p victims of solution s (line 4). The first victim n^* is randomly chosen. After having removed n^*, the closest $p - 1$ victims are removed. We proceed with the repair method (line 4), using the GRA. The neighborhood is defined implicitly by a destroy and a repair method. Solution s' is accepted as current solution s (line 6 and 8) with probability given in (4). We allow a temperature reset during the search in order to escape from local optima.

$$P(s', s) = e^{(f(s)-f(s'))/t}, s' \in N(s) \tag{4}$$

Algorithm 1. Pseudocode of Simulated Annealing with reheat strategy (SA-R)

Require: Feasible solution s
1: $s'' = s$
2: **while** *reheat* **do**
3:      **while** $t > t_f$ **do**
4:          $s' = \text{destroy}(s)$
5:          $s' = \text{repair}(s')$
6:          **if** $rand(0, 1) < P(s', s)$ **then**
7:              $s = s'$
8:          **end if**
9:          **if** $f(s) < f(s'')$ **then**
10:             $s'' = s$
11:         **end if**
12:         $t = updateTemperature()$
13:     **end while**
14: **end while**
15: **return** s''

---

4   Computational Experiments

In this section, the results provided by our metaheuristic approach are presented. The set of 80 instances used in this work were randomly generated considering. Depending on the number of victims and locations each instance is labeled as mxn. Each pair instance-algorithm has been executed 100 times with a maximum computational time adapted to the instance, $5n$ milliseconds. GRA is used in the generation of all initial solutions. The computational experiments were conducted on a computer equipped with Intel Core i7-3632QM CPU @ 2.20 GHz x 8 and 8GB of RAM. With the aim of analyzing the contribution SA-R, the following algorithms are implemented:

1. Variable Neighborhood Search (VNS, [4]) with two environments, *i.e.*, victims swap and victims 2-opt.
2. Simulated Annealing (SA): version described in Sect. 3.2, but without reheat.
3. Large Neighborhood Search (LNS, [11]): it uses the same destroy and repair method as described in Sect. 3.

Parameter configuration is selected based on the Friedman test considering the average objective value. Based on the test, LNS, SA, and SA-R remove 1 percent of victims on every iteration. Otherwise, in SA and SA-R the temperature is progressively decreased from n to 1^{-10}. The cooling schedule follows an exponentially decreasing function.

4.1   SA Computational Results

Table 1 report the minimum (Min.), average (Avg.), and maximum (Max.) objective value obtained during the experiment on large instances: 500, 1000, 1500 and 2000 victims. Since the presented data is grouped by instance identifier $m \times n$, the values correspond to the average. Also, the computational time limit does not depend on the metaheuristic but on the size of the problem instances ($5 \cdot n$ milliseconds), the average computational time thus is not reported.

First, in the results reported in Table 1, it can be seen that VNS for no instance performs better than the other approaches, although it obtains competitive results.

Moreover, it should be noted that LNS and SA approaches implemented are only differentiated by the probability of making the transition from the current state s to a candidate new state s'. Therefore, both approaches exhibit similar performance in terms of solution quality. Furthermore, the average quality of the solutions is better when SA incorporates the reheat strategy than without it. Since all approaches were executed under the same time limit, it can be concluded that SA-R is the most appropriate method for tackling this problem.

Table 1. Overall performance in terms of objective function value of VNS, LNS, SA, and SA-R. Best values among VNS, LNS, SA, and SA-R are shown in boldface type

Instance	GRA	VNS			LNS			SA			SA-R		
$m \times n$	Avg	Min	Avg	Max	Min	Avg	Max	Min	Avg	Max	Min	Avg	Max
10 × 500	2855978.85	2739808.87	2781444.43	2841265.52	2707022.85	2723102.46	2745797.18	2707206.83	**2722701.38**	2744166.39	**2704206.99**	2723037.09	2744438.22
20 × 500	2864564.65	2753129.76	2788354.97	2861673.13	2712009.17	2731087.25	**2751207.88**	2713068.20	**2730964.35**	2752756.14	**2711704.95**	2731043.62	2752659.70
50 × 500	2882140.63	2765855.09	2807526.48	2875829.35	2731627.67	2750625.01	**2770153.52**	2730625.44	2750651.34	2771939.85	**2727105.11**	2750048.45	2770521.39
100 × 500	2798816.32	2692578.72	2727887.45	2804925.29	**2651462.03**	2671447.45	**2691583.32**	2652759.67	2671320.18	2692830.92	2652051.36	**2671159.08**	2692430.97
10 × 1000	10505697.18	10233144.59	10346240.16	10506536.25	10140918.85	10196574.30	10263076.13	**10140339.67**	10197001.97	**10257744.35**	10141749.81	**10194218.98**	10264027.73
20 × 1000	10377501.85	10094603.69	10217089.93	10376578.63	10010907.69	10068280.68	10132804.58	10012296.08	10067000.43	**10125970.73**	**10007005.45**	**10065758.02**	10130255.92
50 × 1000	10391317.14	10114226.37	10236307.84	10406970.15	10023461.25	10079612.18	**10143831.77**	**10021368.80**	10078650.15	10152204.01	10025038.18	**10078561.42**	10148575.62
100 × 1000	10456359.25	10175806.31	10296481.39	10435176.28	10086242.99	10145034.57	**10198344.98**	10092143.93	10145506.33	10204109.29	**10082229.94**	**10143755.35**	10202245.91
10 × 1500	22614206.38	22185849.43	22375997.17	22820131.88	22325727.08	**22214896.93**	22552949.77	**22319402.30**	22426462.75	**22544522.84**	22328878.49	**22422927.51**	22545218.39
20 × 1500	22866662.92	22462764.79	22631322.35	22882694.78	23611988.31	22466634.12	22563924.45	22107606.70	22191189.97	**22324162.70**	22089342.39	22218759.65	22449336.47
50 × 1500	22602553.61	22183481.20	22369850.65	22595984.48	**22065031.24**	22212778.73	22340233.02	22362717.62	22465930.57	22572107.44	22066643.56	**22462527.24**	**22562735.93**
100 × 1500	39187681.46	38655408.13	38901996.83	39161127.75	38600621.58	38817471.48	38989732.37	38651009.77	38816435.76	38992861.37	38590113.90	**22209689.95**	22337659.52
10 × 2000	39493200.69	38978894.51	39206610.39	39521499.76	38965045.25	39123462.51	**39268156.48**	38953846.25	39122245.66	39269202.32	**38948261.91**	**38803994.22**	**38968166.53**
20 × 2000	39734952.79	39211799.22	39460001.33	39777703.95	39220235.22	39368492.42	39538841.70	39206533.40	39374459.53	39546133.30	39204801.14	**39117832.92**	39285008.03
50 × 2000	39215610.63	38688501.09	38933269.23	39259760.15	**38595547.51**	38842455.43	39017656.06	38703941.23	38844426.93	**39012417.74**	38625491.71	**39365165.76**	**39534960.54**
100 × 2000	18854248.59	18520866.16	18666221.29	18863060.44	18454576.29	18552692.26	18643530.04	18467419.88	18552999.72	18643294.79	**18454036.08**	**38834700.88**	39026576.39
Average												**18549573.76**	18644401.08

5   Conclusions

In this work, we have addressed the Emergency k-Location Routing Problem by means of a Simulated Annealing-based approach that incorporates temperature resets (i.e., SA-R) during the search. To contextualize this approach, it was compared with other trajectory-based approaches (i.e., GRA, SA, LNS, and VNS). The computational results show that SA-R performs better in terms of objective function value under the same computational time limitations than the other considered approaches. Furthermore, when assessing the contribution of the reheating strategy, it can be seen that it leads to better solution quality than without using it in the majority of cases. This work thus contributes to the SA state-of-the-art with another example of its benefit for addressing hard combinatorial problems.

As future work, we plan to investigate different reheating schemes for SA in this and other optimization problems.

Acknowledgments. This work has been partially funded by the Spanish Ministry of Economy and Competitiveness (project TIN2015-70226-R).

References

1. Altay, N., Green, W.G.: OR/MS research in disaster operations management. Eur. J. Oper. Res. **175**(1), 475–493 (2006)
2. Balcik, B., Beamon, B.: Facility location in humanitarian relief. Int. J. Logistics **11**(2), 101–121 (2008)
3. Davis, L.: Genetic algorithms and simulated annealing (1987)
4. Hansen, P., Mladenović, N., Pérez, J.: Variable neighbourhood search: methods and applications. Ann. Oper. Res. **175**(1), 367–407 (2010). https://doi.org/10.1007/s10479-009-0657-6
5. Ingber, L.: Very fast simulated re-annealing. Math. Comput. Modell. **12**(8), 967–973 (1989)
6. Ingber, L., Rosen, B.: Genetic algorithms and very fast simulated reannealing: a comparison. Math. Comput. Modell. **16**(11), 87–100 (1992)
7. Luis, E., Dolinskaya, I.S., Smilowitz, K.R.: Disaster relief routing: integrating research and practice. Socio-Econ. Plann. Sci. **46**(1), 88–97 (2012)
8. Michaud, S.: After action review of DREF operation # MDRAF002 - afghanistan floods and landslide. Technical report, International Federation of Red Cross and Afghan Red Crescent Society (2014)
9. Pearson, C.M., Sommer, S.A.: Infusing creativity into crisis management: an essential approach today. Organ. Dyn. **40**(1), 27–33 (2011)
10. Resende, M.G., Ribeiro, C.C.: Greedy randomized adaptive search procedures: advances, hybridizations, and applications. In: Gendreau, M., Potvin, J.Y. (eds.) Handbook of metaheuristics, pp. 283–319. Springer, Boston (2010). https://doi.org/10.1007/978-1-4419-1665-5_10
11. Shaw, P.: Using constraint programming and local search methods to solve vehicle routing problems. In: Maher, M., Puget, J.-F. (eds.) CP 1998. LNCS, vol. 1520, pp. 417–431. Springer, Heidelberg (1998). https://doi.org/10.1007/3-540-49481-2_30

12. Simpson, N., Hancock, P.: Fifty years of operational research and emergency response. J. Oper. Res. Soc. **60**, S126–S139 (2009). https://doi.org/10.1057/jors.2009.3
13. Thomas, A., Humanitarian, L.: Enabling disaster response. Institute Fritz (2007)

Decision Diagram Based Limited Discrepancy Search for a Job Sequencing Problem

Matthias Horn[✉] and Günther R. Raidl

Institute of Logic and Computation, TU Wien, Vienna, Austria
{horn,raidl}@ac.tuwien.ac.at

Abstract. We consider the *Price-Collecting Job Sequencing with One Common and Multiple Secondary Resources* problem. The task is to feasibly schedule a subset of jobs from a given larger set. Each job needs two resources: a common resource for a part of the job's execution time and a secondary resource for the whole execution time. Furthermore each job has one or more time windows and an associated prize. In addition to previous work, we also consider precedence constraints on the jobs. We aim to maximize the total prize over the actually scheduled jobs. To solve large instances heuristically we propose a hybrid of limited discrepancy search and beam search approach that utilize a relaxed decision diagram. We could show that the use of a relaxed decision diagram substantially speed-up the computation times of the search approach.

Keywords: Sequencing problem · Decision diagrams · Limited discrepancy search

1 Introduction

The *Price-Collecting Job Sequencing with One Common and Multiple Secondary Resources* (PC-JSOCMSR) problem without precedence constraints was first introduced by [6,7] and consists of a set of jobs, one common resource, and a set of secondary resources. The common resource is shared by all jobs whereas a secondary resource is shared only by a subset of jobs. Each job has at least one time window and is associated with a prize. A feasible schedule requires that there is no resource used by more than one job at the same time and each job is scheduled within one of its time windows. Due to the time windows it may not be possible to schedule all jobs. The task is to find a subset of jobs that can be feasible scheduled and maximizes the total prize. There are at least two applications [5]. The first is in the field of the daily scheduling of particle therapies for cancer treatments. The second application can be found in the field

This project is partially funded by the Doctoral Program Vienna Graduate School on Computational Optimization, Austrian Science Foundation Project No. W1260-N35.

R. Moreno-Díaz et al. (Eds.): EUROCAST 2019, LNCS 12013, pp. 344–351, 2020.
https://doi.org/10.1007/978-3-030-45093-9_42

of hard real time scheduling of electronics within an aircraft, called avionics, where the PC-JSOCMSR appears as a subproblem.

The PC-JSOCMSR was tackled on the exact side by Horn et al. [6], with an A* based algorithm which is able to solve instances up to 40 jobs to proven optimality. On the heuristic side, Maschler and Raidl [7] applied *decision diagrams* (DDs) to obtain lower and upper bounds for large problem instances with up to 300 jobs. DDs are rooted weighted directed acyclic graphs and provide graphical representations of the solution spaces of combinatorial optimization problems. In particular relaxed DDs represent a superset of the feasible set of solutions and are therefore a *discrete relaxation* of the solution space, providing upper bounds on the objective value. The counterparts are restricted DDs which represent subsets of feasible solutions and therefore provide heuristic solutions. Both types of DDs were investigated in [7] and were compiled with adapted standard methods from the literature. For more details on DDs we refer to [1]. New state of the art results for the PC-JSOCMSR could be obtained by Horn et al. [5] by applying a *beam search* (BS) heuristic that uses a relaxed DD to speed up the search. The relaxed DD is constructed by a novel A*-based construction algorithm.

In particular in the avionic system scenario it often appears that some jobs need to be finished before other jobs may start. To address this aspect, we consider in this work also *precedence constraints*. Thus, there are given relationships between pairs of jobs as additional input such that one job can only be scheduled if the other job is already completely scheduled earlier. These new constraints require an adaption on the algorithmic side of [5] to incorporate the new precedence constraints. The goal is to solve large problem instances of the PC-JSOCMSR with precedence constraints heuristically. Our solution approach builds upon the ideas from [5] but extended them to a *limited discrepancy search* (LDS) combined with BS that exploits structural information contained in a relaxed DD. The usage of the relaxed DD is two-folded: (1) to reduce computation time of the LDS and (2) to provide besides a heuristic solution also an upper bound on the total prize objective.

2 PC-JSOCMSR with Precedence Constraints

The PC-JSOCMSR with precedence constraints is formally defined as follows. Given are a set of n jobs $J = \{1, \ldots, n\}$, a common resource 0 and a set of m secondary resources $R = \{1, \ldots, m\}$. Let $R_0 = \{0\} \cup R$ be the complete set of resources. Each job $j \in J$ needs during its whole execution time $p_j > 0$ one secondary resource $q_j \in R$ and, in addition, after some preprocessing time $p_j^{\mathrm{pre}} \geq 0$ also the common resource 0 for some time $p_j^0 > 0$. Furthermore each job has associated (1) ω_j time windows $W_j = \bigcup_{\omega=1,\ldots,\omega_j} [W_{j\omega}^{\mathrm{start}}, W_{j\omega}^{\mathrm{end}}]$, where $W_{j\omega}^{\mathrm{end}} - W_{j\omega}^{\mathrm{start}} \geq p_j$, $\omega = 1, \ldots, \omega_j$, (2) a set of preceding jobs Γ_j, which must be scheduled before job j can be scheduled w.r.t. the common resource 0, and (3) a prize $z_j > 0$. The task is to find a subset of jobs $S \subseteq J$ which can be feasible scheduled such that the total prize of these jobs is maximized: $Z^* = \max_{S \subseteq J} Z(S)$, $Z(S) = \sum_{j \in S} z_j$. A feasible schedule assigns each job in S a starting time in such a way that all constraints are satisfied. Note that a unique ordered sequence $\pi = (\pi)_{i=1,\ldots,|S|}$ of jobs is implied by each feasible schedule of jobs $S \subseteq J$, since the common resource is required by each job and only one job can use this resource at a time. For each given sequence π of jobs S that can be associated with a feasible schedule, a *normalized schedule* without unnecessary waiting times can be computed greedily (see [6] for further details).

3   Exact/Relaxed Decision Diagrams and Filtering

In order to describe our approach in Sect. 4 we have to introduce some definition and structures beforehand. In our context a DD for the PC-JSOCMSR is a weighted directed acyclic graph $M = (V, A)$ with one root node $\mathbf{r} \in V$, corresponding to the empty schedule and one target node $\mathbf{t} \in V$ corresponding to all feasible schedules that cannot be further extended by any job. Each arc $a = (u, v) \in A$ corresponds to adding a specific job, denoted by $\mathrm{job}(a) \in J$, as the next job after the ones already scheduled up to node u. The length of an arc $a \in A$ is associated width the prize $z_{\mathrm{job}}(a)$. Hence, each path from $\mathbf{r}$ to any node $u \in V$ corresponds to a specific sequence of jobs π and the length of the path is equal to the sum of prizes of jobs in π.

In an exact DD each feasible normalized schedule $S \subseteq J$ has a corresponding path in the exact DD originating from $\mathbf{r}$ and vice versa. The length of such a path corresponds exactly to the total prize $Z(S)$. Therefore a longest path from $\mathbf{r}$ to $\mathbf{t}$ corresponds to an optimal solution of the PC-JSOCMSR. Furthermore, each node $u \in V$ is associated to a state $(P(u), t(u))$, where set $P(u)$ contains all jobs that can be feasibly scheduled next, and vector $t(u) = (t_r(u))_{r \in R_0}$ contains the earliest times from which on each of the resources are available for performing a next job. The transition function to obtain the successor state $(P(v), t(v))$ of state $(P(u), t(u))$ when scheduling job $j \in P(u)$ is

$$\tau((P(u), t(u)), j) = \begin{cases} (P(u) \setminus \{j\}, t(v)), & \text{if } s((P(u), t(u)), j) < T^{\max} \wedge \\ & \qquad P(u) \cap \Gamma_j \neq \emptyset, \quad (1) \\ \hat{0}, & \text{else,} \end{cases}$$

with

$$t_0(v) = s((P(u), t(u)), j) + p_j^{\mathrm{pre}} + p_j^0, \tag{2}$$

$$t_r(v) = s((P(u), t(u)), j) + p_j, \qquad \text{for } r = q_j, \tag{3}$$

$$t_r(v) = t_r(u), \qquad \text{for } r \in R \setminus \{q_j\}, \tag{4}$$

where $\hat{0}$ represents the infeasible state and $s((P(u), t(u)), j)$ corresponds to the earliest start time of job j w.r.t. to state $(P(u), t(u))$ and job j's time windows. If it is not possible to schedule job j feasible then function $s(.,.)$ will return $T^{\max}$. States that are related to exact DDs will be denoted as *exact* states.

Relaxed DDs merge exact states in order to get a more compact DD. Thereby new paths will emerge which correspond to infeasible schedules, denoted as *infeasible* paths. Let merge two nodes $u, v \in V$. The merged state is $(P(u), t(u)) \oplus (P(v), t(v)) = (P(u) \cup P(v), (\min(t_r(u), t_r(v)))_{r \in R_0})$. We compile relaxed DDs with the A*-based compilation (A*C) method from [5] since it could be shown that at least for the PC-JSOCMSR without precedence constraints A*C can produce smaller relaxed DDs in shorter time that represent stronger relaxations than relaxed DDs compiled with standard methods from the literature. Note that we initially ignore the precedence constraints in this compilation of a relaxed DD. Otherwise, we would need to extend the states of the nodes with additional information in order to define a feasible merging rule for two nodes. Preliminary experiments had shown that those larger states cause substantially longer compilation times, which we want to avoid. However, we consider the precedence constraint after the initial construction by applying a respective filtering on the compiled relaxed DD. We try to identify arcs which belong only to infeasible paths. Those arcs can be safely removed from the relaxed DD to reduce the number of infeasible paths without removing paths that correspond to feasible schedules. To identify arcs that violate precedence constraints we adopted the corresponding filter operation suggested by Cire and van Hoeve [2]. Moreover, if we got already a primal solution then we can in addition filter arcs which only belong to paths corresponding to solutions that are worse than this known primal solution. Hence, paths that encode sub-optimal solutions will be removed from the relaxed DD. This cost-based filter operations are adopted from [5].

4   Limited Discrepancy Search

Limited discrepancy search was originally proposed by Harvey and Ginsberg [4] for heuristic binary searches where at each decision point a heuristic $h(.)$ decides between two possibilities to extend the current partial solution. If $h(.)$ would be a perfect heuristic than the algorithm would return the optimal solution as soon as a complete solution is encountered during the search. However, in most cases $h(.)$ will fail at some point and only a non-optimal solution can be returned. To overcome this, LDS allows in a systematic way discrepancies during the search. A *discrepancy* means that at some decision point the algorithm decides against $h(.)$. Hence, if k discrepancies are allowed than LDS will encounter all paths in the search tree where the algorithm exactly decides k times against $h(.)$. To apply LDS on the PC-JSOCMSR we have to consider in general multiple possibilities at each decision point instead of just two, and we do this by counting $i - 1$ discrepancies if we take the i-th best decision according to $h(.)$.

Algorithm 1 shows our LDS-based approach. The search is applied on the exact states defined in Sect. 3. Note that we do not build an exact DD, but we rather keep all not yet expanded nodes in memory and assign to each node v' the so far best encountered partial solution $\pi(v')$. Furthermore, we extend LDS in similar ways as Furcy and Koening [3] by incorporating a BS like approach at each level into LDS. Instead of expanding always one node at each step Algorithm 1 expands at each step β nodes and keeps the $(k + 1)\beta$-best successor nodes according to $h(.)$. As heuristic decision function $h(v')$ for node v' we

---

Algorithm 1. LDSprobe

Input: node set N', relaxed DD $M = (V, A)$, allowed discrepancies k, beam
width β

Output: sequence of jobs π

1 **if** $N' = \emptyset$ **then return** empty sequence;
2 node set $W' \leftarrow \emptyset$; job sequence $\pi_{\text{best}} \leftarrow \emptyset$;
3 **foreach** $u' \in N'$ **do**
4   let $u \in V$ be the node corresponding to u' w.r.t. the path from the root;
5   **foreach** *outgoing arc* $a = (u, v)$ *of node* u **do**
6   **if** $|W'| = (k + 1)\beta \wedge$ *node* v *would be removed from* $W' \cup \{v\}$ **then**
7   continue with next arc;
8   **if** $\tau((P(u'), t(u'), \text{job}(a)) = \emptyset$ **then** continue with next arc;
9   add new node v' to W' and set $(P(v'), t(v')) \leftarrow \tau((P(u'), t(u')), \text{job}(a))$;
10   **if** $|W'| > (k + 1)\beta$ **then** remove worst node from W' according to $h(.)$;
11 **if** $W' = \emptyset$ **then return** $\arg\max_{\pi(u')|u' \in N'} Z(\pi(u'))$;
12 sort W' according to $h(.)$ and split W' into $k + 1$ slices $W'[i]$, $i = 0, \ldots, k$;
13 **foreach** $i = k, \ldots, 0$ **do**
14      $\pi = \text{LDSprobe}(W'[i], M, k - i, \beta)$;
15   **if** $Z(\pi_{\text{best}}) < Z(\pi)$ **then** $\pi_{\text{best}} \leftarrow \pi$;
16 **return** π_{best};

Algorithm 2. LDS+BS

Output: sequence of jobs π

1 compile relaxed DD $M = (V, A)$ by A^*C, ignoring precedence constraints;
2 $\pi_{\text{best}} \leftarrow \text{LDSprobe}(\mathbf{r}, M, 0, 10)$;
3 **for** $k \leftarrow 0$; $k \leq k_{\max} \wedge$ *time limit not exceeded*; $k \leftarrow k + 1$ **do**
4   apply filtering on M;
5      $\pi \leftarrow \text{LDSprobe}(\mathbf{r}, M, k, \beta)$;
6   **if** $Z(\pi_{\text{best}}) < Z(\pi)$ **then** $\pi_{\text{best}} \leftarrow \pi$;
7 **return** π_{best};

---

use the ratio $Z(\pi(v'))/t_0(v')$. In order to quickly identify those $(k + 1)\beta$-best
successor nodes we use the structural information contained in the relaxed DD
$M = (V, A)$ similar as in [5]. For node $u' \in N$ a corresponding node $u \in V$ from
M can be determined by following the job sequence $\pi(u')$ from $\mathbf{r}$ in M. We do
not consider transitions to successor nodes of u' where the corresponding arcs
were removed from the relaxed DD during the filtering step. Furthermore we can
estimate $h(.)$ without creating the successor nodes of u' by using the corresponding nodes in the relaxed DD. Based on these estimation we can decide quickly
if a successor node is a candidate to be one of the β-best successor nodes or
not. Note that for simplification reasons Algorithm 1 shows a recursive version
of LDS, our implementation however, is implemented in an iterative way.

Algorithm 2 gives an overview of the overall approach to tackle the PC-JSOCMSR. First a relaxed DD M is compiled with A*C with the same parameter settings as in [5] and by ignoring the precedence constraints. In order to get quickly an initial primal solution for filtering, Algorithm 1 is applied with the small beam with $\beta = 10$ and no allowed discrepancies. In the main loop we apply first the filtering for sup-optimal paths according to our current best primal solution and precedence constraints violations. Then we apply Algorithm 1 with beam width β and the number of current maximum allowed discrepancies k. After updating the incumbent solution π_{best}, k is increased by one. The algorithm terminates if the maximum allowed discrepancies k_{max} is reached or a certain time limit is exceeded.

5 Computational Study

The LDS-based algorithm for the PC-JSOCMSR with precedence constraints was implemented in C++ using GNU g++ 5.4.1. All tests were performed on a cluster of machines with Intel Xeon E5-2640 v4 processors with 2.40 GHz in single-threaded mode with a memory limit of 16 GB per run. We extended the two sets of benchmark instances for the particle therapy application scenario (denoted as P) and for the avionic system scheduling scenario (denoted as A) from [5] by adding randomly precedence constraints between n pairs of jobs such that circular dependencies between jobs are voided. The instance sets contain 30 instances for each combination of different values of n and m. For further details on the benchmark characteristics we refer to [5].

Figure 1 compares the obtained average total prizes and median computation times between LDS+BS and a standalone variant of LDS+BS without using a relaxed DD dependent on different values of beam width β and different maximum allowed discrepancies k_{max}. The diagrams on the top visualize the obtained average total prizes. There are two main observations regarding the solution quality: First, as expected the solution quality tends to increase with increasing β and/or k_{max}; second, similar results could be obtained from both LDS+BS variants. However, regarding computation times, the LDS+BS approach using the relaxed DD is in almost all cases except for $k_{\text{max}} = 0$ and smaller β substantially faster. Note that we do not show the obtained results from standalone LDS+BS for $k_{\text{max}} = 2$, since the approach exceeded in most cases the time limit of two hours.

Figure 2 compares the LDS+BS approach against a mixed linear integer programing (MIP) approach and a constrained programming (CP) approach. The MIP formulation as well as the CP formulation from [5] were adapted to additionally consider the precedence constraints and are solved with Gurbi Optimizer 7.5.1 and MiniZinc 2.1.7 with backbone solver Chuffed, respectively. All tested approaches use a time limit of 900 s. For LDS+BS the maximum allowed discrepancies k_{max} are set to infinity and β is set to 1000 and 10000 for instance set of type P and A, respectively. The first bar of each group of bars show the obtained average longest path length of the compiled relaxed DD during the

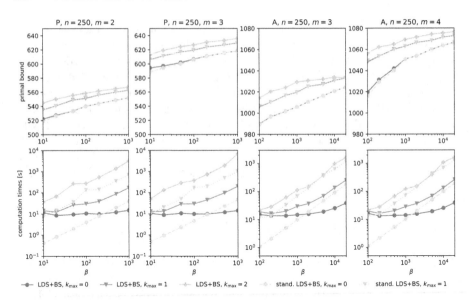

Fig. 1. Comparison between LDS+BS and standalone LDS+BS for middle sized instances with 250 jobs.

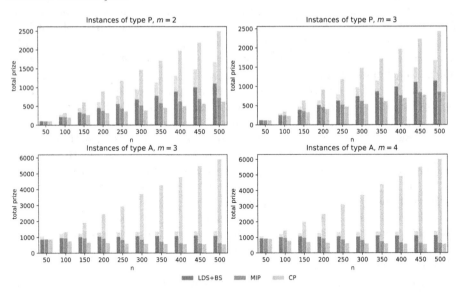

Fig. 2. Primal and Dual Bounds obtainded from LDS+BS, MIP and CP.

LDS+BS approach and the block at the bottom show the obtained average primal bounds. In the same manner, the second bar shows the obtained upper- and primal bounds from the MIP approach. The third bar shows the obtained average primal bounds obtained from the CP approach. On average the LDS+BS approach finds in all considered cases better or equally good solutions than the MIP or the CP solvers. Moreover, LDS+BS is able to return in most cases on average stronger upper bounds than the MIP solver.

6 Conclusion

Exploiting the structural information of relaxed DDs within LDS has following advantages: (1) a substantial speed up of the heuristic search allows to scan larger regions of the search space compared to a standalone LDS approach and (2) a dual bound can be obtained from the relaxed DD. Although we demonstrate this advantages specifically for the PC-JSOCMSR, the general approach also appears promising for other combinatorial optimization problems. Next steps would be to incorporate other filtering techniques to further strengthen the relaxed DD by removing more arcs to speed-up the computation times even more. Another promising research direction would be to apply the general idea of using the structural information of relaxed DDs on further search heuristics and meta-heuristics.

References

1. Bergman, D., Cire, A.A., van Hoeve, W.J., Hooker, J.N.: Decision diagrams for optimization. Artificial Intelligence: Foundations, Theory, and Algorithms. Springer, Cham (2016). https://doi.org/10.1007/978-3-319-42849-9
2. Cire, A.A., van Hoeve, W.: Multivalued decision diagrams for sequencing problems. Oper. Res. **61**(6), 1411–1428 (2013)
3. Furcy, D., Koenig, S.: Limited discrepancy beam search. In: Proceedings of the 19th International Joint Conference on Artificial Intelligence, IJCAI 2005, pp. 125–131. Morgan-Kaufmann (2005)
4. Harvey, W.D., Ginsberg, M.L.: Limited discrepancy search. In: Mellish, C.S. (ed.) Proceedings of the 14th International Joint Conference on Artificial Intelligence, pp. 607–615. Morgan-Kaufmann, Montreal (1995)
5. Horn, M., Maschler, J., Raidl, G., Rönnberg, E.: A*-based construction of decision diagrams for a prize-collecting scheduling problem. Technical report AC-TR-18-011, Algorithms and Complexity Group, TU Wien (2018). submitted
6. Horn, M., Raidl, G.R., Rönnberg, E.: An A* algorithm for solving a prize-collecting sequencing problem with one common and multiple secondary resources and time windows. In: Proceedings of PATAT 2018 - The 12th International Conference of the Practice and Theory of Automated Timetabling, Vienna, Austria, pp. 235–256. (2018)
7. Maschler, J., Raidl, G.R.: Multivalued decision diagrams for a prize-collecting sequencing problem. In: Proceedings of PATAT 2018 - Proceedings of the 12th International Conference of the Practice and Theory of Automated Timetabling, Vienna, Austria, pp. 375–397 (2018)

The Potential of Restarts for ProbSAT

Jan-Hendrik Lorenz[(✉)] and Julian Nickerl

Institute of Theoretical Computer Science, Ulm University, 89069 Ulm, Germany
{jan-hendrik.lorenz,julian.nickerl}@uni-ulm.de

Abstract. This work analyses the potential of restarts for probSAT by estimating its runtime distributions on random 3-SAT instances that are close to the phase transition. We estimate optimal restart times from empirical data, reaching a potential speedup factor of 1.39. Calculating restart times from fitted probability distributions reduces this factor to a maximum of 1.25. We find that the Weibull distribution approximates the runtime distribution well for over 93% of the used instances.

A machine learning pipeline is presented to compute a restart time for a fixed-cutoff strategy to exploit this potential. The presented approach performs statistically significantly better than Luby's restart strategy and the policy without restarts. The structure is particularly advantageous for hard problems.

Keywords: Machine learning · Restart · Runtime distribution

1 Introduction

Stochastic local search (SLS) algorithms achieve the best performance for many hard problems. However, a downside is that the algorithms may get stuck in local optima. A common way to resolve this problem is restarting the probabilistic procedure after a number of search steps. However, choosing a good restart policy is a non-trivial problem.

Luby et al. [14] introduced two theoretically intriguing restart strategies: The fixed-cutoff approach and Luby's universal strategy. The fixed-cutoff approach minimizes the expected runtime for some fixed restart time. However, finding this optimal restart time requires nearly complete knowledge of the algorithm's behavior on the problem instance. Luby's universal strategy does not require any domain knowledge while performing well both in theory and practice [12,14]. Lorenz [13] developed a method to calculate optimal restart times for the fixed-cutoff approach if the runtime behavior is given by a probability distribution. Arbelaez et al. [3] studied the empirical runtime distribution of probSAT, an SLS SAT-solver [5], and found that lognormal distributions describe the runtime behavior well. A recent development is the prediction of runtime distributions (RTD) with machine learning methods [3,8].

A version of this work is available under https://arxiv.org/abs/1904.11757. The authors acknowledge support by the state of Baden-Württemberg through bwHPC.

R. Moreno-Díaz et al. (Eds.): EUROCAST 2019, LNCS 12013, pp. 352–360, 2020.
https://doi.org/10.1007/978-3-030-45093-9_43

Our Contribution: This work analyzes the capability of probability distributions to describe the runtime behavior of probSAT on random 3-SAT instances close to the phase transition. We approximate the RTDs of probSAT with a similar method as presented in [3]. In this work, the lognormal distribution (LD) and the Weibull distribution (WD) are used to describe the runtime behavior. It is observed that the WD is well suited to describe the runtime behavior of probSAT, describing significantly more empirical RTDs than the established LD.

We estimate optimal restart times from empirical data as a baseline to evaluate restart times from the fitted distributions. Applying this restart time to the observed data leads to an average speedup factor of 1.393. The aforementioned fitted distributions are used with Lorenz's [13] method to find theoretically optimal restart times. We found a potential speedup factor of up to 1.253 by applying those restart times. To the best of our knowledge, the potential speedup by restarting at the optimal time according to the respective RTDs has not been systematically studied before.

In the second part of this work, a machine learning pipeline (MP) which predicts the RTD of so far unseen instances is presented. It consists of a random forest predicting the distribution type, and a total of three neural networks which predict the parameters of the distributions. The parameters are used to calculate restart times. This approach differs from the procedure introduced in [10]: The strategies in their portfolio use restart times which are independent of the instance. An average speedup factor of more than 1.21 is observed in hard instances, without worsening the performance on easier ones. An overall speedup factor of 1.06 is achieved. The results are statistically significant (t-test, modified Wilcoxon signed-rank test [11]).

2   The Potential of Restarts in ProbSAT

SLS algorithms commonly employ restart strategies when the number of local search steps exceeds a cutoff value. However, finding a choice for the cutoff value is not trivial, and strategies which work well for one instance might be a bad choice for other instances. However, if the runtime behavior always follows a particular distribution type and only the parameters of the distribution vary, then several conclusions about useful restart strategies can be drawn. Thus, it is useful to study the algorithm's runtime behavior for a specific problem class and model the behavior with RTDs.

The runtime of a randomized algorithm can be interpreted as a random variable. Even though runtimes are discrete values, it is common to model the performance as a continuous process. Following this convention, we considered the LD, the WD, the generalized Pareto distribution, and the exponential distribution to describe the runtime behavior of probSAT. These distributions have been previously regarded as suitable for a similar purpose, see for example [2,3,8]. We observed that the LD and WD suffice to describe the occurring distributions well. In the following, we focus on these two distributions.

Throughout this work we use the so-called fixed-cutoff strategy [14]: The algorithm always restarts after the same number of steps, and the number of possible restarts is unbounded. It is known that there is an optimal fixed restart time t^* which minimizes the expected runtime. If the restart time t^* is used, then the fixed-cutoff strategy is superior to all other restart approaches. However, finding the best possible restart time t^* is a hard task.

We are interested in the calculation of optimal restart times. Lorenz [13] describes a method to find optimal restart times for the LD and the WD. He found that for the WD the optimal approach is either instantly restarting or not restarting at all depending on the parameters. In practice, instantly restarting results in much worse performances. Therefore, the WD is employed with a location parameter that shifts the whole distribution. This resolves the paradox of instantly restarting since the resulting optimal restart times are greater than the location parameter. The used RTDs are obtained from the empirical distribution with maximum likelihood estimations (fits).

We study a version of the SLS SAT-solver [6] probSAT which is known as the break-only-poly-version with $c_b = 2.3$ as suggested in [5] for random 3-SAT formulas. The original version of probSAT does not restart, while the parallel version of probSAT [4] uses one process with a constant restart time regardless of the size and the structure of the instance. The authors did not explain how this restart time was obtained. Thus, in this section, we systematically analyze the runtime behavior and show that there is still much optimizing potential in the probSAT algorithm by using a carefully crafted restart strategy.

Instance Specification: For the experiments, random 3-SAT instances ranging from 1500 to 2500 variables are created using the generator kcnfgen [9]. We use random instances since they are the typical use case of SLS-solvers. The number of clauses was chosen such that the clause to variable ratio is close to the phase transition [6] at about 4.27. We stayed slightly below the phase transition because the instances in this range tend to be "interesting" as they are satisfiable but often hard to solve. Formulas with lower ratios, while almost always satisfiable, are mostly trivial to solve. Above the phase transition, the formulas tend to be unsatisfiable. Most SLS solvers and especially probSAT are so-called incomplete solvers, i.e., if a formula is unsatisfiable, they do not terminate. Thus, the RTDs can only be studied on satisfiable instances. The smallest used ratio was 4.204, the largest 4.272.

The complete instance set consists of 2400 formulas, 1632 of which are satisfiable, ensured by the SAT-solver dimetheus [9]. Note that this can be seen as filtering since formulas on which dimetheus performs well are preferred. Only the 1632 formulas guaranteed to be satisfiable are used for the further steps. We decided to use dimetheus for this task since it is currently one of the best performing solvers on random instances.

Empirical Distributions: For an instance i, the goal is to find a good approximation of its RTD X_i on probSAT. For this, we sample the random variable X_i 300 times to ensure stable results, measuring the number of local search steps until finding a satisfying assignment. This measure is chosen because it is stable and independent of hardware and scheduling. From this, we gain an empirical distribution function $\hat{F}_{300}(X_i)$. The solving is done with the help of Sputnik [16]. The maximum likelihood fits on the empirical distribution yield an estimation of the parameters for an LD and a WD. The Kolmogorov-Smirnov test (KS-test) is applied to compare these new distributions with X_i. Note, that passing the KS-test is not a proof of that distribution being the correct one. We only use it as a goodness-of-fit argument to see how well the actual distribution is described.

Only 14 out of all 1632 instances could not be classified as any of the distributions according to the KS-test at a significance level of 0.05. Table 1 displays how often the distributions passed the KS-test.

Table 1. The number of instances per distribution where the fit passed the KS-test at a significance level of 0.05.

	WD	LD	None
Passed	1529	1273	14
Share	93.7%	78.0%	0.9%

It is noticeable that the WD describes about 93.7% of the observed RTDs well. This value increases to above 97% at a significance level of 0.01. Furthermore, the LD describes 78.0% of the instances. This supports the observations in [2], who report that 389 of their 500 instances (77.8%) are described well by the LD, with the same approach and at a significance level of 0.05.

Calculating the Potential: We approximate an optimal expected runtime using the following approach. Let X be the set of observed runtimes of a fixed instance, and $X_{\leq x}$ the set of elements from X that are smaller than x. Then

$$\hat{opt} = min_{x \in X} \left(\frac{1-p}{p} \cdot x + \frac{\sum_{\{y \in X_{\leq x}\}} y}{|X_{\leq x}|} \right) \tag{1}$$

with $p = \frac{|X_{\leq x}|}{|X|}$. This is the minimum average runtime if restarted at an observed runtime given the measured data.

The speedup is a standard tool to compare the performances of two algorithms. The speedup is defined as the ratio of the expected runtimes of both algorithms on the same instance. As a suitable function to argue about the average speedup on several instances we applied the geometric mean. With this, the average speedup of using the optimal restart time compared to not restarting is 1.393 which serves as a baseline for further comparisons. We interpret this value as the maximum speedup reachable with a fixed-cutoff restart strategy.

The speedup for the different distributions is calculated similarly. However, the considered restart time is now calculated from the parameters of the distribution as proposed in [13]. We calculate the speedup for the LD, the WD, and a combination of both. In the combination, we use the restart time of the distribution with the higher p-value in the KS-test. This is decided for each instance individually. Table 2 shows that either distribution does not reach the potential of the baseline.

Table 2. The calculated speedups for different sets of probability distribution types. In the {LD, WD} column, the distribution was picked to calculate the speedup that performed better in the KS-test.

RTD	Speedup
∅(baseline)	1.393
{LD}	1.154
{WD}	1.174
{LD, WD}	1.253

A reason for the observed discrepancy between the baseline and all other speedups could be that the fitted distributions are too smooth. Describing the runtime behavior with few parameters does not suffice to capture all necessary details. Furthermore, the used sampling method does not provide any specifics on very short runs and the probabilities associated with them.

3   From Theory to Practice

The previous sections argued about the speedup while estimating it from data obtained by running probSAT without any restarts. This section explains how this potential can be exploited. We use an MP to estimate the RTDs of unseen instances. The estimates can be used to find the optimal restart time. Eggensperger et al. [8] propose an MP for this task which we adapt to fit our setting.

The Setup: This part describes the details of the machine learning components. Our implementation uses scikit-learn [15] and Keras [7] with tensorflow backend [1]. The quality of the components is measured as the average potential speedup on a test set with ten iterations of a 10-fold cross-validation. The test set consists of a sample of 162 instances where instances with exceptionally high or low shape parameters are oversampled on purpose. These were sampled from the 1632 instances mentioned in Sect. 2. During training, only the remaining 1470 instances were used. The MP uses 34 features which are obtained by the SATzilla feature extractor [17] with additional feature selection.

There are two steps in the estimation of the parameters: First, a random forest estimates the distribution type. Second, a neural network (NN) trained specifically for that distribution type predicts the distribution's parameters.

The random forest uses the distribution which performed better under the KS-test (higher p-value) as the label. For each distribution type there is a separate NN which predicts the shape and scale parameters of the respective distribution. For the WD, we also use an additional NN which predicts the location parameter. The experiments indicate that the inclusion of the location parameter into the network with the other parameters significantly worsens the overall performance.

For technical reasons, the potential speedup is not a suitable loss function. Therefore, another loss function has to be used. A standard loss function is the root-mean-square error (RMSE) of a prediction w.r.t. the ground truth. It was applied in the location network. However, for the shape and the scale parameter, it is advisable to use a loss function which captures the relationship between both parameters. Eggensperger et al. [8] use the negative log-likelihood function for measuring the quality of a fit. We also experimented with this function but achieved better results by using a different loss function based on the KS-Test.

Our loss function L is given in Eq. 2. Here, σ is the shape and μ the scale parameter. The labels are the value of the maximum-likelihood fit for σ, one observed runtime x and the value of the empirical distribution $emp(x)$. The predictions of the NN, $\hat{\sigma}$ and $\hat{\mu}$, define a cumulative distribution function F:

$$L(\hat{\sigma}, \hat{\mu}) = \left| F(x \mid \hat{\sigma}, \hat{\mu}) - emp(x) \right| + \left| \sigma - \hat{\sigma} \right|. \qquad (2)$$

When each component is evaluated separately, we observed an average potential speedup of 1.218 for the random forest, 1.137 for the WD network and 1.127 for the LD network. The average RMSE of the location network is 0.714. Combining all components and evaluating the whole MP on the test set yields a potential speedup of 1.157.

Experimental Results: We compare three different models: A fixed-cutoff strategy where the restart times are calculated by using the parameters from the MP. Note, that this method might lead to no restarts. In the following evaluation, this approach is denoted by "static restarts." Secondly, Luby restarts [14] are considered. This is a theoretically, approximately optimal strategy and thus a good baseline. Luby's strategy is usually initialized with a term a which is multiplied to the calculated restart times. We use $a = 20n$, where n is the number of variables, this restart time is suggested in [5]. Finally, not restarting the algorithm is considered.

Each of the approaches described above is tested on 100 new satisfiable instances generated with the same settings as defined in Sect. 2. The runtimes for each instance are sampled 100 times, and the average runtimes of each strategy are compared. For this experiment, a timeout of 10^{11} flips was used. Only one instance was affected by the cutoff. For this instance, the static restart strategy found a solution in 72 out of 100 runs, the Luby strategy found a solution in 47 cases and the not restarting approach in 28 cases. The following analysis only considers the remaining 99 instances. Again, we measure the runtime by the number of variable flips until a satisfying assignment is found.

The comparison of the logarithmically scaled runtimes between not restarting, static restarts and Luby restarts is illustrated in Fig. 1. The static restart strategy clearly outperforms the Luby strategy on all but two instances. The comparison with not restarting is harder to interpret. It can be seen that for easy instances there is barely any difference in the performance of the strategies. However, for intermediate and hard instances the static restart strategy outperforms not restarting in most cases. The parameters of probSAT were tuned without restarting with a short timeout. Therefore, it is not surprising that easy instances do not profit from restarting. Hard instances, on the other hand, can show a heavy-tailed behavior, i.e., probSAT's parameters are not chosen optimally for the case without restarts.

The comparison between the static and the Luby strategy yields an average speedup of 2.725. The comparison between the static restart strategy and not restarting yields an overall average speedup of 1.063. This speedup includes instances where the static strategy predicted not restarting. If only instances are considered where the static strategy restarted, then an average speedup of 1.108 is achieved on the remaining 69 instances.

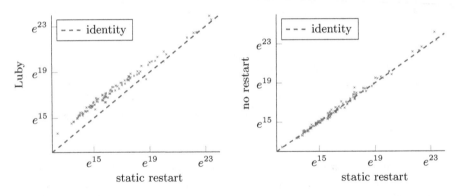

Fig. 1. On the left, the static restart strategy is compared with the Luby strategy. On the right, the static restart strategy is compared with not restarting. The axes are log-scaled average runtimes measured in variable flips.

It is remarkable that the speedup scales well with the hardness of the instance: When considering the 33 instances with the longest runtimes, a speedup of 1.216 is obtained. Finally, the speedups are tested with the t-test and a modified Wilcoxon signed-rank test ([11, Chap. 7.12]) at a significance level of 0.05. In every scenario described above, our approach is statistically, significantly better than both Luby's strategy and the strategy without restarting. We conclude that our system combines the advantages of not restarting on easy instances while also significantly improving the performance on hard problems.

4   Conclusion and Outlook

This work analyzes the potential speedup of restarts for the probSAT algorithm on random 3-SAT instances close to the phase transition. A major result is that if one knew the optimal restart times, the performance of probSAT could be improved by more than 39%. We proceed to approximate the runtime behavior with continuous probability distributions. Two distribution types, the Weibull and the lognormal distribution, are identified which describe the runtime behavior well. The (shifted) Weibull distribution performs best as it describes the runtime distributions of 93.7% instances well. The representations are used to calculate the (theoretically) best restart times; these predictions are used to measure the potential speedup. An average speedup factor of up to 1.253 can be achieved. The original version of probSAT is optimized towards not using restarts [5]. Hence, we find it surprising how much potential it has w.r.t. restarts.

The observations and the approximated runtime distributions are employed to train a random forest and neural networks. The random forest distinguishes between the distributions while the neural networks predict the parameters of the distributions. The predictions are used to decide whether restarts are useful. If they are, then the optimal restart time is calculated, and the fixed-cutoff strategy is used. Otherwise, probSAT is not restarted. The presented approach is statistically significantly better than both Luby's strategy and not restarting. An average speedup of 1.063 is obtained on all instances and an average speedup of 1.216 on the 33 of 99 test instances with the longest runtime where restarts are applied. The observations show that our approach combines the good behavior of probSAT on easy and intermediate instances while considerably improving the performance on hard instances.

Naturally, our approach is not limited to probSAT. Any Las Vegas algorithm can be analyzed and optimized with the same technique. The results imply that well-performing algorithms can be further improved with our approach, especially on hard instances.

References

1. Abadi, M., et al.: TensorFlow: large-scale machine learning on heterogeneous systems (2015). https://www.tensorflow.org/. Software available from tensorflow.org
2. Arbelaez, A., Truchet, C., Codognet, P.: Using sequential runtime distributions for the parallel speedup prediction of SAT local search. Theory Pract. Log. Program. **13**(4–5), 625–639 (2013)
3. Arbelaez, A., Truchet, C., O'Sullivan, B.: Learning sequential and parallel runtime distributions for randomized algorithms. In: ICTAI 2016, San Jose, California, USA, pp. 655–662. IEEE (2016)
4. Balint, A., Schöning, U.: probSAT and pprobSAT. SAT Compet. **2014**, 63 (2014)
5. Balint, A., Schöning, U.: Choosing probability distributions for stochastic local search and the role of make versus break. In: Cimatti, A., Sebastiani, R. (eds.) SAT 2012. LNCS, vol. 7317, pp. 16–29. Springer, Heidelberg (2012). https://doi.org/10.1007/978-3-642-31612-8_3

6. Biere, A., Heule, M., van Maaren, H.: Handbook of Satisfiability. IOS Press, Amsterdam (2009)
7. Chollet, F., et al.: Keras (2015). https://github.com/keras-team/keras
8. Eggensperger, K., Lindauer, M., Hutter, F.: Predicting runtime distributions using deep neural networks. arXiv:1709.07615 [cs] (2017)
9. Gableske, O.: SAT solving with message passing: a dissertation. Ph.D. thesis, Universität Ulm (2016)
10. Haim, S., Walsh, T.: Restart strategy selection using machine learning techniques. In: Kullmann, O. (ed.) SAT 2009. LNCS, vol. 5584, pp. 312–325. Springer, Heidelberg (2009). https://doi.org/10.1007/978-3-642-02777-2_30
11. Hollander, M., Wolfe, D.A., Chicken, E.: Nonparametric Statistical Methods. Wiley, Hoboken (2013)
12. Huang, J.: The effect of restarts on the efficiency of clause learning. In: IJCAI, vol. 7, pp. 2318–2323 (2007)
13. Lorenz, J.-H.: Runtime distributions and criteria for restarts. In: Tjoa, A.M., Bellatreche, L., Biffl, S., van Leeuwen, J., Wiedermann, J. (eds.) SOFSEM 2018. LNCS, vol. 10706, pp. 493–507. Springer, Cham (2018). https://doi.org/10.1007/978-3-319-73117-9_35
14. Luby, M., Sinclair, A., Zuckerman, D.: Optimal speedup of Las Vegas algorithms. Inf. Process. Lett. **47**(4), 173–180 (1993)
15. Pedregosa, F., et al.: Scikit-learn: machine learning in Python. J. Mach. Learn. Res. **12**, 2825–2830 (2011)
16. Völkel, G., et al.: Sputnik: ad hoc distributed computation. Bioinformatics **31**(8), 1298–1301 (2014)
17. Xu, L., et al.: SATzilla2012: improved algorithm selection based on cost-sensitive classification models. In: Proceedings of SAT Challenge, pp. 57–58 (2012)

Hash-Based Tree Similarity and Simplification in Genetic Programming for Symbolic Regression

Bogdan Burlacu[1,2]([✉]), Lukas Kammerer[1,2], Michael Affenzeller[2,3], and Gabriel Kronberger[1,2]

[1] Josef Ressel Centre for Symbolic Regression,
University of Applied Sciences Upper Austria, Softwarepark 11,
4232 Hagenberg, Austria
bogdan.burlacu@fh-hagenberg.at
[2] Heuristic and Evolutionary Algorithms Laboratory,
University of Applied Sciences Upper Austria, Softwarepark 11,
4232 Hagenberg, Austria
[3] Institute for Formal Models and Verification,
Johannes Kepler University, Altenbergerstr. 69, 4040 Linz, Austria

Abstract. We introduce in this paper a runtime-efficient tree hashing algorithm for the identification of isomorphic subtrees, with two important applications in genetic programming for symbolic regression: fast, online calculation of population diversity and algebraic simplification of symbolic expression trees. Based on this hashing approach, we propose a simple diversity-preservation mechanism with promising results on a collection of symbolic regression benchmark problems.

1 Introduction

Tree isomorphism algorithms play a fundamental role in pattern matching for tree-structured data. We introduce a fast *inexact*[1] tree matching algorithm that processes rooted, unordered, labeled trees into sequences of integer hash values, such that the same hash value indicates isomorphism. We define a distance measure between two trees based on the intersection of their corresponding hash value sequences. We are then able to efficiently compute a distance matrix for all trees by hashing each tree exactly once, then cheaply computing pairwise hash sequence intersections in linear time.

Genetic Programming (GP) for Symbolic Regression (SR) discovers mathematical formulae that best fit a given target function by means of evolving a population of tree-encoded solution candidates. The algorithm performs a guided search of the space of mathematical expressions by iteratively manipulating and evaluating a large number of tree genotypes under the principles of natural selection.

---

[1] Inexact due to the possibility of hash collisions causing the algorithm to return the wrong answer. With a reasonable hash function, collision probability is negligible.

R. Moreno-Díaz et al. (Eds.): EUROCAST 2019, LNCS 12013, pp. 361–369, 2020.
https://doi.org/10.1007/978-3-030-45093-9_44

Dynamic properties such as exploratory or exploitative behaviour and convergence speed have a big influence on GP performance [2]. Exploration refers to the ability to probe different areas of the search space, and exploitation refers to the ability to produce improvements in the local neighbourhood of existing solutions. The algorithm's success depends on achieving a good balance between exploration and exploitation over the course of the evolutionary run [11].

Population diversity (measured either at the structural-genotypic or semantic-phenotypic level) is typically used as an indirect measure of the algorithm's state-of-convergence, following the reasoning that the exploratory phase of the search is characterized by high diversity and the exploitative phase is characterized by low diversity. Premature convergence can occur at both genotype or phenotype levels, leading to a large amount of shared genetic material in the final population and a very concentrated set of behaviours [3].

Based on tree node hash values, we propose a new diversity measure defined as an individual's average distance to the rest of the population. We show how the resulting diversity score associated with each individual can be used to shift the focus of selection towards fit-and-diverse individuals.

Section 2 describes the methodology in detail. Section 3 summarizes the empirical results of our diversity-focused algorithmic improvements, and in Sect. 4 we offer our final remarks and conclusions.

2   Methodology

In this section we describe the tree hashing algorithm in detail and introduce a tree distance measure based on the number of common hash values between two tree individuals. The method lends itself well to the efficient computation of distance matrices (for the entire population) since each individual needs only be hashed once and pairwise distances can be subsequently computed using only the corresponding hash value sequences.

An important aspect of our methodology is the ability of the proposed diversity measure to implicitly capture an individual's semantics by hashing the numerical coefficients associated to the tree's leaf nodes. This allows us to differentiate (in terms of tree distance) between individuals with similar structure but different semantics.

Tree Hashing. The hashing procedure shares some common aspects with the earlier algorithm by Merkle [7] where data blocks represented as leaves in the tree are hashed together in a bottom-up manner.

In our approach an associated data block for each tree node is aggregated together with child node data as input to the hash function to a general-purpose non-cryptographic hash function $\oplus$. This concept is illustrated in Fig. 1, where leaf nodes $L_1, ..., L_4$ represent terminal symbols and internal nodes represent mathematical operations. Hash values are computed by the $\oplus$ operator taking as arguments the node's own data hash and its child hash values.

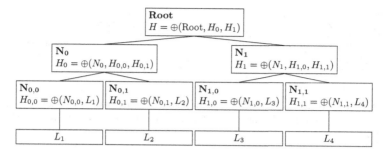

Fig. 1. Example hash tree where each parent hash value is aggregated from its initial hash value and the child node hash values. Leaf nodes $L_1, ..., L_4$ represent data blocks.

The procedure given as pseudocode in Algorithm 1 relies on the linearization of input tree T, such that the resulting array of nodes corresponds to a postorder traversal of T. The benefit of linearization is that subtrees are represented by continuous array regions, thus facilitating indexing and sorting operations. For example, tree node n with postorder index i will find its first child at index $j = i - 1$, its second child at index $k = j - \text{Size}(j)$ and so on, where $\text{Size}(j)$ returns the size of the subtree whose root node has index j.

Sorting the child nodes of commutative symbols is a key part of this procedure. In the general case, this requires a reordering of the corresponding subarrays using an auxiliary buffer. Sorting without an auxiliary buffer is possible when all child nodes are leafs, that is when $\text{Size}(n) = \text{Arity}(n)$ for a parent node n. Child order is established according to the calculated hash values.

We illustrate the sorting procedure in Fig. 2, where a postfix expression with root symbol S at index 9 contains three child symbols with indices 2, 5, 8 and hash values H_2, H_5, H_8, respectively. In order to calculate $\oplus(S)$ the child order is first established according to symbol hash values. Leaf nodes are assigned initial hash values that do not change during the procedure. Assuming that sorting produces the order $\{S_5, S_2, S_3\}$, the child symbols and their respective subarrays need to be reordered in the expression using a temporary buffer. After sorting, the original expression becomes:

$$\{[S_0, S_1, S_2], [S_3, S_4, S_5], [S_6, S_7, S_8], S_9\} \rightarrow \{[S_4, S_5, S_6], [S_1, S_2, S_3], [S_7, S_7, S_8], S_9\}$$

The sorted child hash values are then aggregated with the parent label in order to produce the parent hash value $H_9 = \oplus(\{H_5, H_2, H_8, S_9\})$. The hashing algorithm alternates hashing and sorting steps as it moves from the bottom level of the tree towards the root node. Finally, each tree node is assigned a hash value and the hash value of the root node is returned as the expression's hash value.

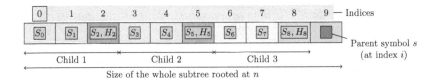

Fig. 2. Example internal node with three child subtrees in postfix representation. Child subarrays are reordered according to priority rules and their respective root hash values H_2, H_5, H_8.

Algorithm 1: Expression hashing

---

    input : A symbolic expression tree T
    output: The corresponding sequence of hash values

1 hashes ← empty list of hash values;
2 nodes ← post-order list of T's nodes;
3 **foreach** *node n in nodes* **do**
4      $H(n)$ ← an initial hash value;
5   **if** *n is a function node* **then**
6   **if** *n is commutative* **then**
7   Sort the child nodes of n;
8   child hashes ← hash values of n's children;
9         $H(n)$ ← ⊕(child hashes, $H(n)$);
10   hashes.append($H(n)$);
11 **return** hashes;

Hash-Based Tree Simplification. Our hash-based approach to simplification enables structural transformations based on tree isomorphism and symbolic equivalence relations. Figure 3 illustrates the simplification of an addition function node with two identical child nodes.

In the expression $E = c_1 x + c_2 yz + c_3 x$ represented in postfix notation, the terms $c_1 x$ and $c_3 x$ are isomorphic and hash to the same value. The simplification algorithm identifies the possibility of folding the constants c_1 and c_3 so that the two terms are simplified to a single term $c_4 x$ where $c_4 = c_1 + c_3$. The simplified expression then becomes $E' = c_4 x + c_2 yz \equiv E$.

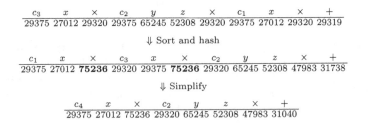

Fig. 3. Expression simplification. The original expression (top) is sorted and hashed according to Algorithm 1. Isomorphic subtrees with the same hash value are simplified.

Hash-Based Population Diversity. Using the algorithm described in Sect. 2, we convert each tree individual into a sequence of hash values corresponding to a post-order traversal of its structure. After this conversion, the distance between two trees can be defined using the intersection between the two sequences[2]:

$$D(T_1, T_2) = \frac{2 \cdot |H_1 \cap H_2|}{|H_1| + |H_2|} \tag{1}$$

where H_1, H_2 represent the hash value sequences of T_1 and T_2, respectively.

Computing the distance matrix for the entire population can be further optimized by hashing all trees in an initial pass, then using the resulting hash value sequences for the calculation of pairwise distances. This leads to the following algorithmic steps:

1. Convert every tree T_i in the population to hash value sequence H_i.
2. Sort each hash value sequence H_i in ascending order.
3. For every pair (H_i, H_j) compute distance using Eq. 1.

Sorting in Step 2 allows us to efficiently compute $|H_i \cap H_j|$ in linear time. As shown in Table 1, in this particular scenario, these optimizations lead to a two order of magnitude improvement in runtime performance over similar methods [10].

Table 1. Elapsed time computing average distance for 5000 trees

Tree distance method	Elapsed time (s)	Speed-up
Bottom-up	1225.751	1.0x
Hash-based	3.677	333.3x

Diversity as an Explicit Search Objective. Diversity maintenance strategies have been shown to improve GP performance [3,11]. A plethora of diversity measures have already been studied: history-based, distance-based, difference-based, entropy-based, etc. [11]. However, due to high computational requirements, tree distances like the tree edit distance have seldomly been used. Our hash-based approach overcomes this limitation making it feasible to compute the population distance matrix every generation.

We define an individual's diversity score as its average distance to the rest of the population. The score is easily computed by averaging the corresponding distance matrix row for a given individual. In effect, this causes selection to favor individuals "farther away" from the crowd, reducing the effects of local optima as attractors in the search space.

We integrate this new objective into a single-objective approach using standard GP and a multi-objective approach using the NSGA-2 algorithm [4].

---

[2] The Sørensen-Dice coefficient (Eq. 1) returns a value in the interval $[0, 1]$.

In the single-objective case, we extent the standard fitness function with an additional diversity term d, such that the new fitness used during selection becomes: $f' = f + d$. Since fitness is normalized between $[0, 1]$ both terms have the same scale and no additional weighting is used.

In the multi-objective case the diversity score is used as a secondary objective along with fitness. The NSGA-2 algorithm performs selection using crowding distance within Pareto fronts of solutions.

3   Empirical Results

We test the proposed approach on a collection of synthetic symbolic regression benchmarks: Vladislavleva [12], Poly-10 [9], Spatial Coevolution [8], Friedman [5] and Breiman [1]. We configure all algorithms to evolve a population of 1000 individuals over 500 generations with a function set consisting of $(+, -, \times, \div, \exp, \log, \sin, \cos, \text{square})$. Different types of mutation (remove branch, replace branch, change node type, one-point mutation) are applied with a probability of 25%. Tree individuals are initialized using the Probabilistic Tree Creator (PTC2) [6].

The results summarized in Table 2 show the benefits of diversity maintenance. Selecting for diversity enables both the GA and NSGA-2 algorithms to better exploit population diversity and achieve better results in comparison with the standard GA approach. The NSGA-2 algorithm in particular is able to avoid overfitting and produce more generalizable models on the Vladislavleva-6 and Vladislavleva-8 problems. In all tested problem instances, both GA Diversity and NSGA-2 outperform standard GA on both training and test data. Figure 4 shows that GA Diversity and NSGA-2 are able to maintain higher diversity and promote smaller, less bloated individuals.

The overhead incurred by tree distance calculation depends on the size of the training data. For large data, this overhead becomes negligible as the algorithm will spend most of its runtime evaluating fitness. In our tests GA Diversity is approximately 20–25% slower than Standard GA. A direct runtime comparison with NSGA-2 is not possible due to different algorithmic dynamics.

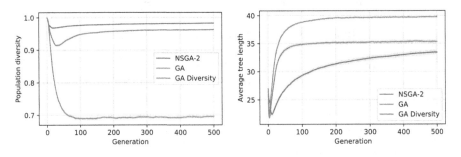

Fig. 4. Evolution of average diversity and average tree length

Table 2. Empirical results expressed as median $R^2\pm$ interquartile range over 50 algorithm repetitions

Problem data	Algorithm	R^2 (Training)	R^2 (Test)	Time (s)
Breiman - I	GA	0.870 ± 0.040	0.865 ± 0.037	1084.5
Breiman - I	GA Diversity	0.875 ± 0.034	0.870 ± 0.029	1207.5
Breiman - I	NSGA-2	$\mathbf{0.885 \pm 0.013}$	$\mathbf{0.879 \pm 0.012}$	1584.0
Friedman - I	GA	0.859 ± 0.007	0.860 ± 0.006	1090.5
Friedman - I	GA Diversity	0.860 ± 0.007	0.860 ± 0.005	1207.0
Friedman - I	NSGA-2	$\mathbf{0.863 \pm 0.002}$	$\mathbf{0.863 \pm 0.003}$	1554.0
Friedman - II	GA	0.944 ± 0.038	0.942 ± 0.041	1092.0
Friedman - II	GA Diversity	0.957 ± 0.023	0.957 ± 0.025	1248.0
Friedman - II	NSGA-2	$\mathbf{0.958 \pm 0.008}$	$\mathbf{0.957 \pm 0.010}$	1474.0
Pagie-1	GA	0.990 ± 0.012	0.889 ± 0.107	442.5
Pagie-1	GA Diversity	0.994 ± 0.005	0.912 ± 0.070	540.5
Pagie-1	NSGA-2	$\mathbf{0.998 \pm 0.002}$	$\mathbf{0.932 \pm 0.075}$	880.0
Poly-10	GA	0.820 ± 0.293	0.764 ± 0.478	405.0
Poly-10	GA Diversity	0.840 ± 0.089	0.838 ± 0.120	518.0
Poly-10	NSGA-2	$\mathbf{0.879 \pm 0.072}$	$\mathbf{0.850 \pm 0.099}$	902.5
Vladislavleva-1	GA	0.999 ± 0.001	0.946 ± 0.120	371.5
Vladislavleva-1	GA Diversity	0.999 ± 0.001	0.980 ± 0.106	450.0
Vladislavleva-1	NSGA-2	$\mathbf{1.000 \pm 0.000}$	$\mathbf{0.987 \pm 0.026}$	784.5
Vladislavleva-2	GA	0.995 ± 0.012	0.987 ± 0.028	355.0
Vladislavleva-2	GA Diversity	0.995 ± 0.012	0.992 ± 0.016	462.0
Vladislavleva-2	NSGA-2	$\mathbf{0.999 \pm 0.001}$	$\mathbf{0.998 \pm 0.002}$	802.0
Vladislavleva-3	GA	0.968 ± 0.062	0.932 ± 0.508	445.0
Vladislavleva-3	GA Diversity	0.979 ± 0.048	0.975 ± 0.053	570.0
Vladislavleva-3	NSGA-2	$\mathbf{0.995 \pm 0.014}$	$\mathbf{0.989 \pm 0.042}$	920.5
Vladislavleva-4	GA	0.951 ± 0.036	0.918 ± 0.053	527.0
Vladislavleva-4	GA Diversity	0.968 ± 0.023	0.936 ± 0.049	651.0
Vladislavleva-4	NSGA-2	$\mathbf{0.982 \pm 0.017}$	$\mathbf{0.966 \pm 0.027}$	973.5
Vladislavleva-5	GA	0.997 ± 0.012	0.933 ± 0.144	407.0
Vladislavleva-5	GA Diversity	0.999 ± 0.002	0.995 ± 0.015	502.5
Vladislavleva-5	NSGA-2	$\mathbf{1.000 \pm 0.000}$	$\mathbf{0.997 \pm 0.023}$	871.0
Vladislavleva-6	GA	0.869 ± 0.155	0.072 ± 0.329	369.0
Vladislavleva-6	GA Diversity	0.939 ± 0.143	0.255 ± 0.978	491.0
Vladislavleva-6	NSGA-2	$\mathbf{1.000 \pm 0.023}$	$\mathbf{1.000 \pm 0.347}$	930.0
Vladislavleva-7	GA	0.895 ± 0.048	0.878 ± 0.100	395.0
Vladislavleva-7	GA Diversity	0.904 ± 0.029	0.892 ± 0.058	501.5
Vladislavleva-7	NSGA-2	$\mathbf{0.917 \pm 0.017}$	$\mathbf{0.899 \pm 0.033}$	843.5
Vladislavleva-8	GA	0.962 ± 0.080	0.541 ± 0.569	362.5
Vladislavleva-8	GA Diversity	0.986 ± 0.033	0.787 ± 0.436	489.5
Vladislavleva-8	NSGA-2	$\mathbf{0.992 \pm 0.012}$	$\mathbf{0.817 \pm 0.412}$	799.0

4   Summary

We described a hashing algorithm for GP trees with applications to distance calculation and expression simplification. The approach is highly efficient, making it feasible to measure diversity on a generational basis, as after an initial preprocessing step tree distance is reduced to a simple co-occurrence count between sorted hash value sequences. With this information new diversity preservation strategies become possible at low computational cost.

A simple strategy illustrated in this work is to bias selection towards individuals that are more distant from the rest of the population. Experimental results using the standard GA and NSGA-2 algorithms showed increased model accuracy on all tested problem instances, correlated with increased diversity and lower average tree size. Although easily integrated with all GA flavours, we conclude empirically that the strategy is more effective in the multi-objective case when diversity is separately considered.

Future work in this area will focus on mining common subtrees in the population based on hash value frequencies and designing more complex diversity preservation strategies.

Acknowledgement. The authors gratefully acknowledge support by the Christian Doppler Research Association and the Federal Ministry of Digital and Economic Affairs within the *Josef Ressel Centre for Symbolic Regression*.

References

1. Breiman, L., Friedman, J., Stone, C., Olshen, R.: Classification and Regression Trees. The Wadsworth and Brooks-Cole Statistics-Probability Series. Taylor & Francis, Routledge (1984)
2. Burke, E.K., Gustafson, S., Kendall, G.: Diversity in genetic programming: an analysis of measures and correlation with fitness. IEEE Trans. Evol. Comput. **8**(1), 47–62 (2004)
3. Burks, A.R., Punch, W.F.: An analysis of the genetic marker diversity algorithm for genetic programming. Genet. Program Evolvable Mach. **18**(2), 213–245 (2016). https://doi.org/10.1007/s10710-016-9281-9
4. Deb, K., Pratap, A., Agarwal, S., Meyarivan, T.: A fast and elitist multiobjective genetic algorithm: NSGA-II. IEEE Trans. Evol. Comput. **6**(2), 182–197 (2002)
5. Friedman, J.H.: Multivariate adaptive regression splines. Ann. Statist. **19**(1), 1–67 (1991)
6. Luke, S.: Two fast tree-creation algorithms for genetic programming. IEEE Trans. Evol. Comput. **4**(3), 274–283 (2000)
7. Merkle, R.C.: A digital signature based on a conventional encryption function. In: Pomerance, C. (ed.) CRYPTO 1987. LNCS, vol. 293, pp. 369–378. Springer, Heidelberg (1988). https://doi.org/10.1007/3-540-48184-2_32
8. Pagie, L., Hogeweg, P.: Evolutionary consequences of coevolving targets. Evol. Comput. **5**(4), 401–418 (1997)
9. Poli, R.: A simple but theoretically-motivated method to control bloat in genetic programming. In: Ryan, C., Soule, T., Keijzer, M., Tsang, E., Poli, R., Costa, E. (eds.) EuroGP 2003. LNCS, vol. 2610, pp. 204–217. Springer, Heidelberg (2003). https://doi.org/10.1007/3-540-36599-0_19

10. Valiente, G.: An efficient bottom-up distance between trees. In: Proceedings of the 8th International Symposium on String Processing and Information Retrieval, pp. 212–219. IEEE Computer Science Press (2001)
11. Črepinšek, M., Liu, S.H., Mernik, M.: Exploration and exploitation in evolutionary algorithms: a survey. ACM Comput. Surv. **45**(3), 35:1–35:33 (2013)
12. Vladislavleva, E.J., Smits, G.F., den Hertog, D.: Order of nonlinearity as a complexity measure for models generated by symbolic regression via Pareto genetic programming. IEEE Trans. Evol. Comput. **13**(2), 333–349 (2009)

Identification of Dynamical Systems Using Symbolic Regression

Gabriel Kronberger$^{(\boxtimes)}$ ⓘ, Lukas Kammerer ⓘ, and Michael Kommenda

Josef Ressel Centre for Symbolic Regression,
Heuristic and Evolutionary Algorithms Laboratory,
University of Applied Sciences Upper Austria, Softwarepark 11,
4232 Hagenberg, Austria
`gabriel.kronberger@fh-hagenberg.at`

Abstract. We describe a method for the identification of models for dynamical systems from observational data. The method is based on the concept of symbolic regression and uses genetic programming to evolve a system of ordinary differential equations (ODE).

The novelty is that we add a step of gradient-based optimization of the ODE parameters. For this we calculate the sensitivities of the solution to the initial value problem (IVP) using automatic differentiation.

The proposed approach is tested on a set of 19 problem instances taken from the literature which includes datasets from simulated systems as well as datasets captured from mechanical systems. We find that gradient-based optimization of parameters improves predictive accuracy of the models. The best results are obtained when we first fit the individual equations to the numeric differences and then subsequently fine-tune the identified parameter values by fitting the IVP solution to the observed variable values.

Keywords: System dynamics · Genetic programming · Symbolic regression

1 Background and Motivation

Modelling, analysis and control of dynamical systems are core topics within the field of system theory focusing on the behavior of systems over time. System dynamics can be modelled using ordinary differential equations (cf. [5]) which define how the state of a system changes based on the current state for infinitesimal time steps.

Genetic programming (GP) is a specific type of evolutionary computation in which computer programs are evolved to solve a given problem. Symbolic regression (SR) is a specific task for which GP has proofed to work well. The goal in SR is to find an expression that describes the functional dependency between a dependent variable and multiple independent variables given a dataset of observed values for all variables. In contrast to other forms of regression

© Springer Nature Switzerland AG 2020
R. Moreno-Díaz et al. (Eds.): EUROCAST 2019, LNCS 12013, pp. 370–377, 2020.
https://doi.org/10.1007/978-3-030-45093-9_45

analysis it is not necessary to specify the model structure beforehand because the appropriate model structure is identified simultaneously with the numeric parameters of the model. SR is therefore especially suited for regression tasks when a parametric model for the system or process is not known. Correspondingly, SR could potentially be used for the identification models for dynamical systems when an accurate mathematical model of the system doesn't exist.

We aim to use SR to find the right-hand sides $f(\cdot)$ of a system of ordinary differential equations such as:

$$\dot{u}_1 = f_{\dot{u}_1}(u_1, u_2, \boldsymbol{\theta}), \dot{u}_2 = f_{\dot{u}_2}(u_1, u_2, \boldsymbol{\theta})$$

We only consider systems without input variables (non-forced systems) and leave an analysis for forced systems for future work.

1.1 Related Work

The vast literature on symbolic regression is mainly focused on static models in which predicted values are independent given the input variable values. However, there are a several articles which explicitly describe GP-based methods for modelling dynamical systems. A straight-forward approach which can be implemented efficiently is to approximate the derivatives numerically [3,4,7,11]. Solving the IVP for each of the considered model candidates is more accurate but also computationally more expensive; this is for instance used in [1].

In almost all methods discussed in prior work the individual equations of the system are encoded as separate trees and the SR solutions hold multiple trees [1,4,7]. A notable exception is the approach suggested in [3] in which GP is run multiple times to produce multiple expressions for the numerically approximated derivatives of each state variable. Well-fitting expressions are added to a pool of equations. In the end, the elements from the pool are combined and the best-fitting ODE system is returned. In this work the IVP is solved only in the model combination phase.

Identification of correct parameter values is critical especially for ODE systems. Therefore, several authors have included gradient-based numeric optimization of parameter values. In [4], parameters are allowed only for scaling top-level terms, to allow efficient least-squares optimization of parameter values. In [11] non-linear parameters are allowed and parameter values are iteratively optimized using the Levenberg-Marquardt algorithm based on gradients determined through automatic differentiation.

Partitioning is introduced in [1]. The idea is to optimize the individual parts of the ODE system for each state variable separately, whereby it is assumed that the values of all other state variables are known. Partitioning reduces computational effort but does not guarantee that the combined system of equations models system dynamics correctly.

Recently, neural networks have been used for modelling ODEs [2] whereby the neural network parameters are optimized using gradients determined via the adjoint sensitivity method and automatic differentiation. The same idea can be used to fit numeric parameters of SR models.

2   Methods

We extend the approach described in [4], whereby we allow numeric parameters at any point in the symbolic expressions. We use an iterative gradient-based method (i.e. Levenberg-Marquardt (LM)) for the optimization of parameters and automatic differentiation for efficient calculation of gradients similarly to [11]. We analyse several different algorithm variations, some of which solve the IVP problem in each evaluation step as in [1,2]. For this we use the state-of-the-art CVODES[1] library for solving ODEs and calculating parameter sensitivities.

We compare algorithms based on the deviation of the IVP solution from the observed data points for the final models. The deviation measure is the sum of normalized mean squared errors:

$$\text{SNMSE}(Y, \hat{Y}) = \sum_{i=1}^{D} \frac{1}{\text{var}(y_{i,.})} \frac{1}{N} \sum_{j=1}^{N} (y_{i,j} - \hat{y}_{i,j})^2 \tag{1}$$

Where $Y = \{y_{i,j}\}_{i=1..D, j=1..N}$ is the matrix of N subsequent and equidistant observations of the D the state variables. Each state variable is measured at the same time points $(t_i)_{i=1..N}$. The matrix of predicted values variables $\hat{Y} = \{\hat{y}_{i,j}\}_{i=1..D, j=1..N}$ is calculated by integrating the ODE system using the initial values $\hat{y}_{.,1} = y_{.,1}$.

2.1   Algorithm Description

The system of differential equations is represented as an array of expression trees each one representing the differential equation for one state variable. The evolutionary algorithm initializes each tree randomly whereby all of the state variables are allowed to occur in the expression. In the crossover step, exchange of sub-trees is only allowed between corresponding trees. This implicitly segregates the gene pools for the state variables. For each of the trees within an individual we perform sub-tree crossover with the probability given by the crossover rate parameter. A low crossover rate is helpful to reduce the destructive effect of crossover.

Memetic optimization of numeric parameters has been shown to improve GP performance for SR [6,10]. We found that GP performance is improved significantly even when we execute only a few iterations (3–10) of LM and update the parameter values whenever an individual is evaluated. This ensures that the improved parameter values are inherited and can subsequently be improved

---

[1] https://computation.llnl.gov/projects/sundials/cvodes.

even further when we start LM with these values [6]. We propose to use the same approach for parameters of the ODEs with a small modification based on the idea of partitioning. In the first step we use the approximated derivative values as the target. In this step we use partitioning and assume all variable values are known. The parameter values are updated using the optimized values. In the second step we use CVODES to solve the IVP for all state variables simultaneously and to calculate parameter sensitivities. The LM algorithm is used to optimize the SNMSE (Eq. 1) directly. Both steps can be turned on individually and the number of LM iterations can be set independently.

2.2   Computational Experiments

Algorithm Configuration. Table 1 shows the GP parameter values that have been used for the experiments.

Table 1. Parameter values for the GP algorithm that have been used for all experiments. The number of generations and the maximum number of evaluated solutions is varied for the experiments.

Parameter	Value
Population size	300
Initialization	PTC2
Parent selection	Proportional (first parent)
	Random (second parent)
Crossover	Subtree crossover
Mutation	Replace subtree with random branch
	Add $x \sim N(0,1)$ to all numeric parameters
	Add $x \sim N(0,1)$ to a single numeric parameter
	Change a single function symbol
Crossover rate	30% (for each expression)
Mutation rate	5% (for the whole individual)
Offspring selection	Offspring must be better than both parents
Maximum selection pressure	$100 <$ # evaluated offspring/population size
Replacement	Generational with a single elite
Terminal set	State variables and real-valued parameters
Function set	$+, *, \sin, \cos$

We compare two groups of different configurations: in the first group we rely solely on the evolutionary algorithm for the identification of parameter values. In the second group we use parameter optimization optimization. Both groups contain three configurations with different fitness functions. In the following we use the identifiers D, I, D + I, D_{opt}, I_{opt}, and $D_{opt} + I_{opt}$ for the six algorithm instances. Configuration D uses the SNMSE for the approximated derivatives for fitness assignment as in [3, 4, 11]; configuration I uses the SNMSE for the solution to the IVP as in [1]; configuration D + I uses the sum of both error measures for fitness evaluation. For the first group we allow maximally 500.000 evaluated solutions and 250 generations; for the second group we only allow 100.000 evaluated solutions and 25 generations. The total number of function evaluations including evaluations required for parameter optimization is similar for all configurations and approximately between 500.000 and 2 million (depending on the dimensionality of the problem).

Problem Instances. We use 19 problem instances for testing our proposed approach as shown in Tables 2 and 3. These have been taken from [4, 7, 8] and include a variety of different systems. The set of problem instances includes simulated systems (Table 2) as well as datasets gathered with motion-tracking from real mechanical systems (Table 3). The simulated datasets have been generated using fourth-order Runge-Kutta integration (RK45). The motion-tracked datasets have been adapted from the original source to have equidistant observations using cubic spline interpolation.

3   Results

Table 4 shows the number of successful runs (from 10 independent runs) for each problem instance and algorithm configuration. A run is considered successful if the IVP solution for the identified ODE system has an SNMSE <0.01. Some of the instances can be solved easily with all configurations. Overall the configuration $I_{opt} + D_{opt}$ is the most successful. With this configuration we are able to produce solutions for all of the 19 instances with a high probability. The beneficial effect of gradient-based optimization of parameters is evident from the much larger number of successful runs.

Notably, when we fit the expressions to the approximated derivatives using partitioning (configurations D and D_{opt}) the success rate is low. The reason is that IVP solutions might deviate strongly when we use partitioning to fit of the individual equations to the approximated derivatives. To achieve a good fit the causal dependencies must be represented correctly in the ODE system. This is not enforced when we use partitioning.

4   Discussion

The results of our experiments are encouraging and indicate that it is indeed possible to identify ODE models for dynamical systems solely from data using GP and SR. However, there are several aspects that have not yet been fully answered in our experiments and encourage further research.

A fair comparison of algorithm configurations would allow the same runtime for all cases. We have use a similar amount of function evaluations but have so

Table 2. The problem instances for which we have generated data using numeric integration.

Instance	Expression	Initial value	N	t_{max}
Chemical reaction [4]	$\dot{y}_1 = -1.4y_1$ $\dot{y}_2 = 1.4y_1 - 4.2y_2$ $\dot{y}_3 = 4.2y_2$	$(0.1, 0, 0)$	100	1
E-Cell [4]	$\dot{y}_1 = -10y_1y_3$ $\dot{y}_2 = 10y_1y_3 - 17y_2$ $\dot{y}_3 = -10y_1y_3 + 17y_2$	$(1.2, 0.0, 1.2)$	40	0.4
S-system [4]	$\dot{y}_1 = 15y_3y_5^{-0.1} - 10y_1^2$ $\dot{y}_2 = 10y_1^2 - 10y_2^2$ $\dot{y}_3 = 10y_2^{-0.1} - 10y_2^{-0.1}y_3^2$ $\dot{y}_4 = 8y_1^2y_5^{-0.1} - 10y_4^2$ $\dot{y}_5 = 10y_4^2 - 10y_5^2$	$(0.1, 0.1, 0.1, 0.1, 0.1)$ $(0.5, 0.5, 0.5, 0.5, 0.5)$ $(1.5, 1.5, 1.5, 1.5, 1.5)$	$3*30$	0.3
Lotka-Volterra (3 species) [4]	$\dot{y}_1 = y_1(1 - y_1 - y_2 - 10y_3)$ $\dot{y}_2 = y_2(0.992 - 1.5y_1 - y_2 - y_3)$ $\dot{y}_3 = y_3(-1.2 + 5y_1 + 0.5y_2)$	$(0.2895, 0.2827, 0.126)$	100	100
Lotka-Volterra (2 species) [3]	$\dot{y}_1 = y_1(0.04 - 0.0005y_2)$ $\dot{y}_2 = -y_2(0.2 - 0.004y_1)$	$(20, 20)$	300	300
Glider [7]	$\dot{v} = -0.05v^2 - \sin(\theta)$ $\dot{\theta} = v - \cos(\theta)/v;$	$(1.5, 1)$	100	10
Bacterial respiration [7]	$\dot{x} = (20 - x - xy)/(1 + 0.5x^2)$ $\dot{y} = (10 - xy)/(1 + 0.5x^2)$	$(1, 1)$	100	10
Predator-prey [7]	$\dot{x} = x(4 - x - y/(1 + x))$ $\dot{y} = y(x/(1 + x) - 0.075y)$	$(1.1, 7.36)$	100	10
Bar magnets [7]	$\dot{\theta}_1 = 0.5\sin(\theta_1 - \theta_2) - \sin(\theta_1)$ $\dot{\theta}_2 = 0.5\sin(\theta_2 - \theta_1) - \sin(\theta_2)$	$(0.7, -0.3)$	100	10
Shear flow [7]	$\dot{\theta} = \cot(\theta)\cos(\phi)$ $\dot{\phi} = (\cos(\phi)^2 + 0.1\sin(\phi)^2)\sin(\phi)$	$(0.7, 0.4)$	100	10
Van der Pol oscillator [7]	$\dot{x} = 10(y - (\frac{1}{3}x^3 - x))$ $\dot{y} = -0.1x$	$(2, 0.1)$	100	10

far neglected the computational effort that is required for numerically solving the ODE in each evaluation step. Noise can have a large effect in the numeric approximation of derivatives. We have not yet studied the effect of noisy measurements. Another task for future research is the analysis of forecasting accuracy of the models. So far we have only measured the performance on the training set. Finally, for practical applications, it would be helpful to extend the method to allow input variables (forced systems) as well as latent variables.

Table 3. The problem instances for which we used the datasets from [9]. For each system type we use two different datasets, one generated via simulation, the other by motion-tracking the real system.

Type	Name	File name	Variables	N
Simulated	Linear oscillator	`linear_h_1.txt`	x, v	512
Motion-tracked	Linear oscillator	`real_linear_h_1.txt`	x, v	879
Simulated	Pendulum	`pendulum_h_1.txt`	θ, ω	502
Motion-tracked	Pendulum	`real_pend_h_1.txt`	θ, ω	568
Simulated	Coupled oscillator	`double_linear_h_1.txt`	x_1, x_2, v_1, v_2	200
Motion-tracked	Coupled oscillator	`real_double_linear_h_1.txt`	x_1, x_2, v_1, v_2	150
Simulated	Double pendulum	`double_pend_h_1.txt`[a]	$\theta_1, \theta_2, \omega_1, \omega_2$	1355
Motion-tracked	Double pendulum	`real_double_pend_h_1.txt`[a]	$\theta_1, \theta_2, \omega_1, \omega_2$	200

[a]First (non-chaotic) configuration only

Table 4. Number of successful runs. A run is successful if the SNMSE for the integrated system is <0.01. Algorithm configurations: numeric differences (D), numeric IVP solution (I), combination of numeric differences and IVP solution (I + D). The configurations using the subscript *opt* include parameter optimization.

Instance	D	I	I + D	D_{opt}	I_{opt}	$I_{opt} + D_{opt}$
Chemical reaction	7	4	2	9	10	10
E-Cell	5	0	4	9	10	10
S-system	0	0	0	10	10	10
Lotka-Volterra (three species)	0	0	0	0	0	8
Bacterial respiration	5	3	3	10	10	10
Bar magnets	3	4	5	10	10	10
Glider	0	0	0	9	10	10
Lotka-Volterra	0	0	0	1	3	10
Predator prey	0	0	0	3	10	10
Shear flow	0	0	0	7	10	10
Van der Pol oscillator	1	0	1	6	10	10
Linear oscillator (motion-tracked)	0	5	9	1	9	10
Linear oscillator (simulation)	0	0	5	0	10	10
Pendulum (motion-tracked)	0	0	0	0	10	10
Pendulum (simulated)	4	9	9	0	10	10
Double oscillator (motion-tracked)	0	0	0	0	0	6
Double oscillator (simulated)	0	0	0	0	0	10
Double pendulum (motion-tracked)	0	0	0	0	0	7
Double pendulum (simulated)	0	0	0	10	0	10
Total	21	16	29	85	122	171

Acknowledgments. The authors gratefully acknowledge support by the Austrian Research Promotion Agency (FFG) within project #867202, as well as the Christian Doppler Research Association and the Federal Ministry of Digital and Economic Affairs within the *Josef Ressel Centre for Symbolic Regression.*

References

1. Bongard, J., Lipson, H.: Automated reverse engineering of nonlinear dynamical systems. Proc. Nat. Acad. Sci. **104**(24), 9943–9948 (2007)
2. Chen, T.Q., Rubanova, Y., Bettencourt, J., Duvenaud, D.K.: Neural ordinary differential equations. In: Bengio, S., Wallach, H., Larochelle, H., Grauman, K., Cesa-Bianchi, N., Garnett, R. (eds.) Advances in Neural Information Processing Systems, vol. 31, pp. 6571–6583. Curran Associates, Inc. (2018). http://papers. nips.cc/paper/7892-neural-ordinary-differential-equations.pdf
3. Gaucel, S., Keijzer, M., Lutton, E., Tonda, A.: Learning dynamical systems using standard symbolic regression. In: Nicolau, M., et al. (eds.) EuroGP 2014. LNCS, vol. 8599, pp. 25–36. Springer, Heidelberg (2014). https://doi.org/10.1007/978-3-662-44303-3_3
4. Iba, H.: Inference of differential equation models by genetic programming. Inf. Sci. **178**(23), 4453–4468 (2008). https://doi.org/10.1016/j.ins.2008.07.029
5. Isermann, R., Münchhof, M.: Identification of Dynamic Systems: An Introduction with Applications. Springer, Heidelberg (2011). https://doi.org/10.1007/978-3-540-78879-9
6. Kommenda, M., Kronberger, G., Winkler, S., Affenzeller, M., Wagner, S.: Effects of constant optimization by nonlinear least squares minimization in symbolic regression. In: Proceedings of the 15th Annual Conference Companion on Genetic and Evolutionary Computation, pp. 1121–1128. ACM (2013)
7. Schmidt, M., Lipson, H.: Data-mining dynamical systems: automated symbolic system identification for exploratory analysis. In: 9th Biennial Conference on Engineering Systems Design and Analysis, Volume 2: Automotive Systems; Bioengineering and Biomedical Technology; Computational Mechanics; Controls; Dynamical Systems, Haifa, Israel. ASME, July 2008
8. Schmidt, M., Lipson, H.: Distilling free-form natural laws from experimental data. Science **324**(5923), 81–85 (2009). https://doi.org/10.1126/science.1165893
9. Schmidt, M., Lipson, H.: Supporting online material for distilling free-form natrual laws from experimental data, April 2009. https://science.sciencemag.org/content/ suppl/2009/04/02/324.5923.81.DC1
10. Topchy, A., Punch, W.F.: Faster genetic programming based on local gradient search of numeric leaf values. In: Proceedings of the 3rd Annual Conference on Genetic and Evolutionary Computation, pp. 155–162. Morgan Kaufmann Publishers Inc. (2001)
11. Worm, T., Chiu, K.: Scaling up prioritized grammar enumeration for scientific discovery in the cloud. In: IEEE International Conference on Big Data, pp. 621–626. IEEE (2014)

Data Aggregation for Reducing Training Data in Symbolic Regression

Lukas Kammerer$^{(\boxtimes)}$![ORCID], Gabriel Kronberger ![ORCID], and Michael Kommenda

Josef Ressel Center for Symbolic Regression,
Heuristic and Evolutionary Algorithms Laboratory,
University of Applied Sciences Upper Austria, Hagenberg, Austria
`lukas.kammerer@fh-hagenberg.at`

Abstract. The growing volume of data makes the use of computationally intense machine learning techniques such as symbolic regression with genetic programming more and more impractical. This work discusses methods to reduce the training data and thereby also the runtime of genetic programming. The data is aggregated in a preprocessing step before running the actual machine learning algorithm. K-means clustering and data binning is used for data aggregation and compared with random sampling as the simplest data reduction method. We analyze the achieved speed-up in training and the effects on the trained models' test accuracy for every method on four real-world data sets. The performance of genetic programming is compared with random forests and linear regression. It is shown, that k-means and random sampling lead to very small loss in test accuracy when the data is reduced down to only 30% of the original data, while the speed-up is proportional to the size of the data set. Binning on the contrary, leads to models with very high test error.

Keywords: Symbolic regression · Machine learning · Sampling

1 Introduction

One of the first tasks in data-based modeling of systems is collection and selection of data with which a meaningful model can be learned. One challenge is to provide the right amount of data – w.r.t. both instances and features. There should be enough data to compensate noise and train a sufficiently complex model, but not too much to unnecessarily slow down the training. With the growing volume of data, especially the latter becomes more and more an issue when working with computationally intensive algorithms like genetic programming (GP).

An intuitive idea to keep the training data small is outlined in Fig. 1. Instead of using all training data, only a few representative instances are extracted first and then used for the training. Ideally, these few instances retain all information that are necessary to train a well-generalizing model. This idea has been already applied for support vector machines for classification [9] and regression tasks

R. Moreno-Díaz et al. (Eds.): EUROCAST 2019, LNCS 12013, pp. 378–386, 2020.
https://doi.org/10.1007/978-3-030-45093-9_46

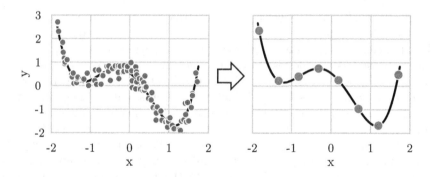

Fig. 1. Schematic outline of extracting a data set of representative instances (right) out of original noisy data (left).

[3]. In this previous work, the authors heuristically selected those instances, which are likely to determine the support vectors and therefore the SVM model. Another proposed approach for speeding up GP's evaluation is to use only a small random sample in every evaluation [5]. Kugler et al. [6] suggested to aggregate similar instances together before training a neural network. Grouping together instances should cancel out noise and shrink the data set, which is similar to the initially mentioned idea.

This paper builds on the idea of Kugler et al. [6] and uses clustering algorithms for aggregating and reducing training instances. The applied methods are random sampling, data binning and k-means clustering. The new, aggregated data are then used for training with different machine learning algorithms, with a focus on symbolic regression with GP. The trade-off between speed-up and loss in prediction accuracy due to potential removal of information is analyzed. We test how much we can reduce data so that we still can train accurate and complex models.

2   Aggregation Methods

All three aggregation methods require a predefined number of instances, which should be generated out of the original data. The first one, random sampling without replacement, serves as lower baseline. It will show, how much data can be removed without losing relevant information for modeling.

In data binning, the training data are aggregated based on the target variable. Similar to a histogram, instances with target values in fixed ranges are grouped together into bins. The range is determined by the minimum/maximum of the original target values and the predefined number of bins. To reduce the instances in a bin to one single instance, we use the median for all of the features. Binning should provide an equal target value distribution and reduce noise and variance [6]. However, binning also implies a "many-to-one mapping" between features and target variable, which incurs a loss of information in cases where interactions of features are relevant.

K-means clustering searches k cluster centroids that minimize the sum of the Euclidean distance between each point and its closest centroid. In this work, the calculated centroids are used as new training data, while in usual applications, k-means is used as unsupervised learning algorithm to separate the data into k groups. We deliberately consider both features and target values when aggregating the data because it is assumed that information about variable interactions is retained in contrast to data binning. The most common algorithm for determining the centroids is the heuristic Lloyd's algorithm [7], is infeasible for larger data sets with high k, since its runtime complexity is linear to the number of instances and k – especially when the goal is actually to speed up the overall modeling process. For our experiments, we use the mini-batch k-means algorithm [10], which reduces the runtime of the preprocessing from days to a few minutes with comparable results.

3   Experimental Setup

This work focuses on the effect of the described aggregation methods on symbolic regression with GP. We use the algorithm framework described by Winkler [13], in which the prediction error of mathematical formulas in syntax tree representation as individuals is minimized. The algorithm implementation applies strict offspring selection with gender-specific selection [1], a separate numerical optimization of constants in the formula [5] and explicit linear scaling [4]. Since we focus on the effect of preprocessing on symbolic regression, we use a standard parameter setting which has shown in our experience to provide good results. GP's maximum selection pressure was set to 100, the mutation rate to 20% and the population size to 300. The trees are build with a grammar of arithmetic and trigonometric symbols, as well as exponential and logarithm functions. The tree size is limited at most 50 symbols and a maximum depth of 30. The crossover operator is subtree swapping and mutation operators are point mutation, tree shaking, changing single symbols and replacing/removing branches [1]. We use for all experiments the *HeuristicLab* framework[1] [11].

Random forests (RF) [2] and linear regression (LR) are run on the same data sets in order to provide comparability of the achieved results. Linear regression serves as a lower bound, as the resulting linear models have low complexity and all relations in the data sets are nonlinear. Random forests are used as a rough indicator, which accuracy values are achievable on different data sets, as this algorithm has shown to be both fast and often very accurate. The settings for RF were set train 50 trees and sample from 30% of instances and 50% of features for each tree.

This work uses four real-world data sets: The *Chemical-I* data set (711 training instances, 57 features) [12], the *Tower* data set [12] (3136 training instances, 25 features), the *SARCOS* data set [8] (44 500 training instances, 21 features) and the *puma8NH* data set (6144 training instances, 8 features) from the Delve

---

[1] https://dev.heuristiclab.com.

repository[2]. The data sets are split into training and test data set according to the cited papers. The data sets are normalized to a mean of zero and a standard deviation of 1 in the training data, which is especially important for the k-means algorithm, which uses the Euclidean distance.

To analyze the tradeoff between accuracy and speed-up, the training data are reduced stepwise, starting from a reduction to one percent up to 50% of the original training data size. For each of these steps, ten reduced training data sets are generated. The three machine learning algorithms are then run on each of these reduced data sets. Figure 2 outline the estimation of the generalization error, where we evaluate the trained models on the not-aggregated test data. The generalization error is measured with Pearson's R^2 coefficient, which describes the correlation between actual target values predictions in the interval $[0, 1]$.

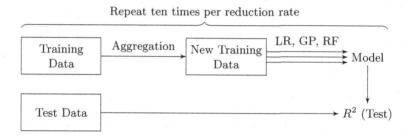

Fig. 2. Experiment workflow of reducing data for each data set and reduction rate.

4   Results

Figure 3 shows, that the speed-up of k-means clustering and sampling is proportional to the reduction rate – how much the data has been reduced relative to the original size. This is expected, as GP spends most of its time evaluating individuals. The computational effort of evaluations increases proportionally with larger numbers of training instances, because every instance's prediction has to be calculated in every evaluation. However, the proportional speed-up also indicates, that algorithm dynamics such as the earlier convergence due to preprocessing can be ruled out. The runtime for the preprocessing step itself is neglected, because it made up at most three minutes (for the large Sarcos data set) per run, which is only a very small fraction of the runtime of GP. Figure 3 also excludes the runtime for RF and LR because both methods took only seconds to finish in all experiments.

---

[2] https://www.dcc.fc.up.pt/ltorgo/Regression/puma.html.

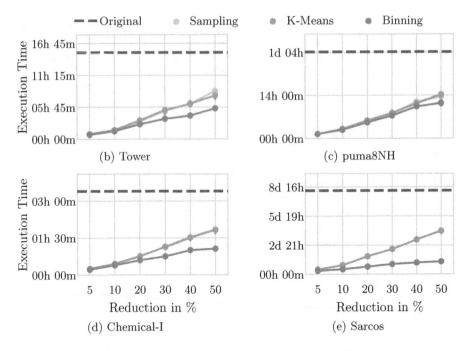

Fig. 3. Median execution of GP for different rates of reduction of the original data set.

While data binning and k-means led to similar execution times, GP runs with preceding data binning were slightly faster. This is most likely due to a loss of information about the relations in the data. Fewer relations in the training data make the search for a model, which fits the training data well, easier and therefore faster although it degrades its test errors. Table 1 shows that data binning yielded throughout worse models regarding test accuracy.

K-means clustering and random sampling produced very similar results, as listed in Table 1. In both cases, the loss in prediction accuracy is small when the data is reduced to 30%, 40% or 50% of the original data in comparison to runs with the original data. The only difference is the more stable behavior of k-means preprocessing when the training data is reduced to a size of 20% or less of the original data set size. The small difference between preprocessing with k-means and data binning can be explained with the small number of instances, with which each centroid is computed – e.g. if the training data is reduced to 50%, each centroid is the center of only two similar instances on average. This leaves little space for improvements over random sampling. However, all methods failed to yield meaningful models when the training data are reduced to only 1% of their original size.

Table 1. Test R^2 median and interquartile range for symbolic regression with GP.

		Chemical-I	Sarcos	Tower	puma8NH
Original sampling		.863 (.02)	.949 (.00)	.937 (.01)	.684 (.00)
	1	.002 (.00)	.928 (.01)	.390 (.31)	.369 (.06)
	5	.431 (.36)	.946 (.00)	.907 (.02)	.572 (.04)
	10	.396 (.36)	.947 (.00)	.919 (.02)	.632 (.02)
	20	.728 (.21)	.948 (.00)	.925 (.01)	.657 (.00)
	30	.787 (.08)	.948 (.00)	.933 (.00)	.668 (.01)
	40	.836 (.01)	.947 (.00)	.934 (.01)	.675 (.01)
	50	.834 (.02)	.949 (.00)	.936 (.00)	.677 (.01)
K-means	1	.018 (.04)	.943 (.01)	.729 (.14)	.480 (.11)
	5	.523 (.15)	.948 (.00)	.918 (.01)	.596 (.07)
	10	.704 (.13)	.948 (.00)	.926 (.00)	.596 (.09)
	20	.806 (.06)	.948 (.00)	.933 (.01)	.656 (.05)
	30	.828 (.03)	.948 (.00)	.932 (.00)	.665 (.01)
	40	.853 (.04)	.948 (.00)	.932 (.00)	.672 (.01)
	50	.848 (.02)	.947 (.00)	.931 (.00)	.680 (.00)
Binning	1	.009 (.01)	.504 (.45)	.378 (.37)	.021 (.03)
	5	.253 (.30)	.866 (.01)	.313 (.35)	.189 (.31)
	10	.345 (.21)	.893 (.01)	.700 (.27)	.532 (.05)
	20	.578 (.13)	.892 (.01)	.806 (.05)	.530 (.07)
	30	.673 (.05)	.899 (.01)	.853 (.03)	.576 (.05)
	40	.699 (.05)	.898 (.01)	.872 (.05)	.630 (.03)
	50	.718 (.08)	.897 (.01)	.883 (.02)	.648 (.01)

When compared to RF and LR, GP achieves in most experiments similar accuracy as random forests. However, when the data is strongly reduced to only 30% or less of the original data, GP tends to have less loss in test accuracy than RF. Figure 4 describes the median R^2 results of all three algorithms with differently reduced data for each data set. Modeling results on the original data set are shown on the left. The results for data, that was reduced to 1%, as well as the binning results are not shown in Fig. 4 since they would degrade the axis scale.

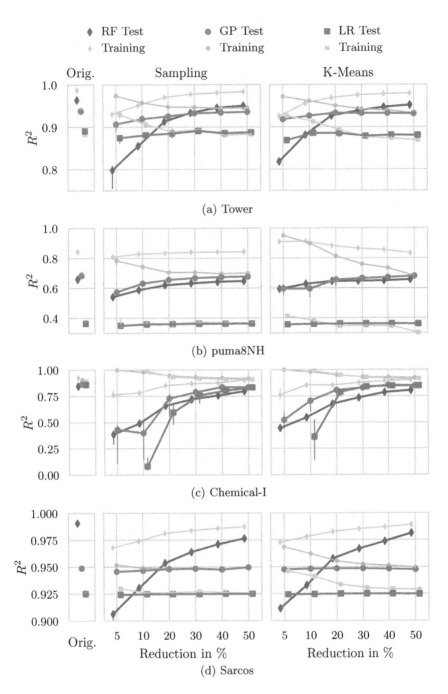

Fig. 4. Median R^2 values of the generated models.

5 Conclusion

The experimental results show, that the original training data can be reduced to only 30% of the original size while still achieving only slightly worse test accuracy. While data binning led to unusable models regarding test accuracy, there is no noticeable difference in the resulting modeling accuracy between k-means and random sampling. Depending on the actual application, whether a higher loss in accuracy is acceptable, k-means might only be useful if the data is reduced to 5–20%, as less variance among test errors compared to random sampling was observed in this range. Otherwise, it is more convenient to use random sampling instead of the additional effort of implementing k-means clustering in the modeling process. The impact of data reduction was smaller for GP than for RF, which underlines the stability and the generalization capabilities of symbolic regression.

The speed-up for GP is proportional to the data reduction ratio, how much the original data was reduced to. While the proposed reduction methods might not be suitable for a final model in most applications due to the (even slight) loss in accuracy, random sampling as preprocessing step might be a suitable tool for speeding up early experimental phases and meta parameter tuning.

Acknowledgements. The authors gratefully acknowledge support by the Austrian Research Promotion Agency (FFG) within project #867202, as well as the Christian Doppler Research Association and the Federal Ministry of Digital and Economic Affairs within the *Josef Ressel Centre for Symbolic Regression*.

References

1. Affenzeller, M., Winkler, S., Wagner, S., Beham, A.: Genetic Algorithms and Genetic Programming - Modern Concepts and Practical Applications, Numerical Insights, vol. 6. CRC Press, Chapman & Hall, Boca Raton (2009)
2. Breiman, L.: Random forests. Mach. Learn. **45**(1), 5–32 (2001)
3. Guo, G., Zhang, J.S.: Reducing examples to accelerate support vector regression. Pattern Recogn. Lett. **28**(16), 2173–2183 (2007)
4. Keijzer, M.: Scaled symbolic regression. Genet. Program Evolvable Mach. **5**(3), 259–269 (2004). https://doi.org/10.1023/B:GENP.0000030195.77571.f9
5. Kommenda, M., Kronberger, G., Affenzeller, M., Winkler, S., Feilmayr, C., Wagner, S.: Symbolic regression with sampling. In: 22nd European Modeling and Simulation Symposium EMSS, pp. 13–18 (2010)
6. Kugler, C., Hochrein, T., Dietl, K., Heidemeyer, P., Bastian, M.: Softsensoren in der Kunststoffverarbeitung: Qualitätssicherung für die Compoundierung und Extrusion. Shaker Verlag GmbH, SKZ - Forschung und Entwicklung (2015)
7. Lloyd, S.: Least squares quantization in PCM. IEEE Trans. Inf. Theory **28**(2), 129–137 (1982)
8. Pagie, L., Hogeweg, P.: Evolutionary consequences of coevolving targets. Evolutionary Comput. **5**(4), 401–418 (1997)
9. Rychetsky, M., Ortmann, S., Ullmann, M., Glesner, M.: Accelerated training of support vector machines. In: Proceedings of the International Joint Conference on Neural Networks, IJCNN 1999, (Cat. No. 99CH36339). vol. 2, pp. 998–1003. IEEE (1999)

10. Sculley, D.: Web-scale k-means clustering. In: Proceedings of the 19th International Conference on World Wide Web, pp. 1177–1178. ACM (2010)
11. Wagner, S., Affenzeller, M.: HeuristicLab: a generic and extensible optimization environment. In: Ribeiro, B., Albrecht, R.F., Dobnikar, A., Pearson, D.W., Steele, N.C. (eds.) Adaptive and Natural Computing Algorithms, pp. 538–541. Springer, Vienna (2005). https://doi.org/10.1007/3-211-27389-1_130
12. White, D.R., et al.: Better GP benchmarks: community survey results and proposals. Genet. Program Evolvable Mach. **14**(1), 3–29 (2013). https://doi.org/10.1007/s10710-012-9177-2
13. Winkler, S.: Evolutionary system identification: modern concepts and practical applications. Schriften der Johannes Kepler Universität Linz, Universitätsverlag Rudolf Trauner (2009)

Genetic Programming Based Evolvement of Models of Models

Mariia Semenkina[1,2(✉)], Bogdan Burlacu[1,2], and Michael Affenzeller[1,2]

[1] Heuristic and Evolutionary Algorithms Laboratory (HEAL), University of Applied Sciences Upper Austria, Softwarepark 11, 4232 Hagenberg, Austria
Mariia.Semenkina@fh-hagenberg.at
[2] Johannes Kepler University, Altenberger Straße 69, 4040 Linz, Austria

Abstract. The main idea of this paper is to use Simple Symbolic Formulas generated offline with the help of the deterministic function extraction algorithm as building blocks for Genetic Programming. This idea comparison to Automatically Defined Functions approach was considered. A possibility to take into consideration an expert's knowledge about the problem in hand has been reviewed. In this work a map of building block's set is generated by means of clustering. All distances between blocks are calculated offline by using a special metric for symbolic expressions. A mutation operator in Genetic Programming was modified for work with this kind of nodes. The effectiveness of this approach was evaluated on benchmark as well as on real world problems.

Keywords: Genetic Programming · Symbolic Regression · Models of Models

1 Introduction

Symbolic regression (SR) is one of the tasks that can be solved with Genetic Programming (GP) [1]. Genetic Programming can generate solutions of arbitrary form using components from functional ant terminal sets. Genetic Programming for symbolic regression has always been one of the topics discussed at workshops. However, GP also has some limitations for using it for symbolic regression in practice. Genetic Programming (GP) typically performs a long search for relatively simple bundles of variables.

The idea to use some "good parts of tree" as parts of new generated trees during the GP run for GP boosting was first time considered in [1]. One of the most common variants is to use so called Automatically Defined Functions (ADF) [2]. Most researches in this direction focuses on using some kind of coevolution of GP trees and ADFs or independently train ADFs and then use them inside of GP trees [3]. In both variants ADF inherits disadvantages and problems of GP (such as bloat, one phenotype for different genotypes, and small changes in tree that lead to big changes in formula and sufficiently influent on the kind of the model behavior) and introduce additional complexity (in particular, when performing evaluation of trees with ADFs). Fast Function Extraction (FFX) algorithms [4, 5] do not suffer from these disadvantages, because they are deterministic and provide simpler models with more generalization ability.

© Springer Nature Switzerland AG 2020
R. Moreno-Díaz et al. (Eds.): EUROCAST 2019, LNCS 12013, pp. 387–395, 2020.
https://doi.org/10.1007/978-3-030-45093-9_47

A main idea of the concrete contribution is to use Simple Symbolic Formulas generated offline with help of Fast Function Extraction Algorithm like in [5] as a building blocks for GP.

The structure of this article contains a description of the approach and analysis of the experiments results to evaluate the method effectiveness. In Sect. 2, we explain how map of models can be created and used during the GP run. Section 3 shows general parameter settings, datasets that will be used along the experiments and results of investigation. In Sect. 3, the approach effectiveness comparison with traditional GP and GP using buildings blokes without map. Lastly, our findings and draw conclusions are discussed in Sect. 4.

2   Genetic Programming Based Evolvement of Models of Models

Genetic Programming based Evolvement of Models of Models (EMM-GP) is the approach that is discussed in this paper. As the first step of each problem solving a set of models in form of symbolic expression with help of algorithm like in [5] should be created and then they will be used as a leafs in GP trees. Certain difficulties arise on the step of mutation, because the most obvious way of models replacement (replace on a random one from the set) can provide very big changes in phenotype that contradicts to idea of mutation. In this work, we decided to build a map of building block's set by means of clustering [6] and all distances between blocks were calculated by using metric based on [7]. In this case, on the stage of mutation operator works every model can be randomly replaced only on model from the same cluster that should be close enough.

The main scheme of algorithm is presented below and the most interesting steps description are considered in Sects. 2.1–2.3.

EMM-GP Algorithm scheme
Initialize model set (offline)
Create map of models (offline)
Genetic programming algorithm run:
Initialization (using additional terminal nodes from model set)
Fitness function calculation
Selection
Crossover
Mutation (with special operations for nodes from models set)
Survivor selection

This decision will not affect the running time of a GP because all preparation works (sub-models extraction, distances calculation, clustering and map creation) can be done offline (before GP should be started) and only one time for each problem (algorithm for simple models creation [5] is deterministic). In addition, for this approach, it is possible to take into consideration an expert's knowledge about a problem in hand for reducing building block's set and search space, or for inclusion of expert views on the type of model (for example, some previously existed models or hypotheses about behavior of object in hand).

2.1 Model Sets Description

As it was written above, all sub-models in this paper were created with help of algorithm from [5]. The main problem with using this deterministic models extraction algorithm for our goal is the multiple possible choice of answer on the question: How complicated should be sub-models that will be used as building blocks?

In this paper we consider several possible variants: Complete System and Simple System. The Complete System is understood as a set of models that was created with using all available elements of a functional set as components (*sin, cos, exp* and so on). This kind of models set provides for GP more features and greater flexibility on the one hand, but on the other hand, this significantly increases the amount of search space (in case of two dimension more than 1200 possible sub-models) and may lead to additional problems. The Simple System is understood as a set of models that was created with using only basic elements of a functional set as components (+, −, *, /) and contains only of 120 models. It is obviously that the Complete System contains the Simple System. The examples from both models sets are presented in Table 1.

Table 1. Examples of models from examples from complete system and simple system models sets.

Complete system	1. $(0 + (1/(0 + ((SIN((0 + ('X' * 1))) * SIN((0 + ('Y' * 1)))) * 1)) * 1))$
	2. $(0 + (1/(0 + ((LOG((0 + ('X' * 1))) * SIN((0 + ('Y' * 1)))) * 1)) * 1))$
	3. $(0 + (1/(0 + ((LOG((0 + ('X' * 1))) * EXP(('Y' * 1))) * 1)) * 1))$
	4. $(0 + ((1/(0 + (LOG((0 + ('Y' * 1))) * 1)) * EXP((('X' * 'X') * 1))) * 1))$
	5. $(0 + (1/(0 + (((EXP(('X' * 1)) * EXP(('Y' * 1))) * SIN((0 + ('Y' * 1)))) * 1)) * 1))$
	and so on.
Simple system	1. $((0 + ((1/(0 + ('X' * 1)) * 'X') * 1)) + ('X' * 1))$
	2. $((0 + ('X' * 1)) + ('Y' * 1))$
	3. $(0 + (1/((0 + (('X' * 'Y') * 1)) + ('X' * 1)) * 1))$
	4. $((0 + (('Y' * 'Y') * 1)) + ('Y' * 1))$
	and so on

Second important parameter is a way of constants and coefficients using in sub-models. In this paper we consider 3 possible variants: all coefficients is equal 1 (Simple-1, Complete-1), models without free constants (with 0 on it places) (Simple-2, Complete-2) and models with randomly generated coefficients (Simple, Complete).

From the view point of Genetic Programming models in such model sets are considered like third possible type of terminal node. During initialization Model Terminal Node can be generated randomly with probability equal 1/3. The exact value (exact model) for this node is randomly selected from the entire set of models. On the step of mutation we face a new problem: How should algorithm choose a new model?

2.2 Mutation Operators

In this paper we consider 3 variants of mutation operator: Change Node Type Manipulation, Multipoint mutation and Multipoint mutation that saves terminal node type. Change Node Type Manipulation randomly generates a new node from suitable type for the parent node. Multipoint mutation randomly generates a new node from the same type like previous node value (terminal node, binary function, unary function and so on). Second variant of Multipoint mutation saves not only common information about node type like terminal or functional, but also in addition it saves type of terminal node (variable, constant or model).

All 3 mutation variants have common problem with new model node generation and the same possible way for solving it:

1. Absolutely randomly select a one new model from model set (ZeroMap);
2. Randomly select a one new model from preliminary chosen subset (IslandMap).

Efficiency for GP with both variants of models map were evaluated. The results are considered in Sect. 3.

2.3 Island Map

A concept of Island Map is based on the idea that mutation should provide small changes on genotype and phenotype level. Obviously, that random selection of a new model for Model Terminal Node can dramatically change the whole structure of the tree. According Island Map model's set is dividing on isolated islands (by clustering) according distance matrix. During mutation or local search each sub-model can be exchange only for model from the same island. Island Map is based on clustering K-means algorithm. Let's take a closer look.

In this investigation we used K-means clustering algorithm for discrete structures, that means that initial centroids can not be created absolutely random, but they should be randomly selected from existed model's set. On the second step we should calculate distance matrix according distance measure between two trees based on the intersection of their corresponding hash value sequences. Each tree is converted into a sequence of hash values corresponding to a post-order traversal of its structure. The distance between two trees can be denned using the intersection between the two sequences:

$$D(T_1, T_2) = \frac{2 \cdot |H_1 \cap H_2|}{|H_1| + |H_2|},$$

where H_1, H_2 represent the hash value sequences of T_1 and T_2, respectively. On the third step the closest one centroid should be found for each tree, so first version of clusters will be created. On the next step, new centroids should be found in the way that total distance of all trees in cluster to the centroid became minimal. Then we should repeat last two steps until during the next repetition the clusters will remain unchanged. A K-means algorithm result example is presented on Fig. 1.

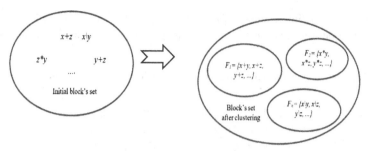

Fig. 1. An example of models set before and after clustering.

3 Empirical Results

Investigation of Genetic Programming based Evolvement of Models of Models effectiveness was performed on a several test problem and one real world problem (Prognoses of turbine vibration characteristics). All characteristics for problems that are considered bellow in details are presented in Table 2.

Table 2. Problems description.

Problem name	Test function view	Dimension
Spatial co-evolution	$F(x, y) = 1/(1 + x^{(-4)}) + 1/(1 + y^{(-4)})$	2
Keijzer-10	$F(x, y) = x^y$	2
Keijzer-11	$F(x, y) = xy + \sin((x - 1)(y - 1))$	2
Keijzer-12	$F(x, y) = x^4 - x^3 + y^2 / 2 - y$	2
Prognoses of turbine vibration characteristics	1000 elements in data set	11 inputs

In this paper we consider influence of model set type, map type, mutation type on efficiency of Genetic Programming based Evolvement of Models of Models, all other parameters of GP was fixed. We use Standard GP and Standard GP with coefficient optimization like base line for our comparison of effectiveness. During all test each algorithm made 10000 fitness function calculations. In tables bellow all results are presented for test part of data set in term of Pearson R^2 correctness measure. All results are presented for 100 algorithm runs.

The example of impact, that clusters number have on efficiency, is presented in Table 3. It was proved by statistical tests that any reasonable for model set size number of clusters can be used without any influence on algorithm performance. The example of mutation operator comparison is presented in Table 4. Easy to see that Median values for all three variants are almost equal. So, Multipoint mutation and two values of cluster number (k = 10 for Simple System and k = 50 for Complete System) were used in all other tests below.

Efficiency evaluation results for four test problems that are considered in details and all possible model set's and map's variation are presented in Tables 5, 6, 7 and 8.

Table 3. Efficiency comparison for different number of clusters for spatial co-evolution problem.

k	Min	Max	Median	Average	SD
5	0,0632	0,9948	0,8142	0,7987	0,1191
10	0,0357	0,9963	0,8151	0,7981	0,1259
15	0,0981	0,9946	0,8155	0,8017	0,1160
20	0,1099	0,9955	0,8154	0,8051	0,1187

From the results anybody can see that the type of the models set is more important than the Map type. In majority cases Genetic Programming based Evolvement of Models of Models that did not use coefficient optimization outperform Standard GP and can compete with Standard GP with coefficient optimization. Advantage of Genetic Programming based Evolvement of Models of Models is the more noticeable for tasks that are more difficult. As it turned out, the biggest problem for the Island Map is the randomness of the choice of initial centroids. Usual behavior of this variant of algorithm is to decrease a variability of the using in trees models, as a rule at the end population contain only models from one cluster (see Fig. 2). Generally in case using Island Map, that was created before successful GP run, for next GP run increase essentially probability of it's success.

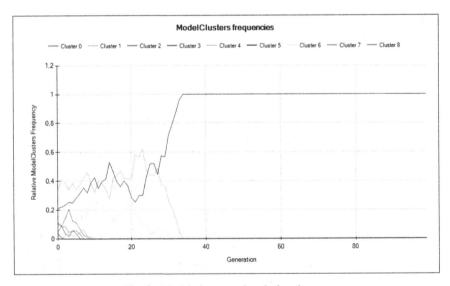

Fig. 2. Model clusters using during the run.

Table 4. Efficiency comparison for different mutation operator types for spatial co-evolution problem.

Mutation	Min	Max	Median	Average	SD
Change node type manipulation	0,0594	0,9967	0,8151	0,8167	0,1356
Multipoint mutation	0,0756	0,9946	0,8158	0,7984	0,1027
Multipoint mutation + Node type saving	0,0357	0,9854	0,8146	0,7875	0,1174

Table 5. Efficiency comparison for different model subsets for spatial co-evolution problem.

Model set	Map	Min	Max	Median	Average	SD
Simple-1	Island	0,0163	0,9746	0,8382	*0,7945*	0,1464
	Zero	0,0081	0,9786	0,8403	0,8178	0,1193
Simple-2	Island	0,0067	0,9696	0,8146	*0,7926*	0,1482
	Zero	0,0088	0,9522	0,8146	*0,8129*	0,0985
Simple	Island	0,4171	0,9516	0,8290	0,7868	0,1192
	Zero	0,0015	0,9509	0,8097	0,7803	0,1377
Complete	Island	0,2272	0,9546	0,7450	0,7057	0,1674
	Zero	0,2857	0,9546	0,7166	0,6856	0,1648
Standard GP		0,2320	0,8218	0,5981	0,5965	0,1164
Standard GP + Opt.		0,7778	0,9876	0,8500	***0,8446***	0,0299

Table 6. Efficiency comparison for different model subsets for Keijzer-10 Problem.

Model set	Map	Min	Max	Median	Average	SD
Simple-1	Island	0,2010	0,9998	0,9638	0,9016	0,2012
	Zero	0,0301	0,9972	0,9636	0,8927	0,2170
Simple-2	Island	0,0227	0,9773	0,9490	0,9134	0,1652
	Zero	0,1184	0,9996	0,9488	0,9132	0,1648
Simple	Island	0,1345	0,9862	0,9527	0,9119	0,1734
	Zero	0,1386	0,9939	0,9507	0,9322	0,1158
Complete	Island	0,2330	0,9994	0,9471	0,9176	0,1400
	Zero	0,2372	0,9986	0,9464	0,9386	0,0731
Standard GP + Opt.		0,9117	0,9990	0,9584	0,9594	0,0140

Table 7. Efficiency comparison for different model subsets for Keijzer-11 Problem.

Model set	Map	Min	Max	Median	Average	SD
Simple-1	Island	0,0310	0,9565	0,9313	0,7773	0,3082
	Zero	0,0245	0,9565	0,9287	0,7151	0,3472
Simple-2	Island	0,0426	0,9583	0,9544	0,8492	0,2710
	Zero	0,0533	0,9583	0,9538	*0,8957*	0,1998
Simple	Island	0,0159	0,9603	0,9374	0,8048	0,3152
	Zero	0,0185	0,9588	0,9366	*0,8708*	0,2187
Complete	Island	0,0048	0,9580	0,9526	*0,8665*	0,2503
	Zero	0,0326	0,9584	0,9529	0,8325	0,2986
Standard GP		0,0251	0,9614	0,9427	0,7706	0,3543

Generalized for all test problems results are presented in Table 9. In first two rows results are presented in terms of reliability (proportion of successful runs to total run's number). In the last row results for real world problem are presented like percent of correct prognoses. Minimal and maximal values for all runs for all problems are presented in brackets, and below average value is presented.

It is predictable that for more complex test problems a Complete System of models was more efficient.

Table 8. Efficiency comparison for different model subsets for Keijzer-12 Problem.

Model set	Map	Min	Max	Median	Average	SD
Simple-1	Island	0,0028	0,9992	0,0465	0,2953	0,4104
	Zero	0,0007	0,9971	0,05290	0,3131	0,4121
Simple-2	Island	0,0015	0,9985	0,9646	0,8291	0,2676
	Zero	0,0228	0,9998	0,9618	0,8430	0,2494
Simple	Island	0,0151	0,9927	0,9697	*0,8901*	0,2009
	Zero	0,0036	0,9914	0,9638	*0,9111*	0,1747
Complete	Island	0,1291	0,9946	0,9618	*0,9153*	0,1408
	Zero	0,0030	0,9948	0,9665	*0,9174*	0,1479
Standard GP		0,0447	0,9923	0,9866	*0,8997*	0,2602

Table 9. The common efficiency comparison.

Problem	Standard GP	Complete system	Simple system
Simple benchmark problems	[0.324, 0.653] 0.576	[0.213, 0.624] 0.498	**[0.345, 0.701] 0.603**
More complex benchmark problems	[0.276, 0.543] 0.368	**[0.301, 0.674] 0.412**	[0.287, 0.576] 0.398
Real world problem	[64,52; 77,42] 70,97	[53,85; 69,23] 61,54	**[61,54; 92,31] 76,92**

4 Conclusion

Genetic Programming based evolvement models of models work quicker than conventional GP. In majority cases Genetic Programming based Evolvement of Models of Models that did not use coefficient optimization outperform Standard GP and can compete with Standard GP with coefficient optimization. Advantage of Genetic Programming based Evolvement of Models of Models is the more noticeable for tasks that are more difficult. Unfortunately, it adds new parameters for algorithm, so the best type of model's set is not obvious for a problem in hand. Effective Island Map creation requires additional algorithm run for map evaluation. Probably, other variants of using distance information in map can be useful.

References

1. Koza J.: Simultaneous discovery of reusable detectors and subroutines using genetic programming. In: Proceedings of the 5th International Conference on Genetic Algorithms, ICGA 1993, pp. 295–302. Morgan Kaufmann, University of Illinois at Urbana-Champaign (1993)
2. Koza, J.: Genetic Programming II: Automatic Discovery of Reusable Programs. MIT Press, Cambridge (1994)
3. Niimi A., Tazaki, E.: Genetic programming combined with association rule algorithm for decision tree construction. In: Proceedings of Fourth International Conference on Knowledge-based Intelligent Engineering Systems and Allied Technologies, Brighton, UK, pp. 746–749 (2000)
4. McConaghy, T.: FFX: fast, scalable, deterministic symbolic regression technology. In: Riolo, R., Vladislavleva, E., Moore, J. (eds.) Genetic Programming Theory and Practice IX, pp. 235–260. Springer, New York (2011). https://doi.org/10.1007/978-1-4614-1770-5_13
5. Kammerer, L., Kronberger, G., Burlacu, B., Winkler, S.M., Kommenda, M., Affenzeller, M.: Symbolic regression by exhaustive search – reducing the search space using syntactical constraints and efficient semantic structure deduplication. In: Banzhaf, W., Goodman, E., Sheneman, L., Trujillo, L., Worzel, B. (eds.) Genetic Programming Theory and Practice XVII. Springer, New York (2019). https://doi.org/10.1007/978-3-030-39958-0
6. Zojaji, Z., Ebadzadeh, M.M.: Semantic schema modeling for genetic programming using clustering of building blocks. Appl. Intell. **48**(6), 1442–1460 (2017). https://doi.org/10.1007/s10489-017-1052-7
7. Valiente, G.: An efficient bottom-up distance between trees. In: Proceedings of International Symposium on String Processing and Information Retrieval, Laguna de San Rafael, Chile (2001)

"Incremental" Evaluation for Genetic Crossover

Erik Pitzer[1(✉)] and Michael Affenzeller[1,2]

[1] Department Software Engineering, University of Applied Sciences Upper Austria,
Softwarepark 11, 4232 Hagenberg, Austria
{erik.pitzer,michael.affenzeller}@fh-hagenberg.at
[2] Institute for Formal Models and Verification, Johannes Kepler University,
Altenbergerstr 68, 4040 Linz, Austria

Abstract. Incremental evaluation is a big advantage for trajectory-based optimization algorithms. Previously, the application of similar ideas to crossover-based algorithms, such as genetic algorithms did not seem appealing as the expected benefit would be marginal. We propose the use of an immutable data structure that stores partial evaluation results inside of the solution representation, and composing new solution from parts of previously evaluated candidates, which can speed up re-evaluation. The application of this idea to the knapsack problem shows promising results hinting at logarithmic complexity in case all genetic operators can be adapted accordingly.

1   Introduction

Recombination of genetic information in the form of nature-inspired crossover operators has proven a worthwhile addition to the arsenal of metaheuristic optimization methods and is successful in a wide range of applications [8] including genetic algorithms [9]. On the other hand, trajectory-based methods such as Tabu Search [7], have also shown impressive results. A big advantage of these methods is the applicability of incremental evaluation. When a solution candidate changes only slightly, it is possible to evaluate the change's influence on fitness rather than re-evaluation the whole solution candidate. Especially for high-dimensional optimization tasks this can be a significant advantage. Unfortunately, it is not straightforward to combine incremental evaluation with crossover, as the changes made to the parent individuals from one generation to the next can be substantial in size and are often difficult to track, especially, when the recombination results need to be *repaired* [2].

Recent developments in immutable data structures that allow structure sharing of solution candidates have enabled complex operations to be made with few actual changes to the data representation, even during repeated crossover operations [13]. The same idea can be taken even further to incorporate *evaluation hints* that can be re-used to achieve fast re-evaluation even in the presence of large changes, as long as changes have been evaluated elsewhere and the evaluation hints have been preserved and are still applicable. In this paper, we present

© Springer Nature Switzerland AG 2020
R. Moreno-Díaz et al. (Eds.): EUROCAST 2019, LNCS 12013, pp. 396–404, 2020.
https://doi.org/10.1007/978-3-030-45093-9_48

a simple extension to the ideas described previously [13], that allows re-use of previous evaluation results to achieve evaluation speed-ups for crossover-based algorithms, similar to incremental evaluation in trajectory-based algorithms.

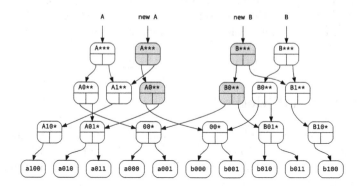

Fig. 1. Crossover of two RRB-Trees [13]

2   State of the Art

Persistent data structures are as old as the Information Processing Language, presented in 1957 [11], where a persistent linked list was used. The idea is simple: While every suffix of a linked list cannot be changed, it can be re-used as a suffix of any other list, *because* it cannot be changed. Therefore, prepending an element to a linked list does not invalidate the old list. Recent stagnation of processor speeds have spurred multi-core processing [15] which in turn promoted functional languages [6]. Classic data structures have been adapted to functional – i.e. immutable – variants [12]. In particular, arrays have received interesting pendants in the form of Array Mapped Tries (AMTs) [3] and Relaxed Radix Balanced Trees (RRB-Trees) [4]. These data structures allow single-point modifications and even slicing and concatenation of conceptual arrays in logarithmic time. Only a narrow *seam* of nodes from the root to the changed index needs to be created [13]. In Fig. 1, an example of a single-point crossover using persistent data structures is shown, where the conceptual arrays A and B are cut in half and recombined re-using many nodes [13].

Incremental evaluation has been incorporated into many metaheuristics [16], and has been a contributing factor to the success of trajectory-based algorithms. Equation 1 shows a simple fitness function that is just the sum of all alleles. When only a single element of x changes, it would be wasteful to calculate the whole sum again. Instead, it is much more efficient to subtract the old value x_i at the changed position i and add the new value x'_i to arrive at the new *fitness* in constant time (Eq. 2).

$$f(x) := \sum_{i=1}^{n} x_i \tag{1}$$

$$f(x') = f(x) - x_i + x'_i \text{ with change at index } i \tag{2}$$

When we try to apply this scheme to solution candidates produced by recombination, which typically consist of more or less equal contribution from all parents, we would get a calculation similar to Eq. 3, which contains of two steps with linear time complexity. This scheme is hardly advantageous compared to directly summing over the whole solution candidate z in the first place.

$$f(z) = f(x) - \sum_{i=j}^{n} x_i + \sum_{i=j}^{n} y_i \tag{3}$$

3 Materials and Methods

When we generalize the simple fitness function of Eq. 1 to the sum of alleles each of which is located at a position $i \in I$, we can formulate a partial fitness function over only some of these positions. The full fitness is then simply any sum over a partition $\mathcal{I}$ of positions (Eq. 4).

$$f_I(x) := \sum_{i \in I} x_i \qquad f(x) := \sum_{I \in \mathcal{I}} f_I(x) \tag{4}$$

While this generalization is initially more cumbersome to handle, it generates a set of new leverage points to facilitate incremental evaluation. We can generalize these partial evaluation results $f_I(x)$ even further, using increasing partition sizes, re-using smaller sub-partitions. This idea has been nicely captured in the *Map Reduce* calculation scheme [5]. First the initial position-based fitness is *mapped* into an intermediate representation that we will call *residual fitness*, which can be a single value or tuple of values to facilitate further aggregation. These residuals, will then be combined or *reduced* into more aggregated fitness residuals of the same form. Finally, once all positions' residuals have been aggregated into a single residual, a step called finalization will convert it into the actual fitness value, e.g. a real number as shown in Eq. 5: The mapping step is conducted with the help of a function μ or "map" that applies another function m individually to each allele and produces a corresponding set of residual fitness values. A second function ρ or "reduce" applies another function r repeatedly to combine two residuals. The mapper function $m : \chi_i \rightarrow F_r$ maps a single allele or a subsequence of alleles to a residual fitness type F_r which can vary between different problems. The reducer $r : F_r \times F_r \rightarrow F_r$ combines two (or more) residual fitness values. The finalizer $g : F_r \rightarrow \mathbb{R}$ takes the final, most aggregated residual and transforms it into a fitness value, e.g. a real number. Using this calculation scheme, the initial example using the simple sum of alleles of Eq. 1 can be broken down into the map-reduce scheme shown in Eq. 6.

$$\mu(m, x) := [m(x_1), m(x_2), \ldots, m(x_n)]$$
$$\rho(r, x) := f(f(f(\ldots f(f(x_1, x_2), x_3), \ldots, x_{n-2}), x_{n-1}), x_n) \tag{5}$$
$$f(x) := g(\rho(r, \mu(m, x)))$$

$$m(x_i) := x_i \qquad r(r_a, r_b) := r_a + r_b \qquad g(r_x) := r_x \tag{6}$$

We can apply the same idea to the Knapsack problem [10], where the fitness function is the sum of all items, with the constraint that the sum of all weights of the same items must stay below a certain capacity. In a relaxed formulation, a violation of this constraint simply reduces the fitness instead of invalidating it, which can be used to guide a metaheuristic towards valid solutions in the initial phase of the optimization process, as shown in Eq. 7, where $V(x)$, and $W(x)$ give the sum of values and weights of all selected items in x, and p gives a penalty for exceeding the capacity C.

$$f(x) := V(x) - \begin{cases} 0 & \text{if } W(x) < C \\ p(W(x) - C) & \text{otherwise} \end{cases} \qquad (7)$$

The knapsack's fitness function can be decomposed into mapper, reducer and finalizer functions as shown in Eq. 8. In this case the residual fitness is a tuple in $\mathbb{R}^2$ consisting of the partial sums of all values and weights. The mapper simply returns the value $v(i)$ and weight $w(i)$ of an individual item if it has been selected, while the reducer simply calculates their sums independently. Eventually, the finalizer function g discerns the two cases using the overall sums of weights and values.

$$\begin{aligned} m(x_i) &:= (x_i \cdot v(i), x_i \cdot w(i)) \\ r((v_a, w_a), (v_b, w_b)) &:= (v_a + v_b, w_a + w_b) \\ g((v_x, w_x)) &:= v_x - \begin{cases} 0 & \text{if } w_x < C \\ p(w_x - C) & \text{otherwise} \end{cases} \end{aligned} \qquad (8)$$

For the traveling salesman problem [14] (Eq. 9), where usually a permutation is used to describe the solution candidates, the formulation becomes more complex. Here, a distance function $d : \mathbb{N} \times \mathbb{N} \to \mathbb{R}$ can be used to return the distance between two locations. The fitness is simply the sum over the distances of the locations in the order specified by the permutation x.

$$f(x) := d(x_n, x_1) + \sum_{i=1}^{n-1} d(x_i, x_{i+1}) \qquad (9)$$

This problem requires the calculation of distances between consecutive elements of the solution candidates. Therefore, in addition to the sum of the consecutive distances between the locations of a partial solution, the first and last location need to be propagated as well, as they need to be joined when combining subsequent partial evaluation results. The residual fitness becomes a tuple containing the first location index, the last location index and the sum of consecutive distances between the contained indexes. The mapper m initializes the distance to zero and propagates the single element as its first and last location. The reducer r returns the first location of the first residual x_{f_b} and the last location of the second residual x_{l_b} and calculates the sum of the smaller residuals r_a and r_b plus the distance between the newly joint locations in between. While this yields a relatively elegant solution for incremental evaluation, in this case, the problem comes with genetic operators that need to maintain permutations as described below.

$$m(x_i) := (x_i, x_i, 0)$$
$$r((x_{f_a}, x_{1_a}, r_a), (x_{f_b}, x_{1_b}, r_b)) := (x_{f_a}, x_{1_b}, r_a + r_b + d(x_{1_a}, x_{f_b})) \tag{10}$$
$$g((x_f, x_1, r)) := r + d(x_f, x_1)$$

To enable incremental evaluation, the existing implementation of RRB-Trees can be extended to include additional meta-information (MI) in each internal node in the tree. This MI then consists of the residual fitness values as shown in Fig. 2 and is immutable once it has been added. New nodes start out with empty MI and the evaluation process successively adds MI from the leaves to the root. As the nodes below each node are immutable, the MI is still valid for fragments that have been incorporated into another tree, assuming it has been devised accordingly.

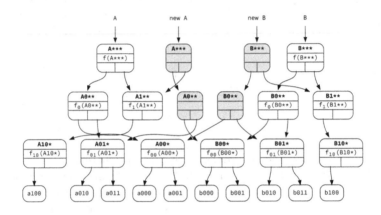

Fig. 2. Example of crossover retaining many residual fitness values

The additional code required to add recursive annotation is only a few lines, as shown in Listing 1. The function `visit_all_nodes` traverses the tree using a visitor function before descending and another one after re-ascending to the child nodes. This enables canceling the traversal once a node is found that already contains MI, possibly saving quite some computation time.

```
def annotate(rrb, mapper, reducer):
  visit_all_nodes(
    tree = rrb,
    pre_visitor = lambda(node, cancel): if mi(node): cancel()
    post_visitor = lambda(node): set_mi(node,
      leaf(node) ? mapper(node) : reducer(subjacent_mis(node))))
```

Listing 1: Pseudo-Code for Adding Meta-Info Annotations to RRB-Trees

To be able to exploit existing MI, all solution candidate manipulations must be able to preserve or re-use parts of the tree structure. For localized mutation operations this usually means replacing a narrow branch from the root to the changed leaf, re-using most of the partial evaluation results. For crossover, however, this becomes more difficult. Slicing operations, such as single-point or multi-point crossovers or even reversals can re-use long stretches of the predecessors' tree structures. These operations have logarithmic runtime with respect to the size of the solution candidates. Using RRB-Trees with a radix of 32 and 64-bit unsigned integer index, only seven levels have to be recursively modified at most, for any of these operations, making many practical examples quasi-constant (with a relatively large constant, however) [4]. For position-based encodings, such as for the knapsack problem, where each item's selection can be represented by a Boolean value, this is rather easy. Both, modification and evaluation can be evaluated in logarithmic, or quasi-constant time, as all operations will automatically create valid solution candidates. For the traveling salesman problem, however, while these operations are theoretically logarithmic, they could destroy the permutation by having duplicate or missing indexes and have to be repaired by checking which locations have been used more or less than once, which is the case for many successful crossover operations such as the edge recombination operator [17]. This requires linear checking of the constituents, and hence, increases the computation effort for crossover. So, while the evaluation could be performed logarithmically, the crossover would be more costly. For problems, where the evaluation is expensive, this could still pay off, as large parts could be re-used, however, due to the more complex data structure, part of the benefit would be lost.

4   Experimental Setup and Results

We have implemented a simple offspring-selection genetic algorithm (OSGA) [1] to solve random knapsack problems for different problem sizes. The algorithm settings have been kept constant for all experiments: The population size was set to 100, and the maximum number of evaluations was limited to one million. In these experiments, the resulting quality was not important, but we made sure the results are meaningful. To be able to directly compare computation speed the random seed has been fixed and all results have been checked for exact numerical equality with an array-based implementation. In the analysis we have instrumented the code to measure the different phases, like selection, mutation, crossover or evaluation, of the algorithm. Table 1 shows the breakdown into the most time-consuming phases for problems of different size. It can be seen, that for small problems the array-based implementation is much faster. For larger problems, the runtime is completely dominated by evaluation and crossover, but more interestingly, the tree-based implementation is much faster, even though it produces exactly the same results. The different time complexities are clearly visible in Fig. 3 where it is shown per data structure for different problem sizes. While the array implementation follows, as expected, a linear pattern, the tree-based method appears to be even sub-logarithmic.

Table 1. Runtimes in minutes of different phases of an OSGA on Knapsack problems of exponentially increasing size

Size [log10]	Array			RRB		
	Crossover	Evaluation	Mutation	Crossover	Evaluation	Mutation
2	0.005	0.004	0.004	0.069	0.009	0.005
3	0.104	0.111	0.024	0.629	0.074	0.024
4	4.200	4.523	0.097	3.134	0.392	0.103
5	40.768	44.957	0.136	3.928	0.489	0.095
6	408.428	455.226	0.516	4.378	0.566	0.102

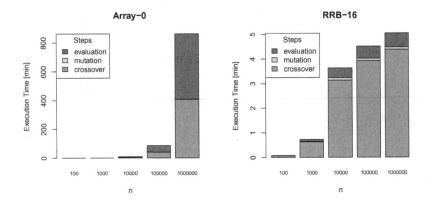

Fig. 3. Runtime development for different problem sizes

Finally, we can calculate the speedup in comparison to the array-based implementation for the individual phases of the algorithm an a knapsack problem with one million dimensions. For different radixes, crossover is between 50 and 90 times faster, mutation is about two to five times faster and evaluation achieves a speedup between 500 and 800. At this point it should also be noted that the radix 2 variant did not complete as it ran out of memory due to doubling the memory requirement for the trees in comparison to the array-based implementation.

5   Conclusion

Using immutable data structures with partial evaluation shows a large potential in speeding up recombination-based optimization algorithms. On the other hand, transformation into a map-reduce pattern alone is not straightforward, and especially the manipulation operators can be quite difficult to adapt. The results are rather directly applicable to knapsack problems which covers a wide range of applications. This method can be a powerful weapon when dealing with high-dimensional problems. It remains to be seen, however, how easy other problem domains can be adapted to make them susceptible to this approach. Future research will have to show to the applicability to real-world high-dimensional problems, for example of the knapsack problem. In addition, other optimization problems will have to be reformulated and suitable manipulation and recombination operators have to be devised that preserve partial evaluation results and do not need to examine solution candidates as a whole.

Acknowledgments. The work described in this paper was done within the project "Smart Factory Lab" which is funded by the European Fund for Regional Development (EFRE) and the country of Upper Austria as part of the program "Investing in Growth and Jobs 2014–2020".

Europäische Union Investitionen in Wachstum & Beschäftigung. Österreich.

References

1. Affenzeller, M., Winkler, S., Wagner, S., Beham, A.: Genetic Algorithms and Genetic Programming - Modern Concepts and Practical Applications. Numerical Insights. CRC Press, Boca Raton (2009)
2. Bäck, T., Fogel, D.B., Michalewicz, Z.: Handbook of Evolutionary Computation. CRC Press, Boca Raton (1997)
3. Bagwell, P.: Ideal hash trees. Technical report, École Polytechnique Fédéerale de Lausanne (2001)
4. Bagwell, P., Rompf, T.: RRB-trees: efficient immutable vectors. Technical report, École Polytechnique Fédéerale de Lausanne (2011)
5. Dean, J., Ghemawat, S.: MapReduce: simplified data processing on large clusters. Commun. ACM **51**(1), 107–113 (2008)
6. Eyler, P.: The rise of functional languages. Linux J. (2007). https://www.linuxjournal.com/content/rise-functional-languages
7. Glover, F., Laguna, M.: Tabu Search. Kluwer, Alphen aan den Rijn (1997)
8. Goldberg, D.E.: Simple genetic algorithms and the minimal deceptive problem. In: Davis, L. (ed.) Genetic Algorithms and Simulated Annealing, pp. 74–88. Pitman, London (1987)
9. Holland, J.H.: Adaptation in Natural and Artificial Systems. University of Michigan Press, Ann Arbor (1975)
10. Martello, S.: Knapsack Problems: Algorithms and Computer Implementations. Wiley-Interscience Series in Discrete Mathematics and Optimization. Wiley, New York (1990)
11. Newell, A., Shaw, J.: Programming the logic theory machine. In: Proceedings of the Western Joint Computer Conference, pp. 230–240 (1957)
12. Okasaki, C.: Purely Functional Data Structures. Cambridge University Press, Cambridge (1999)
13. Pitzer, E., Affenzeller, M.: Facilitating evolutionary algorithm analysis with persistent data structures. In: Moreno-Díaz, R., Pichler, F., Quesada-Arencibia, A. (eds.) EUROCAST 2017. LNCS, vol. 10671, pp. 416–423. Springer, Cham (2018). https://doi.org/10.1007/978-3-319-74718-7_50

14. Rosenkrantz, D.J., Stearns, R.E., Lewis II, P.M.: An analysis of several heuristics for the traveling salesman problem. SIAM J. Comput. **6**(3), 563–581 (1977)
15. Sutter, H.: The free lunch is over: a fundamental turn toward concurrency in software. Dr. Dobbs J. **30**(3), 16–20 (2005)
16. Talbi, E.G.: Metaheuristics: From Design to Implementation. Wiley, Hoboken (2009)
17. Whitley, D., Starkweather, T., Fuquay, D.: Scheduling problems and traveling salesman: the genetic edge recombination operator. In: International Conference on Genetic Algorithms, pp. 133–140 (1989)

A Model-Based Learning Approach for Controlling the Energy Flows of a Residential Household Using Genetic Programming to Perform Symbolic Regression

Kathrin Kefer[1(✉)], Roland Hanghofer[2], Patrick Kefer[3], Markus Stöger[1],
Michael Affenzeller[2], Stephan Winkler[2], Stefan Wagner[2], and Bernd Hofer[1]

[1] Fronius International GmbH, Günter-Fronius-Straße 1, 4600 Thalheim, Austria
{kefer.kathrin-maria,stoeger.markus,hofer.bernd}@fronius.com
[2] University of Applied Sciences Upper Austria Campus Hagenberg, Softwarepark 11,
4232 Hagenberg, Austria
{Roland.Hanghofer,Michael.Affenzeller,Stephan.Winkler,
Stefan.Wagner}@fh-ooe.at
[3] University of Applied Sciences Upper Austria, Ringstraße 43a, 4600 Wels, Austria
Patrick.Kefer@fh-ooe.at

1 Introduction

In recent years, renewable energy resources have become increasingly important. Due to the fluctuating and changing environment, these energy sources are not permanently available. At certain times, e.g. a photovoltaic (PV) power plant can only generate little or no electricity at all. This is why energy management systems (EMS), which store, use and distribute the available energy as optimally as possible, have been strongly promoted and further developed recently. However, current EMS are often easy and non-optimal or computationally very intensive but then optimal controllers. Conventional approaches developed by experts like model predictive controls (MPCs) [1,2] require a great amount of computing time and resources. That is why they can currently just be executed at fixed intervals like once per hour. This prevents them from being able to instantly react to changes in the environment. Additionally, such controllers are usually tailored to one specific system. For each new system, the respective logic must be written anew. Recent works in this research area also use linear programming algorithms [3,4] or meta heuristic algorithms [5,6]. However, they were used to optimize only small systems with a limited number of household appliances and few energy sources so far. Therefore it would be favourable to automatically learn controllers that are able to adapt themselves to possible changes in the system e.g. new controllable appliances, from time to time. As a first step towards this goal, a new model-based optimization approach using heuristic optimization techniques was developed in the course of this work.

© Springer Nature Switzerland AG 2020
R. Moreno-Díaz et al. (Eds.): EUROCAST 2019, LNCS 12013, pp. 405–412, 2020.
https://doi.org/10.1007/978-3-030-45093-9_49

2   Model-Based Optimization Approach

In this new model-based optimization approach, the software *HeuristicLab* [11] is extended to work together closely with *MATLAB* and *MATLAB Simulink*. It already contains all necessary algorithms for performing symbolic regression [10] using genetic programming (*GP*). But there is no optimization problem available that is able to optimize a *MATLAB Simulink* model. Therefore, two new *HeuristicLab* optimization problems were implemented - one single objective and one multi objective problem. They communicate with *MATLAB* via the *MATLAB API* for *.NET*, which provides all necessary functions for executing *MATLAB* scripts and reading and writing variables to and from the *MATLAB* workspace.

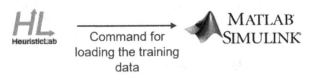

(a) The first step of the training process is to load the needed training data into the *MATLAB* workspace.

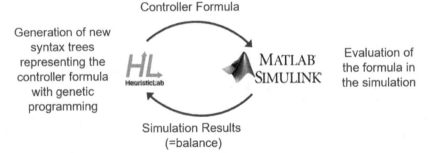

(b) The second step is the training of the controllers. *HeuristicLab* creates the syntax trees which are converted to the controller formula that is inserted into the *MATLAB Simulink* model. The simulation is run to evaluate the current solution candidate. The result is returned to *HeuristicLab* via the *MATLAB* workspace.

Fig. 1. The newly developed model-based two-step optimization approach.

The optimization approach consists of two main steps (Fig. 1). First, *HeuristicLab* initiates the loading of the training data into the *MATLAB* workspace by starting a data loading script in *MATLAB*. Then, the actual optimization begins by starting the *GP* algorithm. This algorithm runs in the *HeuristicLab* and creates new syntax trees that represent the solution candidates. Each syntax tree is converted into a mathematical formula, which is put as a variable into the *MATLAB* workspace. The *MATLAB* evaluation script replaces the old formula in the simulation model with the new one and starts the simulation. The result of the simulation is the balance of the system at the end. This quality value is then read by *HeuristicLab* from the *MATLAB* workspace and is assigned as fitness value to the respective solution candidate. Based on this fitness value, the algorithm evaluates the solution candidates and generates the new generation.

3 Simulation Model

The simulation model used for training and evaluating the heuristic controllers was developed by Kirchsteiger et al. [9]. It is implemented in *MATLAB Simulink* and consists of the trained controller, the power inverter, the battery storage and the evaluator which calculates the balance of the system. The only adaption on the model done for this work, was to replace the linear optimizer by a *MATLAB Simulink Fcn* block to be able to run and evaluate the trained controllers.

4 Data Basis

As data basis real world data from a single family house in Upper Austria with a PV system and a battery storage is used. The data is collected from 10th February 2017 until 31st December 2017 and consists of the electrical power and voltage from the PV system and the electrical power consumed by the household. The values are measured in five minute intervals and the household load is linearly interpolated to one second as the simulation interval is set to this. Also missing PV power values are interpolated linearly. The dynamic feed-in tariff and the grid salary which are also needed as input for the simulation, are provided by *aWATTar*[1] in form of their monthly tariff.

The whole data set is split up into multiple different parts for training and evaluation. First, it is separated into two main parts, being a combined training and testing data set running from 10th February 2017 until 31st August 2017. This data set is then used in a first version to train controllers with one day training data, being the 10th of February 2017, and to test them with the directly following period starting at the 11th of February 2017. The second version of the data set is used to train controllers with 30 days training data, which is the time period from 10th February 2017 until 11th March 2017. Here again, the directly following period starting at the 12th of March 2017 is used as test data. The second main data set, running from 1st September 2017 until 31st December 2017 is only used as test data for comparing all these controllers with each other.

5 Evaluation

To test and evaluate the developed optimization approach, two different genetic algorithms (*GAs*) are used to perform symbolic regression to calculate the controllers for the model. The first one is the *Offspring Selection Genetic Algorithm*

---

[1] https://www.awattar.com/tariffs/monthly.

(OSGA) developed by Affenzeller and Wagner [7]. It is a single-objective GA that aims at maximizing the balance. It works with a *selection pressure*, which lets the algorithm end if no sufficient number of offsprings (solution candidates) that outperform their own parents [7] can be found. This should lead to better results and shorter training times. The second GA is the *Non-dominated Sorting Genetic Algorithm-II (NSGA-II)*, which is a multi-objective algorithm with a two dimensional Pareto Front [8]. The aim of the first dimension is to maximize the balance just like the *OSGA* does. Additionally, the second dimension aims at minimizing the complexity of the syntax trees, which can help to prevent overfitting. This is achieved by favouring mathematical functions with a lower complexity like addition or subtraction.

With each of these algorithms, five controllers were trained with one day training data and 30 days training data using the data sets described in Sect. 4 and a simulation interval of one second. For both algorithms, the maximum number of generations was set to 100, the population size to 500 and the mutation probability to 15%. As selectors, a *Tournament Selector* with a group size of six was used for the controllers trained with the *OSGA* and one day of training data, whereas a *Gender Specific Selector* was used for the ones trained with the *OSGA* and 30 days training data. For all controllers trained with the *NSGA-II*, a *Crowded Tournament Selector* also with a group size of six was used as selector.

All trained controllers are compared to the linear optimizer (*LO*) developed by Kirchsteiger et al. [9] and the ownconsumption optimizer (*OCO*) implemented by Fronius International GmbH. The ownconsumption optimizer follows a zero feed-in strategy. First of all, the household loads are supplied with the power from the *PV* system. When there is more power available than needed for supplying the loads, this surplus energy is fed into the battery until it is full. Finally, only when all loads are supplied and the battery is fully charged, the remaining surplus power is fed into the grid. First, the controllers are evaluated with the directly following timespan, being from the 11th of February 2017 onwards for the controllers trained with one day of training data and from the 12th of March 2017 onwards for the ones trained with 30 days training data. This is done by inserting the respective controller formula into the *MATLAB Simulink* model and running the simulation for 120 days with a simulation interval of one second. Additionally, all controllers independently of their training data length are also evaluated with the second main evaluation data set starting at the 1st of September 2017.

6 Results

After 120 simulation days, the own consumption optimizer achieves a balance of −83.33€. The linear optimizer reached a balance of −90.24€, which is 8.28% worse than the own consumption optimizer. For this test data, none of the trained controllers achieved better results than the own consumption optimizer and also no controller trained with the *OSGA* algorithm was better than the linear optimizer. However, four of the five controllers trained with the *NSGA-II* algorithm achieved better results than the linear optimizer. The best of them is only 0.96%

worse than the own consumption optimizer. Compared to the linear optimizer, this best controller was able to save 6.09€ or 7.24%, while the worst of the good controllers still saves 2.50€ or 2.85%. The worst of all the trained controllers (one of the *OSGA* controllers) achieves a balance of −106.07€, which is 17.55% worse than the linear optimizer and 27.29% worse than the own consumption optimizer. Figure 2(a) shows these results.

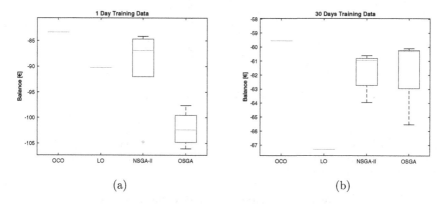

Fig. 2. A comparison of the controllers trained with (a) one day and (b) 30 days training data to the own consumption optimizer (*OCO*) and the linear optimizer (*LO*) [9] for 120 days of evaluation data from the continuing data sets starting at the 11th of February 2017 and the 12th of March 2017.

The controllers trained with 30 days training data (Fig. 2(b)) achieved similar results as the ones trained with one day training data when being evaluated on the continuing data set starting at the 12th of March 2017. None of them was better than the own consumption optimization, which achieved a balance of −59.56€. But three of the controllers trained with the *OSGA* algorithm resulted in balances between −60.11€ and −60.30€, which is only between 0.55%–0.74% worse than the own consumption optimizer but between 10.40%–10.68% better than the linear optimizer. The worst controller was trained with the *OSGA* algorithm and achieved a balance of −65.53€, which is 10.02% worse than the own consumption optimizer. However, all these controllers achieved better results than the linear optimizer. The three best *NSGA-II* algorithm trained controllers achieved balances between −60.94€ and −60.60€, which means an improvement between 9.44%–9.94% compared to the linear optimizer. All others achieved balances that improved the linear optimizer by 2.63%–7.75%.

Finally, both optimizers and all trained controllers are evaluated on the independent second main test data set reaching from 1st September 2017 until 31st December 2017 (Fig. 3). The own consumption optimizer achieves a balance of −142.03€. The linear optimizer [9] performs 4.23% worse with a balance of −148.31€ compared to that. Three of the controllers trained with one day training data (two of the *NSGA-II* controllers and one of the *OSGA* controllers)

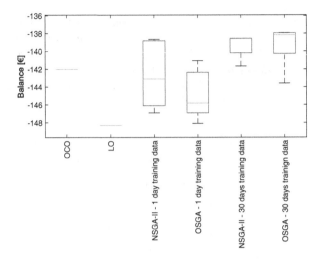

Fig. 3. A comparison of all controllers to the own consumption optimizer (OCO) and the linear optimizer (LO) [9] with the second main test data set reaching from 1st September 2017 to 31st December 2017.

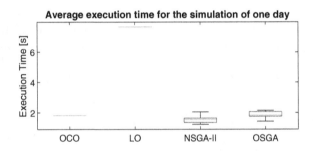

Fig. 4. The visualization of the execution times from the evaluation of all controllers on the second main test data set compared to the own consumption optimizer (OCO) and the linear optimizer (LO) [9].

achieve between 0.69%–2.43% better results than the own consumption optimizer. From the controllers trained with 30 days training data, all except one (a controller trained with the $OSGA$) have a better balance than the own consumption optimizer. The one controller performing worse than the own consumption optimizer achieves a balance of $-143.58€$, which is 1.08% more. All other controllers achieve improvements between 0.26%–2.96% by resulting in balances between $-141.66€$ and $-137.94€$. Additionally, all twenty trained controllers perform better than the linear optimizer by achieving balances between $-148.10€$ and $-137.94€$, which means an improvement between 0.14%–7.51%. Eight of the controllers that were trained with 30 days training data (four of the $NSGA$-II controllers and four of the $OSGA$ controllers) achieve improvements of at least 6.15% compared to the linear optimizer. From the controllers that were trained with one day training data, two of the *NSGA-II* controllers and one of the *OSGA* controllers yield improvements of at least 5.14% and up to 6.95%.

All controllers trained with this new approach achieve big execution time savings compared to the linear optimizer [9] (Fig. 4). Due to running the simulation in each optimization step, the execution of this optimizer takes on average 7.66 s for one day of simulation, whereas the own consumption optimizer and all the trained controllers achieve an average execution time of 1–3 s.

7   Conclusions and Future Work

Summing up, the newly developed model-based optimization approach is a promising step towards the future of energy management systems. The controllers trained with this approach have a similarly low execution time as the ownconsumption optimizer has. But this is at least twice as fast as the linear optimizer. Taking a closer look on the results, one can find a big difference between the controllers trained with one day training data and the ones trained with 30 days. This is not surprising due to the most probably better generalization when training the controllers with a bigger training data set. However, training the controllers with more than 30 days of training data is currently not useful due to the long training times of just one controller. This is why it is planned to speed up the training process by exporting the model to C/C++ code and then not simulating the model but instead running the code of the model. Additionally, the influence of the simulation interval on the training time and the results should be examined. It is expected that the training runs faster with a bigger simulation interval. However, the effects on the results need to be taken into account, so that the model does not lose to much of its precision. Once the training of the controllers is faster, it is also planned to perform more trainings with longer training data sets containing at least half a year or even a year of data, which should further improve the results of the trained controllers.

Acknowledgements. This project is co-financed by the European Regional Development Fund and the Province of Upper Austria. It was carried out by Fronius International GmbH together with partner researches from the University of Applied Sciences Upper Austria.

Europäische Union Investitionen in Wachstum & Beschäftigung. Österreich.

References

1. Chen, C., Wang, J., Heo, Y., Kishore, S.: MPC-based appliance scheduling for residential building energy management controller. IEEE Trans. Smart Grid **4**(3), 1401–1410 (2013)
2. Kothare, M.V., Balakrishnan, V., Morari, M.: Robust constrained model predictive control using linear matrix inequalities. Automatica **32**(10), 1361–1379 (1996)

3. De Angelis, F., Boaro, M., Fuselli, D., Squartini, S., Piazza, F., Wei, Q.: Optimal home energy management under dynamic electrical and thermal constraints. IEEE Trans. Industr. Inf. **9**(3), 1518–1527 (2013)
4. Chen, Z., Wu, L., Fu, Y.: Real-time price-based demand response management for residential appliances via stochastic optimization and robust optimization. IEEE Trans. Smart Grid **3**(4), 1822–1831 (2012)
5. AlRashidi, M.R., El-Hawary, M.E.: A survey of particle swarm optimization applications in electric power systems. IEEE Trans. Evol. Comput. **13**(4), 913–918 (2009)
6. Pedrasa, M.A.A., Spooner, T.D., MacGill, I.F.: Coordinated scheduling of residential distributed energy resources to optimize smart home energy services. IEEE Trans. Smart Grid **1**(2), 134–143 (2010)
7. Affenzeller, M., Wagner, S., Winkler, S., Beham, A.: Genetic Algorithms and Genetic Programming. Chapman and Hall/CRC, New York (2009)
8. Deb, K., Pratap, A., Agarwal, S., Meyarivan, T.A.M.T.: A fast and elitist multi-objective genetic algorithm: NSGA-II. IEEE Trans. Evol. Comput. **6**(2), 182–197 (2002)
9. Kirchsteiger, H., Rechberger, P., Steinmaurer, G.: e & i Elektrotechnik und Informationstechnik **133**(8), 371–380 (2016). https://doi.org/10.1007/s00502-016-0447-1
10. Riolo, R., Vladislavleva, E., Ritchie, M., Moore, J.H.: Genetic Programming Theory and Practice X. Genetic and Evolutionary Computation. Springer, New York (2013). https://doi.org/10.1007/978-1-4614-6846-2
11. Wagner, S., et al.: HeuristicLab 3.3: a unified approach to metaheuristic optimization. In: Actas del séptimo congreso español sobre Metaheurísticas, Algoritmos Evolutivos y Bioinspirados (MAEB 2010), p. 8, September 2010

Understanding and Preparing Data of Industrial Processes for Machine Learning Applications

Philipp Fleck[1(✉)], Manfred Kügel[2], and Michael Kommenda[1]

[1] Heuristic and Evolutionary Algorithms Laboratory,
University of Applied Sciences Upper Austria, Softwarepark 11,
4232 Hagenberg i.M., Austria
{philipp.fleck,michael.kommenda}@fh-hagenberg.at
[2] Primetals Technologies Austria GmbH, Turmstraße 44, 4031 Linz, Austria
manfred.kuegel@primetals.com

Abstract. Industrial applications of machine learning face unique challenges due to the nature of raw industry data. Preprocessing and preparing raw industrial data for machine learning applications is a demanding task that often takes more time and work than the actual modeling process itself and poses additional challenges. This paper addresses one of those challenges, specifically, the challenge of missing values due to sensor unavailability at different production units of nonlinear production lines. In cases where only a small proportion of the data is missing, those missing values can often be imputed. In cases of large proportions of missing data, imputing is often not feasible, and removing observations containing missing values is often the only option. This paper presents a technique, that allows to utilize all of the available data without the need of removing large amounts of observations where data is only partially available. We do not only discuss the principal idea of the presented method, but also show different possible implementations that can be applied depending on the data at hand. Finally, we demonstrate the application of the presented method with data from a steel production plant.

Keywords: Machine learning · Data preprocessing · Missing values

1 Introduction

In industrial production processes, items are often processed in nonlinear production layouts with different sensor types equipped at different production stages, leading to sparse sensor data caused by the different routes items take through the production facility. Considering the sample production layout in Fig. 1, all processed items will pass station A and E; therefore, data from those station will always be available. Station B1 and B2 are of the same type with the same sensors, thus, data from B is always available, regardless if an item was processed

© Springer Nature Switzerland AG 2020
R. Moreno-Díaz et al. (Eds.): EUROCAST 2019, LNCS 12013, pp. 413–420, 2020.
https://doi.org/10.1007/978-3-030-45093-9_50

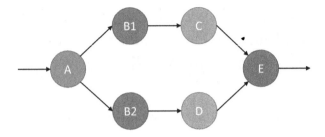

Fig. 1. A nonlinear production layout where an item passes through different processing units. As a result, sensor data for a specific item is only available if it passes a certain processing unit.

at B1 or B2. However, depending on the route, only sensor data from station C or D is available, never both.

Most machine learning algorithms struggle with sparse data due to their inability to handle missing values themselves. For such cases, missing values are often *imputed* [4] with the help of existing data. Imputing missing values is often feasible in scenarios, where missing data occurs infrequently and at random. However, imputing missing values can be challenging if large proportions of the data is missing, especially if related signals that could be used to reconstruct the missing values are also missing at the same time. This is often the case, because related signals from the same production unit are also completely missing if the product did not pass this production unit. Instead, observations or even entire signals that contain missing values are often removed due to the lack of alternatives [3]. However, considering a production line with two exclusive paths, removing observations that contain any missing values would lead to the removal of the entire data. Alternatively, removing all signals for the exclusive paths, leads to a considerable loss of information. As a conclusion, neither imputing nor removing missing values are desirable options.

Figure 2 shows an overview of available and missing data for the sample production layout from Fig. 1. The full block A symbolizes that this data is always available for all observations, which is also true for E. B is always available, either in the form of B1 or B2. C is only available for the top half of the observations and D only for the lower half. Note that the grouping of C and D in the top and bottom half was done for illustration, this is not necessarily true in real world cases. If we remove either column C or D, potentially valuable signals are lost. Alternatively, if we remove all observations where D is not available, then half the available data from A, B, and E is lost.

Some machine learning algorithms are capable of dealing with missing values on their own [5]. For example, random forests can be extended, so that for each decision within the a tree, there is a default choice in case a missing value is at hand. In this paper, however, we present a method on a higher level, that can be used in conjunction with different modeling techniques that are not required to handle missing values themselves.

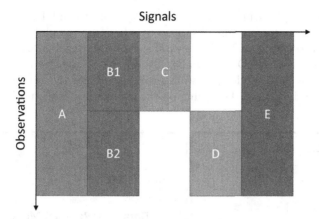

Fig. 2. This chart shows available data, grouped by the processing units based on the production layout in Fig. 1.

As an exemplary use case for this paper, we use the data of a production plant producing steel coils, where missing values in the data are caused by the different routes that the coil takes through the production plant. For this use case, we will demonstrate and discuss in detail how we applied the proposed methods.

2   Methods

We propose a method for dealing with missing values due to nonlinear production lines, that can utilize all available data for training a model, without the need of imputation or removal of missing values. This method consists of the following three steps:

(1) Create multiple subsets of the data that do not contain any missing values.
(2) Create an ensemble model [6], based on the different data subsets.   .
(3) Predict, using only models of the ensemble, where the corresponding input variables are available.

The basic idea of the proposed method is to create multiple, missing-value-free datasets based on the actual dataset. Then, multiple models are trained on these new datasets. When a prediction is made, only models are selected where all corresponding input variables are available, and the model results are combined. Typically, we create a base model based on features that are always available, and multiple residual models that are used as additional correction factors. In other words, the residual models are building upon the predictions of the base model, and do only predict the error of the base model. This way, the residual models can reuse knowledge the base model has already discovered.

An overview and example of this method is shown in Fig. 3. Typically, we create a base model based on signals that are always available, for instance, signals A, E and, B (which is the combination of B1 and B2). The data for residual model 1 only contains observations where C is available, as well as A, B1 and E for those observations. Likewise, the data for residual model 2 contains only observations where D is available. To obtain a prediction value of a new observation, the base model is always evaluated while the application of the residual models 1 and 2 is depending on the availability of C and D.

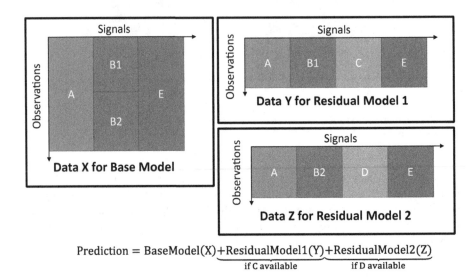

$$\text{Prediction} = \text{BaseModel}(X) + \underbrace{\text{ResidualModel1}(Y)}_{\text{if C available}} + \underbrace{\text{ResidualModel2}(Z)}_{\text{if D available}}$$

Fig. 3. An overview and example of our proposed method that splits the data into multiple subparts, trains an ensemble model on those subparts, and uses selected models of the ensemble for a final prediction.

The challenge of this method is to anticipate, which signals will commonly be available together, so a data subset can be created accordingly. Additionally, all three steps are interdependent and also dependent on the data, therefore selecting appropriate strategies is important. In the following sections, we describe each section in detail and also discuss advantages and disadvantages of the presented strategies for each step.

2.1   Data Subset Strategies

The first step is to create subsets of the dataset that do not contain any missing values. Those subsets define which models of the ensemble are available, therefore, should be selected based on signals that are commonly available together. As a base, there are two common strategies that we encountered:

Subset by Grouped Signals: In this strategy, the data is split based on signals from the same processing unit. This is done, for example, in Fig. 3, where we split whether C or D is available, thus we obtain datasets for a base model and for two residual models. For this strategy there are two options concerning which signals are included in the data subset: If relations between the signals for a single processing unit (C or D) and the other data (A, B, or C) are suspected, both should be included. If no relationship is to be expected, only the signals from the processing unit (C or D) should be used in the data subset.

Subset by Common Routes: If the production layout graph is complex and contains many branches, creating all potential combinations of data subsets would result in a large number of subsets and residual models. Instead, it is advised to split the data by common production routes. For example, even if the production layout would allow for a high number of potential routes in theory, there is often only a small number of routes that are used predominantly. In such case, the data can be split along those routes with the signals used that are available for the specific routes. Uncommon routes can either be grouped together with only their common signals remaining, or completely dropped. We will demonstrate such a splitting strategy later in Sect. 3.

2.2   Ensemble Modeling

For modeling, there are two considerations: the actual modeling method and the aggregation of the individual models' results. As for modeling methods, there are no restrictions; thus, different methods can be used, also for the separate models. For aggregating the models' results, there are two common strategies for ensembles:

Bagging (bootstrap aggregating) [1] is combining all models equally, e.g. averaging all predictions for a regression ensemble. In this case, each model predicts the target itself, and all applicable models' outputs are simply averaged. A downside of this approach is that models cannot built upon results that are already achieved by other models.

Boosting involves an incremental modeling method that builds upon the results of previous models [2]. Instead of each model trying to predict the target itself, a model only tries to predict the residuals from a previous model, allowing the residual models to build upon knowledge that the base model (or previous residual models) has already discovered. This is a core element of our proposed method, because it allows the residual models to build upon knowledge that was created from different data than the residual model has available. This additional data corresponds to the observations that were removed from the original data when building the residuals data subset due to missing values and, therefore, would not be available if the model was trained in a conventional manner.

2.3   Ensemble Prediction

When a prediction is made based on the ensemble model, only models of the ensemble are used where all the input variables are available. In other words, based on the available data, all applicable models are selected, and their prediction combined. In case of a bagging model, the models' average is determined; in case of a boosting model, the base model is evaluated and the residual models' predictions (i.e. the correction factors) are added.

3   Application

In this section, we demonstrate how we applied the proposed method, especially how we partitioned the data into missing-value-free subsets, and which ensemble modeling strategy we applied. As an exemplary use case for this paper, we use the data of a production plant producing steel coils, where missing values are caused by different routes through the production plant, shown in Fig. 4.

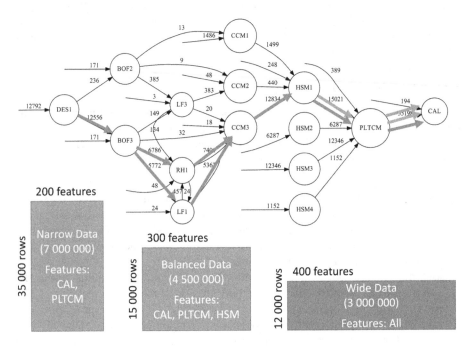

Fig. 4. Layout with route-segments for splitting the data. The numbers in the graph show the number of coils that passed this route. On the bottom, the subsets of the data are shown with their number of values and included signal groups. (Color figure online)

For building data subsets, one option would be to split the data by processing unit, e.g. one partition for all processing units of the same type (one for BOF, one for CCM, etc.). However, since there is a single predominated route that most of the coils followed, we opted for splitting the data by common routes. More specifically, we split by route-segments that a steel coil must pass in order to be considered for a data subset. For example, we define a short route-segment (orange in Fig. 4) only spanning PLTCM and CAL, and all coils that passed this route-segment are selected for our first data subset, which we call *narrow data*. As for the signals, only PLTCM and CAL signals are used. Next, the green route, forming the route-segment HSM1-PLTCM-CAL is used for the *balanced data*, which contains less coils than the previous route (i.e. coils from HSM2-4 are removed), but additionally also contain HSM signals. Only HSM1 is used for this route-segment, because there is no sensor data available for the other HSMs. Finally, the blue route, spanning the entire production layout, is used for the *wide data* and contains even less coils than the other routes, but in exchange, all signals are available. Although there is a branch within the blue route that would usually lead to missing values, we consider RH and LF to be identical since they both have the same signals.

For the ensemble modeling, we opted for a boosting strategy with incremental residual models. As a start, we use the *narrow data* that contains the most observation as a base model. Next, the first residual model uses the *balanced data* and predicts the residuals from the base model. And finally, a second residual model uses the *wide data* and predicts the residuals of the first residual model. This sequential application of the residual model is possible, because all the signal of a model's dataset are also available in the subsequent model's dataset, i.e. all signals of the *narrow data* are also available at the *balanced data*.

As a result, if a coil only passed PLTCM and CAL, the base model is used for prediction. In case a coil also passed HSM1, a correcting term based on the first residual model is applied. In case the coil passed the long, blue route from DES1 to CAL, also the second residual model is applied.

Table 1 shows the accuracy of the ensemble model of our proposed method, and a conventional model as comparison where observations containing missing values were removed. We can clearly observe that our method yields higher accuracy (in terms of lower mean absolute error and higher R^2) than the conventional model. Since the individual modeling methods had the same inputs and observations available, the results indicate, that the proposed method was successful in creating better models due to its ability to incorporate data that would otherwise has been removed.

Table 1. Model accuracies on the test partition for our proposed method and a conventional model were observations with missing values were removed.

	Narrow		Balanced		Wide	
	MAE	R^2	MAE	R^2	MAE	R^2
Proposed method	0.285	0.549	0.233	0.717	0.183	0.811
Conventional model	0.285	0.549	0.284	0.585	0.315	0.450

For this paper, the main focus was the application of the proposed method. Thus, only minimal parameter optimization and method tuning was performed.

However, both the proposed method and the conventional method used the same training data as well as same algorithm parameters for a fair comparison.

4   Conclusion

The presented method enables working with data that contains many missing values if they do occur systematically. One typical case is industry data, where the production layout is nonlinear and therefore, a single item often does not pass all sensors.

The idea of our method is to build subsets of the data without missing values, based on signals that are available together. Then, we train a model for each data subset, similar to ensemble modeling. For a prediction with the ensemble model, the applicable models are selected based on the available data and the models' required input features, and then the predictions of the models are combined. We showed an application using a boosting approach, where we build residual models based on common routes within the production layout.

Although the methodology was designed for industrial applications of machine learning, the presented method is also applicable in general if missing values occur. However, the method strongly relies on an underlying reason why groups of signals are missing together, for example due to sensors only available at certain processing units. In cases where missing values occur infrequently and at random, the proposed method is not recommended.

In this paper, we have proposed a new method for dealing with missing values that appear systematically due to the production layout. It would be interesting to see how the different approaches (e.g. boosting, bagging) compare to each other across multiple datasets in a comprehensive comparison. For instance, the proposed subsetting strategies (by grouped signals or common routes) are only basic ideas that could be analyzed in more detail and developed further. In that sense, the propose method is still very elementary, and many variants could be further developed and explored.

References

1. Breiman, L.: Bagging predictors. Mach. Learn. **24**(2), 123–140 (1996). https://doi.org/10.1007/BF00058655
2. Friedman, J.H.: Stochastic gradient boosting. Comput. Stat. Data Anal. **38**(4), 367–378 (2002)
3. García, S., Luengo, J., Herrera, F.: Data Preprocessing in Data Mining. Intelligent Systems Reference Library. Springer, Cham (2015). https://doi.org/10.1007/978-3-319-10247-4
4. Royston, P.: Multiple imputation of missing values. Stata J. **4**(3), 227–241 (2004)
5. Tierney, N.J., Harden, F.A., Harden, M.J., Mengersen, K.L.: Using decision trees to understand structure in missing data. BMJ Open **5**(6), e007450 (2015)
6. Zhou, Z.H.: Ensemble Methods: Foundations and Algorithms. Chapman and Hall/CRC (2012)

Genetic Algorithm Applied to Real-Time Short-Term Wave Prediction for Wave Generator System in the Canary Islands

C. Hernández$^{(\boxtimes)}$, M. Méndez, and R. Aguasca-Colomo

Instituto Universitario de Sistemas Inteligentes (SIANI), Universidad de Las Palmas de Gran Canaria (ULPGC), Las Palmas de Gran Canaria, Spain
`carlos.hernandez119@alu.ulpgc.es`,
`{maximo.mendez,ricardo.aguasca}@ulpgc.es`

Abstract. In island territories, as is the case of the Canary Islands, renewable energies mean greater energy independence, in these cases wave and wind energy favour this independence, all the more so when the generation of these types of energy is optimised. The increase in wave energy extracted from the waves requires knowledge of the future wave incident on the energy converters. A prediction system is presented using Genetic Algorithm to optimize the parameters that govern an autoregressive model, model necessary for the prediction of the incident wave. The comparison of the Yule-Walker equations with that of the Genetic Algorithm will provide us with a knowledge of the prediction technique that offers the best results, for the sake of its application. All this under the restriction of limited execution times, less than the periods of the waves to be predicted, and a demanding precision through distant prediction horizons, with reduced training datasets.

Keywords: Genetic algorithms · Yule-Walker method · Forecasting methods · Wave energy

1 Introduction

The oceans contain a large amount of energy. This energy potential has hardly been exploited, although in recent years the trend has changed, and there is increasing interest in the installation of devices for collecting renewable energy at sea. Proof of this is the remarkable growth of offshore wind energy in recent years, especially in Europe [14]. Also in the Canary Islands, the region where this article focuses, which already 5 MW has installed and more are foreseen in the coming years.

Wave energy, which uses the energy resource contained in the waves, is another way to extract the potential of the oceans, however, it is not as developed as wind energy. However, it is estimated that wave energy will achieve remarkable growth in the coming decades as technology advances and has the necessary years of development [1,4,7].

© Springer Nature Switzerland AG 2020
R. Moreno-Díaz et al. (Eds.): EUROCAST 2019, LNCS 12013, pp. 421–428, 2020.
https://doi.org/10.1007/978-3-030-45093-9_51

Fig. 1. Artistic representation of the W2Power wind-wave hybrid offshore platform (Courtesy of Pelagic Power AS [12]).

The growth associated with wave devices will be linked to the reduction of capital and maintenance costs, as well as to the capture of the greatest amount of available resources. The reduction of capital and maintenance costs go, among many other possibilities, by the integration of the devices themselves in platforms installed for wind power use, see Fig. 1. Costs are reduced by sharing a single evacuation line, the same infrastructure and the same anchoring system, as well as the annual hours of joint generation are increased [11].

With the intention of optimizing the generation of wave energy in this article, a real-time wave prediction system with autoregressive model is proposed, a model that has proven to be one of the best results in terms of accuracy of future sea state values [5], and on which the obtaining of the parameters that describe it is obtained through a Genetic Algorithm. This in turn will be compared with the Yule-Walker equations, in performance, precision and execution time.

The system will only require the past states of the sea to make predictions, which simplifies the system and makes unnecessary the use and spatial distribution of measurement buoys around the structure for prediction. All this with the limitation of some execution times of the algorithms limited by the wave periods that are intended to be predicted.

With this, it is sought, as has been mentioned, the increase in the technical and economic viability of wave energy, which, together with wind power, can provide energy independence to insular territories, as is the case of the Canary Islands, and a reduction in energy prices in those territories highly dependent on fossil fuels.

In the Canary Islands, the installed electric power dependent on petroleum products stood at 2,697.8 MW in 2015, while only 11.8% of the total installed power, about 361.1 MW corresponded to renewable generation sources [2]. With this situation of dependence on hydrocarbons and their derivatives, the average cost of energy was placed in 2015 in the Canary Islands at 160€/MWh [2]. Meanwhile, the cost of wind energy in floating platforms, necessary to overcome the difficult insular bathymetries, is between 82–236€/MWh [10], depending on

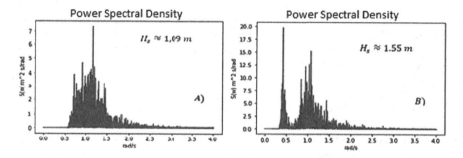

Fig. 2. Wave spectrum of the sample dataset: (A) 1 January 2015 (11:00 h–23:00 h); (B) 15 January 2015 (00:00 h–11:00 h).

many parameters such as depth and distance at coast, and the cost of wave energy between 330–630€/MWh [9], also dependent on many parameters.

However, the cost of wave energy is expected to be between 113–226€/MWh by the year 2030 [9], which, with a presumable decrease in offshore wind energy costs, the combination of these technologies, sharing costs and increasing their productivity, will generate an average cost of energy lower than those given by energy sources that use hydrocarbons. To achieve this reduction and reach competitive costs, optimization measures such as the one presented here are needed, especially designed for the use of wave energy in insular territories and incorporated in multipurpose floating platforms.

2   Available Data

The island of Gran Canaria offers a wave energy that is between 25 and 30 kW/m [8], lower than sites of the Atlantic European coast but equally significant for its exploitation, with the highest concentration of energy in the waves of heights between 1.5 m and 3.5 m and periods between 7 s and 13 s [6].

The data used comes from real observations provided by the Triaxys Las Palmas East measurement buoy operated by the state agency Puertos del Estado, whose identification is 1414, and is located at approximately 28.05° N and 15.39° W, over a depth of 30 m, on the eastern slope of the island. The buoy has a sampling frequency of 8.04 rad/s (1.28 Hz), it takes data every 0.78 s, in intervals of 20 min each hour of measurement.

The two sets of samples that are counted belong to the data collected by the measurement buoy on January 1, between 11:00 and 23:00 h (A), and January 15, 2015, between 00:00 h and 11:00 h (B), to have different days and alternate schedules, or what is the same, to have different wave spectra, spectra that are presented in Fig. 2.

The main characteristic that is derived from the analysis of the data, and that can be seen in its spectral representation (Fig. 2), is that the vast majority of the energy contained in the waves is concentrated at low frequencies, low frequency waves that they are more regular and less affected by non-linearity [5]. It is in this way, with the waves of low frequency, with which it is possible to assume a linearity between the past and future wave values, since the past values are closely correlated with the coming ones and have a smaller amount of white noise contained in the waves.

3   Prediction Model

The prediction model to be used is an autoregressive AR that is one that imposes a linear dependence between the past and future values, therefore, the rise of the wave $\eta(k)$ will depend on its n past values, n being the order of the model.

$$\eta(k) = \sum_{i=1}^{n} a_i \eta(k - i) + \xi(k) \tag{1}$$

where the innovation term $\xi(k)$ has been included, and the parameters a_i are estimated, through the Genetic Algorithm (GA) and the Yule-Walker (YW) equations, and the noise is considered Gaussian and white, so that the best prediction with l advance steps from instant k, is estimated as shown in (2).

$$\widehat{\eta}(k + l|k) = \sum_{i=1}^{n} \widehat{a}_i \widehat{\eta}(k + l - i|k) \tag{2}$$

The parameters estimated with the GA and YW will give us different results depending on whether they are able to adjust to the real parameters that govern the phenomenon. The parameters are estimated from a number N of values that make up the training sample, from minimizing the error, in the case of GA and solving the equations given by YW.

3.1   Genetic Algorithm

The Genetic Algorithm (GA) has been tested as a good method for estimating and predicting time series [13]. The GA aims to optimize the fundamental parameters described by the AR autoregressive model. It is therefore, that the different variables can be improved than previously described in (2) and that they are described by the term $\widehat{a}_i(k)$ for each time interval. Say, that is the order of the autoregressive model is 32, that is, the 32 states of the sea will be used to establish the next and the continuous ones.

The GA will have an initial population of 100 individuals counting the elitists, who will make up a total of 5, as well as the probability of selection will be 70% and the cross of 80%. The probability of mutation will be 0.2%. Finally, it should be noted that the number of generations will be 100, seeking not to have a robust and inefficient algorithm, with restrictive execution times given the resource to be predicted, the waves and their elevation. The algorithm has been optimized to these values, after testing different configurations.

Finally, establish the objective function that is pursued and seeks to optimize, which is nothing more than the reduction of the error in the prediction, therefore the objective function is defined by the error and is expressed as shown in (3).

$$f_{objective} = \sqrt{\sum_{i=1}^{n} \eta(k) - \widehat{\eta}(k)} \tag{3}$$

The GA and its objective function, as well as its results have been obtained through the Matlab software tool.

3.2 Yule-Walker Equations (YW)

The Yule-Walker equations [3], in this case, is used to obtain the different parameters that govern the autoregressive model described previously in (1).

The parameters of the Yule-Walker equations can be obtained through the covariance function, or through the correlation function (and autocorrelation), which is shown in (4).

$$Corr[y_t, y_{t-k}] = \frac{E[y_t - \mu][y_{t-k} - \mu]}{E[y_t - \mu]^2} = \frac{y_k}{y_0} = \rho_k \tag{4}$$

From the autocorrelation function we can estimate and calculate the parameters that define the autoregressive model AR through the Yule-Walker equation defined in matrix form in (5)

$$\begin{bmatrix} \rho_1 \\ \rho_2 \\ \vdots \\ \rho_{p-1} \\ \rho_p \end{bmatrix} = \begin{bmatrix} 1 & \rho_1 & \cdots & \rho_{p-2} & \rho_{p-1} \\ \rho_1 & 1 & \cdots & \rho_{p-3} & \rho_{p-2} \\ \vdots & & & & \\ \rho_{p-2} & \rho_{p-3} & \cdots & 1 & \rho_1 \\ \rho_{p-1} & \rho_{p-2} & \cdots & \rho_1 & 1 \end{bmatrix} \begin{bmatrix} a_1 \\ a_2 \\ \vdots \\ a_{p-1} \\ a_p \end{bmatrix} \tag{5}$$

Solving the system of equations, we can estimate the values of the parameters a_i that define the model that we will impose on our wave data, so that we can estimate the values of the incident wave. The parameters are 32, given the order of the model equated to that used in the GA.

The Yule Walker equations, as well as their results have been obtained through the software tool Matlab.

4 Results

The autoregressive model and the optimization methodologies of the parameters that are used have been tested and trained with real data. These real data have been previously filtered with a low pass filter of 0.4 Hz cut-off frequency. This frequency has been taken, according to the spectrum given by the available data samples, under a rough estimate, given that the optimization of this and the

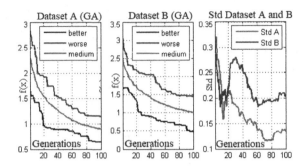

Fig. 3. Evolution of the error by generation, better, worse and average after 30 executions, dataset A, left and dataset B, center, and deviations of the two datasets by generation, right.

adaptation of the cut-off frequency to the needs are beyond the scope of this study. The filter has been developed in Python.

The effect of the filter can be greater or lesser depending on the state of the sea, that is, according to the spectrum of the waves given in each situation. Therefore they require a detailed study with the idea of generating an adaptive low pass filter to the sea conditions.

With the filter the focus is on the low frequency waves, however, they are distributed in different ways.

The first results refer to the average error given by the difference between the real and predicted values. This difference is estimated in the function of the two methodologies used. In the case of the GA, the error corresponds to the average error after 30 executions, that is, the average of the average error of each one of the executions.

With these methodologies, GA and YW, the result is shown in Table 1.

Table 1. Error values with GA and YW

Algorithm	Data set	Mean error	Std
GA	A	0.92	0.161
GA	B	1.16	0.175
YW	A	3.25	–
YW	B	4.71	–

Looking at the results presented in Table 1, we can see that the error originated using GA is less than the error originated using YW. Also, in Fig. 3 we show the evolution by generation of the best, worst and average error in 30 runs achieved by the GA. It is observed how the error in dataset B has a lower slope throughout the generations, and how it is situated with a higher average error, partly due to the characteristics of the waves and therefore of the data. The

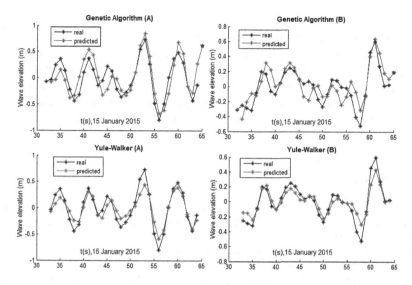

Fig. 4. Real vs predicted series given by GA (data set A: top left image) (data set B: top right image) and YW (data set A: lower left image) (data set B: lower right image).

deviation also decreases throughout the generations with greater promptitude in dataset A, which shows the ability to reach the optimum more accurately, with less uncertainty. Finally, in Fig. 4 we present the graphical results of the real vs predicted series (data sets A and B) obtained with the GA and YW techniques.

5 Conclusions

The results show first how the behaviour of the two methodologies used differs depending on the data set used, which indicates that a specific result will be obtained depending on the spectrum and the characteristics of the waves. In this sense, the low pass filter applied has a remarkable influence, so it must be adjusted to the sea conditions so that the models can make predictions with precision.

Regarding the results shown by the different methodologies, we find that the GA presents better results than the YW equations. All this, in terms of predictions with a prediction horizon of a time step in advance, and with a training dataset of 32 measurements, for a model order of the same number, equated in both models.

Demonstrated that the GA presents better results we must say that there are many lines of research that must be addressed. First, an analysis of the behaviour of the filter with different sea states, and the possibility of performing an adaptive one.

Secondly, analyse the prediction capabilities of the GA, the best model, with prediction horizons higher than a step of time, that is, prediction horizons higher than 0.78 s, so that it can be predicted with 5–6 s of In advance, which is the average period of waves in the area.

In third and last place, it is possible to analyse the execution times when implementing the prediction algorithm in a real wind-wave hybrid platform. The analysis of the execution times should respond to whether it is better to install the predictor system in situ on the platform or if it is convenient to use external servers that send the predictions in real time to the platform to fit the incident wave.

References

1. Carballo, R., Iglesias, G.A.: A methodology to determine the power performance of wave energy converters at a particular coastal location. Energy Convers. Manag. **61**, 8–18 (2012)
2. Consejería de Economía, Industria, Comercio y Conocimiento: Gobierno de Canarias. Anuario Energético de Canarias (2015)
3. Eshel, G.: The Yule Walker Equations for the AR Coefficients. University of South Carolina. http://www.stat.sc.edu/vesselin/STAT520_YW.pdf. Accessed 27 May 2018
4. Falcão, A.F. de O.: Wave energy utilization: a review of the technologies. Renew. Sustain. Energy Rev. **14**(3), 899–918 (2010)
5. Fusco, F., Ringwood, J.V.: Short-term wave forecasting for real-time control of wave energy converters. IEEE Trans. Sustain. Energ. **1**(2), 99–106 (2010)
6. Gonçalves, M., Martinho, P., Guedes Soares, C.: Assessment of wave energy in the Canary Islands. Renewable Energy **68**, 774–784 (2014)
7. Guedes Soares, C., Bhattacharjee, J., Tello, M., Pietra, L.: Review and classification of wave energy converters. In: Guedes, S.C., Garbatov, Y., Sutulo, S., Santos, T.A. (eds.) Maritime Engineering and Technology, pp. 585–594. Taylor & Francis, Milton Park (2012)
8. Hernández-Brito, J.J., Monagas, V., González, J., Schallenberg, J., linás, L.: Vision for marine renewables in the Canary Islands. In Proceedings of 4th International Conference on Ocean Energy, Dublin (2012)
9. IRENA. https://www.irena.org/documentdownloads/publications/wave-energy_v4_web.pdf. Accessed 27 May 2018
10. Myhr, A., Bjerkseter, C., Agotnes, A., Nygaard, T.A.: Levelised cost of energy for offshore floating wind turbines in a life cycle perspective. Renewable Energy **66**, 714–728 (2014)
11. Pérez-Collazo, C., Greaves, D., Iglesias, G.: A review of combined wave and offshore wind energy. Renew. Sustain. Energy Rev. **42**, 141–153 (2015)
12. W2POWER. http://www.pelagicpower.no/. Accessed 27 May 2018
13. Yang, C.X., Zhu, Y.F.: Using genetic algorithms for time series prediction. In: IEEE 2010 Sixth International Conference on Natural Computation (ICNC), vol. 8, pp. 4405–4409 (2010)
14. Wind Europe Business Intelligence. https://windeurope.org/wp-content/uploads/files/about-wind/statistics/WindEurope-Annual-Offshore-Statistics-2017.pdf. Accessed 27 May 2018

A Heuristic Approach for Solving
the Longest Common Square
Subsequence Problem

Marko Djukanovic[1(✉)], Günther R. Raidl[1], and Christian Blum[2]

[1] Institute of Logic and Computation, TU Wien, Vienna, Austria
{djukanovic,raidl}@ac.tuwien.ac.at
[2] Artificial Intelligence Research Institute (IIIA-CSIC), Campus UAB, Bellaterra,
Barcelona, Spain
christian.blum@iiia.csic.es

Abstract. The longest common square subsequence (LCSqS) problem, a variant of the longest common subsequence (LCS) problem, aims at finding a subsequence common to all input strings that is, at the same time, a square subsequence. So far the LCSqS was solved only for two input strings. We present a heuristic approach, based on randomized local search and a hybrid of variable neighborhood search and beam search, to solve the LCSqS for an arbitrary set of input strings. The beam search makes use of a novel heuristic estimation of the approximated expected length of a LCS to guide the search.

Keywords: Longest common subsequence problem · Reduced Variable Neighborhood Search · Beam Search · Hybrid metaheuristic · Expected value

1 Introduction

A *string* s is a finite sequence of symbols from a finite alphabet Σ. The length of a string s, denoted by $|s|$, is defined as the number of symbols in s. Strings are data structures for storing words or even complete texts. In the field of bioinformatics, strings are used to represent DNA and protein sequences. As a consequence, many computational problems in bioinformatics are phrased in terms of so-called string problems. These problems usually present measures of similarities for different molecular structures. Each string obtained from a string s by removing zero or more characters is a *subseqence* of s. A fundamental measure of similarity among molecules is the length of the *longest common subsequence* (LCS): given an arbitrary set of m input strings, $S = \{s_1, \ldots, s_m\}$, we aim at finding the longest subsequence that is common to all input strings [7]. The classical variant of the LCS for $m = 2$ has been studied for almost 50 years. The general LCS for $m \geq 2$ has been tackled for the first time by Huang et al. [4], and later by Blum et al. [1], Mousavi and Tabataba [9], and others.

© Springer Nature Switzerland AG 2020
R. Moreno-Díaz et al. (Eds.): EUROCAST 2019, LNCS 12013, pp. 429–437, 2020.
https://doi.org/10.1007/978-3-030-45093-9_52

Recently, *the longest common square subsequence* (LCSqS) problem, a variant of the LCS, was proposed by Inoue et al. [5]. It requires that the resulting LCS is at the same time a square subsequence. A string s is a *square* iff $s = s' \cdot s' = s'^2$, for some string s', where "$\cdot$" denotes the string concatenation. The length of the LCSqS can be seen as a measure of similarity between disjunctive parts of each of the compared molecules. Therefore, it can give more insight into the internal similarity of the compared molecules than when just considering a LCS. Moreover, the information about those parts of the molecules that are similar to each other is obtained by identifying a LCSqS. Inoue et al. [5] proved that the LCSqS problem is $\mathcal{NP}$-hard for an arbitrary set of input strings and proposed two approaches for the case of two input strings: (1) a *Dynamic Programming* (DP) approach running in $O(n^6)$ time (n denotes the length of the largest input string), and (2) a sparse DP-based approach, which makes use of a special geometric data structure. It can be proven that, if m is fixed, the LCSqS is polynomially solvable by DP in $O(n^{3m})$ time which is not practical already for small input sizes. To the best of our knowledge, no algorithm has yet been proposed for solving the LCSqS problem for an arbitrary number $m \geq 2$ of input strings. The main contributions of this paper are as follow:

- A transformation of the LCSqS problem to a series of the standard LCS problems is described.
- An approach based on a randomized local search and a hybrid of *a Reduced Variable Neighborhood Search* (RVNS) [8] and *a Beam Search* (BS) are proposed for solving the general LCSqS problem.
- An approximation for the expected length of a LCS is derived and incorporated into the BS framework to guide its search.

Organization of the Paper. The paper is organized as follows. Section 2 describes basic solution approaches for the LCS known from the literature. Section 3 gives a basic reduction from the LCSqS to the LCS problem and two approaches to solve the LCSqS. Section 4 presents computational results, and Sect. 5 outlines some research questions and directions for future work.

2   Solution Approaches for the LCS Problem

For $i \leq j$, let $s[i,j] = s[i] \cdots s[j]$ be a continuous (sub)string of a string s which starts from index i and ends at index j. If $i > j$, $s[i,j]$ is the empty string ε. For $p^{\mathrm{L}} \in \mathbb{N}^m$, which is called the left position vector, $S[p^{\mathrm{L}}] := \{s_i[p_i^{\mathrm{L}}, |s_i|] \mid i = 1, \ldots, m\}$ denotes the set of the remaining parts of the input strings of S w.r.t. p^{L}.

The Best-Next Heuristic (BNH) for the LCS was proposed by Huang et al. [4]. This heuristic starts with an empty partial LCS solution $s^{\mathrm{p}} = \varepsilon$ which is then iteratively extended by a feasible letter. If there exists more than one candidate to extend s^{p}, the decision which one to choose is made by a *greedy* heuristic, calculating for each of the candidate letters a greedy value. The BNH works in detail as follows. We initialize s^{p} to the empty string ε, and the left pointers p^{L} to

$(1, \ldots, 1)$, indicating that the complete input strings are still relevant for finding extensions of s^{P}. Each letter $a \in \Sigma$ which appears at least once in all strings from $S[p^{\mathrm{L}}]$ is considered as a *feasible* candidate to extend s^{P}. Let us denote by $p_{a,i}^{\mathrm{L}}$ the position of the first occurrence of letter a in $s_i[p_i^{\mathrm{L}}, |s_i|]$. Among the feasible letters, *dominated* ones may occur. We say that letter a *dominates* letter b iff $p_{a,i}^{\mathrm{L}} \leq p_{b,i}^{\mathrm{L}}, \forall i = 1, \ldots, m$. Non-dominated letters are preferred, since the choice of a dominated letter will lead to a suboptimal solution. Denote a set of feasible, non-dominated letters by $\Sigma_{p^{\mathrm{L}}}^{\mathrm{nd}} \subseteq \Sigma$. The letter given by

$$a^* = \arg \min_{a \in \Sigma_{p^{\mathrm{L}}}^{\mathrm{nd}}} \left(g(p^{\mathrm{L}}, a) = \sum_{i=1}^{m} \frac{p_{a,i}^{\mathrm{L}} - p_i^{\mathrm{L}}}{|s_i| - p_i^{\mathrm{L}} + 1} \right)$$

is chosen to extend s^{P}, and we update $s^{\mathrm{P}} := s^{\mathrm{P}} \cdot a^*$, and $p_i^{\mathrm{L}} := p_{a^*,i}^{\mathrm{L}} + 1$, $i = 1, \ldots, m$, accordingly. We repeat the steps with the new s^{P} and p^{L} until $\Sigma_{p^{\mathrm{L}}}^{\mathrm{nd}} = \emptyset$ is met, constructing a complete greedy solution s^{P}. Let us label this procedure by $\mathrm{BNH}(S)$, taking an instance S as input and returning the derived greedy solution.

Beam Search is known as an incomplete tree search algorithm which expands nodes in a breadth-first search manner. It maintains a collection of nodes called *beam*. The $\beta > 0$ most promising nodes of these expansions are further used to create the beam of the next level. This step is repeated level by level until the beam is empty. We will consider several ways to determine the most promising nodes to be kept at each step in Sect. 2.1. A BS for the LCS has been proposed by Blum et al. [1]. Each node v of the LCS is defined by a left position vector $p^{\mathrm{L},v}$ which corresponds to the set $S[p^{\mathrm{L},v}]$ relevant for further extension of v, and an l^v-value, denoting the length of the partial solution represented by the node. Initially, the beam contains the (root) node $r := ((1, \ldots, 1), 0)$. In order to expand a node v, the corresponding set $S[p^{\mathrm{L},v}]$ is used to find all *feasible*, non-dominated letters $\Sigma_{p^{\mathrm{L},v}}^{\mathrm{nd}}$, and for each $a \in \Sigma_{p^{\mathrm{L},v}}^{\mathrm{nd}}$, the positions $p_{a,i}^{\mathrm{L},v}$ are determined, and further used to create all successor nodes $v' = (p_a^{\mathrm{L},v} + 1, l^v + 1)$ of v, where $p_a^{\mathrm{L},v} + 1 = \{p_{a,i}^{\mathrm{L},v} + 1 \mid i = 1, \ldots, m\}$. If $\Sigma_{p^{\mathrm{L},v}}^{\mathrm{nd}} = \emptyset$, a *complete* node has been reached. If the l^v-value of a complete node is greater than the length of the current incumbent solution, we derive the respective solution and store it as the new incumbent. We emphasize that directly storing partial solutions within the nodes is not necessary. For any node of the search tree, the respective partial solution can be derived in a backward manner by iteratively identifying predecessors in which the l^v-values always decrease by one. Let us denote this procedure by $\mathrm{BS}(S, \beta)$, taking a set S and a beam size β as input and returning the best solution found by the BS execution.

2.1   Estimating the Length of the LCS Problem

For each node of the search tree, an upper bound on the number of letters that might further be added—i.e., the length of a LCS of $S[p^{\mathrm{L},v}]$—is given by $\mathrm{UB}(v) = \mathrm{UB}(S[p^{\mathrm{L},v}]) = \sum_{i=1}^{m} c_a$, where $c_a = \min\{|s_i[p_i^{\mathrm{L},v}, |s_i|]|_a \mid i = 1, \ldots, m\}$, and $|s|_a$ is the number of occurrences of letter a in s; see [10]. Unfortunately, this upper bound is not tight in practice.

We develop here a novel estimation based on the approximated expected length for uniform random strings. Mousavi and Tabataba [9] came up with a recursion which calculates the probability that a specific string s of length k is a subsequence of a uniform random string t of length q in the form of a table $\mathcal{P}(|s|,|t|) = \mathcal{P}(k,q)$. Let us assume that the following holds: (1) all strings from S are mutually independent and (2) for any sequence of length k over Σ, the event that the sequence is a subsequence common to all strings in S is independent of the corresponding events for other sequences. By making use of the table $\mathcal{P}$ and some basic laws from probability theory, we can derive the estimation for the expected length as

$$\text{EX}(v) = \text{EX}(S[p^{\text{L},v}]) = l_{\max} - \sum_{k=1}^{l_{\max}} \left(1 - \prod_{i=1}^{m} \mathcal{P}(k, |s_i| - p_i^{\text{L},v} + 1) \right)^{|\Sigma|^k}, \quad (1)$$

where $l_{\max} = \min_{i=1,\dots,m}\{|s_i| - p_i^{\text{L},v} + 1\}$. EX provides, in practice, a much better approximation than the afore-mentioned upper bound UB or the heuristic from [9] which is also of limited use since it cannot be used to compare nodes from different levels of the search tree. Formula (1) is numerically calculated by decomposing the power $|\Sigma|^k = \underbrace{|\Sigma|^p \cdots |\Sigma|^p}_{\lfloor k/p \rfloor} \cdot |\Sigma|^{(k \mod p)}$, since intermediate values would otherwise be too large for a commonly used floating point arithmetic. Moreover, the calculation of (1) can be run in $O(m \log(n))$ on average by determining $v_k = \left(1 - \prod_{i=1}^{m} \mathcal{P}(k, |s_i| - p_i^{\text{R},v} + 1) \right)^{|\Sigma|^k} \in (\epsilon, 1 - \epsilon)$ using the divide-and-conquer principle exploiting the fact that $\{v_k\}_{k=1}^{l_{\max}}$ is a monotonic sequence. We set $\epsilon = 10^{-10}$ in our implementation. If the product which appears in (1) is close to zero, it can cause stability issues resolved by replacing v_k by an approximation derived from the Taylor expansion $(1 - x)^\alpha \approx 1 - \alpha x + \binom{\alpha}{2}x^2 + o(x^2)$ which approximates v_k well. This estimation was developed by following the same idea for the palindromic case of the LCS problem; see [3] for more details.

3   Algorithms for Solving the LCSqS Problem

Let us denote by $\mathbb{P} := \{(q_1, \dots, q_m) : 1 \le q_i \le |s_i|\} \subset \mathbb{N}^m$ all possibilities for partitioning the strings from S each one into two consecutive substrings. For each $q \in \mathbb{P}$, we define the left and right partitions of S by $S^{\text{L},q} = \{s_1[1,q_1], \dots, s_m[1,q_m]\}$ and $S^{\text{R},q} = \{s_1[q_1+1,|s_1|], \dots, s_m[q_m+1,|s_m|]\}$, respectively. Let $S^q := S^{\text{L},q} \cup S^{\text{R},q}$ be the joint set of these partitions. Finding an optimal solution s_{lcsqs}^* to the LCSqS problem can then be done as follows. First, an optimal LCS $s_{\text{lcs},q}^*$ must be derived for all S^q, $q \in \mathbb{P}$. Let $s_{\text{lcs}}^* = \arg\max\{|s_{\text{lcs},q}^*| : q \in \mathbb{P}\}$ Then, $s_{\text{lcsqs}}^* = s_{\text{lcs}}^* \cdot s_{\text{lcs}}^*$. Unfortunately, the LCS problem is already $\mathcal{NP}$-hard [7], and the size of $\mathbb{P}$ grows exponentially with the instance size. This approach is, therefore, not practical. However, we will make use of this decomposition approach in a heuristic way as shown in the following.

3.1 Randomized Local Search Approach

In this section we adapt and iterate BNH in order to derive approximate LCSqS solutions in the sense of a randomized local search (RLS).

We start with the $q = \left(\lfloor \frac{|s_1|}{2} \rfloor, \ldots, \lfloor \frac{|s_m|}{2} \rfloor \right)$ and execute BNH on the corresponding set S^q to produce an initial approximate LCSqS solution $s_{\text{lcsqs}} = (\text{BNH}(S^q))^2$. At each iteration, q is perturbed by adding to each q_i, $i = 1, \ldots, m$, a random offset sampled from the discretized normal distribution $\lceil \mathcal{N}(0, \sigma^2) \rceil$ with a probability $destr \in (0, 1)$, where the standard deviation is a parameter of the algorithm. BNH is applied to the resulting string set S^q for producing a new solution. A better solution is always accepted as new incumbent solution s_{lcsqs}. The whole process is iterated until a time limit $t_{\max}$ is exceeded. Note that if s_{lcsqs} is the current incumbent, only values in $\{|s_{\text{lcsqs}}|/2 + 1, \ldots, |s_i| - |s_{\text{lcsqs}}|/2 - 1\}$ for q_i can lead to better solutions. We therefore iterate the random sampling of each q_i until a value in this range is obtained.

3.2 RVNS and BS Approach

As an alternative to the RLS described above we consider a variable neighborhood search approach [8]. More precisely, we use a version of the VNS with no local search method included, known as *Reduced* VNS (RVNS).

For a current vector $q \in \mathbb{P}$, we define a move in the k-th neighborhood, $k = 1, \ldots, m$, by perturbing exactly k randomly chosen positions as above by adding a discretized normally distributed sampled random offset. Again, we take care not to choose meaningless small or large values. We then evaluate q by the following 3-step process. We first calculate $ub_q = 2 \cdot \text{UB}(S^q)$, and if $ub_q \leq |s_{\text{lcsqs}}|$, q cannot yield an improved incumbent solution and q is discarded. Otherwise, we perform a fast evaluation of q by applying BNH which yields a solution $s = (\text{BNH}(S^{q'}))^2$. If $|s| > \alpha \cdot |s_{\text{lcsqs}}|$, where $\alpha \in (0, 1)$ is a threshold parameter of the algorithm, we consider q promising and further execute BS on S^q, yielding solution $s^{\text{bs}} = (\text{BS}(S^{q'}, \beta))^2$. Again, the incumbent solution s_{lcsqs} is updated by any obtained better solution. If an improvement has been achieved, the RVNS&BS always continues with the first neighborhood, i.e. $k := 1$, otherwise with the next neighborhood, i.e. $k := k + 1$ until $k = m + 1$ in which case k is reset to 1. To improve the performance, we store all partitionings evaluated by BS, together with their evaluations, in a hash map and retrieve these values in case the corresponding partitionings are re-encountered.

4   Computational Experiments

The algorithms are implemented in C++ and all experiments are performed on a single core of an Intel Xeon E5-2640 with 2.40 GHz and 8 GB of memory.

We used the set of benchmark instances provided in [2] for the LCS problem. This instance set consists of ten randomly generated instances for each combination of the number of input strings $m \in \{10, 50, 100, 150, 200\}$, the length of the input strings $n \in \{100, 500, 1000\}$, and the alphabet size $|\Sigma| \in \{4, 12, 20\}$. This makes a total of 450 problem instances. We apply each algorithm ten times to each instance, with a time limit of 600 CPU seconds.

Table 1. Selected results for the LCSqS problem.

n	m	$|\Sigma|$	RVNS&BS		RLS& BS		RVNS& Dive		RLS	
			$|\overline{s}|$	$\overline{t}_{best}[s]$	$|\overline{s}|$	$\overline{t}_{best}[s]$	$|\overline{s}|$	$\overline{t}_{best}[s]$	$|\overline{s}|$	$\overline{t}_{best}[s]$
100	10	4	**27.08**	67.71	26.54	44.94	26.96	51.20	26.42	34.40
	10	20	3.84	0.02	**4.00**	1.66	3.96	0.05	**4.00**	4.44
	50	4	**18.54**	10.53	18.16	24.12	**18.54**	45.81	18.04	19.43
	50	20	0.20	0.01	**0.46**	4.77	0.20	0.00	0.40	0.01
	200	4	**14.00**	4.88	**14.00**	8.68	**14.00**	1.36	13.94	24.12
	200	20	**0.00**	0.05	**0.00**	0.00	**0.00**	0.00	**0.00**	0.00
500	10	4	**156.58**	143.70	156.14	146.08	149.78	160.69	149.24	110.09
	10	20	**35.78**	53.31	35.12	48.07	34.54	50.42	34.56	71.16
	50	4	**124.30**	52.66	124.12	160.39	120.32	86.33	120.12	109.37
	50	20	**21.14**	78.62	20.52	34.12	20.64	66.00	20.68	61.19
	200	4	**109.86**	152.55	108.78	102.03	106.22	66.72	104.94	90.66
	200	20	**14.48**	62.08	14.26	35.50	14.04	3.51	14.10	12.82
1000	10	4	**321.14**	206.16	320.94	193.50	304.48	186.65	304.34	161.08
	10	20	**76.84**	126.40	76.66	141.40	73.80	118.86	73.72	76.98
	50	4	**261.52**	127.81	260.82	135.14	252.94	131.88	249.84	153.18
	50	20	**49.78**	116.89	49.76	188.74	48.12	54.48	48.70	74.04
	200	4	**235.50**	213.72	234.44	202.34	230.10	135.37	222.66	145.99
	200	20	38.04	132.86	**38.12**	165.15	38.00	59.74	38.02	24.07

From preliminary experiments we noticed that the behavior of our algorithms mostly depends on the length n of the input strings. Therefore, we tuned the algorithms separately for instances with string length 100, 500, and 1000. The *irace* tool [6] was used for this purpose. For RLS, we obtained $destr = 0.2$ and $\sigma = 5$ (for $n = 100$), $destr = 0.3$ and $\sigma = 10$ (for $n = 500$), and $destr = 0.3$ and $\sigma = 20$ (for $n = 1000$). For RVNS&BS, we obtained $\alpha = 0.9$ and $\beta = 100$ (for $n = 100$), $\alpha = 0.9$ and $\beta = 200$ (for $n = 500$), and $\alpha = 0.9$ and $\beta = 200$ (for $n = 1000$). For σ of the RVNS&BS, *irace* yielded the same values as for the RLS. Moreover, EX was preferred over UB as a guidance for BS.

We additionally include here results for RVNS&Dive, which is RVNS&BS with $\beta = 1$. In this case BS reduces to a simple greedy heuristic (or dive). This was done for checking the impact of a higher beam size. Moreover, RLS&BS refers to a version of RLS in which BNH is replaced by BS with the same beam size as in RVNS&BS. Selected results are shown in Table 1. For each of the algorithms we present the avg. solution quality and the avg. median time when the best solution was found. From the results we conclude the following:

- RVNS&BS produces solutions of significantly better quality then the other algorithms on harder instances.
- The rather high beam size is apparently useful for finding approximate solutions of higher quality.
- Concerning the computation time for harder instances, the times of the RVNS&BS are usually higher than those of the RLS. It seems harder for BNH to help to improve solution quality in later stages of the RLS than for the BS in RVNS&BS.
- From Fig. 1 we can see that, for smaller instances with larger alphabet sizes, stronger jumps in the search space are in essence preferred. This is because a small number of feasible solutions is distributed over the search space, and to find them it is convenient to allow large, random jumps in the search. When n is larger, choosing to do larger jumps in the space is not a good option (see the bar plot on the right). This can be explained by the fact that already the vector q that is defined by the middle of all input strings (which are generated uniformly at random) yields a promising solution, and many promising partitions are clustered around this vector. By allowing larger jumps, we move further away from this middle vector quickly, which yields usually in weaker solutions.

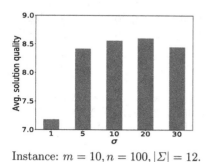

Instance: $m = 10, n = 100, |\Sigma| = 12$.

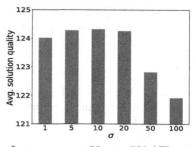

Instance: $m = 50, n = 500, |\Sigma| = 4$.

Fig. 1. The impact of parameter σ on the solution quality of RVNS&BS.

Fig. 2 provides box plots showing the relative differences between the quality of the solutions obtained by RVNS&BS using EX and RVNS&BS using UB ($\beta = 200$). The figure shows a clear advantage of several percent when using EX over the classical upper bound UB as search guidance.

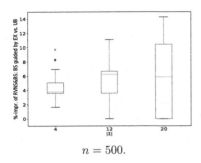
$n = 500.$

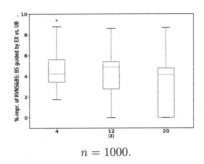
$n = 1000.$

Fig. 2. Improvements of solution quality when using EX instead of UB for guiding BS in RVNS&BS.

5 Conclusions and Future Work

This article provides the first heuristic approaches to solve the LCSqS problem for an arbitrary set of input strings. We applied a reduction of the LCSqS problem to a series of standard LCS problems by introducing a partitioning of the input strings as a first-level decision. Our RVNS framework explores the space of partitionings, which are then tackled by BNH and, if promising, by BS. Hereby, BS is guided by a heuristic which approximates the expected length of a LCS. Overall, RVNS&BS yields significantly better solutions that the also proposed, simpler RLS.

In future work we want to solve smaller instances of the LCSqS problem to optimality. To achieve this, we aim for an A* search that is also based on the described reduction to the classical LCS problem.

Acknowledgments. We gratefully acknowledge the financial support of the project by the Doctoral Program "Vienna Graduate School on Computational Optimization" funded by the Austrian Science Foundation (FWF) under contract no. W1260-N35.

References

1. Blum, C., Blesa, M.J., López-Ibáñez, M.: Beam search for the longest common subsequence problem. Comput. Oper. Res. **36**(12), 3178–3186 (2009)
2. Blum, C., Festa, P.: Metaheuristics for String Problems in Bio-Informatics. Wiley, Hoboken (2016)
3. Djukanovic, M., Raidl, G., Blum, C.: Anytime algorithms for the longest common palindromic subsequence problem. Technical report AC-TR-18-012, Algorithms and Complexity Group, TU Wien (2018)
4. Huang, K., Yang, C.-B., Tseng, K.-T., et al.: Fast algorithms for finding the common subsequence of multiple sequences. In: Proceedings of the International Computer Symposium, pp. 1006–1011. IEEE Press (2004)
5. Inoue, T., Inenaga, S., Hyyrö, H., Bannai, H., Takeda, M.: Computing longest common square subsequences. In: Proceedings of CPM 2018–29th Annual Symposium on Combinatorial Pattern Matching. Schloss Dagstuhl-Leibniz-Zentrum für Informatik, Dagstuhl Publishing (2018)

6. López-Ibáñez, M., Dubois-Lacoste, J., Pérez Cáceres, L., Stützle, T., Birattari, M.: The irace package: iterated racing for automatic algorithm configuration. Oper. Res. Perspect. **3**, 43–58 (2016)
7. Maier, D.: The complexity of some problems on subsequences and supersequences. J. ACM (JACM) **25**(2), 322–336 (1978)
8. Mladenović, N., Hansen, P.: Variable neighborhood search. Comput. Oper. Res. **24**(11), 1097–1100 (1997)
9. Mousavi, S.R., Tabataba, F.: An improved algorithm for the longest common subsequence problem. Comput. Oper. Res. **39**(3), 512–520 (2012)
10. Wang, Q., Pan, M., Shang, Y., Korkin, D.: A fast heuristic search algorithm for finding the longest common subsequence of multiple strings. In: Proceedings of AAAI 2010-24th AAAI Conference on Artificial Intelligence, pp. 1287–1292 (2010)

Investigating the Dynamic Block Relocation Problem

Sebastian Raggl[1]([✉]), Andreas Beham[1,2], and Michael Affenzeller[1,2]

[1] Heuristic and Evolutionary Algorithms Laboratory,
University of Applied Sciences Upper Austria, Hagenberg, Austria
{`sebastian.raggl,andreas.beham,michael.affenzeller`}`@fh-hagenberg.at`
[2] Institute for Formal Models and Verification,
Johannes Kepler University, Linz, Austria

Abstract. The dynamic block relocation problem is a variant of the BRP where the initial configuration and retrieval priorities are known but are subject to change during the implementation of an optimized solution. This paper investigates two kinds of potential changes. The exchange of assigned priorities between two blocks and the arrival of new blocks. For both kind of events we present algorithms that can adjust an existing solution to the changed situation. These algorithms are combined with a branch and bound based solver to enable online optimization with look-ahead. Our experiments show that the algorithms enable finding better solutions in a shorter time after a event occurs.

Keywords: Dynamic block relocation problem · Changing priorities

1 Introduction

The block relocation problem is an important problem in both container logistics and the steel industry. Steel slabs as well as containers are stored in stacks and have to be retrieved from the storage yard in a given order using the minimal number of crane movements possible. Solvers are based on heuristics [3,6] mixed-integer programming [7] or branch and bound [4,5].

Using a BRP solver to control a real-life crane is a three step process. First we observe the current state of the environment and generate a BRP instance. Then we use a BRP solver to find a optimized crane schedule. Finally the crane executes one move after another until all blocks are delivered. Of course this only works if the delivery priorities never change and no new blocks are added while the plan is executed. There are several approaches for dealing with such uncertainties discussed in the literature. The online BRP avoids the problem by not planning ahead at all and just deciding one move at a time. There is a

The work described in this paper was done within the project *Logistics Optimization in the Steel Industry (LOISI)* (#855325) within the funding program Smart Mobility 2015, organized by the Austrian Research Promotion Agency (FFG) and funded by the Governments of Styria and Upper Austria.

© Springer Nature Switzerland AG 2020
R. Moreno-Díaz et al. (Eds.): EUROCAST 2019, LNCS 12013, pp. 438–445, 2020.
https://doi.org/10.1007/978-3-030-45093-9_53

leveling heuristic with a good and known competitive ratio but using planning data when available enables finding better solutions [8]. There is work on a stochastic BRP where items are grouped and the delivery order of groups is assumed to be known but the order within a group is determined at random in an online fashion. It is shown that such partial information can be used to improve solution quality [2]. Both of these works assume that some information is just not known, in contrast the work in this paper assumes that the initial state and priorities are known but can change over time. When such a change is observed the plan must change to accommodate it and the simplest way to achieve this is to rerun the optimizer. However this is wasteful as the problems are clearly related but the solver is unable to take advantage of this fact. We show that in the case of two kinds of unplanned events we can efficiently update an existing solution. If the original solution was optimal the update algorithms can under certain conditions even guarantee that the resulting updated solution remains optimal.

The rest of this paper is structured as follows. We start in Sect. 2 by defining the dynamic BRP we are dealing with and describing our general approach. In Sect. 2.1 we describe the problem of new items arriving over time and describe an algorithm for updating solutions to incorporate the newly arrived items. Section 2.2 is similar but for events where the priorities of two blocks are swapped. We evaluate the performance of our algorithms in comparison to a reoptimization scheme on problem instances from the literature in Sect. 3 and finally present conclusions and an outlook on further research in Sect. 4.

2   Dynamic Block Relocation Problem

The state of a BRP denoted as θ consists of a set of blocks with distinct delivery priorities $n \in N$. The blocks are stacked onto a number of stacks $s \in S$. There is a crane that can perform two different kinds of moves. If the block that has to be delivered next is on top of a stack the crane performs a `removal` move and delivers the block. Otherwise the crane has to perform a `relocation` move by taking the top block of one stack and placing it on top of a different stack. The goal is to deliver all blocks in order of the delivery priorities using the minimal number of crane movements. Such an optimized crane schedule Ω is therefore a list of moves the crane has to executed.

The BRP variant that this paper is concerned with is dynamic meaning that the problem changes over time. We observe the current state θ and use a branch and bound based solver to obtain a near optimal solution ($\texttt{solve}(\theta) \rightarrow \Omega$). Then the crane movements are executed according to plan until an event happens that invalidates the current plan. We then apply the adjustment algorithms described below in order to regain a valid plan and there are two possible outcomes. Either the current solution could be adjusted by inserting the minimal number of `relocation` and `removal` moves or it needs to be changed more extensively. In the latter case the same optimizer used to find the initial solution is applied to a partial solution to search for the best changed solution. After we have a new adjusted or changed solution we can if there is more time call the optimizer again to continue searching for even better solutions. This search can use the new solution as an upper bound which enables it to be more efficient.

2.1   Arrival of a New Item

In addition to the normal stacks we introduce so called arrival locations. When new items arrive they are added to the last position of one of the arrival locations. The crane can only take the first item of a arrival location and it can never put items back to a arrival location.

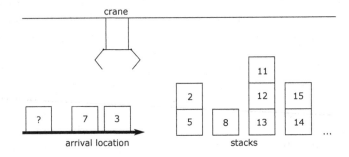

Fig. 1. New item arriving at arrival location.

Items that are added can have any removal priority, this is indicated by the question mark in the example shown in Fig. 1. So depending on the priorities of the items at the arrival location the introduction of an additional item can necessitate additional relocation moves. There are three different possibilities of when the new item must be delivered:

1. Deliver after all items at the arrival location. (e.g. ? = 9)
2. Deliver before all items at the arrival location. (e.g. ? = 1)
3. Deliver before some of the items at the arrival location. (e.g. ? = 4)

We define R as the list of items for which we have to introduce additional relocations. In the first case R is empty because we can simply wait until the items are delivered and than remove the new item. In the second case $R = [7, 3]$ contains all items at the arrival location, because in order to remove the new item the others must be put aside. In case the new item has priority 4 in the example of Fig. 1 R would contain the item with priority 7 but not the one with priority 3. If we imagine that the priorities 3 and 7 were exchanged R would be empty because even though 7 must be delivered after 4 every valid solution already contains the relocation of item 7 in order to remove item 3. Therefore by the time item 4 must be removed both 7 and 3 are no longer at the arrival location. For every item in R we need at least one additional relocation. Where

Data: Ω, n_{new}, s_{arr}
Result: adjusted solution
$(R, i_{min}) \leftarrow$ **needs_relocation**$(\Omega_0, n_{new}, s_{arr})$;
$i_{max} \leftarrow$ **removal_index**(Ω, n_{new});
insert_removal$(\Omega, i_{max}, n_{new}, s_{arr})$;
foreach $n \in R$ **do**
   | **delete_all_moves_of**(Ω, n);
   | $sucess \leftarrow false$;
   | $i_{rem} \leftarrow$ **removal_index**(Ω, n);
   | **for** $i_{rel} \leftarrow i_{max}$ **to** i_{min} **do**
   | | **if** $\exists s \in S$ *is_well_placed*$(n, s, \Omega, i_{rel}, i_{rem})$ **then**
   | | | **insert_relocation**$(\Omega, i_{rel}, n, s_{arr}, s)$;
   | | | **insert_removal**(Ω, i_{rem}, n, s);
   | | | $i_{max} \leftarrow i_{reloc}$;
   | | | $sucess \leftarrow true$;
   | | | break;
   | | **end**
   | **end**
   | **if** *not sucess* **then**
   | | **return** *solve*(Ω_0);
   | **end**
end
return Ω;

Algorithm 1. Adjusting solution when a new item arrives

this relocation can be inserted into the existing solution Ω is limited by the blocks above and below.

Algorithm 1 describes how this can be done. It takes a solution to the original problem Ω and the arrival location s_{arr} and delivery priority of the new block n_{new}. It first uses the function **needs_relocation** to determine which blocks need to be relocated R because of the new block and when the earliest relocation can take place i_{max}. For every item in R it first deletes all existing moves using **delete_all_moves_of** and then tries to insert only a single relocation and removal move per item such that the solution stays valid. The index where the remove move must be inserted is calculated using the **removal_index** function. The algorithm tries to find an index in the solution where a block can be relocated to a stack and remain there until it can be removed. To check this the algorithm calls **is_well_placed** which takes a block n a stack s the current solution Ω as well as the proposed relocation and removal indices. It checks that the item does not block anything in the stack when it is placed, that nothing blocks it when it needs to be removed and that no relocation in between violates the height restriction of the stack. With the appropriate preprocessing both **is_well_placed** and **removal_index** can be made to run in constant time which makes the entire algorithm very fast as we will show in Sect. 3. If the algorithm can incorporate all blocks in R using only a single relocation the resulting solution will be optimal given that the solution was optimal before the new item arrived. When any block in R would need more than one relocation the algorithm simply runs the solver again.

2.2 Swapping Delivery Priorities

Another event that can happen is that the delivery priorities of the blocks changes over time. If the order changes entirely the best course of action is to simply restart the optimization. On the other hand if only a small change in delivery order occurs it might be advantageous to adjust the current solution. Such a possible small change can be that two blocks exchange their delivery priorities. Algorithm 2 shows how to adjust a solution after such a priority swap has occured. It works by comparing the planned moves with the possible moves performing removal moves as soon as possible and skipping all moves of already removed blocks. As long as the planned moves are in the set of possible moves the solution remains valid. If this is no longer the case we apply the solver starting from the current state.

Data: Ω
Result: adjusted solution
$\Omega_{new} \leftarrow \emptyset$;
$\theta \leftarrow \Omega_0$;
foreach *planned* $\in \Omega$ **do**
    $B \leftarrow$ possible_moves(θ);
    if $\exists \beta \in B$ *is_remove_move*(β) **then**
        | $\theta \leftarrow$ apply$(\Omega_{new}, \theta, \beta)$;
    end
    if *not_already_removed*(θ, β) **then**
        if *planned* $\in$ *possible* **then**
            | $\theta \leftarrow$ apply$(\Omega_{new}, \theta, planned)$;
        else
            | **return** *solve*(θ);
        end
    end
end
return Ω_{new};

Algorithm 2. Adjusting solution when priorities are swapped

3 Experiments

The question we want to answer with our experiments is does the techniques described above really enable us to find better solutions faster after a event occured that made the current solution invalid? All benchmarks are implemented using the Rust language and run on a Dell Latitude E6540 with an Intel i7 4810MQ CPU @2.80 GHz and 16 GB RAM running Windows 10. We used problem instances BRP instances from the literature [1] and simulated the events described in Sect. 2. We start by searching for near optimal solutions for each problem instance and then we apply the change produced by a single event. The optimizer we use is an iterative variant of the rake search algorithm with a width limit of 100 and a depth limit of 1000 [6]. When a event occurs we have two possibilities, we can either treat the resulting problem instance as entirely separate and just start the solver from scratch or we use the algorithms described above to try to adapt the current solution to the change. We call the former the offline approach and the latter the online approach. By recording the times and qualities of the found solutions and comparing them we can get a idea of the relative performance of these two approaches. The problem instances vary in both the number of stacks and the number of items per stack from instances with three stacks of height 3 to instances of 10 stacks each with 10 items. We choose to focus on the 40 largest 10×10 instances because the small instances can be solved so quickly by the offline approach that improving that time is not particularly interesting.

In order to simulate arrival of new items one stack of each problem instance was randomly choosen to serve as the arrival location and for every items on the arrival location the arrival is simulated. This means we have 10 events for each of our 40 instances totaling 400 simulated arrival events. Figure 2 shows a empirical cumulative distribution function (ECDF), it compares the time it takes for a certain percentage of algorithm runs to reach the best solution between the online and offline approach. Out of all plans 68% could be adjusted when a new block arrived at the arrival location and for 2% additional improvements could be found after adjusting the previous solution. After 13% of events the plans were changed and no further improvement could be reached while in 17% of cases further improvements could be achieved starting from the changed solution. Algorithm 1 enables the online approach to both find good solutions very quickly while still allowing improvements over time.

For the test instances with 100 blocks there are $\binom{100}{2} = 4950$ possible swap events per problem instance. We chose to investigate swap events where the

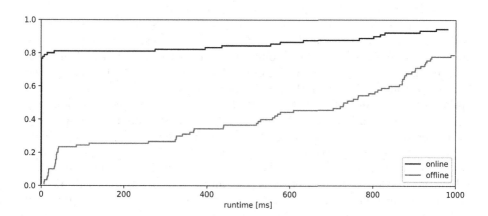

Fig. 2. Percentage of solutions reaching the best quality over time after a new block is added.

difference between the swapped priorities is less then 5. Out of all plans 18% could be adjusted after a swap event and 82% needed to be changed. These percentages vary depending on the priority of the items involved in the swap. The further into the future the removal of the affected items is the more likely it is that the solution can be adjusted. Similarly the the closer the priorities are to each other the more likely it as that adjustment is possible. Figure 3 shows that after swapping item priorities the online approach enables finding solutions faster than the offline approach. The in graph on the right where the affected blocks have to be delivered later the online curve starts out higher because of the higher proportion of adjustable solutions.

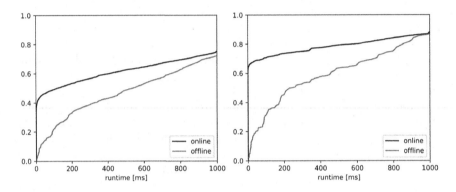

Fig. 3. Percentage of solutions reaching the best quality over time after two blocks exchange priorities. Swapped priorities <50 shown left and swapped priorities >50 shown right.

4   Conclusion and Outlook

Using the procedures described in this paper it is possible to adjust existing optimized solutions when unplanned events occur. The adjustment procedures are fast enough that even if the solution could not be adjusted and must be changed the online approach outperforms starting from scratch. This is achieved by starting the search only on the part of the solution that needs to change. If the optimal solution has not been found jet the adjusted or changed solutions can also be used as an initial upper bound for the continued search. The reason our online optimizer performs better than simply calling the offline solver repeatedly is that it is able to preserve some of the search state. This state is the current best solution that gets adjusted, changed and improved over time. Of course a branch and bound based optimizer has much more state information. One direction for further research is to investigate if performance can be improved by keeping and updating more of that information.

In this paper we looked at handling individual events but the long term goal of this work is to enable a solver that can run indefinitely improving and adjusting the solution as time goes on. This is relevant because both adjustment algorithm greedily choose the first adjustment they can find but it might be the case that some adjustments enable or prohibit future adjustments and therefore influence overall performance. Extending the algorithms to consider multiple alternative is relatively simple but in order to evaluate the performance of such extensions we are working on a simulator and appropriate benchmark scenarios and metrics.

References

1. Caserta, M., Voß, S.: A corridor method-based algorithm for the pre-marshalling problem. In: Giacobini, M., et al. (eds.) EvoWorkshops 2009. LNCS, vol. 5484, pp. 788–797. Springer, Heidelberg (2009). https://doi.org/10.1007/978-3-642-01129-0_89
2. Galle, V., Manshadi, V.H., Boroujeni, S.B., Barnhart, C., Jaillet, P.: The stochastic container relocation problem. Transp. Sci. **52**(5), 1035–1058 (2018). https://doi.org/10.1287/trsc.2018.0828
3. Kim, K.H., Hong, G.P.: A heuristic rule for relocating blocks. Comput. Oper. Res. (2006). https://doi.org/10.1016/j.cor.2004.08.005
4. Tanaka, S., Mizuno, F.: An exact algorithm for the unrestricted block relocation problem. Comput. Oper. Res. (2018). https://doi.org/10.1016/j.cor.2018.02.019
5. Tanaka, S., Takii, K.: A faster branch-and-bound algorithm for the block relocation problem. IEEE Trans. Autom. Sci. Eng. (2016). https://doi.org/10.1109/TASE.2015.2434417
6. Tricoire, F., Scagnetti, J., Beham, A.: New insights on the block relocation problem. Comput. Oper. Res. (2018). https://doi.org/10.1016/j.cor.2017.08.010
7. Zehendner, E., Caserta, M., Feillet, D., Schwarze, S., Voß, S.: An improved mathematical formulation for the blocks relocation problem. Eur. J. Oper. Res. (2015). https://doi.org/10.1016/j.ejor.2015.03.032
8. Zehendner, E., Feillet, D., Jaillet, P.: An algorithm with performance guarantee for the Online Container Relocation Problem. Eur. J. Oper. Res. (2017). https://doi.org/10.1016/j.ejor.2016.09.011

Multi-objective Evolutionary Approaches for Solving the Hybrid Flowshop Scheduling Problem

Fabricio Niebles-Atencio[1]($\boxtimes$), Miguel Rojas-Santiago[2],
and Daniel Mendoza-Casseres[3]

[1] Institute of Information Systems, Universität Hamburg, Von-Melle-Park 5,
20146 Hamburg, Germany
fabricio.niebles.atencio@uni-hamburg.de
[2] Industrial Engineering Department, Universidad del Norte,
Km 5 Ant. Via Puerto Colombia, Barranquilla, Colombia
miguelrojas@uninorte.edu.co
[3] Facultad de Ingenieria, Universidad del Atlantico, Km 7 Ant. Via Puerto
Colombia, Barranquilla, Colombia
danielmendoza@mail.uniatlantico.edu.co

Abstract. In this paper we schedule a set of jobs on a production system with more than one stage and several machines in parallel per stage, considering multiple objectives to be optimized. This problem is known as the Flexible or Hybrid Flowshop Scheduling problem (HFSP), which is NP-hard even for the case of a system with only two processing stages where one stage contains two machines and the other stage contains a single machine. In that sense, it is possible to find an optimal solution for this problem with low computing resources, only for small instances which, in general, do not reflect the industrial reality. For that reason, the use of meta-heuristics as an alternative approach it is proposed with the aim to determine, within a computational reasonable time, the best assignation of the jobs in order to minimize the makespan, total tardiness and the number of tardy jobs simultaneously. In this regard, a Multi-Objective Ant Colony Optimization algorithm (MOACO) and the Non-dominated Sorting Genetic Algorithm-II (NSGA-II) are used for solving this combinatorial optimization problem. Results show the effectiveness of the approaches proposed.

Keywords: Hybrid flow shop · Meta-heuristics · Multi-objective optimization

1 Introduction

Scheduling is one of the hard optimization problems found in real industrial contexts. Generally speaking, scheduling is a form of decision-making that plays a crucial role in manufacturing and service industries. According to [1] scheduling

© Springer Nature Switzerland AG 2020
R. Moreno-Díaz et al. (Eds.): EUROCAST 2019, LNCS 12013, pp. 446–453, 2020.
https://doi.org/10.1007/978-3-030-45093-9_54

problems deal with the allocation of resources to tasks over given time periods and its goal is to optimize one or more objectives. In this regard, flow-shop scheduling (FS) problems are among the most studied combinatorial optimization problems in the literature. These scheduling problems frequently can be found in many real world applications such as manufacturing, transportation systems and service companies [2]. In this paper, an extension of the classical FS is studied, in which a set of jobs on a production system with more than one stage and several machines in parallel per stage is scheduled, considering one or multiple objectives to be optimized. This problem is known as the Flexible or Hybrid Flowshop Scheduling problem (HFSP), which can be reduced to the FS problem when each stage has only one machine. Furthermore, HFS adds more flexibility per stage than the classical Flowshop by increasing the overall capacity, avoiding bottle necks if some operations are too long [3]. Since most extensions of the FS problems are NP-hard, thus, HFSP is also NP-hard even for the case of a system with only two processing stages where one stage contains two machines and the other stage contains a single machine [4].

According to above, this paper considers the multi-objective hybrid flowshop scheduling problem with the aim of minimizing makespan, total tardiness and number of tardy jobs at the same time. This problem has not been studied previously in the literature since most works consider mainly two of these objectives simultaneously. Our solution approach is based on a Ant Colony System Optimization (ACO) Algorithm and the Non-dominated Sorting Genetic Algorithm-II (NSGA-II). The rest of this paper is organized as follows: Sect. 2 is devoted to present a review of literature related to the HFSP. Section 3 explains in detail the problem under study. In addition, Sect. 4 presents the proposed ACO and NSGA-II algorithms, while Sect. 5 is devoted to computational experiments and the analysis of results. Finally, this paper ends in Sect. 6 by presenting some concluding remarks and suggestions for further research.

2   Literature Review

The scientific literature regarding the utilization of meta-heuristics for solving scheduling problems is overwhelming. Although most of these works have focused on single objective applications, we have some works that involve the optimization of two or more objectives simultaneously, especially in shop scheduling problems [5]. In addition, with respect to the Hybrid Flow Shop scheduling problem, there is a considerable amount of papers which address mono-objective versions of the problem. Nevertheless, there are not so many works which deal with multi-objective version of this problem in comparison to another scheduling problems. For more details we recommend the works of [2] and [6], where comprehensive surveys are provided.

On the other hand, there are several examples of meta-heuristic approaches for addressing the HFSP with multiple objectives. In that sense, genetic algorithms [7,8], ant colony system algorithms [9,10] and simulated annealing and tabu search [11], are among the most common approaches. However, less used methods such as: migrating birds optimization algorithm [12] and energy consumption algorithm [13], have also been used recently. Since the previous studies included mostly two objectives, the multi-objective function with three objectives studied in this paper has not been considered by any other research.

3   Problem Formulation

Formally, the problem under study can be described as follows. In accordance with Grahams' notation, given a set of n jobs to be processed on a set of s stages organized in series, each one containing a set of m_s machines in parallel, the problem can be stated as: $FH_s(Pm_1^{(t)}, Pm_2^{(t)}, ..., Pm_n^{(t)})\|C_{max}, \sum T, \sum U$. In this notation, $Pm_1^{(t)}, Pm_2^{(t)}, ..., Pm_n^{(t)}$ correspond to $m^{(t)}$ homogeneous or identical machines in parallel at stage $t = 1, 2, ... , s$; i.e., at each stage t, the number of machines $m^{(t)}$ varies, $1 \leq m^{(t)} \leq M$. In addition, the processing route of all jobs is identical; i.e., jobs are consecutively moved from stage 1 to stage s. Moreover, at each stage a job j has to be processed on only one machine, and once the processing of a job is started, it cannot be interrupted (e.g., preemption is not allowed). The processing time of job j on any machine of stage s is denoted as p_{js}. All jobs are available at the beginning of the time horizon (e.g., all jobs have the same release date). The due date is denoted as d_j; the makespan or C_{max} is equal to $\max\{0, C_j\}$, where C_j is the completion time of job j; the tardiness of job j, T_j, is equal to $\max\{0, C_j - d_j\}$. Thus, the total tardiness is denoted as $\sum T$. Finally, the number of tardy jobs is equal to $\sum U$, where U_j is 1 when a job j is delayed, that is: $(C_j > d_j)$, and 0 otherwise.

4   Approaches Proposed

In this section, the structures of the approaches used for the problem under consideration are described. Therefore, MOACO and NSGA-II procedures are showed as well as the way that were implemented in the problem under study.

Multi-objective Ant Colony Optimization Algorithm. Ant colony optimization (ACO) is a meta-heuristic which was initially proposed by [14] and later on improved in further studies (e.g. see [15] and [17]). The common pattern of all variants of ant-based algorithms, consists on the emulation of real ants when they find the optimal path between their nest and a food source. Several studies have applied ACO to solve different discrete and continuous optimization problems [16]. One of these applications involves scheduling problems as stated by [18] in a comprehensive survey. ACO has experimented an evolution through eight variants or extensions which have been used to different problems.

In this work, the Ant Colony System (ACS) variant is used for the HFSP. In this regard, a model to represent pheromones, a mechanism to update the pheromone trail, and heuristic function are defined [19]. These elements are employed to guide the selection of a job to be executed at a given time. This also impacts the system behavior. Moreover, and with the aim of obtaining feasible solutions, job routing sequence has to be respected at each step when building a solution. This is ensured by using a restricted candidate list of all jobs that may be carried out at a given time of the schedule. The structure of ACS proposed, is presented in Algorithm 1:

Algorithm 1. Pseudocode of MOACO proposed

1: **Initialize parameters ()**
2: **while** Termination condition not reached **do**
3:     **for** Each ant k **do**
4:         **for** Each stage s **do**
5:             Select the next processing job applying transition state rule
6:             Assign a position to selected processing job
7:             Update pheromone trails locally
8:         **end for**
9:     **end for**
10:     Evaluate solution()
11:     Update pheromone trails globally
12: **end while**

Non-dominated Sorting Genetic Algorithm-II. The non-dominated sorting genetic algorithm (NSGA) was initially proposed by [20], and has been applied to diverse optimization problems. The main feature of a non-dominated sorting procedure, is that a ranking selection method is performed in order to characterize non-dominated points of the objective function as well as establishing a subset of fixed populations around those values [20]. Nevertheless, as stated by [21], there have been a number of criticisms due to issues such as: High complexity in the non-dominated sorting procedure, a lack of elitism and the necessity of a high parameterization. Therefore, an improved version of NSGA, named NSGA-II was proposed in order to alleviate all the above difficulties.

Concerning the application of NSGA-II for the problem under study, the population is randomly generated. At each iteration a sorting procedure is applied to rank the solutions according its dominance over the total population. Then, a solution is said to be better if it is not dominated by any other solution in the population. In addition, the procedure for producing new solutions is performed by the application of Partially Mapped Crossover (PMX). Mutation with a fixed probability is also considered. The algorithm stops after a predefined number of iterations. The structure of NSGA-II used, is presented in Algorithm 2, which was adapted from [21].

5   Computational Experiments

The computational experiments were carried out on a PC Intel Core i7, 2.9 GHz with 8 GB of RAM. The proposed meta-heuristics were coded using Visual Basic

6.0. Data sets used in our experiments were extracted from the OR-Library [22]. Instances with 20, 50 and 100 jobs and shops with 2, 5 and 8 stages were considered. For each meta-heuristic, non-dominated solutions (i.e., the set of Pareto-optimal solutions) were registered. The results obtained included the set of non-dominated solutions for each meta-heuristic for the selected instances, as well as the evaluation of distance measures between two fronts, the coverage of the solutions of both meta-heuristics, the deviation with respect to a single objective, the deviation with respect to the best initial solution and the computational time. Nevertheless, this last aspect was excluded from the results showed since it presented no statistical or significant differences among both meta-heuristics. An overview of these results is presented in Tables 1, 2 and 3.

Algorithm 2. Pseudocode of the NGSA-II used

1: **for** each $p \in P$ **do**
2:     $S_p \leftarrow \emptyset$
3:     $n_p \leftarrow \emptyset$
4:     **for** each $q \in P$ **do**
5:         **if** $(p \prec q)$ **then**
6:             $S_p \leftarrow S_p \bigcup \{q\}$
7:         **else if** $(q \prec p)$ **then**
8:             $n_p \leftarrow n_p + 1$
9:         **end if**
10:         **if** $(n_p = 0)$ **then**
11:             $p_{rank} \leftarrow 1$
12:             $F_1 \leftarrow F_1 \bigcup \{q\}$
13:         **end if**
14:     **end for**
15:     $i \leftarrow 1$
16:     **while** $F_i \neq \emptyset$ **do**
17:         $Q \leftarrow \emptyset$
18:         **for** each $p \in F_i$ **do**
19:             **for** each $q \in S_p$ **do**
20:                 $n_p \leftarrow n_p - 1$
21:                 **if** $(n_p = 0)$ **then**
22:                     $q_{rank} \leftarrow i + 1$
23:                     $Q \leftarrow Q \bigcup \{q\}$
24:                 **end if**
25:             **end for**
26:         **end for**
27:         $i \leftarrow i + 1$
28:         $F_i \leftarrow Q$
29:     **end while**
30: **end for**

For example, regarding the performance of both meta-heuristics, results show that for small instances, the performance of MOACO is better than the corresponding one of NGSA-II in 2 of the 3 objective functions evaluated. Nevertheless, if the number of jobs is increased, NSGA-II outperforms MOACO in each one of the objective functions considered. Therefore, while for MOACO, the quality of solutions is affected when the number of jobs to be scheduled is increased, this is not the case of NSGA-II. In fact, this algorithm is much more stable in terms of the behavior of solutions. Therefore, for the problem under consideration, NSGA-II presents more reliability and robustness.

Table 1. Average values of Cmax, $\sum T$ and $\sum U$ for 20 jobs instances

Approach	20 jobs					
	Cmax	%dev	$\sum Tj$	%dev	$\sum Uj$	%dev
MOACO	649	–	4146	–	14	–
NGSA-II	663	−2%	3825	8%	15	1%

Table 2. Average values of Cmax, $\sum T$ and $\sum U$ for 50 jobs instances

Approach	50 jobs					
	Cmax	%dev	$\sum Tj$	%dev	$\sum Uj$	%dev
MOACO	1709	–	21423	–	43	–
NGSA-II	1525	−2%	21235	12%	41	1%

Table 3. Average values of Cmax, $\sum T$ and $\sum U$ for 100 jobs instances

Approach	100 jobs					
	Cmax	%dev	$\sum Tj$	%dev	$\sum Uj$	%dev
MOACO	3879	–	162245	–	85	–
NGSA-II	3565	9%	121335	34%	83	34%

6   Conclusions

This paper studied the Flexible or Hybrid Flowshop scheduling problem (HFSP) considering multiple objectives. A multi-objective ant colony optimization algorithm (MOACO) and the Non-dominated Sorting Genetic Algorithm-II (NSGA-II) were proposed for solving this complex combinatorial optimization problem. Computational experiments were carried out using data sets from the literature. Results showed that very good solutions can be found using NSGA-II algorithm in comparison with the MOACO algorithm, especially for larger instances. Therefore, NSGA-II seems to be the most suitable approach for this problem, due to its reliability and capability of improving the quality of solutions once the number of jobs is increased, whereas the variability of solutions remains stable, which constitutes a measure of robustness.

For further research, MOACO and NSGA-II algorithms can be combined each other in order to improve much more their individual performances when solving larger instances of HFSP. Moreover, other opportunities for further research would be based on adapting MOACO and NSGA-II algorithms for solving different versions of the multi-objective hybrid flowshop scheduling problem. Such versions of the problem could include, for instance, cases considering specific time windows for jobs arrivals and stochastic or deteriorating processing times, among others.

References

1. Pinedo, M.L.: Scheduling: Theory, Algorithms, and Systems, 3rd edn. Springer, Heidelberg (2008). https://doi.org/10.1007/978-0-387-78935-4
2. Ruiz, R., Vázquez-Rodríguez, J.A.: The hybrid flow shop scheduling problem. Eur. J. Oper. Res. **205**(1), 1–18 (2010)
3. Khalouli, S., Ghedjati, F., Hamzaoui, A.: An ant colony system algorithm for the hybrid flow-shop scheduling problem. Int. J. Appl. Metaheuristic Comput. (IJAMC) **2**(1), 29–43 (2011)
4. Gupta, J.N.: Two-stage, hybrid flowshop scheduling problem. J. Oper. Res. Soc. **39**(4), 359–364 (1988)
5. Atencio, F.N., Prasca, A.B., Rodado, D.N., Casseres, D.M., Santiago, M.R.: A comparative approach of ant colony system and mathematical programming for task scheduling in a mineral analysis laboratory. In: Tan, Y., Shi, Y., Niu, B. (eds.) Advances in Swarm Intelligence. LNCS, vol. 9712, pp. 413–425. Springer, Cham (2016). https://doi.org/10.1007/978-3-319-41000-5_41
6. Lei, D., Wu, Z.: Multi-objective production scheduling: a survey. Int. J. Adv. Manuf. Technol. **43**(9–10), 926–938 (2009)
7. Montoya-Torres, J., Vargas-Nieto, F.: Solving a bi-criteria hybrid flowshop scheduling problem occurring in apparel manufacturing. Int. J. Inf. Syst. Supply Chain Manage. **4**, 42–60 (2011)
8. Jun, S., Park, J.: A hybrid genetic algorithm for the hybrid flow shop scheduling problem with nighttime work and simultaneous work constraints: a case study from the transformer industry. Expert Syst. Appl. **42**(15–16), 6196–6204 (2015)
9. Solano-Charris, E., Montoya-Torres, J., Paternina-Arboleda, C.: Ant colony optimization algorithm for a bi-criteria 2-stage hybrid flowshop scheduling problem. J. Intell. Manuf. **22**, 815–822 (2011)
10. Niebles Atencio, F., Solano-Charris, E.L., Montoya-Torres, J.R.: Ant colony optimization algorithm to minimize makespan and number of tardy jobs in flexible flowshop systems. In: Proceedings of IEEE XXXVIII Conferencia Latinoamericana en Informática (CLEI), pp. 1–10 (2012)
11. Janiak, A., Kozan, E., Lichtenstein, M., Oğuz, C.: Metaheuristic approaches to the hybrid flow shop scheduling problem with a cost-related criterion. Int. J. Prod. Econ. **105**(2), 407–424 (2007)
12. Zhang, B., Pan, Q.K., Gao, L., Zhang, X.L., Sang, H.Y., Li, J.Q.: An effective modified migrating birds optimization for hybrid flowshop scheduling problem with lot streaming. Appl. Soft Comput. **52**, 14–27 (2017)

13. Li, J.Q., Sang, H.Y., Han, Y.Y., Wang, C.G., Gao, K.Z.: Efficient multi-objective optimization algorithm for hybrid flow shop scheduling problems with setup energy consumptions. J. Clean. Prod. **181**, 584–598 (2018)
14. Colorni, A., Dorigo, M., Maniezzo, V.: Distributed optimization by ant colonies. In: Proceedings of the First European Conference on Artificial Life, vol. 142, pp. 134–142 (1992)
15. Dorigo, M., Maniezzo, V., Colorni, A.: Ant system: optimization by a colony of cooperating agents. IEEE Trans. Syst. Man Cybern. B Cybern. **26**(1), 29–41 (1996)
16. Dorigo, M., Stützle, T.: Ant Colony Optimization. Massachusetts Institute of Technology, Cambridge (2004)
17. Stützle, T., Hoos, H.H.: MAX-MIN ant system. Future Gener. Comput. Syst. **16**(8), 889–914 (2000)
18. Neto, R.T., Godinho Filho, M.: Literature review regarding Ant Colony Optimization applied to scheduling problems: guidelines for implementation and directions for future research. Eng. Appl. Artif. Intell. **26**(1), 150–161 (2013)
19. Blum, C., Sampels, M.: Ant colony optimization for FOP shop scheduling: a case study on different pheromone representations. In: Proceedings of the 2002 Congress on Evolutionary Computation (CEC 2002), vol. 2, pp. 1558–1563. IEEE (2002)
20. Srinivas, N., Deb, K.: Multiobjective optimization using nondominated sorting in genetic algorithms. Evol. Comput. **2**(3), 221–248 (1994)
21. Deb, K., Pratap, A., Agarwal, S., Meyarivan, T.: A fast and elitist multiobjective genetic algorithm: NSGA-II. IEEE Trans. Evol. Comput. **6**(2), 182–197 (2002)
22. OR-Library.   http://people.brunel.ac.uk/mastjjb/jeb/orlib/multiflowinfo.htm. Accessed 4 Oct 2017

Model-Based System Design, Verification and Simulation

Models of the Heat Exchanger Assuming the Ideal Mixing

Anna Czemplik[⊠]

Faculty of Electronics, Wroclaw University of Science and Technology, Wroclaw, Poland
anna.czemplik@pwr.edu.pl

Abstract. The paper presents models of heat exchangers assuming the ideal mixing of medium. The models are used to describe exchangers while designing control strategies in heating, ventilation and air conditioning systems. This approach was tested on the exchangers with a parallel-flow and counter-flow.

Keywords: Heat exchanger · Ideal mixing · HVAC

1 Introduction

Effective work of complex heating, ventilation and air conditioning (HVAC) systems requires a development of appropriate control strategies. Computer Aided Control System Design (CACSD) tools give a significant support during design and testing of control systems. For their proper operation, it is necessary to employ models allowing to determine reactions of the plant to various control variables and disturbances [1].

A typical element of HVAC systems is a heat exchanger and plants may contain many such exchangers. Typical models of the exchangers [2, 3] can be divided into three categories:

- the most accurate models based on the partial differential equations (PDE) – dedicated to design of exchanger structures, very accurate but also too complicated for studies of complex systems;
- the simplest models, in a form of transfer functions (TF) – used in control system research but only for constant flows of medium in an exchanger;
- the most popular models, in a form of balance equations with a logarithmic mean temperature difference (L-model) – static models dedicated to design a system with exchangers (for example to design of heating substations).

In this paper, an alternative type of model is proposed – a model assuming the ideal mixing (M-model), which means that there are no temperature gradients in a given volume of fluid (i.e. changes of temperature are spreading immediately in the entire volume), therefore the entire volume may be described by one temperature [4]. This is a very simple model and it is usually believed that so simple model is not appropriate to describe heat exchangers. The following analysis shows that the model has appropriate

© Springer Nature Switzerland AG 2020
R. Moreno-Díaz et al. (Eds.): EUROCAST 2019, LNCS 12013, pp. 457–464, 2020.
https://doi.org/10.1007/978-3-030-45093-9_55

Table 1. Categories of exchanger models and their properties

PDE	TF	L-model	M-model
Distributed Parameters	**Lumped Parameters**	**Lumped Parameters**	**Lumped Parameters**
Nonlinear	Linear	**Nonlinear**	**Nonlinear**
Static/dynamic	Dynamic	Static	**Static/dynamic**

properties to design of control strategies – lumped parameters and an ability to describe a dynamics of nonlinear plants (Table 1).

Two typical types of heat exchangers was taken into consideration:

- parallel-flow (PFE) – a hot and cold medium flow in the same direction,
- counter-flow (CFP) – a hot and cold medium flow in the opposite directions.

Temperature profile along PFE and CFP exchangers and typical nominal values of inlet and outlet temperatures (T_{1in}, T_{2in}, T_{1out}, T_{2out}) are shown in Fig. 1.

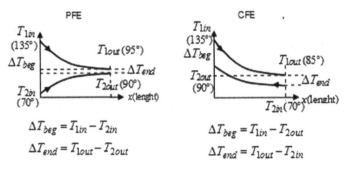

Fig. 1. Two types of heat exchangers

The reference point for the performed studies of PFE and CFE exchangers are accurate PDE models based on the following balance equations in micro-scale

$$PFE: \begin{cases} C_{v1}dx\dfrac{dT_1(t,x)}{dt} = m_1(t)(T_1(t,x) - T_1(t,x+dx)) - k(T_1(t,x) - T_2(t,x))dx \\ C_{v2}dx\dfrac{dT_2(t,x)}{dt} = k(T_1(t,x) - T_2(t,x))dx - m_2(t)(T_2(t,x+dx) - T_2(t,x)) \end{cases} \quad (1)$$

$$CFE: \begin{cases} C_{v1}dx\dfrac{dT_1(t,x)}{dt} = m_1(t)(T_1(t,x) - T_1(t,x+dx)) - k(T_1(t,x) - T_2(t,x))dx \\ C_{v2}dx\dfrac{dT_2(t,x)}{dt} = k(T_1(t,x) - T_2(t,x))dx - m_2(t)(T_2(t,x) - T_2(t,x+dx)) \end{cases} \quad (2)$$

where, T_1 and T_2 are temperatures of hot and cold medium along the exchanger, m_1 and m_2 are functions of media flows $\left(m_1 = c_p\rho f_1, m_2 = c_p\rho f_2, \text{ W/K}\right)$, k is a thermal conductance (W/K), C_{v1} and C_{v2} are a heat capacity, dx is an elementary section of an exchanger (a length of an exchanger is scaled to the range $0 \div 1$). Typical initial conditions for CACSD are steady state, i.e. a solution of the system (1) and (2) assuming that the time derivatives are equal to zero. In this way, it is possible to determine the media temperature profile along an exchanger in the form shown in Fig. 1, as well as a temperature difference between hot and cold media defined as

$$\Delta T(x) = T_1(x) - T_2(x) = \Delta T_{beg} e^{-mkx} \tag{3}$$

where $m = 1/m_1 + 1/m_2$ for PDE and $m = 1/m_1 - 1/m_2$ for CFE. Temperature profiles are unnecessary in CACSD, since the theory and techniques of control design are developed for lumped systems.

2   The Static M and L Models

The static models with lumped parameters (M and L models) are based on heat balance equations in macro-scale and have the same form for both types of the heat exchangers [4], as follows

$$\begin{cases} 0 = m_1(T_{1in} - T_{1out}) - k\,\Delta T \\ 0 = k\,\Delta T - m_2(T_{2out} - T_{2in}) \end{cases} \tag{4}$$

where, T_{1in} and T_{2in} are inlet temperatures, T_{1out} and T_{2out} are outlet temperatures and ΔT is a mean temperature difference between both media. Definitions of ΔT for the M and L models of both exchangers (PFE, CFE) are shown in Table 2.

Table 2. Definition of the mean ΔT in Eq. (3) for model M ($\Delta T_{(M)}$) and L ($\Delta T_{(L)}$)

M-model, $\Delta T_{(M)}$	L-model, $\Delta T_{(L)}$	
$T_{1out} - T_{2out}$	Logarithmic mean difference	Form with a exponential function
	$\dfrac{\Delta T_{end} - \Delta T_{beg}}{\ln\left(\Delta T_{end}/\Delta T_{beg}\right)}$	$\dfrac{1-e^{-mk}}{mk}\Delta T_{beg} = a_L\,\Delta T_{beg}$
The same for PFE and CFE	PFE: $\Delta T_{beg} = T_{1in} - T_{2in}$, $\Delta T_{end} = T_{1out} - T_{2out}$, $m = 1/m_1 + 1/m_2$ CFE: $\Delta T_{beg} = T_{1in} - T_{2out}$, $\Delta T_{end} = T_{1out} - T_{2in}$, $m = 1/m_1 - 1/m_2$	

As can be seen from the table above, the simple definition of the mean ΔT for the proposed M-model results directly from the assumption about the ideal mixing. Whereas ΔT for the L-model is calculated based on the initial condition (3) and has two forms. The first form, with the logarithmic mean temperature difference, based on the temperature difference at the beginning and end of the exchanger (ΔT_{beg}, ΔT_{end}), is the most commonly used but it is suitable mainly for design of exchangers. The second form, with an exponential function is more appropriate for control system design and it has been used in further transformations.

Solutions of balance equations (4) with different definitions of ΔT (from Table 2) are outlet temperatures (T_{1out}, T_{2out}) and the mean ΔT as functions of inlet temperatures (T_{1in}, T_{2in}) and media flows (m_1, m_2). These solutions (output variables) for the M and L models have the same general form

$$T_{1out} = a_{11}T_{1in} + a_{12}T_{2in}, \quad T_{2out} = a_{21}T_{1in} + a_{22}T_{2in}, \quad \Delta T = a_\Delta(T_{1in} - T_{2in}) \quad (5)$$

but differ in coefficients ($a_{11} \div a_{22}, a_\Delta$). The coefficients depend on media flows and are presented in Table 3. The M and L models are linear due to the temperatures and nonlinear due to the flows.

Table 3. Coefficients for temperatures T_{1out}, T_{2out} and ΔT in functions (5)

	M-model for PFE/CFE	L-model for PFE	L-model for CFE
$a_{11}(m_1, m_2)$	$\dfrac{m_1(m_2+k)}{m_1m_2+k(m_1+m_2)}$	$\dfrac{m_1+m_2e^{-km}}{m_1+m_2}$	$\dfrac{(m_2-m_1)e^{-km}}{m_2-m_1e^{-km}}$
$a_{12}(m_1, m_2)$	$\dfrac{m_2k}{m_1m_2+k(m_1+m_2)}$	$\dfrac{m_2-m_2e^{-km}}{m_1+m_2}$	$\dfrac{m_2-m_2e^{-km}}{m_2-m_1e^{-km}}$
$a_{21}(m_1, m_2)$	$\dfrac{m_1k}{m_1m_2+k(m_1+m_2)}$	$\dfrac{m_1-m_1e^{-km}}{m_1+m_2}$	$\dfrac{m_1-m_1e^{-km}}{m_2-m_1e^{-km}}$
$a_{22}(m_1, m_2)$	$\dfrac{m_2(m_1+k)}{m_1m_2+k(m_1+m_2)}$	$\dfrac{m_2+m_1e^{-km}}{m_1+m_2}$	$\dfrac{m_2-m_1}{m_2-m_1e^{-km}}$
$a_\Delta(m_1, m_2)$	$\dfrac{m_1m_2}{m_1m_2+k(m_1+m_2)}$	$\dfrac{m_1m_2}{k}\dfrac{1-e^{-km}}{m_1+m_2}$	$\dfrac{m_1m_2}{k}\dfrac{1-e^{-km}}{m_2-m_1e^{-km}}$

In order to compare the M and L models, nominal values of a thermal power q_N as well as temperatures at the inlet and outlet the PFD and CFD exchangers were assumed, and values of flows and model coefficients (according to the type of model and exchanger) were calculated. The values are collected in Table 4.

As the goal of the analysis is not to compare PFE i CFE exchangers or their design but to compare the M and L models, so the negative value of coefficient k in M-model of CFE exchanger does not matter. Although, coefficients $a_{11N}, a_{12N}, a_{21N}, a_{22N}$ in Table 4 were determined for nominal values m_{1N}, m_{2N} on the basis of different formulas (according to Table 3), the same values were obtained for M and L models. This means that functions T_{1out} and T_{2out} (5) for m_{1N}, m_{2N} are identical for the both models. The outlet temperatures characteristics (T_{1out}, T_{2out}) as a function of media flows m_1, m_2 are presented in Fig. 2, where dotted lines (♦♦) refer to the M-model and solid lines (—) to the L-model. As can be seen, the M and L models are quite similar. The differences are bigger for CFE, but not very significant for purpose of testing the control system strategy.

Table 4. Values of variables and coefficients for the M and L models

$q_N = 500$ kW	PFE exchanger		CFE exchanger	
T_{1inN}; T_{1outN}; T_{2inN}; T_{2outN}, °C	135; 95; 90; 70		135; 85; 90; 70	
	M-model	L-model	M-model	L-model
ΔT_N, °C	5.0	23.4	−5.0	27.3
k, W/°C	100 000	21 376	−100 000	18 310
m_{1N}, W/°C	12 500		10 000	
m_{2N}, W/°C	25 000			
a_{11N}; a_{12N}; a_{21N}; a_{22N}	0.3846; 0.6154; 0.3077; 0.6923		0.2308; 0.7692; 0.3077; 0.6923	
$a_{\Delta N}$	0.0769	0.3599	−0.0769	0.4201

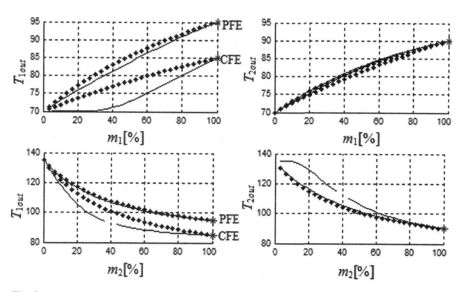

Fig. 2. The outlet temperatures characteristics as a functions of media flows for PFE and CFE

3 The M Model as an Approximation of the L Model

All expressions in the L-models (Table 3) contain a characteristic exponential function. The function was approximated with $e^x = 1 + x$ in two variants (Table 5). The variant Xf concerns approximation of the function m and the variant Xe is approximation of elements m_1 and m_2.

The approximation of the L-model can be obtained doing calculations both, on the output variables (Table 3) and also on the definition of ΔT (Table 2). Obtained results are the same. The relationship between models and approximations is shown in Fig. 3, where O1 and O2 are mathematical operations used when solving the balance Eqs. (3) with the ΔT definition (Table 2), and X is the Xf or Xe variants approximation of the L-model.

Table 5. Two variants of approximation (Xf, Xe)

	PFE	CFE
	$e^{-mk} = \dfrac{1}{\exp((1/m_1 + 1/m_2)k)}$	$e^{-mk} = \dfrac{1}{\exp((1/m_1 - 1/m_2)k)}$
Xf	$e^{-mk} = \dfrac{1}{e^{mk}} \approx \dfrac{1}{1+mk} = \dfrac{1}{1 + \left(\frac{1}{m_1} + \frac{1}{m_2}\right)k}$	$e^{-mk} = \dfrac{1}{e^{mk}} \approx \dfrac{1}{1+mk} = \dfrac{1}{1 + \left(\frac{1}{m_1} - \frac{1}{m_2}\right)k}$
Xe	$e^{-mk} = \dfrac{1}{e^{k/m_1} e^{k/m_2}} \approx \dfrac{1}{\left(1 + \frac{k}{m_1}\right)\left(1 + \frac{k}{m_2}\right)}$	$e^{-mk} = \dfrac{e^{k/m_2}}{e^{k/m_1}} \approx \dfrac{1 + k/m_2}{1 + k/m_1}$

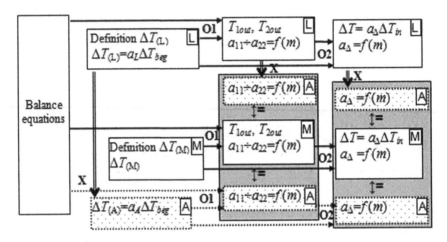

Fig. 3. The relationships between the M, L and A models

It has been found that the M-model is the approximation of L-model since the M-model and the approximated L-model (i.e. named A-model) have the same forms of coefficient a_{11}, a_{12}, a_{21}, a_{22}, a_Δ if only the Xf variant was used for PFE exchanger and the Xe variant for CFE exchanger. However, the A and M models are not the same, because the value of the thermal conductance k in the A-model is inherited from the L-model and differs from the conductance in M-model ($k_{(A)} = k_{(L)} \neq k_{(M)}$). The value of k can be transformed according to the dependency $k_{(L)}\Delta T_{(L)} = k_{(M)}\Delta T_{(M)} = q_N$, which will ensure that A and M models will be identical.

4   The Dynamic Models M and L

Construction of lumped dynamics models of heat exchangers was based on the balance Eqs. (4) and taking into account a heat accumulation. It leads to the following ordinary differential equations (ODE),

$$\begin{cases} C_{v1}\dot{T}_{1out}(t) = m_1(t)(T_{1in}(t) - T_{1out}(t)) - k\,\Delta T(t) \\ C_{v1}\dot{T}_{2out}(t) = k\,\Delta T(t) - m_2(t)(T_{2out}(t) - T_{2in}(t)) \end{cases} \tag{6}$$

The application of T_{1out}, T_{2out} temperatures to describe the heat accumulation in the macro-scale results from the assumption of ideal mixing analogically as describing the accumulation in the micro-scale in the PDE Eqs. (1, 2). On this basis, three variants of blocks PFE and CFE (i.e. for the PFE and CFE exchanger) were constructed in the Matlab/Simulink. The variants rely on the use of various forms for the difference $\Delta T(t)$ in (6), as follows

1. the definition for M-model (Table 2): $\Delta T(t) = T_{1out}(t) - T_{2out}(t)$,
2. the definition for L-model (Table 2): $\Delta T(t) = a_L \Delta T_{beg}(t)$,
3. the function of input variables for L-model (Table 3): $\Delta T(t) = a_\Delta (T_{1in}(t) - T_{2in}(t))$.

The models were used to determine the step response for each of the inputs (T_{1in}, T_{2in}, m_1, m_2). The reference point for these characteristics should be the step responses of partial Eqs. (1, 2), but such simulations are difficult [5]. Therefore, reference models of exchangers with distributed parameters are implemented as a PFE or CFE block chain, using blocks with $\Delta T(t) = a_L \Delta T_{beg}(t)$, as shown in Fig. 4.

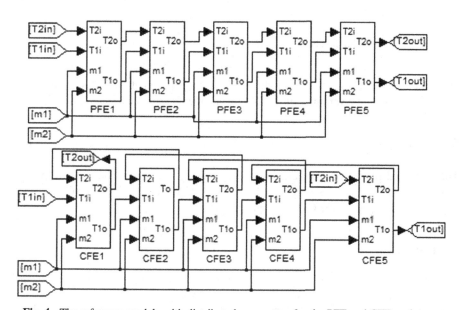

Fig. 4. The reference models with distributed parameters for the PFE and CFE exchanger

The Figs. 5 and 6 show the comparison of models based on selected step responses – i.e. responses of T_{1out} to step change in flow (m_1, m_2), where dotted lines (♦♦) refer to the M-model (variant 1), thin lines to the L-model (variant 2: solid —, variant 3: dashed --) and thick solid lines (■■) to the reference model with distributed parameters.

Differences in steady state of responses result from static characteristics (Fig. 2). Step responses are intended for comparing dynamic properties such as setting time. Unfortunately, the M-model for the CFE exchanger is unstable, due to the negative value

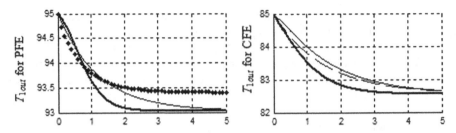

Fig. 5. The response T_{1out} to step change in flow m_1.

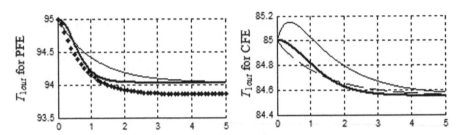

Fig. 6. The response T_{1out} to step change in flow m_2.

of coefficient k (Table 4). Both L-models (i.e. variant 2 and 3) of the PFE exchanger are identical, whereas the variant 3 of the CFE exchanger is more accurate than variant 2.

5   Conclusions

The presented results confirm the suitability of analyzed models that assume perfect mixing – their accuracy is sufficient to use with CACSD. The only problem is the M-model instability for the CFE exchanger.

References

1. Åström, K., Hägglund, T.: Advanced PID Control. ISA (Instrumentation, Systems, and Automation Society) (2006). ISBN-13: 978-1-55617-942-6
2. Chmielnicki, W.: Energy Control in Buildings Connected to Urban Heating Sources. PAN, Warszawa (1996)
3. Recknagel, H., Spenger, E., Hönmann, W., Schramek, E.: Taschenbuch für Heizung und Klimatechnik 92/93. Oldenbourg Verlag GmbH, München (1994)
4. Czemplik, A.: Simple models of central heating system with heat exchangers in the quasi-static conditions. In: Moreno-Díaz, R., Pichler, F., Quesada-Arencibia, A. (eds.) EUROCAST 2015. LNCS, vol. 9520, pp. 597–604. Springer, Cham (2015). https://doi.org/10.1007/978-3-319-27340-2_74
5. Schiesser, W.E., Griffiths, G.W.: A compendium of Partial Differential Equation Models. Cambridge University Press, New York (2009)

Algorithm to Express Lie Monomials in Ph. Hall Basis and Its Practical Applications

Ignacy Duleba$^{(\boxtimes)}$ (iD) and Iwona Karcz-Duleba (iD)

Electronics Faculty, Wroclaw University of Science and Technology,
Janiszewski Street 11/17, 50-372 Wroclaw, Poland
{ignacy.duleba,iwona.duleba}@pwr.edu.pl

Abstract. In this paper an algorithm is presented to expand any Lie monomial in the Ph. Hall basis. The algorithm exploits an algorithm for the optimal generation of Ph. Hall basis and it can be used to express $\log(\exp X \cdot \exp Y)$ in this basis. An explicit formula for the logarithm includes the Ph. Hall elements up to the 7th degree. The Ph. Hall expansion is useful in the theory of Lie algebras and in robotics for nonholonomic motion planning.

Keywords: Algorithm · Lie monomial expansion · Hall basis

1 Introduction

Lie monomials are quite general objects that span a Lie algebra without discovering its nature. From a practical point of view it is better to work with minimal sub-sets of Lie monomials, i.e. bases, without missing any desirable property they offer. In the literature a few bases are known: Chibrikov [1], Lyndon [7], Shirshov [13] and Ph. Hall [10,12]. The latter one will be used throughout this paper. In some formulas (like (generalized) Campbell-Baker-Hausdorff-Dynkin [11]), an infinite series of Lie monomials appear rather than items of their bases. Only a low degree truncation of the series can be processed by hand, therefore an automatic algorithm is needed to process the series when higher degree terms are also required. The Campbell-Baker-Hausdorff-Dynkin (CBHD) formula allows one to predict a net motion of concatenated elementary flows [11,14]

$$\log(\exp X \cdot \exp Y) = \sum_{m=1}^{\infty} \frac{(-1)^{m-1}}{m} \sum_{\substack{(p_1,q_1,\ldots,p_m,q_m) \\ p_k+q_k \geq 1}} \frac{ad_X^{q_m} \, ad_Y^{p_m} \ldots ad_X^{q_1} \, ad_Y^{p_1}}{\left(\sum_{i=1}^{m}(p_i+q_i)\right) \prod_{i=1}^{m} p_i! \, q_i!}$$

$$(1)$$

© Springer Nature Switzerland AG 2020
R. Moreno-Díaz et al. (Eds.): EUROCAST 2019, LNCS 12013, pp. 465–473, 2020.
https://doi.org/10.1007/978-3-030-45093-9_56

where indexing vectors $(p_1, q_1, \ldots, p_m, q_m)$ satisfy $\forall k \in \{1, \ldots, m\}$ $p_k, q_k \geq 0$, $p_k + q_k \geq 1$, and either $p_1 = 1$ or $p_1 = 0$ and $q_1 = 1$, and

$$ad_X^{q_m} ad_Y^{p_m} \ldots ad_X^{q_1} ad_Y^{p_1} =$$

$$[\underbrace{X, [X, \ldots [X,}_{q_m} [\underbrace{Y, [Y, [\ldots [Y,}_{p_m} \ldots [\underbrace{X, [X, \ldots [X,}_{q_1} [\underbrace{Y, [Y, [\ldots [Y, Y]}_{p_1} \ldots] \ldots] \ldots] \ldots] \ldots]$$

CBHD formula (1) can be treated purely algebraically in the scope of the free Lie algebra generated by X and Y. But recently, it has attracted considerable attention in robotics in the scope of nonholonomic motion planning, where X and Y become vector fields spanning the controllability Lie algebra of a nonholonomic system, [3,5]. In nonholonomic motion planning CBHD formula allows one to plan locally (around a given point in the state space of a system) control actions producing a motion in a direction defined by higher degree vector fields. In this paper CBHD formula will be considered in the pure algebraic sense. Although formula (1) is presented in a closed form but this form is not the simplest possible. Clearly, the anti-symmetry of Lie brackets and the Jacobi identity may be applied to simplify it. The simplest form of CBHD formula will be obtained when the right hand side of Eq. (1) is expressed as a linear combination of basis elements of the free Lie algebra spanned by X and Y.

In this paper, being extended version of [4], we will present an algorithm computing Ph. Hall basis expansion of any Lie monomial. The algorithm will be used to expand the CBHD formula up to any prescribed degree. It processes Lie monomials in different order than suggested by Eq. (1) (first fixing m, and sequentially increasing it by one). By fixing m, generated Lie monomials differ in their degrees. Therefore, to obtain a fixed degree Ph. Hall basis expansion, more than required Lie monomials should be produced. We will examine sequences $(p_1, q_1, \ldots, p_m, q_m)$ satisfying $\sum_k (p_k + q_k) = r = const$ with r increasing. In such a case, all Lie monomials of degree equal to r are generated simultaneously. After generating all Lie monomials of degree r, a preliminary simplification is performed aimed at: (1) neglecting those sequences which produce zero Lie monomial; (2) coupling those monomials differing only by their coefficients. In the last step of the algorithm each Lie monomial left is subject to the Ph. Hall basis expansion. By summing up resulting coefficients appropriately a final form is obtained of the Ph. Hall basis expansion of $\log(\exp X \cdot \exp Y)$ up to a prescribed degree. The above considerations allow us to reformulate Eq. (1) and render it the form

$$\log(\exp X \cdot \exp Y) = X + Y +$$
$$\sum_{r=1}^{\infty} \frac{1}{r} \sum_{i=1}^{r-1} \sum_{\substack{(p_1, q_1, \ldots, p_m, q_m) \\ p_k + q_k \geq 1 \\ \sum_k p_k = i, \ \sum_k q_k = r-i}} \frac{(-1)^{m-1}}{m \prod_{k=1}^{m} p_k! q_k!} ad_X^{q_m} ad_Y^{p_m} \ldots ad_X^{q_1} ad_Y^{p_1}. \quad (2)$$

The expansion of $\log(\exp X \cdot \exp Y)$ is known in mathematics for many years [6,8,9] although it was derived with different approaches (algorithms).

This paper is organized as follows. In Sect. 2 a terminology is introduced and simple lemmas are formulated being a base of the algorithm construction.

In Sect. 3 the algorithm generating Ph. Hall expansion of $\log(\exp X \cdot \exp Y)$ is given. Its steps are exemplified and discussed. Section 4 concludes the paper.

2   Terminology

– the Lie bracket is denoted as $[\cdot, \cdot]$,
– Lie monomials (Mon) is a collection of generators (X and Y in our case) and all nested Lie brackets of generators, e.g. $[X, Y], [Y, [Y, X]] \in$ Mon, $-[X, Y], X + Y \notin$ Mon,
– the Lie bracket is bi-linear $[k_1 U + k_2 W, Z] = k_1 [U, Z] + k_2 [W, Z]$, and homogeneous: $[k_1 U, k_2 W] = k_1 k_2 [U, W]$, $k_1, k_2 \in \mathbb{R}$, $U, W, Z \in$ Mon,
– for any Lie monomial its degree can be defined recursively as follows:

$$\deg(Z) = 1 \qquad \text{for generators}$$
$$\deg([A, B]) = \deg(A) + \deg(B) \text{ for compound Lie monomials,}$$

– index of a Lie monomial is a vector whose components indicate how many times every generator appears in this Lie monomial: $\mathrm{idx}(Z \in$ Mon$) = (\#X(Z), \#Y(Z))$, denotes the number of occurrences of generators X and Y in the Lie monomial Z. E.g. $\mathrm{idx}([Y, [Y, X]]) = (1, 2)$,
– a minimal set of Lie monomials that spans the free Lie algebra is called a basis of the algebra. Later on Ph. Hall basis will be exploited, H. elements of the basis $H = (H^1, H^2, H^3, \ldots)$ are ordered (it is denoted as $\overset{H}{<}$) thus $i < j \Rightarrow H^i \overset{H}{<} H^j$.
– Elements of the free Lie algebra spanned by generators X, Y are not independent of each other due to the anti-symmetry property and the Jacobi identity

$$[U, W] = -[W, U], \quad [U, [W, Z]] + [W, [Z, U]] + [Z, [U, W]] = 0.$$

The properties are encoded in the form of operators acting on an ordered pair (c, L) composed of a coefficient $c \in \mathbb{R}$ and a Lie monomial L
  • anti-symmetry: $O_1((c, [U, W])) = (-c, [W, U])$,
  • the Jacobi identity: $O_2((c, [U, [W, Z]])) = \{(c, [W, [U, Z]]), (-c, [Z, [U, W]])\}$,
– layers of Lie monomials within the Ph. Hall basis are defined as follows

$$H_i = \{X \mid X \in H, \; degree(X) = i\}.$$

Easy-to-prove and useful lemmas state that:

Lemma 1: Resulting Lie monomials after (multi-) actions of O_1, O_2 inherit the index of the operators' argument.

Lemma 2: In the Ph. Hall expansion of a given Lie monomial Z only such Ph. Hall basis elements may appear that have the same index as Z as only O_1, O_2 can be applied to transform Z.

3   Ph. Hall Expansion of $\log(\exp X \cdot \exp Y)$

In a preliminary step, the Ph. Hall basis $H = \cup_{i=1}^{r_{max}} H_i = (H^1, H^2, \ldots)$ elements up to required degree r_{max} are generated with the optimal algorithm presented in [2]. The basis for $r_{max} = 7$ spanned by generators X, Y is collected in Table 1.

Table 1. The Ph. Hall basis generated by X, Y up to the 7th degree

X,	[X,[X,[X,[X,[X,Y]]]]],	[Y,[Y,[Y,[Y,[Y,[X,Y]]]]]],
Y,	[Y,[X,[X,[X,[X,Y]]]]],	[[X,Y],[X,[X,[X,[X,Y]]]]],
[X,Y],	[Y,[Y,[X,[X,[X,Y]]]]],	[[X,Y],[Y,[X,[X,[X,Y]]]]],
[X,[X,Y]],	[Y,[Y,[Y,[X,[X,Y]]]]],	[[X,Y],[Y,[Y,[X,[X,Y]]]]],
[Y,[X,Y]],	[Y,[Y,[Y,[Y,[X,Y]]]]],	[[X,Y],[Y,[Y,[Y,[X,Y]]]]],
[X,[X,[X,Y]]],	[[X,Y],[X,[X,[X,Y]]]],	[[X,Y],[[X,Y],[X,[X,Y]]]],
[Y,[X,[X,Y]]],	[[X,Y],[Y,[X,[X,Y]]]],	[[X,Y],[[X,Y],[Y,[X,Y]]]],
[Y,[Y,[X,Y]]],	[[X,Y],[Y,[Y,[X,Y]]]],	[[X,[X,Y]],[X,[X,[X,Y]]]],
[X,[X,[X,[X,Y]]]],	[[X,[X,Y]],[Y,[X,Y]]],	[[X,[X,Y]],[Y,[X,[X,Y]]]],
[Y,[X,[X,[X,Y]]]],	[X,[X,[X,[X,[X,[X,Y]]]]]],	[[X,[X,Y]],[Y,[Y,[X,Y]]]],
[Y,[Y,[X,[X,Y]]]],	[Y,[X,[X,[X,[X,[X,Y]]]]]],	[[Y,[X,Y]],[X,[X,[X,Y]]]],
[Y,[Y,[Y,[X,Y]]]],	[Y,[Y,[X,[X,[X,[X,Y]]]]]],	[[Y,[X,Y]],[Y,[X,[X,Y]]]],
[[X,Y],[X,[X,Y]]],	[Y,[Y,[Y,[X,[X,[X,Y]]]]]],	[[Y,[X,Y]],[Y,[Y,[X,Y]]]],
[[X,Y],[Y,[X,Y]]],	[Y,[Y,[Y,[Y,[X,[X,Y]]]]]],	[[Y,[X,Y]],[Y,[Y,[X,Y]]]]

Algorithm 1. (Ph. Hall basis expansion of a given Lie monomial processed simultaneously with its real coefficient 1).

Step 1. Read in a Lie monomial L, put it into a list to process, $List \leftarrow (1, L)$ and clear out a result, $R \leftarrow 0$

Step 2. if $List \neq \emptyset$, then process with Step 3, otherwise output the result R.

Step 3. Take the first item from the List: $(c, I) \leftarrow \text{First}(List)$ and exclude it from the list: $List \leftarrow \text{Drop}(\text{First}((List))$

Step 4. Applying O_1 only (possibly many times) to I, transform it into a canonical form, defined as follows:

($\star$) Generators are in a canonical form by definition. ($\star\star$) For compound Lie monomials, the canonical form is characterized by the condition (satisfied for any of sub-Lie bracket $[U, W]$ of the Lie monomial Z):

$\text{struct}(U) < \text{struct}(W) \quad \Leftrightarrow$

1. $\deg(U) < \deg(W)$ or
2. $\text{struct}(\text{left}(U)) < \text{struct}(\text{left}(W))$ or
3. $\text{struct}(\text{right}(U)) < \text{struct}(\text{right}(W))$

with $\text{left}(A = [A_l, A_r]) = A_l$ and $\text{right}(A = [A_l, A_r]) = A_r$. When none of the conditions is decidable then $\text{struct}(U) = \text{struct}(W)$ and a lexicographic order is applied to establish precedence between U and W. The lexicographic order is defined as $U \overset{lex}{<} W \Leftrightarrow \text{flat}(U) \overset{lex}{<} \text{flat}(W)$, where $\text{flat}(A)$ removes all symbols of Lie brackets $[,]$ from A.

Note that in Step 4 coefficient $\tilde{c}$ of the resulting Lie monomial is the same as input Lie monomial when even number of O_1 operations is performed or $-c$ otherwise, consequently $(\tilde{c}, I) \leftarrow (c, \text{canon}(I))$.

Step 5. If $I \in H$ then $R \leftarrow R + \tilde{c}I$ and go to Step 2. If I is a zero Lie monomial then go to Step 2.

Step 6. Here $I \notin H$, thus in I there exists such a sub-Lie bracket G that $G = [U, [W, Z]]$, and U, W, Z belong to the Ph. Hall basis, and $U \overset{H}{<} W \overset{H}{<} Z$. To G sub-tree O_2 is applied: two copies of I are created, in the first G is replaced with $[W, [U, Z]]$, in the second with $[Z, [U, W]]$. The two Lie monomials paired with coefficient inherited from I are added to *List* and go to Step 2.

Comment: it is easy to check that the Lie monomials $[W, [U, Z]]$ and either $[Z, [U, W]]$ or $[[U, W], Z]$ when processed with O_1 belong to the Ph. Hall basis. Moreover, O_2 moves a Lie monomial rightward along the ordered sequence of Lie monomials. For the Lie monomial $[X, [Y, [X, Y]]]$ the push is under-braced in the series (3). Because there exists an upper bound for Lie monomials of a prescribed degree expressed in the canonical form (e.g. $[Y, [Y, Y]]$ and $[[Y, Y], [Y, Y]]$ for degrees 3 and 4 respectively), therefore the procedure of transforming a Lie monomial into a canonical form and applying then O_2 converges. Those Lie monomials in a canonical form which do not admit O_2 belong to the Ph. Hall basis.

Example 1: Structure comparisons: when 1/2/3 conditions presented are decidable
(1) $\text{struct}([\cdot, [\cdot, \cdot]]) < \text{struct}([\cdot[\cdot, [\cdot, \cdot]]])$, (2) $\text{struct}([\cdot[\cdot, [\cdot, \cdot]]]) < \text{struct}([[\cdot, \cdot], [\cdot, \cdot]])$,
(3) $\text{struct}([[\cdot, \cdot], [\cdot, [\cdot, [\cdot, \cdot]]]]) < \text{struct}([[\cdot, \cdot], [[\cdot, \cdot], [\cdot, \cdot]]])$.
Example 2: A lexicographic comparison:
$\text{flat}([X, [X, Y]]) = (XXY) \overset{lex}{<} (YXY) = \text{flat}([Y, [X, Y]])$
Example 3: Lie monomials generated by X, Y canonically ordered (using O_1 only)

$$X, Y, [X, X], [X, Y], [Y, Y], [X, [X, X]], [X, [X, Y]], \dots, [Y, [Y, Y]],$$
$$[X, [X, [X, X]]], \dots, \underbrace{[X, [Y, [X, Y]]]}, \dots \underbrace{[Y, [X, [X, Y]]]}, \dots [Y, [Y, [Y, Y]]],$$
$$[[X, X], [X, X]], \dots, \underbrace{[[X, Y], [X, Y]]}, \dots [[Y, Y], [Y, Y]], \dots, \tag{3}$$

Example 4: The Ph. Hall expansion of Lie monomial $[X, [X, [X, [Y, [Y, X]]]]] = -2[[X, Y], [X, [X, [X, Y]]]] - [Y, [X, [X, [X, [X, Y]]]]]$. Underbraces point out subbrackets to which O_2 is applied.

$$[X, [X, [X, [Y, [Y, X]]]]]$$
$$\downarrow$$
$$-[X, [X, \underbrace{[X, [Y, [X, Y]]]}]] \longrightarrow [X, [X, [[X, Y], [X, Y]]]] = 0$$
$$\downarrow$$
$$-[X, [\underbrace{X, [Y, [X, [X, Y]]]}]]$$
$$-\underbrace{[X, [Y, [X, [X, [X, Y]]]]]}   [X, [[X, [X, Y]], [X, Y]]]$$
$$\downarrow$$
$$-[Y, [X, [X, [X, [X, Y]]]]] \; [[X, [X, [X, Y]]], [X, Y]] \quad -\underbrace{[X, [[X, Y], [X, [X, Y]]]]}$$
$$\downarrow$$
$$-[[X, Y], [X, [X, [X, Y]]]]   -[[X, Y], [X, [X, [X, Y]]]][[X, [X, Y]], [X, [X, Y]]] = 0$$

Algorithm 1 is particularly well suited to be implemented in any symbolic language (LISP, Wolfram's Mathematica) because Lie monomials are naturally represented as binary trees, cf. Fig. 1 and the languages have got a lot of functions to manipulate on binary trees and lists.

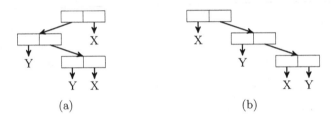

(a)   (b)

Fig. 1. Lie monomial $[[Y, [Y, X]], X]$ in a binary tree form (a), the monomial in the canonical form (b)

Equation (2) and Lemmas 1 and 2 serve as a base for the algorithm of Ph. Hall expansion of $\log(\exp X \cdot \exp Y)$. **Algorithm 2** proceeds as follows: (it is assumed that the Ph. Hall basis has been generated already)

Step 1. For degree $r = 1$ with step 1 until $r = r_{max}$ do Steps 2–9
Step 2. Split degree r into an index pair (i, j) such that $i + j = r$, $i, j \geq 1$.
Step 3. for the pair (i, j) do Steps 4–9
Step 4. Select a set of admissible indexing vectors $(p_1, q_1, \ldots, p_m, q_m)$ satisfying $\sum_k p_k = i$, $\sum_k q_k = j$. Any indexing vector determines a Lie monomial uniquely. Observe that the same Lie monomial can be produced by many indexing vectors. With each indexing vector (Lie monomial) associate its coefficient equal to $(-1)^{m-1}/(m \left(\sum_{i=1}^{m}(p_i + q_i)\right) \prod_{i=1}^{m} p_i! q_i!)$
Step 5. Exclude from the set vectors that correspond to zero Lie monomials (in the form either $[\ldots [X, X] \ldots]]$ or $[\ldots [Y, Y] \ldots]]$). The vectors are characterized by conditions either $p_1 = 0$ ($q_1 = 1$) and $p_2 = 0$, or $q_1 = 0$ ($p_1 = 1$) and $p_2 \neq 0$. For each indexing vector define uniquely a corresponding Lie monomial.

Step 6. Non-zero Lie monomials may be duplicated. Thus leave only one representative monomial for each group of identical Lie monomials and assign to it a coefficient being a sum of coefficients of Lie monomials forming the group. Non-duplicated Lie monomials are collected in the set $S_{(i,j)}$. Finally, form a linear combination of Lie monomials

$$comb_{(i,j)} = \sum_l \beta_l Z_l, \quad \beta_l \in \mathbb{R}, \quad Z_l \in S_{(i,j)}. \tag{4}$$

Most of Z_l do not belong to the Ph. Hall basis.

Step 7. From the Ph. Hall basis select its subset $H_{(i,j)}$ including all elements with index (i,j): $H_{(i,j)} = \{Z \in H_{i+j} \mid index(Z) = (i,j)\}$.

Step 8. Using Algorithm 1, expand any element in $S_{(i,j)}$ over elements of $H_{(i,j)}$:

$$S_{(i,j)} \ni Z_l = \sum_k \alpha_k H^k \quad H^k \in H_{(i,j)}, \quad \alpha_k \in \mathbb{R}. \tag{5}$$

Step 9. Collect results

$$comb_{(i,j)} = \sum_l \beta_l \left(\sum_k \alpha_k H^k \right) = \sum_k \gamma_k H^k \quad \gamma_k \in \mathbb{R} \tag{6}$$

and write them down into resulting Ph. Hall basis expansion.

Remarks on Algorithm 2:

- The behavior of Algorithm 2 will be exemplified on determining coefficients of the Ph. Hall elements with index $(i,j) = (2,2)$.
- In Steps 4–6 purely combinatoric calculations are performed. All indexing vectors $p_1, q_1, \ldots, p_m, q_m$ generating Lie monomial with index equal to $(2,2)$ are given below:
(1 0 0 2 1 0)   (1 1 1 0 0 1)   (0 1 1 1 1 0)   (1 0 0 1 1 1)
(1 1 0 1 1 0)   (0 1 1 0 1 1)   (1 0 0 1 1 0 0 1) (1 2 1 0)
(0 1 1 0 1 0 0 1) (1 0 0 1 0 1 1 0) (0 1 2 1)   (0 1 1 0 0 1 1 0)
(1 1 1 1)   (0 1 2 0 0 1)
These indexing vectors yield a combination (Step 8)

$$comb_{(2,2)} = -\frac{1}{48}[X,[Y,[X,Y]]] + \frac{1}{48}[Y,[X,[Y,X]]].$$

Note that indexing vectors $(1 0 1 2)$, $(1 0 1 1 0 1)$, $(1 0 1 0 0 1 0 1)$, $(1 0 1 0 0 2)$, $(0 1 0 1 2 0)$, $(0 1 0 1 1 0 1 0)$ do not appear because they produce vanishing Lie monomials.
- There is only one element of Ph. Hall basis $[Y,[X,[X,Y]]]$ with index $(2,2)$ (Step 7), cf. Table 1.
- The most demanding step of Algorithm 2 is Step 8. For this particular case, it is a simple exercise to show that $[X,[Y,[X,Y]]] = [Y,[X,[X,Y]]]$ and $[Y,[X,[Y,X]]] = -[Y,[X,[X,Y]]]$. Generally, in this step Algorithm 1 is invoked.

Algorithm 2 run for $r_{max} = 7$ produces the Ph. Hall expansion of $\log(\exp X \cdot \exp Y)$:

$$
\begin{aligned}
\log(\exp X \cdot \exp Y) = {} & X + Y + \tfrac{1}{2}[X,Y] + \tfrac{1}{12}[X,[X,Y]] - \tfrac{1}{12}[Y,[X,Y]] + \\
& - \tfrac{1}{24}[Y,[X,[X,Y]]] - \tfrac{1}{720}[X,[X,[X,[X,Y]]]] - \tfrac{1}{120}[[X,Y],[X,[X,Y]]] + \\
& - \tfrac{1}{180}[Y,[X,[X,[X,Y]]]] + -\tfrac{1}{360}[[X,Y],[Y,[X,Y]]] + \tfrac{1}{180}[Y,[Y,[X,[X,Y]]]] + \\
& + \tfrac{1}{720}[Y,[Y,[Y,[X,Y]]]] + \tfrac{1}{1440}[Y,[X,[X,[X,[X,Y]]]]] - \tfrac{1}{240}[[X,[X,Y]],[Y,[X,Y]]] + \\
& + \tfrac{1}{240}[[X,Y],[Y,[X,[X,Y]]]] + \tfrac{1}{360}[Y,[Y,[X,[X,[X,Y]]]]] + \\
& + \tfrac{1}{720}[[X,Y],[Y,[Y,[X,Y]]]] + \tfrac{1}{1440}[Y,[Y,[Y,[X,[X,Y]]]]] + \\
& + \tfrac{1}{30240}[X,[X,[X,[X,[X,[X,Y]]]]]] + \tfrac{1}{5040}[[X,[X,Y]],[X,[X,[X,Y]]]] + \\
& + \tfrac{1}{2016}[[X,Y],[X,[X,[X,[X,Y]]]]] + \tfrac{1}{5040}[Y,[X,[X,[X,[X,[X,Y]]]]]] + \\
& + \tfrac{13}{15120}[[Y,[X,Y]],[X,[X,[X,Y]]]] - \tfrac{1}{10080}[[X,[X,Y]],[Y,[X,[X,Y]]]] + \\
& + \tfrac{1}{1260}[[X,Y],[[X,Y],[X,[X,Y]]]] + \tfrac{23}{15120}[[X,Y],[Y,[X,[X,[X,Y]]]]] + \\
& + \tfrac{1}{3780}[Y,[Y,[X,[X,[X,[X,Y]]]]]] - \tfrac{1}{1120}[[Y,[X,Y]],[Y,[X,[X,Y]]]] + \\
& + \tfrac{1}{1680}[[X,[X,Y]],[Y,[Y,[X,Y]]]] + \tfrac{1}{5040}[[X,Y],[[X,Y],[Y,[X,Y]]]] + \\
& + \tfrac{1}{5040}[[X,Y],[Y,[Y,[X,[X,Y]]]]] - \tfrac{1}{3780}[Y,[Y,[Y,[X,[X,[X,Y]]]]]] + \\
& - \tfrac{1}{5040}[[Y,[X,Y]],[Y,[Y,[X,Y]]]] - \tfrac{1}{10080}[[X,Y],[Y,[Y,[Y,[X,Y]]]]] + \\
& - \tfrac{1}{5040}[Y,[Y,[Y,[Y,[X,[X,Y]]]]]] - \tfrac{1}{30240}[Y,[Y,[Y,[Y,[Y,[X,Y]]]]]] \cdots
\end{aligned}
$$

$$(7)$$

The Ph. Hall expansion of CBHD given in Formula (7) reveals that

- not all Ph. Hall basis elements appear in the formula. In fact $[X,[X,[X,Y]]]$, $[Y,[Y,[X,Y]]]$, $[X,[X,[X,[X,[X,Y]]]]]$, $[Y,[Y,[X,[X,[X,Y]]]]]$, $[Y,[Y,[Y,[Y,[X,Y]]]]]$, $[[X,Y],[X,[X,[X,Y]]]]$ are absent,
- the elements in the expansion are given not as they appear in the Ph. Hall basis, cf. Table 1, but as they were generated by the algorithm.

4   Conclusions

In this paper an algorithm of the Ph. Hall expansion of any Lie monomial is presented. It can be used to effectively construct the Ph. Hall basis expansion of Campbell-Baker-Hausdorff-Dynkin formula, up to a prescribed degree. An explicit expansion up to the 7th degree has been provided. The algorithm can be extended without any additional effort to the case of more than two generators of the CBHD formula. The formula is important both for the theory of Lie algebras and for practical robotic applications in nonholonomic motion planning.

References

1. Chibrikov, E.S.: A right normed basis for free Lie algebras and Lyndon-Shirshov words. J. Algebra **302**(2), 593–612 (2006)
2. Duleba, I.: Checking controllability of nonholonomic systems via optimal Ph. Hall basis generation. In: IFAC SyRoCo Conference, Nantes, pp. 485–490 (1997)

3. Duleba, I.: Locally optimal motion planning of nonholonomic systems. J. Robot. Syst. **14**(11), 767–788 (1997)
4. Duleba, I., Karcz-Duleba, I.: On the Ph. Hall expansion of lie monomials. In: Quesada-Arencibia, A., Rodriguez, J.-C., Moreno-Diaz, R., Moreno-Diaz Jr., R., de Blasio, G., Garcia, C.R. (eds.) The 17th EUROCAST Conference, Las Palmas, pp. 128–129 (2019). (extended abstract)
5. Hermes, H.: On the synthesis of a stabilizing feedback control via Lie algebraic method. SIAM Journ. Control Opt. **18**(6), 352–361 (1980)
6. Goldberg, K.: The formal power series for $\text{Log}(e^x e^y)$. Duke Math. J. **23**, 13–21 (1956)
7. Jacob, G. : Lyndon discretization and exact motion planning. In: European Control Conference, Grenoble, pp. 1507–1512 (1991)
8. Macdonald, I.G.: On the Baker-Campbell-Hausdorff series. Aust. Math. Soc. Gaz. **8**, 69–95 (1981)
9. Newman, M., Thompson, R.C.: Numerical values of Goldberg's coefficients in the series for $\log(e^x e^y)$. Math. Comput. **48**, 265–271 (1987)
10. Serre, J.-P.: Lie Algebras and Lie Groups. W.J. Benjamin, New Jork (1965)
11. Strichartz, R.S.: The Campbell-Baker-Hausdorff-Dynkin formula and solutions of differential equations. J. Funct. Anal. **72**, 320–345 (1987)
12. Varadarayan, V.S.: Lie Groups, Lie Algebras and Their Representations. Prentice-Hall Inc., Englewood Cliffs (1974)
13. Viennot, G.: Algèbres de Lie Libres et Monoïdes Libres. LNM, vol. 691. Springer, Heidelberg (1978). https://doi.org/10.1007/BFb0067950
14. Wojtynski, W.: Lie Groups and Algebras. Polish Scientific Publishers, Warsaw (1986). (in Polish)

Simulation Infrastructure for Automated Anesthesia During Operations

Martin Hrubý[1]([✉]), Antonio Gonzáles[2], Ricardo Ruiz Nolasco[2], Ken Sharman[3], and Sergio Sáez[3]

[1] Brno University of Technology, Brno, Czech Republic
hrubym@fit.vutbr.cz
[2] RGB Medical Devices, Madrid, Spain
{agonzalez,rruiznolasco}@rgb-medical.com
[3] Instituto Tecnológico de Informática, Valencia, Spain
{ken,ssaez}@iti.es

Abstract. This paper deals with a simulation infrastructure for comprehensive testing of safety, security and performance of a specific medical device (TOF-*Cuff* Controller) being developed by RGB Medical Devices. The controller is designed to monitor and regulate patient's blood pressure and muscle relaxation during operations. The controller is still at the laboratory testing level of development and needs to be fully accredited by national health-care agencies before practical deployment. By having a simulation infrastructure, we can study the Controller's behaviour in various pre-defined test scenarios.

Keywords: Simulation in the loop · Control system · Medical device · Anesthesia

1 Introduction

Anesthesia is a complex of monitoring and regulation of various vital functions of patients during surgeries and post-operation treatments in Intensive Critical Care Units. Number of all medical devices installed in operation rooms rises and controlling all of them becomes difficult. Our goal is to develop a safe and reliable medical device called TOF-*Cuff* (Controller) which automatises two common components of anesthesia: *controlled hypotension*, i.e. regulation of a patient's blood pressure (BP) in order to lower it and *muscle relaxation* (NMT – Neuromuscular Transmission). Monitoring and regulation of both components is essential. Decreased blood pressure causes the patient to bleed less. The state of muscle relaxation softens patient's muscles and so it simplifies the surgery.

Developing such a Controller requires plenty of laboratory experiments which obviously cannot be done using a living patient. We simulate the entire *medical system of anesthesiologic regulation* consisting of infusion pumps for BP/NMT, BP/NMT transducers and the patient. In this paper, we describe details of the simulator focused at blood pressure monitoring and regulation.

© Springer Nature Switzerland AG 2020
R. Moreno-Díaz et al. (Eds.): EUROCAST 2019, LNCS 12013, pp. 474–481, 2020.
https://doi.org/10.1007/978-3-030-45093-9_57

This work is a part of the EU research project AQUAS which is oriented towards creating a general methodology for developing cuber-physical devices which are critical from security, safety and performance points of view.

2   Principle of Controlled Hypotension

By definition, *Mean Arterial Pressure* (MAP) is an average blood pressure measured during a single cardiac cycle, and estimated as $\frac{1}{3}SP + \frac{2}{3}DP$ where SP denotes Systolic Pressure and DT denotes Diastolic Pressure. Mean arterial pressure is widely referred in situations when expressing the blood pressure as a single number is preferred. Sensors for measuring BP/MAP are commonly used in operation rooms. They are implemented as autonomous electronic devices inter-connectable via LAN or point-to-point to other medical systems.

Saying blood pressure regulation, we mostly mean lowering blood pressure (controlled hypotension). The goal is to drive patient's BP towards some desired target level p_t [mmHg] and hold it there as long as possible. The patients body reacts in opposite way though. The controlled hypotension is based on infusion of *vasoactive drugs* into the patient's bloodstream [8]. For this purpose, we employ medical devices called *Infusion Pump*. Infusion pumps are also autonomous electronic devices that are LAN/point-to-point remotely connectable. They get configured with a certain dosage rate [ml/h] and then infuse continuously the drug to the patients body.

So called *vasoactive drugs* can change physical parameters of patient's blood vessels in the sense that the blood flow rises, and consequently, the blood pressure drops. Sodium Nitroprusside (SNP) is one of the commonly chosen vasodilators, often mentioned in the literature [1,2,6] and also being a referential drug in this paper.

Figure 1 depicts the hypotensive effect of SNP dosage. At the beginning, patients's arterial blood pressure is at some initial level p_0. Then we infuse a short impulse D_i of SNP [ml/h]. After 1–2 min of *initial transport delay*, SNP gets distributed through patient's body and we can measure an almost instant decrease of blood pressure. The effect is temporary though, since SNP decomposes quickly into a variety of other (toxic) chemicals. After a short period of time, patient's blood pressure rises back to its initial p_0 value.

Hypotensive effect of SNP is certain and, theoretically, with a continuous infusion D_i [ml/h] of SNP, the patient's MAP should remain at p_t as long as we need during the surgery. However, there are some influencers and symptoms. At first, patients differ in their *drug sensitivity*, i.e. certain dosage rate D_i causes distinct MAP decreases ([mmHg]) across different patients. Patient's drug sensitivity is usually unknown, since there is no intact preoperative method of estimating that. Practically, anesthesiologists investigate patient's drug sensitivity experimentally, starting with some safe *initial dosage*. After that, they may adjust the dosage rate in order to bring the patient towards the target p_t MAP.

Measured patient's blood pressure fluctuates during surgery thanks to many physiological and technical reasons, so some kind of permanent regulation is

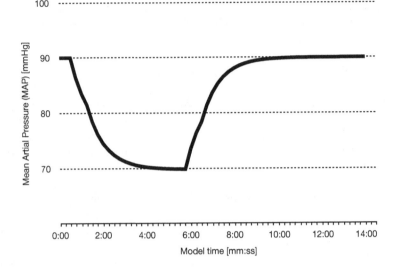

Fig. 1. Body response to a unit infusion of Sodium Nitroprusside (SNP).

necessary. There are numerous papers describing automated regulation mechanisms based on PID regulators [3,5], Fuzzy regulators [9,10] and even Genetic algorithms [4]. TOF-*Cuff* Controller is constructed on Fuzzy regulation principles [9].

Equation (1) shows three elements considered to influence the measured $MAP(t)$:

- p_0 is an initial patient's preoperative MAP. It is the reference value of patient's blood pressure considered to remain constant in duration of the operation.
- $p_n(t)$ is an time variable complex of all random influencers to blood pressure, such as natural fluctuations of blood pressure, errors in measurement, etc.
- $P_\Delta(t)$ is the hypotensive effect of SNP that we model in this paper.

$$MAP(t) = p_0 + p_n(t) - P_\Delta(t) \tag{1}$$

System of automatic blood pressure regulation works in discrete time cycles. In every step, the controller estimates a new level of infusion dosage rate [ml/h] and reconfigures the infusion pumps. The patient receives the infusion. Blood pressure transducer measures the resulting patient's BP/MAP and passes that information to the controller.

Formally speaking, the objective of blood pressure regulation working n discrete time steps ahead is to determine such future $D_0, D_1, ..., D_n$ dosage rates that the outcoming time sequence $MAP(0), MAP(1), ...$ of blood pressure measurements were close to p_t. Number of cycles where $|MAP(t) - p_t| < Limit$ (for some pre-defined $Limit$ [mmHg]) gives us statistical overview about regulator's performance. Let us emphasise that hypotensive effects comes with some time delay and also includes some unpredictable $p_n(t)$ stochastic element. An eventual *over-dosing* causing too deep hypotension might be extremely dangerous.

3   Simulation Infrastructure

The entire system of monitoring and regulation of BP/NMT represents a typical closed-loop regulation system that, in our case, consists of a measurement sensor, infusion pumps, the patient and Controller. Figure 2 depicts its architecture in real and simulated schemes. The simulation scheme (Fig. 2b) combines Hardware-in-the-loop and Simulation-in-the-loop techniques. All real and simulated components are interconnected using A2K - a versatile platform for simulation of cyber-physical systems.

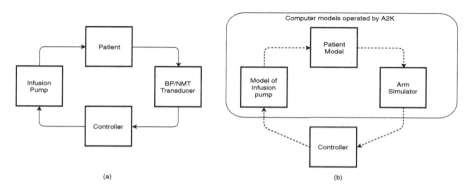

Fig. 2. (a) Scheme of the practical closed-loop BP/MNT regulation system, (b) Scheme of the simulation infrastructure based on computer models and A2K.

4   Patient Model

So called *Patient model* is a software simulation model describing *pharmaco-dynamic* and *pharmacokinetic* dynamic response of patient's body to a given infusion of a particular drug. It is a component of our simulation infrastructure depicted in Fig. 2b.

Pharmacodynamic of a drug defines its influence into a patient's body, whilst pharmacokinetics describes how the body recirculates this substance through the cardiovascular system. The hypotensive effect of a vasodilator is given by *pharmacodynamic* and *pharmacokinetic* features of a particular drug which is Sodium Nitroprusside in our case.

The elementary mathematical model of Sodium Nitroprusside comes from the original work of Slate [7,8] and is cited in many scientific papers. We implement this model in discrete time domain with 15 s time step. The model itself is described in Eq. (2).

$$P_\Delta(Z) = \frac{Z^{-d_i}(\beta_0 + \beta_1 Z^{-d_c})}{1 - \alpha Z^{-1}} \tag{2}$$

$$P_\Delta(k) = \alpha \cdot P_\Delta(k-1) + \beta_0 \cdot I(k-d_i) + \beta_1 \cdot I(k-d_i-d_j) \tag{3}$$

In the model, Z^{-1} represents a unit time delay (15 s) and Z^{-d} accordingly. Numerical coefficients model pharmacodynamic and pharmacokinetic behaviour of Sodium Nitroprosside, where β_0 specifies the strength of SNP's effect, d_i determines *initial time delay*, d_c denotes *recirculation time delay* and recirculation fraction (β_1). Equation (3) represents the model in the discrete time domain, where k denotes the discrete time steps and $I(k)$ is the input infusion dosage rate [ml/h].

We have already mention that estimating patient's *sensitivity* to SNP is generally difficult. At least, we can classify patients into three classes (normal/nominal sensitivity, low sensitivity, high sensitivity) of *SNP sensitivity*. See Table 1 for particular values of model's coefficients. Figure 3 demonstrates a complex regulation of BP starting with an initial dose and later increases of SNP dosage rate.

Table 1. Coefficients for Sodium Nitroprusside model, Eq. (2)

Parameter	Low sensitivity	Normal sensitivity	High sensitivity
Time coefficients d_i, d_c	2	3	5
Response coefficient (α)	0.606	0.741	0.779
Drug sensitivity (β_0)	0.053	0.187	3.546
Recirculation fraction (β_1)	0.01	0.075	1.418

5   Deployment of the System

There are two regimes of intended deployment of TOF-*Cuff* Controller System. In the first regime, TOF-*Cuff* is intended for use in operation rooms where all its components (Infusion pump, Controller itself, BP/NMT Transducer) are physically present and point-to-point directly interconnected (Fig. 2a). In the second regime, TOF-*Cuff* is considered to work in Intensive Critical Care Units (ICCU). In this scenario, infusion pumps and measurement devices are physically present in the room, just Controller connects them remotely via LAN.

In both scenarios, TOF-*Cuff* is supposed to be LAN connected to hospital's Anesthesia/Patient Information Management System. Local network connection to the hospital information system allows the user to access relevant anesthesiologic data as well as to record the entire anesthesiology for further evidence and analysis.

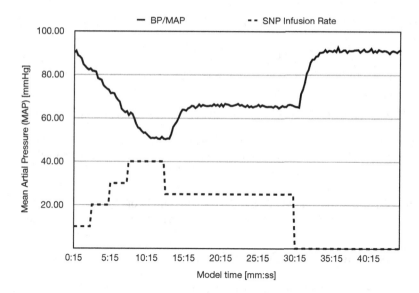

Fig. 3. Regulation of BP/MAP using Sodium Nitroprusside (SNP).

6 Test Scenarios and Experiments

Our mission at the AQUAS project is to prepare TOF-*Cuff* Controller for accreditation in USA/EU national health agencies. Practically, we need to provide a sequence of tests regarding operational safety, security and robustness of this medical device. Safety tests are intended to inspect hardware failures of the entire infrastructure and also tests oriented at human factor failures. Security tests aim at LAN connections of Controller in the context of its target deployments (operation room, ICCU). Since safety and security tests are out of the scope of this paper, let us rather comment the robustness tests.

Tests of Controller Robustness

The simulation testbed consists of TOF-*Cuff* Controller, simulated infusion pump, simulated patient (with various sensitivity coefficients) and simulated blood pressure sensor – all connected via A2K platform as described in Fig. 2. Since we insert the real TOF-*Cuff* Controller in the loop, all simulations has to work in the real time, making our experiments very time consuming (1–2 h per experiment).

From the robustness point of view, Controller must guarantee its correct behaviour under all thinkable circumstances, i.e. Controller must not put *any patient* in danger by over-dosing or ignoring unusual measurements. For this reason, we decided to experiment on patients that are *generated totally random*.

Test experiments are supposed to run in a loop. In each experiment, we randomly sample the space of eventual patient's sensitivity coefficients with the following scheme:

- Patient's sensitivity s is uniformly generated from $\{low, normal, high\}$.
- Let $(\alpha, \beta_0, \beta_1)$ are sensitivity coefficients from Table 1 appropriate to s.
- Patient's sensitivity coefficients $(\alpha', \beta_0', \beta_1')$ are randomly generated using Normal distribution with $mean = \alpha$ (resp. β_0, β_1), and standard deviation 10% of α (resp. β_0, β_1).

In every experiment, we record Controller's behaviour with a special care on:

- The controller does not cause over-dosing of the patient, i.e. patient's simulated MAP does not drop under critical level (e.g. 40 mmHg MAP).
- The controller outcomes *conservative* decisions, i.e. estimated dosage rate does not fluctuate.
- The controller alarms when dangerous conditions occur, e.g. MAP measurements are out of expected boundaries or MAP fluctuations are too high.

7   Conclusions and Future Work

We presented our simulation infrastructure constructed as a *simulation-in-the-loop* allowing us to intensively experiment with TOF-*Cuff* Controller under various simulated conditions. We also defined a part of test scenarios for rather large testing upcoming in our future work.

Acknowledgements. The work was supported by the H2020 ECSEL project Aquas (reg. no. 737475), the IT4IXS: IT4Innovations Excellence in Science project (LQ1602), and the FIT BUT internal project FIT-S-17-4014. Partners contributing to the Aquas project: RGB Medical Devices, Instituto Tecnológico de Informática, City University of London, Brno University of Technology, Trustport, All4Tec, Tecnalia.

References

1. Behbehani, K., Cross, R.R.: A controller for regulation of mean arterial blood pressure using optimum nitroprusside infusion rate. IEEE Trans. Biomed. Eng. **38**(6), 513–521 (1991)
2. Genc, S.: Prediction of mean arterial blood pressure with linear stochastic models. In: 2011 Annual International Conference of the IEEE Engineering in Medicine and Biology Society, pp. 712–715 (2011)
3. Isaka, S., Sebald, A.V.: Control strategies for arterial blood pressure regulation. IEEE Trans. Biomed. Eng. **40**(4), 353–363 (1993)
4. Kumar, H., Kumar, R., Yadav, J., Rani, A., Singh, V.: Genetic Algorithm based PID controller for blood pressure and Cardiac Output regulation. In: IEEE 1st International Conference on Power Electronics, Intelligent Control and Energy Systems, Delhi (2016)
5. Liu, G.Z., Wang, L., Zhang, Y.T.: A robust closed-loop control algorithm for mean arterial blood pressure regulation. In 2009 Sixth International Workshop on Wearable and Implantable Body Sensor Networks, pp. 77–81 (2009)
6. Ma, J., Zhu, K.Y., Krishnan, S.M.: Automatic postoperative blood pressure control. In: Proceedings of the 22nd Annual International Conference of the IEEE Engineering in Medicine and Biology Society, vol. 2, pp. 817–820 (2000)

7. Slate, J.B.: Model-based design of a controller for infusing sodium nitroprusside during postsurgical hypertension. Ph. D. dissertation. University of Wisconsin-Madison, no. 8028208 (1980)
8. Slate, J.B., Sheppard, L.C.: Automatic control of blood pressure by drug infusion. Phys. Sci. Measur. Instrum. Manage. Educ. Rev. IEE Proc. A. **129**, 639–645 (1983)
9. Ruiz, R., Borches, D., González, A., Corral, J.: A new sodium-nitroprusside-infusion controller for the regulation of arterial blood pressure. Biomed. Instrum. Technolol. **27**(3), 244–251 (1993)
10. Ying, H., Sheppard, L.C.: Regulating mean arterial pressure in postsurgical cardiac patients. A fuzzy logic system to control administration of sodium nitroprusside. IEEE Eng. Med. Biol. Mag. **13**(5), 671–677 (1994)

Approximating Complex Arithmetic Circuits with Guaranteed Worst-Case Relative Error

Milan Češka jr., Milan Češka, Jiří Matyáš[✉], Adam Pankuch, and Tomáš Vojnar

Faculty of Information Technology, Centre of Excellence IT4Innovations, Brno University of Technology, Brno, Czech Republic
imatyas@fit.vutbr.cz

Abstract. We present a novel method allowing one to approximate complex arithmetic circuits with formal guarantees on the *worst-case relative error*, abbreviated as WCRE. WCRE represents an important error metric relevant in many applications including, e.g., approximation of neural network HW architectures. The method integrates SAT-based error evaluation of approximate circuits into a verifiability-driven search algorithm based on Cartesian genetic programming. We implement the method in our framework ADAC that provides various techniques for automated design of arithmetic circuits. Our experimental evaluation shows that, in many cases, the method offers a superior scalability and allows us to construct, within a few hours, high-quality approximations (providing trade-offs between the WCRE and size) for circuits with up to 32-bit operands. As such, it significantly improves the capabilities of ADAC.

1 Introduction

In the recent years, reduction of power consumption of computer systems and mobile devices has become one of the biggest challenges in the computer industry. *Approximate computing* has been established as a new research field aiming at reducing system resource demands by relaxing the requirement that all computations are always performed correctly. Approximate computing can be conducted at different system levels with *arithmetic circuit approximation* being one of the most popular as such circuits are frequently used in numerous computations. Approximate circuits exploit the fact that many applications, including image and multimedia processing, machine learning, or neural networks, are *error resilient*, i.e., produce acceptable results even though the underlying computations are performed with a certain error. Chippa et al. [3] claims that almost 80% of runtime is spent in procedures that could be approximated.

This work has been partially supported by the Brno Ph.D. Scholarship Program, the Czech Science Foundation (project No. 19-24397S), the IT4Innovations Excellence in Science (project No. LQ1602), and the FIT BUT internal project FIT-S-17-4014.

© Springer Nature Switzerland AG 2020
R. Moreno-Díaz et al. (Eds.): EUROCAST 2019, LNCS 12013, pp. 482–490, 2020.
https://doi.org/10.1007/978-3-030-45093-9_58

Circuit approximation can be formulated as an optimisation problem where the error and non-functional circuit parameters (such as power consumption or chip area) are conflicting design objectives. Designing complex approximate circuits is a time-demanding and error-prone process, and its automation is challenging too since the design space is huge and evaluating candidate solutions is computationally demanding, especially if formal guarantees on the error have to be ensured. In our previous work [10], we proposed a scalable evolutionary circuit optimisation algorithm integrating a SAT-based circuit evaluation method that provides formal guarantees on the *worst-case absolute error* (WCAE).

In this paper, we extend the algorithm towards the *worst-case relative error* (WCRE) that represents another important error metric capturing the worst-case behaviour of the approximate circuits. Bounds on WCRE, in contrast to WCAE, require that the approximate circuits provide results that are close to the correct values even for small input values. This is essential for many application domains including, e.g., approximation of neural network hardware architectures [5]. Designing approximate circuits with WCRE bounds, however, represents a more challenging problem (when compared to WCRE) as the approximation has to preserve a larger part of the circuit logic and the circuit evaluation requires a more complicated procedure. To mitigate these challenges, we propose a novel construction of an auxiliary circuit (so-called miter) enabling an efficient SAT-based circuit evaluation against WCRE bounds. We integrate this evaluation procedure into the verifiability-driven circuit optimisation [10] implemented in our tool ADAC [1] and thus significantly extend the existing capabilities of automated techniques for the circuit approximation. Our experiments on circuits with up to 32-bit operands show that, in many cases, the proposed approach offers a superior scalability compared to alternative methods and allows us to construct, within a few hours, high-quality approximate circuits.

2   Search-Based Circuit Approximation

This section briefly summarises state-of-the-art methods for functional approximation with the focus on search-based approaches with formal error guarantees.

In *functional approximation*, the original system is replaced by a less complex one which exhibits some errors but reduces power consumption, delay, etc. Functional approximation can then be formulated as an optimisation problem where the error and energy efficiency/performance are conflicting design objectives. The approximation process either (1) tries to build an approximate solution from scratch or (2) tries to gradually modify the original system. The goal of the design process is to obtain an approximate solution with the best trade-off between the approximation error and resource savings.

Functional approximation can be performed manually by experts, but the current trend is to develop fully automated functional approximation methods that can be integrated into computer-aided design tools for digital circuits. There exist systematic approaches such as SALSA [11] or SASIMI [12], however, their drawback is an inability to generate novel logic structures. Search-based approximation techniques overcame this problem, and existing literature shows that this approach offers good performance and scalability [7].

Search-based approximation techniques typically iterate over two basic steps until a certain termination criterion is satisfied. The first step is the generation of candidate approximate solutions, and the error of these solutions is evaluated during the second step. Eventually, the method produces a solution (or a set of solutions) providing a good trade-off between the approximation error and resource consumption.

Our search-based approach builds, in particular, on the *Cartesian genetic programming* [6]—a specialised version of evolutionary algorithms suitable for circuit approximation [8]. The circuits are represented as an oriented acyclic graph where nodes are located in a fixed-size two-dimensional matrix. New solutions are obtained from existing ones by simply changing the functionality and interconnections of the nodes.

To obtain a near-optimal solution, search-based techniques typically have to explore and evaluate a high number of candidate approximations [8]. Therefore, the efficiency of the methods used to evaluate the approximation error of the candidates is essential for the performance of the overall approach.

There exist several *error metrics* characterising different types of errors such as the worst-case error, the mean error, or the error rate. In this work, we primarily focus on the worst-case error that is essential when guarantees on the worst behaviour of the approximate circuits are required. For arithmetic circuits, the worst-case behaviour is typically captured either by the *worst-case absolute error* (WCAE) or by the *worst-case relative error* (WCRE), defined as follows.

For an original golden circuit G computing a function f_G and its approximation C computing a function f_C, we define:

$$\mathrm{WCAE}(G, C) = \frac{\max_{x \in \{0,1\}^n} |\mathrm{int}(f_G(x)) - \mathrm{int}(f_C(x))|}{2^m} \tag{1}$$

$$\mathrm{WCRE}(G, C) = \max_{n \in \mathbb{N}} \frac{|f_G(n) - f_C(n)|}{f_G(n)} \tag{2}$$

Figure 1 illustrates the difference between the approximation process targeting at WCAE and WCRE. This difference, in fact, motivates our work. It shows two sets of circuits approximating 8-bit multipliers optimised for WCRE (green squares) and WCAE (red circles), respectively. The plots show the trade-off between the circuit area (directly effecting the power consumption) and WCRE (left) and WCAE (right), respectively. First, we observe that circuits optimised for WCAE have very bad WCRE (red dots left) and vice versa (green squares right). Second, the plots demonstrate that when optimising 8-bit multipliers circuits for WCAE, we achieve about 50% area reduction with WCAE = 1% while we need to set WCRE = 40% to obtain similar area improvements when optimising for WCRE. This is indeed caused by the fact that a larger part of the circuit logic has to be preserved to obtain approximations with low WCRE.

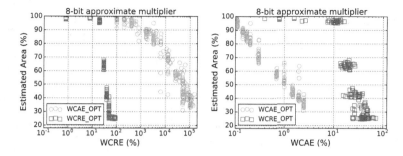

Fig. 1. A comparison of 8-bit multipliers approximated for WCRE and WCAE. (Color figure online)

Methods evaluating the approximation error have a crucial impact on the performance of the approximation process. A popular class of methods employs *circuit simulation* on a set of inputs to evaluate the error. Such methods typically suffer from low scalability (when an exhaustive simulation is applied) or a lack of guarantees (when the circuits are simulated for a subset of the possible inputs only). In order to provide guarantees on the approximation error and scale to complex circuits at the same time, various *formal verification* techniques, such as model-checking, SAT solving, or BDDs have recently been integrated to the approximation process [4,9]. They typically employ auxiliary circuits, so-called *miters*, that combine the original circuit and the approximate circuit and evaluate the error [2]. In our previous work [10], we proposed a miter construction allowing one to subsequently use an efficient SAT-based procedure to check whether the approximate circuit satisfies a given WCAE bound. Moreover, we proposed a verifiability-driven search-strategy that drives the search towards promptly verifiable approximate circuits. The strategy introduces a limit on resources that the underlying SAT solver can use to prove that the WCAE bound is met. This approach currently provides the best performance for the circuit approximation with WCAE guarantees.

3   SAT-based WCRE Evaluation

To evaluate whether the given approximate circuit meets the required bound on WCRE, we adapt and extend the miter we designed for WCAE [10]. As shown in Fig. 2 (left), the miter interconnects the golden circuit G and the candidate circuit C that both share identical inputs. The subtractor and absolute value blocks allow us to quantify the approximation error between C and G. Finally, the error is compared to a given threshold value T, and the output of the comparator is set to logical *true* if and only if the threshold T is violated. Thus the miter construction allows us to evaluate whether $WCAE(C, G) > T$ in a single SAT query. Note that, for a given approximation scenario, the threshold T is constant and can therefore be built into the structure of the comparator.

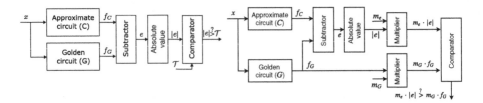

Fig. 2. The approximation miter used for WCAE evaluation (left) and the novel construction for WCRE evaluation (right).

3.1 A Generic WCRE Miter

To obtain a WCRE miter, we extend the WCAE miter by adding some components. Recall that we need to check the satisfiability of the following formula:

$$\max_{n \in \mathbb{N}} \frac{|f_G(n) - f_C(n)|}{f_G(n)} > T. \tag{3}$$

Note that we do not need to find the maximum of the left-hand side of the formula, but rather determine if there exists a single input combination for which the bound T is violated. Therefore, we can replace Formula 3 by the following constraint

$$\exists n \in N : |f_G(n) - f_C(n)| * m_e > f_G(n) * m_G \tag{4}$$

where $T = m_G/m_e$. Based on this formula, we build a general WCRE miter using two multipliers by a constant and a generic comparator, see Fig. 2 (right).

3.2 Variants of the WCRE Miter

Observe that the resulting WCRE miter is larger and more complex than the WCAE miter. This indeed slows down its evaluation and thus reduces the overall performance of the approximation process. To improve the performance and scalability with respect to the circuit complexity, we simplify the general WCRE miter and propose three variants of the miter that are smaller but can be used for certain values of the bound T only.

As we work with the binary representation of integers, multiplication by the powers of 2 is identical to a bit shift operation. Thus, each of the constants m_G and m_e can be expressed using two values: namely, mc_x denoting a multiplicative constant and bs_x denoting a number of shifted bits. The original values of m_G and m_e are then computed as:

$$m_G = mc_G * 2^{bs_G} \qquad\qquad m_e = mc_e * 2^{bs_e}$$

In combinational circuits, a shift by a constant number of bits is represented by a reconnection of wires only and does not contain any logical gates. This setting allows us to remove one or even both of the constant multiplications for a subset of target WCRE error bounds T. If we restrict the values of m_g and

m_e to powers of two, we suffice with utilising bit shifts only. This restricts the obtainable values of T to $1/2^{bs_e}$, e.g., 50%, 25%, or 12.5%. Adding one multiplier by a constant significantly broadens the range of supported target values. These can be expressed by one of the formulas: $2^{bs_G}/(mc_e * 2^{bs_e})$ or $(mc_G * 2^{bs_G})/2^{bs_e}$. However, the constants should be kept small. Using higher values leads to larger bit widths representing the compared numbers, and therefore a more complex comparator, thus negating the contribution of this optimisation.

4   Experimental Evaluation

We have integrated the proposed WCRE miters into our tool ADAC [1] and evaluated its performance on a benchmark of circuit approximation problems.

4.1   Comparison of the WCRE Miters

In Table 1, we compare the size of the proposed WCRE miters. We select three target WCRE bounds T for the bit-shift variant and four target error values for one and two multiplier miter designs. The table shows the average sizes of the different variants of the miter obtained for the three chosen bit widths in adder and multiplier approximation. The size is measured in the number of nodes in the AIG graph representation of the miter. Note that AIG is a basic representation of circuits in ADAC and is directly used as the input for the SAT solving procedure. A larger size of AIGs negatively affects the performance of the solver. We can see that the bit-shift variant is about a factor 2 smaller than the general construction using two multipliers. For multipliers, the differences in the size between the variants are less significant as the circuits themselves form a bigger part of the miter. Note that the average size of the WCAE miter for a 32-bit adder and 12-bit multiplier is 810 and 2437 AIG nodes, respectively, which is smaller than even the bit-shift variant of the WCRE miters for the corresponding circuits. This clearly indicates that the evaluation against WCRE is considerably harder.

Table 1. Numbers of AIG nodes for different miters and WCRE bounds T.

	Bit shifts			One multiplier				Two multipliers			
T [%]	12.5	25.0	50.0	10.0	33.3	66.7	80.0	30.0	42.9	71.4	85.7
add8	226	228	233	324	327	342	360	447	510	506	519
add16	497	501	502	770	755	773	756	1079	1225	1181	1220
add32	1120	1090	1114	2074	2116	2084	2106	3125	3249	3315	3354
mult4	268	267	273	335	347	356	347	464	499	492	513
mult8	1175	1177	1183	1393	1421	1436	1414	1685	1833	1803	1841
mult12	2617	2621	2622	3032	3057	3060	3051	3512	3748	3726	3756

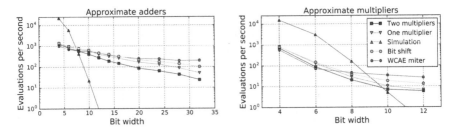

Fig. 3. Performance of the circuit evaluation using different WCRE miters, the WCAE miter, and simulation. Left: Adders. Right: Multipliers.

Figure 3 illustrates how the size of the miters affects the performance of the candidate circuit evaluation. In particular, it shows the average number of evaluations per second (taken from 20 independent runs) when the approximation of adders (left) and multipliers (right) with different bit-widths (the x-axis) is performed. We also compare the miter-based methods with full simulation and WCAE miter evaluation.

We can observe that the simulation is considerably faster for small bit-widths, however, its performance significantly drops for circuits with operands larger than 10-bits. The proposed-SAT based approach scales much better. For the adders, it provides very good performance (around 100 evaluations per second) even for 32-bit operands. For the multipliers (representing structurally more complex circuits), the performance is much lower and drops to 10 evaluations per second for 12-bit operands. As expected, the speed of the miter evaluation slows down with increasing miter complexity—the bit-shift variant is the fastest while the version with two multipliers is the slowest. The difference in the evaluation speed is negligible for smaller circuits but becomes more significant for larger bit-widths. Note that the evaluation of the WCAE miters is significantly faster due their smaller sizes (e.g. 4-times smaller for the 32-bit adders and 1.5-times smaller for the 12-bit multipliers in comparison to the two multiplier implementation).

For larger miters, the bounds on the SAT solver resources get applied, and a small number of circuit evaluation tasks is skipped (e.g., for the WCRE mitters, 0.7% for the 32-bit adders and 6% for the 12-bit multipliers). The idea is to skip candidates for which the evaluation takes too long because their successors typically have the same problem and thus they reduce the performance of the overall approximation process. For more details, see our previous work [10], where we introduced this, so-called verifiability-driven strategy.

4.2   Circuit Approximation

In this section, we study how the proposed SAT-based circuit evaluation can be leveraged in circuit approximation. Recall that we integrate the evaluation procedure into the verifiability-driven circuit approximation based on Cartesian genetic programming. The optimisation is formulated as a single-objective optimisation, i.e., for a given threshold on the WCRE bound T, the approximation

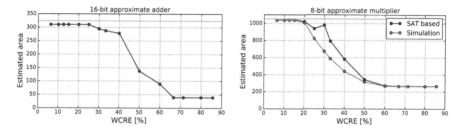

Fig. 4. The median circuit area of approximate adders (left) and multipliers (right). The red line indicates the area of the original circuit. (Color figure online)

seeks for a circuit satisfying the bound and having the smallest circuit area[1]. For every value of T, we run a 2-hours-long approximation process. To take into account the randomness of the evolutionary optimisation, we report the median of the circuit area obtained from 20 independent runs. Figure 4 illustrates the results of the approximation process, in particular, the obtained approximate circuits for the 16-bit adders (left) and the 8-bit multipliers (right). The circuits form a Pareto front that captures the trade-offs between the area and the approximation error. The red line shows the area of the golden circuit.

For the *adders*, the proposed approximation method works very well and is able to successfully approximate circuits up to 32-bit operands (not presented here). Figure 4 (left) shows that, for 16-bit adders, the most interesting solutions in the terms of accuracy and area savings are located in the interval between 30% and 60% WCRE. For smaller target error values, the reduction of the circuit size is negligible. On the other hand, the solutions with larger approximation errors do not feature further improvements. We can also observe a dramatic area reduction between 40% and 50% WCRE.

Approximation of the *multipliers* represents a significantly harder problem. Recall that the size of multipliers (and thus also of the miters) grows quadratically with respect to their bit-widths. Therefore, the design space is larger and candidate evaluation is more complicated as discussed in the previous section. Figure 4 (right) compares the approximate 8-bits multipliers obtained using the simulation-based and SAT-based evaluation procedure. The SAT-based approach slightly lags behind the simulation mainly in the interval between 25% and 40% WCRE. This can be explained by the worse performance of the SAT-based evaluation on the 8-bit multipliers (recall Fig. 3 (right)).

As the performance of the simulation-based evaluation is very low beyond 10-bit multipliers, the approximation process is not able to provide a good approximation of these circuits within a 2-hours-long run. Although the SAT-based approach (namely the bit-shift solution) is able to evaluate around 10 candidates per second (for the 12-bit multipliers), the approximation process also

---

[1] We estimate the area as the sum of sizes of the gates (in the target 45 nm technology) used in the circuit. The estimation tends to be accurate and also adequately captures the circuit power consumption [9].

fails to provide good Pareto sets. This is probably caused by the candidates that are skipped during the evaluation due to the resource limits on the underlying SAT solver. Note that this behaviour was not observed for the WCAE approximation that works very well even for 16-bit multipliers despite many skipped solutions [10]. This again indicates that the WCRE approximation is very challenging, and future research is necessary in this area.

References

1. Češka, M., Matyáš, J., Mrazek, V., Sekanina, L., Vasicek, Z., Vojnar, T.: ADAC: automated design of approximate circuits. In: Chockler, H., Weissenbacher, G. (eds.) CAV 2018. LNCS, vol. 10981, pp. 612–620. Springer, Cham (2018). https://doi.org/10.1007/978-3-319-96145-3_35
2. Chandrasekharan, A., et al.: Precise error determination of approximated components in sequential circuits with model checking. In: DAC 2016. IEEE (2016)
3. Chippa, V.K., et al.: Analysis and characterization of inherent application resilience for approximate computing. In: DAC 2013. ACM (2013)
4. Ciesielski, M., et al.: Verification of gate-level arithmetic circuits by function extraction. In: DAC 2015. ACM (2015)
5. Judd, P., Albericio, J., et al.: Proteus: exploiting numerical precision variability in deep neural networks. In: ICS 2016, pp. 1–12 (2016)
6. Miller, J.F., Thomson, P.: Cartesian genetic programming. In: Poli, R., Banzhaf, W., Langdon, W.B., Miller, J., Nordin, P., Fogarty, T.C. (eds.) EuroGP 2000. LNCS, vol. 1802, pp. 121–132. Springer, Heidelberg (2000). https://doi.org/10.1007/978-3-540-46239-2_9
7. Mrazek, V., et al.: Library of approximate adders and multipliers for circuit design and benchmarking of approximation methods. In: DATE 2017 (2017)
8. Vasicek, Z., Sekanina, L.: Circuit approximation using single- and multi-objective cartesian GP. In: Machado, P., et al. (eds.) EuroGP 2015. LNCS, vol. 9025, pp. 217–229. Springer, Cham (2015). https://doi.org/10.1007/978-3-319-16501-1_18
9. Vasicek, Z., et al.: Towards low power approximate DCT architecture for HEVC standard. In: DATE 2017, pp. 1576–1581. IEEE (2017)
10. Češka, M., et al.: Approximating complex arithmetic circuits with formal error guarantees: 32-bit multipliers accomplished. In: ICCAD 2017, pp. 416–423 (2017)
11. Venkataramani, S., et al.: SALSA: systematic logic synthesis of approximate circuits. In: DAC 2012, pp. 796–801. ACM (2012)
12. Venkataramani, S., et al.: Substitute-and-simplify: a unified design paradigm for approximate and quality configurable circuits. In: DATE 2013. EDA (2013)

Solving String Constraints with Approximate Parikh Image

Petr Janků$^{(\boxtimes)}$ and Lenka Turoňová

Faculty of Information Technology, Brno University of Technology, Brno, Czech Republic
{ijanku,ituronova}@fit.vutbr.cz

Abstract. In this paper, we propose a refined version of the Parikh image abstraction of finite automata to resolve string length constraints. We integrate this abstraction into the string solver SLOTH, where on top of handling length constraints, our abstraction is also used to speed-up solving other types of constraints. The experimental results show that our extension of SLOTH has good results on simple benchmarks as well as on complex benchmarks that are real-word combinations of transducer and concatenation constraints.

1 Introduction

Strings are a fundamental data type in many programming languages, especially owing to the rapidly growing popularity of scripting languages (e.g. JavaScript, Python, PHP, and Ruby) wherein programmers tend to make heavy use of string variables. String manipulations could easily lead to unexpected programming errors, e.g., cross-site scripting (a.k.a. XSS), which are ranked among the top three classes of web application security vulnerabilities by OWASP [11]. Some renowned companies like Google, Facebook, Adobe and Mozilla pay to whoever (hackers) finds a web application vulnerability such as cross-site scripting and SQL injection in their web applications [1], e.g., Google pays up to $10,000.

In recent years, there have been significant efforts on developing solvers for string constraints. Many rule-based solvers (such as Z3STR2 [15], CVC4 [8], S3P [12]) are quite fast for the class of simple examples that they can handle. They are sound but do not guarantee termination. Other tools for dealing with string constraints (such as NORN [1], SLOTH [6], OSTRICH [4]) are based on automata. They use decision procedures which work with fragments of logic over string constraints that are rich enough to be usable in real-world web applications.

This work has been supported by the Czech Science Foundation (project No. 17-12465S), the IT4Innovations Excellence in Science (project No. LQ1602), and the FIT BUT internal projects FIT-S-17-4014 and FEKT/FIT-J-19-5906.
We thank you to Lukáš Holík for all the support and encouragement he gave us and also the time he spent with us during discussions.
[1] For more information, see https://www.netsparker.com/blog/web-security/google-increase-reward-vulnerability-program-xss/.

R. Moreno-Díaz et al. (Eds.): EUROCAST 2019, LNCS 12013, pp. 491–498, 2020.
https://doi.org/10.1007/978-3-030-45093-9_59

They are sound and complete. SLOTH was the first solver that can handle string constraints including transducers, however, unlike NORN and OSTRICH it is not able to handle length constraints yet. Moreover, these tools are not efficient on simple benchmarks as the rule-based solvers above.

Example 1. The following JavaScript snippet is an adaptation of an example from [2,7]:

```
var x = goog.string.htmlEscape(name);
var y = goog.string.escapeString(x);
nameElem.innerHTML = '<button onclick= "viewPerson(\'' + y +
    '\')">' + x + '</button>';
```

This is a typical example of string manipulation in a web application. The code attempts to first sanitise the value of name using the sanitization functions htmlEscape and escapeString from the Closure Library [5]. The author of this code accidentally swapped the order of the two first lines. Due to this subtle mistake, the code is vulnerable to XSS, because the variable y may be assigned an unsafe value. To detect such mistakes, we have to first translate the program and the safety property to a string constraint, which is satisfiable if and only if y can be assigned an unsafe value. However, if we would add the length constraints (e.g. `x.length == 2*y.lenght;`) to the code, none of SLOTH, OSTRICH, or NORN would be able to handle them.

The length constraints are quite common in programs like this. Hence, in this paper, we present how to extend the method of SLOTH to be able to cope with them. Our decision procedure is based on the computation of Parikh images for automata representing constraint functions. Parikh image maps each symbol to the number of occurrences in the string regardless to its position.

For one nondeterministic finite automaton, one can easily computate the Parikh image by standard automata procedures. However, to compute an exact Parikh image for a whole formula of contraints is demanding. The existing solution proposes first to compute the product of the automata representing the subformulea and then compute the Parikh image of their product. Unfortunately, the exact computation of the Parikh images is computationally far too expensive. Even more importantly, the resulting semilinear expressions become exponential to the number of automata.

We therefore propose a decision procedure which computes an over-approximation of the exact solution that is sufficiently close to the exact solution. We first compute the membership Parikh images of the automata representing the string constraints. Then we use concatenation and substitution to compute the over-approximation of the Parikh image of the whole formula. However, we will not get the same result as with the previous approach since the Parikh image forgets the ordering of the symbols in the world. This causes that we could accept even words that are not accepted by the first approach. But even though our method does not provide accurate results, it is able to handle the lenght constraints and solve also real-world cases.

Outline. Our paper is organized as follow. In Sect. 1, we introduce relevant notions from logic and automata theory. Section 3 presents an introduction to a string language. Section 4 explains the notion of Parikh image and operations on Parikh images. Section 5 presents the main decision procedure. In Sect. 6, the experimental results are presented.

2 Preliminaries

Bit Vector. Let $\mathbb{B} = \{0, 1\}$ be a set of Boolean values and V a finite set of bit variables. *Bit vectors* are defined as functions $b : V \rightarrow \mathbb{B}$. In this paper, bit vectors are described by conjunctions of literals over V. We will denote the set of all bit vectors over V by $\mathbb{P}(V)$ and a set of all formulae over V by $\mathbb{F}_V$.

Further, let $k \geq 1$, and let $V\langle k \rangle = V \times [k]$ where $[k]$ denotes the set $\{1, \dots, k\}$. Given a word $w = b_1^k \dots b_m^k \in \mathbb{P}(V\langle k \rangle)^*$ over bit vectors, we denote by $b_j^k[i] \times \{i\} = b_j^k \cap (V \times \{i\})$, $1 \leq j \leq m$ where $b_j^k[i] \in \mathbb{P}(V)$ the j-th bit vector of the i-th track. Further, $w[i] \in \mathbb{P}(V)^*$ such that $w[i] = b_1^k[i] \dots b_m^k[i]$ is the word which keeps the content of the i-th track of w only. For a bit vector $b \in \mathbb{P}(V)$, we denote by $\{b\}$ the set of variables in the vector.

Automata and Transducers. A *succinct nondeterministic finite automaton* (NFA) over bit variables V is a tuple $\mathcal{A} = (V, Q, \Delta, q_0, F)$ where Q is a finite set of *states*, $\Delta \subseteq Q \times \mathbb{F}_V \times Q$ is *transition relation*, $q_0 \in Q$ is an *initial state*, and $F \subseteq Q$ is a finite set of *final states*. $\mathcal{A}$ accepts a word w iff there is a sequence $q_0 b_1^k q_1 \dots b_m^k q_m$ where $b_i^k \in \mathbb{P}(V)$ for every $1 \leq i \leq m$ such that $(q_i, \varphi_i, q_{i-1}) \in \Delta$ for every $1 \leq i \leq m$ where $b_1^k \models \varphi_i$, $q_m \in F$, and $w = b_1^k \dots b_m^k \in \mathbb{P}(V)^*$, $m \geq 0$, where each b_i^k, $1 \leq i \leq m$, is a bit vector encoding the i-th letter of w. The *language* of $\mathcal{A}$ is the set $L(\mathcal{A})$ of accepted words.

A k-track *succinct finite automaton* over V is an automaton $\mathcal{R}\langle k \rangle = (V\langle k \rangle, Q, \Delta, I, F)$, $k \geq 1$. The relation $R(\mathcal{R}\langle k \rangle) \subseteq (\mathbb{P}(V)^*)^k$ *recognised* by $\mathcal{R}$ contains a k-tuple of words $(x_1, \dots, x_k)$ over $\mathbb{P}(V)$ iff there is a word $w \in L(\mathcal{R})$ such that $x_i = w[i]$ for each $1 \leq i \leq k$. A *finite transducer* (FT) $\mathcal{R}$ is a 2-track automaton.

Strings and Languages. We assume a finite alphabet Σ. Σ^* represents a set of finite words over Σ, where the empty word is denoted by ϵ. Let x and y be finite words in Σ^*. The concatenation of x and y is denoted by $x \circ y$. We denote by $|x|$ the length of a word $x \in \Sigma^*$. A language is a subset of Σ^*. The concatenation of languages L, L' is the language $L \circ L' = x \circ x' | x \in L \wedge x' \in L'$, and the iteration L^* of L is the smallest language closed under $\circ$ and containing L and ϵ.

3 String Language

Let $\mathbb{X}$ be a set of variables and x, y be *string variables* ranging over Σ^*. A *string formula* over string terms $\{t_{str}\}^*$ is a Boolean combination of *word equations* $x =$

t_{str} whose right-hand side t_{str} might contain the concatenation operator, *regular constraints* P, *rational constraints* R and arithmetic inequalities:

$$\varphi ::= x = t_{str} \mid P(x) \mid R(x,y) \mid t_{ar} \geq t_{ar} \mid \varphi \wedge \varphi \mid \varphi \vee \varphi \mid \neg\varphi$$
$$t_{str} ::= x \mid \epsilon \mid t_{str} \circ t_{str}$$
$$t_{ar} ::= k \mid |t_{str}| \mid t_{ar} + t_{ar}$$

In the grammar, x ranges over string variables, $R \subseteq (\Sigma^*)^2$ is assumed to be a binary rational relation on words of Σ^*, and $P \subseteq \Sigma^*$ is a regular language. We will represent regular languages by succinct automata and tranducers denoted as $\mathcal{R}$ and $\mathcal{A}$, respectively. The arithmetic terms t_{ar} are linear functions over term lengths and integers, and arithmetic constraints are inequalities of arithmetic terms. The set of word variables appearing in a term is defined as follows: $Vars(\epsilon) = \emptyset$, $Vars(c) = \emptyset$, $Vars(u) = \{u\}$ and $Vars(t_1 \circ t_2) = Vars(t_1) \cup Vars(t_2)$.

To simplify the representation, we do not consider *mixed* string terms t_{str} that contain, besides variables of $\mathbb{X}$, also symbols of Σ. This is without loss of generality since a mixed term can be encoded as a conjunction of the pure terms over $\mathbb{X}$ obtained by replacing every occurrence of a letter $a \in \Sigma$ by a fresh variable x, and adding a regular membership constraint $\mathcal{A}_a(x)$ with $L(\mathcal{A}_a) = \{a\}$.

Semantics. A formula φ is interpreted over an *assignment* $\iota : \mathbb{X}_\varphi \to \Sigma^*$ of its variables $\mathbb{X}_\varphi$ to strings over Σ^*. ι is extended to string terms by $\iota(t_{s_1} \circ t_{s_2}) = \iota(t_{s_1}) \circ \iota(t_{s_2})$ and to arithmetic terms by $\iota(|t_s|) = |\iota(t_s)|$, $\iota(k) = k$ and $\iota(t_i + t_i') = \iota(t_i) + \iota(t_i')$. We formalize the satisfaction relation for word equations, regular constraints, rational constraints, and arithmetic inequalities, assuming the standard meaning of Boolean connectives:

$$x = t_{str} \text{ iff } \iota(x) = \iota(t_{str})$$
$$\iota(P(x)) = \top \text{ iff } \iota(x) \in P$$
$$\iota(\mathcal{R}(x,y)) = \top \text{ iff } (\iota(x), \iota(y)) \in \mathcal{R}$$
$$\iota(t_{i_1} \leq t_{i_2}) = \top \text{ iff } \iota(t_{i_1}) \leq \iota(t_{i_2})$$

The truth value of Boolean combinations of formulae under ι is defined as usual. If $\iota(\varphi) = \top$ then ι is a *solution* of φ, written $\iota \models \varphi$. The formula φ is *satisfiable* iff it has a solution, otherwise it is *unsatisfiable*.

The unrestricted string logic is undecidable, e.g., one can easily encode Post Correspondence Problem (PCP) as the problem of checking satisfiability of the constraint $\mathcal{R}(x,x)$ for a rational transducer $\mathcal{R}$ [10]. Therefore, we restrict the formulae to be in so-called *straight-line form*. The definition of *straight-line fragment* as well as a linear-time algorithm for checking whether a formula φ falls into the straight-line fragment is defined in [9].

4   Parikh Image

The Parikh image of a string abstracts from the ordering in the string. Particularly, the Parikh image of a string x maps each symbol a to the number of its occurrences in the string x (regardless to their position). Parikh image of a given language is then the set of Parikh images of the words of the language.

In this chapter, we present a construction of the Parikh image of a given NFA $\mathcal{A} = (V, Q, \Delta, q_0, F)$. The algorithm is modified version of the algorithm from [13] which computes the Parikh image for a given context-free grammar G. This algorithm contains a small mistake that has been fixed by Barner in a 2006 Master's thesis [3]. Since for every regular grammar there exists a corresponding NFA, we can easily customize the algorithm for NFA such that one can compute an existential Presburger formula $\phi_{\mathcal{A}}$ which characterizes the Parikh image of the language $L(\mathcal{A})$ recognized by $\mathcal{A}$ in the following way.

Let us define a variable $\#_\varphi$ for each $\varphi \in \mathbb{F}_V$, y_t for each $t \in \Delta$, and u_q for each $q \in Q$, respectively. The free variables of $\phi_{\mathcal{A}}$ are variables $\#_\varphi$ and we write $Free(\phi_{\mathcal{A}})$ for the set of all free variables in the formula $\phi_{\mathcal{A}}$. The formula $\phi_{\mathcal{A}}$ is the conjunction of the following three kinds of formulae:

- $u_q + \sum_{t=(q',\varphi,q)\in\Delta} y_t - \sum_{t=(q,\varphi,q')\in\Delta} y_t = 0$ for each $q \in Q$, where the variable u_q is restricted as follows: $u_{q_0} = 1$, $u_{q_F} \in \{0, -1\}$ for $q_F \in F$, and $u_q = 0$ for all other $q \in Q \setminus (\{q_0\} \cup F)$.
- $y_t \geq 0$ for each $t \in \Delta$ since the variable y_t cannot be assigned a negative value.
- $\#_\varphi = \sum_{t=(q,\varphi,q')\in\Delta} y_t$ for each $\varphi \in \mathbb{F}$ to ensure that the value x_φ are consistent with the y_t.
- To express the connectedness of the automaton, we use an additional variable z_q for each $q \in Q$ which reflects the distance of q from q_0 in a spanning tree on the subgraph of $\mathcal{A}$ induced by those $t \in \Delta$ with $y_t \geq 0$. To this end, we add for each $q \in Q$ a formulae $z_q = 1 \wedge y_t \geq 0$ if q is an initial state, otherwise $(z_q = 0 \wedge \bigwedge_{t\in\Delta_q^+} y_t = 0) \vee \bigvee_{t\in\Delta_q^+}(y_t \geq 0 \wedge z_{q'} \geq 0 \wedge z_q = z_{q'} + 1)$ where $\Delta_q^+ = \{(q', \varphi, q) \in \Delta\}$ is a set of ingoing transitions.

The resulting existential Presburger formula is then $\exists z_{q_1}, \ldots, z_{q_n}, u_{q_1}, \ldots, u_{q_n}, y_{t_1}, \ldots, z_{t_m} : \phi_{\mathcal{A}}$ where n is the number of states and m is the number of transitions of the given automaton. This algorithm can be directly applied to transducers where the free variables are $\#_\varphi$ such that $\varphi \in \mathbb{F}_{V\langle 2\rangle}$.

4.1   Operations on Parikh Images

In our decision procedure, we will need to use *projection* of the Parikh image of transducers and *intersection* of Parikh images. We have to find a way how to deal with alphabet predicates of transducers since our version of the intersection of Parikh images works only with alphabet predicates over a non-indexed set of bit variables. The intersection of Parikh images is needed since the alphabet predicates of one automaton can represent a set of symbols which may contains common symbols for more than one automaton. These operations can be implemented in linear space and time.

Projection. Let $\mathcal{R} = (V\langle 2\rangle, Q, \Delta, I, F)$ be a transducer representing a constraint $R(x, y)$ and let $\varphi \in \mathbb{F}_{V\langle 2\rangle}$ be a formula over $\{b^k\} \in 2^{V\langle 2\rangle}$ where $b^k \in \mathbb{P}(V\langle 2\rangle)$. We write $\varphi[x]$ to denote a alphabet projection of φ where $\varphi[x]$ is the subformula of φ such that only contains bits from $b^k[i]$ and i is the position of x in R. Given the Parikh image $\phi_\mathcal{R}$ of $\mathcal{R}$, we denote by $\phi_\mathcal{R}[x]$ a *projection* of $\phi_\mathcal{R}$ where the set of free variables is $Free(\phi_\mathcal{R}[x]) = \{\#_{\varphi[x]} \mid \#_\varphi \in Free(\phi_\mathcal{R})\}$. Further, we need to introduce the auxiliary function λ that assigns to each variable $\#_{\varphi[x]}$ a set $\{\#_{\varphi'} \mid \varphi'[x] = \varphi[x]\}$. The resulting formula of projection $\phi_\mathcal{R}$ has then the form $\phi_\mathcal{R}[x] = \exists \#_{\varphi_1}, \ldots, \#_{\varphi_n} : \phi_\mathcal{R} \wedge \bigwedge_{\#_{\varphi[x]} \in Free(\phi_\mathcal{R}[x])} \left(\#_{\varphi[x]} = \sum_{\#_\varphi \in \lambda(\#_{\varphi[x]})} \#_\varphi\right)$ where $\#_{\varphi_i} \in Free(\phi_\mathcal{R})$ for $1 \leq i \leq n$.

Intersection. We assume that both Parikh images have alphabet predicates $\mathbb{F}_V$ over the same set of bit variables V. Given two Parikh images ϕ_1 and ϕ_2, their intersection $\phi_\curlywedge = \phi_1 \curlywedge \phi_2$ can be constructed as follows. First, we compute a set of fresh variables $\mathcal{I} = \{\#_{\varphi_1 \curlywedge \varphi_2} \mid \#_{\varphi_1} \in Free(\phi_1) \wedge \#_{\varphi_2} \in Free(\phi_2) \wedge \exists b \in \mathbb{P}(V) : b \models \varphi_1 \wedge \varphi_2\}$ representing the number of common symbols for ϕ_1 and ϕ_2. Next, we define for each Parikh image ϕ_i a function $\tau_i : Free(\phi_i) \to 2^{\mathcal{I}}$ such that $\tau_1(\#_{\varphi_1}) = \{\#_{\varphi_1 \curlywedge \varphi_2} \in \mathcal{I}\}$ and $\tau_2(\#_{\varphi_2}) = \{\#_{\varphi_1 \curlywedge \varphi_2} \in \mathcal{I}\}$. Finally, the intersection is define as $\phi_\curlywedge = \phi_1 \wedge \phi_2 \wedge \bigwedge_{\#_{\varphi_1} \in Free(\phi_1)} \left(\#_{\varphi_1} = \sum_{\#_{\varphi'_1} \in \tau_1(\#_{\varphi_1})} \#_{\varphi'_1}\right) \wedge \bigwedge_{\#_{\varphi_2} \in Free(\phi_2)} \left(\#_{\varphi_2} = \sum_{\#_{\varphi'_2} \in \tau_2(\#_{\varphi_2})} \#_{\varphi'_2}\right)$.

5   Decision Procedure

Our decision procedure is based on computation of the Parikh images of the automata representing string constraints. Let $\varphi := \varphi_{cstr} \wedge \varphi_{eq} \wedge \varphi_{ar}$ be a formula in straight-line form where φ_{cstr} is a conjunction of regular constraints (or their negation) and rational constraints, φ_{eq} is a conjunction of word equations of the form $x = y_1 \circ y_2 \circ \cdots \circ y_n$, and φ_{ar} is a conjunction of arithmetic inequalities. The result of the decision procedure is an existential Presburger formula ϕ_φ which represents an over-approximation of the Parikh image of φ.

We assume that each variable $x \in Vars(\varphi)$ is restricted by an automaton or a transducer. Note that the function $Vars(\varphi)$ denotes a set of variables appearing in the formula φ. We write $\mathbb{T}$ to denote a set of Parikh images. The procedure is divided into three steps as follows.

- **Step 1:** First, we compute Parikh images of automata and transducers representing the constraints from φ_{cstr} using the algorithm from Sect. 4. We define a mapping $\rho_{cstr} : Vars(\varphi_{cstr}) \to \mathbb{T}$ that maps each string variable $x \in Vars(\varphi_{cstr})$ to the over-approximation of its Parikh image. Let $P_1(x), \ldots, P_n(x)$ and $R_1(x, y), \ldots, R_m(x, y)$ be constraints from φ_{cstr} restricting x. A formula ϕ_x representing the Parikh image of x is then computed using the algorithm from Sect. 4.1 as $\phi_x = \phi_{\mathcal{A}_x} \curlywedge \phi_{\mathcal{A}_1} \curlywedge \ldots \curlywedge \phi_{\mathcal{A}_n} \curlywedge \phi_{\mathcal{R}_1}[x] \curlywedge \ldots \curlywedge \phi_{\mathcal{R}_m}[x]$ where $\phi_{\mathcal{A}_i}$, $0 \leq i \leq n$, is the Parikh image of the automaton $\mathcal{A}_i$ representing $P_i(x)$ and $\phi_{\mathcal{R}_j}$, $0 \leq j \leq m$, is the Parikh image of the transducer $\mathcal{R}_j$ representing $R_j(x, y)$.

- **Step 2:** We define a mapping $\rho_{eq} : Vars(\varphi_{eq}) \to \mathbb{T}$ that maps each string variable $x \in Vars(\varphi_{eq})$ to the over-approximation of its Parikh image as $\phi_x = (\bigwedge_{i=1}^{k} \rho_{cstr}(y_i) \wedge \bigwedge_{j=k+1}^{n} \rho_{eq}(y_j)) \curlywedge \rho_{cstr}(x)$. We assume that $Free(y_1) \cap \cdots \cap Free(y_n) = \emptyset$. This can be done by adding double negation to the alphabet predicates which helps to distinguish free variables of individual y_i. Parikh image does not preserve the ordering of the symbols in the string, therefore, we can reorder the right side of the equation $y_1 \circ \cdots \circ y_k \circ \cdots \circ y_n$ such that $\forall 1 \le i \le k : y_i \in \varphi_{cstr}$ and $\forall k \le j \le n : y_j \in \varphi_{eq}$. Moreover, the reordering can be done in such a way that each variable on the right side of the equation is already defined since φ falls into the straight-line fragment.
- **Step 3:** Finally, we build the final formula ϕ_φ using mappings ρ_{cstr} and ρ_{eq}. Let $\mathbb{X}_{eq} \subseteq Vars(\varphi_{eq})$ be a set of all variables that are on the left side of the equations. The resulting formula ϕ_φ is then a conjunction $\phi_\varphi = (\curlywedge_{x \in \mathbb{X}_{eq}} \rho_{eq}(x)) \curlywedge (\curlywedge_{x \in Vars(\varphi) \setminus \mathbb{X}_{eq}} \rho_{cstr}(x))$.

6 Experiments

We have implemented our decision procedure extending the method of SLOTH [6] as a tool, called PICoSo. SLOTH is a decision procedure for the straight-line fragment and acyclic formulas. It uses succinct alternating finite-state automata as concise symbolic representation of string constraints. Like SLOTH, PICoSo was implemented in Scala.

To evaluate its performance, we compared PICoSo against SLOTH. We performed experiments on benchmarks with diverse characteristics.

The first set of benchmarks is obtained from Norn group [1] and implements string manipulating functions such as the Hamming and Levenshtein distances. It consists of small test case that is combinations of concatenations, regular constraints, and length constraints. The second set SLOG

Table 1. Performance of PICoSo in comparison to SLOTH.

		SLOTH	PICoSo
Norn (1027)	sat (sec)	**314 (545)**	313 (566)
	unsat (sec)	353 (624)	**356 (602)**
	timeout	0	0
	error/un	360	**358**
SLOG (3392)	sat (sec)	922 (**5526**)	923 (5801)
	unsat (sec)	2033 (5950)	**2080 (4382)**
	timeout	437	**389**
	error/un	0	0
SLOG-LEN (394)	sat (sec)	0	0
	unsat (sec)	266 (659)	**296 (773)**
	timeout	4	15
	error/un	124	**83**

[14] is derived from the security analysis of real web applications. It contains regular constraints, concatenations, and transducer constraints such as `Replace` but no length constraints. The last set is obtained from SLOG by selecting 394 examples containing `Replace` operation. It was extended by `RelaceAll` operation and since in practice, it is common to restrict the size of string variables in web applications, we added length constraints of the form $|x| + |y| R n$, where $R \in \{=, <, >\}$, $n \in \{4, 8, 12, 16, 20\}$, and x, y are string variables. The summary of the experiments is shown in Table 1. All experiments were executed on a computer with Intel Xeon E5-2630v2 CPU @ 2.60 GHz and 32 GiB

RAM. The time limit was 30 s was imposed on each test case. The rows indicate the number of times the solver returned satisfiable/unsatisfiable (sat/unsat), the number of times the solver ran out of 30-s limit (timeout), and the number of times the solver either crashed or returned unknown (error/un).

The results show that PICoSo outperforms Sloth on all of unsat examples. Sloth is however slightly better in case of sat examples due to the addition computation of the over-approximation of the Parikh image. Sloth timed out on 441 cases while PICoSo run out of time only in 404 cases. This shows that our proposed procedure is efficient in solving not only length constraints, but also other types of constraints.

References

1. Abdulla, P.A., et al.: String constraints for verification. In: CAV 2014, pp. 150–166 (2014)
2. Barceló, P., Figueira, D., Libkin, L.: Graph logics with rational relations. Proc. ACM Program. Lang. **9**, 30 (2013)
3. Barner, S.: H3 mit gleichheitstheorien. Master's thesis, Technical University of Munich, Germany (2006)
4. Chen, T., Hague, M., Lin, A.W., Rümmer, P., Wu, Z.: Decision procedures for path feasibility of string-manipulating programs with complex operations. Proc. ACM Program. Lang. **3**, 49:1–49:30 (2019)
5. G. co. 2015. Google closure library (referred in Nov 2015) (2015). https://developers.google.com/closure/library/
6. Holík, L., Janků, P., Lin, A.W., Rümmer, P., Vojnar, T.: String constraints with concatenation and transducers solved efficiently. PACMPL **2**(POPL), 1–32 (2018)
7. Kern, C.: Securing the tangled web. ACM **57**, 38–47 (2014)
8. Liang, T., Reynolds, A., Tinelli, C., Barrett, C., Deters, M.: A DPLL(T) theory solver for a theory of strings and regular expressions. In: CAV 2014 (2014)
9. Lin, A.W., Barceló, P.: String solving with word equations and transducers: towards a logic for analysing mutation XSS. In: POPL, pp. 123–136 (2016)
10. Morvan, C.: On rational graphs. In: Tiuryn, J. (ed.) FoSSaCS 2000. LNCS, vol. 1784, pp. 252–266. Springer, Heidelberg (2000). https://doi.org/10.1007/3-540-46432-8_17
11. OWASP: The ten most critical web application security risks (2013). https://www.owasp.org/images/f/f8/OWASP_Top_10_-_2013.pdf
12. Trinh, M., Chu, D., Jaffar, J.: Progressive reasoning over recursively-defined strings. In: CAV 2016, pp. 218–240 (2016)
13. Verma, K.N., Seidl, H., Schwentick, T.: On the complexity of equational horn clauses. In: Nieuwenhuis, R. (ed.) CADE 2005. LNCS (LNAI), vol. 3632, pp. 337–352. Springer, Heidelberg (2005). https://doi.org/10.1007/11532231_25
14. Wang, H.-E., Tsai, T.-L., Lin, C.-H., Yu, F., Jiang, J.-H.R.: String analysis via automata manipulation with logic circuit representation. In: Chaudhuri, S., Farzan, A. (eds.) CAV 2016. LNCS, vol. 9779, pp. 241–260. Springer, Cham (2016). https://doi.org/10.1007/978-3-319-41528-4_13
15. Zheng, Y., et al.: Z3str2: an efficient solver for strings, regular expressions, and length constraints. Formal Meth. Syst. Des. **50**(2–3), 249–288 (2014)

Verification of Architectural Views Model 1+5 Applicability

Tomasz Górski[✉][iD]

Institute of Naval Weapons and Computer Science, Polish Naval Academy,
Gdynia, Poland
t.gorski@amw.gdynia.pl
https://amw.gdynia.pl/index.php/en/

Abstract. The paper presents architectural views model 1+5 which has
been proposed for designing integration solutions of collaborating software systems. The author has introduced modeling extensions of Unified
Modeling Language (UML) in form of UML profiles. The author has proposed an *Integration flow diagram* which is special form of UML activity
diagram. The diagram arranges mediation mechanisms from *UML Profile for Integration Flows* into an integration flow. The paper presents
transformations of model-to-model and model-to-code types which automate design of integration platform. The 1+5 was successfully applied
in Service-Oriented Architecture. The approach reveals its potential in
Domain-Driven Design, Micro-services and blockchain solutions.

Keywords: Software architecture description · Service-Oriented
Architecture · Domain-Driven Design · Micro-services · Blockchain

1 Introduction

The concept of the model crossed the author's mind after participation as a
Chief Analyst in project for Polish Border Guards. The first important issue
was getting of business context. The author stepped into running project with
almost 200 use cases without strict business justification. So, business modeling
helped to identify essential business processes and resulting from them about 30
use cases. The second crucial element was communication between systems. For
proper functioning, Border Guards system required collaboration with external
ones. And the third one are non-functional requirements, especially performance
and security. For example, when you cross border, Border guard officer checks
your passport. It usually takes about 20 s. Meanwhile, Border Guards system has
to communicate with several external systems, e.g.: Visa Information System,
Central Register of Issued and Canceled Passport Documents.

Having that experience in mind the author has proposed architectural views
model 1+5 for solutions, in which integration with external systems is required,
[1,2]. Three views are completely new: *Integrated processes*, *Integrated services*,
Contracts.

© Springer Nature Switzerland AG 2020
R. Moreno-Díaz et al. (Eds.): EUROCAST 2019, LNCS 12013, pp. 499–506, 2020.
https://doi.org/10.1007/978-3-030-45093-9_60

Integrated processes view presents business processes. *Integrated services* view describes communication between software systems. *Contracts* view defines collaborating parties, rules which should be fulfilled and gathers non-functional requirements, e.g.: performance, security.

The paper is structured as follows. The second section contains description of architectural views model 1+5. The third section presents UML extensions for architecture modeling of integration solution. The next section describes transformations of model-to-model and model-to-code types to automate design of integration platform. The fifth section presents two case studies where the model was applied in the context of ISO/IEC/IEEE 42010:2011 standard. The sixth section reveals part of development process for building integration solution. The last one summarizes the paper and outlines directions for further works.

2   Architectural Views Model 1+5

The model consists of the following architectural views: Integrated processes, Use cases, Logical, Integrated services, Contracts, Deployment (see Fig. 1).

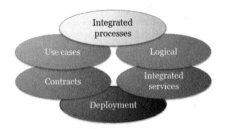

Fig. 1. Architectural views model 1+5.

The purpose of *Integrated processes view* is the identification of business processes which should be supported by software systems. Usually those processes activate many systems to cooperate. There are mainly two kinds of task within business process which require support: human task and automated task. First of them is task realized by human that can be supported. The second type is task that become service which fully automates execution of actions without interaction with human being. Processes are presented in a Business Process Modeling Notation business process diagram or UML activity diagram.

The *Use cases view* defines use cases which application should realize. The scope of an application is presented in UML Use case diagram. In case of collaboration among systems we need to distinguish actors who represent external ones. So, a new stereotype ≪IntegratedSystem≫ was proposed. A stereotype is one of three types of extension mechanisms in Unified Modeling Language. The other two are tagged values and constraints. The Use case diagram (see Fig. 2) presents scope of functionality of e-Prescription application. External system e-Pharmacy can invoke "Get prescriptions" use case.

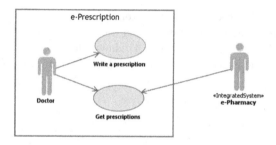

Fig. 2. Use cases of e-Prescription application.

The *Logical view* presents realizations of the use cases identified in the *Use case* view. For this purpose UML Sequence, Communication, and Class diagrams are applied. The view also presents, in UML class diagrams, the structure of business entities used in human tasks in *Integrated processes view*.

The *Contracts view* presents contracts imposed on collaborating parties. Provider is the component implementing the service and Consumer is a component that uses the service. ≪Consumer≫ and ≪Provider≫ are stereotypes from Service-Oriented Architecture Modeling Language (SoaML). Contracts are presented in the UML component diagram. In order to add further details the contract can be shown in UML composite structure diagram. Furthermore, we can use contract to specify non-functional requirements imposed of communication between systems. For example, response time in case of performance or required protocol to communication (https) in case of security.

The *Integrated services view* concentrates on communication between service providers and service consumers which are presented as components with applied ≪Capability≫ stereotype. The central part of an UML component diagram is Enterprise Service Bus (ESB) component. So as to clearly identify the bus, a new stereotype ≪ESB≫ was proposed. The UML component diagram (see Fig. 3) shows providers and consumers of services. *Prescription* is provided by e-Prescription application. *PrescriptionRealization* is provided by e-Pharmacy. The main component that is responsible for communication between those two systems is marked with ≪ESB≫ stereotype.

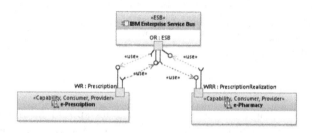

Fig. 3. The UML component diagram shows services integrated through ESB.

Communication between software systems usually requires the use of mediation mechanisms arranged into integration flow. Such flow is executed by ESB. The Integration flow diagram (see Fig. 4) shows flow for getting prescriptions with applied mediation mechanism from *UML Profile for Integration Flows.*

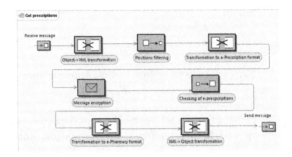

Fig. 4. The Integration flow diagram for getting prescriptions from e-Prescription.

The *Deployment view* defines the physical runtime environment for the solution. It encompasses: hardware environment, execution environment required to run developed software, the mapping of the software onto the runtime nodes that execute them. UML offers deployment diagram to express that view.

3   UML Extensions for Architecture Modeling

3.1   UML Profiles

Proposed new stereotypes have been grouped into two UML profiles: *UML Profile for Integration Platform* and *UML Profile for Integration Flows*, [5]. The first one encompasses stereotypes which represent elements of integration platform, e.g.: ≪ESB≫, ≪IntegratedSystem≫. Software systems can differ significantly. Those differences may relate to, e.g.: communication protocols, data formats, data structures. So, successful communication usually requires the use of mediation mechanisms. So, the *UML Profile for Integration Flows* contains stereotypes which represent elements of integration flows, e.g.: ≪ContentEnricher≫, ≪ContentFilter≫, ≪Translator≫. The latter has been developed in form of profile, EIP.epx file, which can be included in design of integration flow, [13]. The profile encompasses 40 mediation mechanisms and each of them has its own icon to be easily distinguished.

3.2   Integration Flow Diagram

An UML activity diagram was extended, and its special form was obtained for modeling integration flows on ESB. This special form of UML activity diagram was called *Integration flow diagram*. Having prepared the UML profile and defined appropriate diagram we can model integration flows (see Fig. 4).

4   Automation of Integration Platform Design

4.1   Transformations of Model-to-model Type

So, as to automate the process of architecture description of integration solution model-to-model transformations have been proposed, [6]. Those transformations, [14], encompass generation of modeling elements between the following views: Integrated processes and Use cases (BPMN2UC), Use cases and Logical (UC2Logical), Use cases and Integrated services (UC2IS) and Use cases and Contracts (UC2Contracts). Figure 5 depicts rules of the last transformation which generates UML component diagram from UML use case diagram. The diagram shows part of railway system (contract is presented as a collaboration).

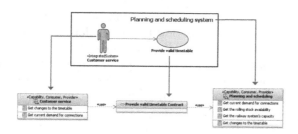

Fig. 5. Transformation from Use cases view to Contracts view.

4.2   Transformations of Model-to-code Type

In order to automate design of integration solution, model-to-code transformations were implemented for both Java [8] and WS-BPEL [9] based integration flows. Integration2Java transformation generates Enterprise ARchive file installed on ESB which realizes modeled integration flow. The transformation was realized in form of plug-in which generates integration flow application, installs and starts it on IBM ESB. It was developed only for selected mediation mechanisms: Message channel, Content enricher and Publish-subscribe. The second transformation (Integration2BPEL) translates UML model stored in XML file into WS-BPEL file. Because both files are of extensible markup language format it is easier to implement such transformation and it is also less prone to errors. Transformed file can be executed in open source ESB, e.g., OpenESB.

5   Case Studies

The ISO/IEC/IEEE 42010:2011 standard defines an architectural description's content requirements in terms of its elements, [12]. The standard requires that architecture description must identify their architectural concerns and enumerates the following ones: Functionality, Performance, Security, Feasibility.

The section shows two examples of application and the same way verification of solutions feasibility which were designed according to 1+5 model. Both examples encompassed description of functionality in Use cases view. Performance was area of interest of the second case study.

5.1   Circulation of Electronic Prescription

When you receive a prescription from a doctor, you have to go with this prescription to the pharmacy in order to realize it. A pharmacist realizes prescription in full. So, my goal was to build applications for a doctor and a pharmacist and ensure electronic circulation of prescription and its realization. Both applications have been realized in Java Server Faces technology in IntelliJ IDEA and source code is available on Github repositories for e-Prescription, [15], and e-Pharmacy, [16], respectively. The e-Prescription implements functions for Doctor who must be able to write prescriptions and view them. The latter is also available for e-Pharmacy application in order to find specific prescription for its realization (see Fig. 2). The e-Pharmacy implements functions for Pharmacist, who must be able to realize prescriptions issued by doctors and view realizations of prescriptions. The UML use case diagram depicts functions of e-Pharmacy (see Fig. 6).

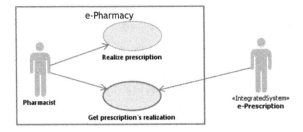

Fig. 6. Use cases of e-Pharmacy application.

Moreover, use cases *Get prescription* and *Get prescription's realization* were implemented as services and exposed onto ESB (see Fig. 3). In each application there are slightly different data structures to store prescriptions and their realizations but exchange format is the same. The e-Prescription is the source of prescriptions while e-Pharmacy is the source of realizations. Each service execution invokes integration flow. The Integration flow diagram (see Fig. 4) shows the flow for getting prescriptions. As far as scenario of using the platform is concerned Doctor writes prescription in e-Prescription. The Pharmacist realizes the prescription in e-Pharmacy. After that Doctor is able to view realization of the prescription in e-Prescription. In that way, the business goal was achieved and circulation of electronic prescription was provided.

The model proved to be useful in designing that integration solution, [3].

5.2 ARMA and AFQI Communication

The subject was communication between Agricultural and Food Quality Inspection (AFQI) and The Agency for Restructuring and Modernization of Agriculture (ARMA). ARMA is the competent authority supervising quality of agricultural and food products in Poland. The certification bodies carry out inspections and issue and revoke certificates in the field of organic farming. A list of organic producers is transmitted by the certification bodies to AFQI and ARMA, as well. In case of irregularities in the list, the ARMA submit request to AFQI to punish the certification body. Such files, concerning one case of irregularity, may contain 700 MB of data. Due to the large volume of data it was simply burned at DVD disc and send by post mail. So, it usually took 2 days to get the request to punish the certification body. An architectural decision was to design communication between Agency and Inspection using ESB. The 1+5 views model was applied and *Send request to penalize* use case was designed.

The main challenge was to send such a large documentation expeditiously. Size of documentation for testing was 675.71 MB, which contained photos and sketches of plots. I looked at the solution like at queuing system with queue and service node. Furthermore, thanks to the Little's Law length of the queue and required number of service nodes were determined, [10]. So, execution environment was divided into two parts: queue (RabbitMQ Message Broker) and service node (Mule ESB). Finally, application of architectural pattern *Decoupled Invocation* in combination with *Service Instance* pattern allowed on effectively dealing with a short-term heavy load, [11]. Moreover, each file from the documentation was sent as a separate message (169 files).

The solution allows on transferring request to penalize in 2 min, [7].

6 Integration Platform Development Process

As far as ISO/IEC/IEEE 42010:2011 standard is concerned it does not require definition of software development process but recommends identification of stakeholders. According to Software Process Engineering Metamodel Specification there are three basic elements in process description: role, work product and task. Preliminary configuration of newly proposed *Integration* discipline was defined which gathers tasks connected with integration design, [4]. Tasks for Integration Architect and role's responsibility have been configured (see Fig. 7).

Fig. 7. The definition of tasks and work products for Integration Architect role.

7   Conclusions and Further Work

Architectural views model 1+5 was presented which is devoted for designing solutions of collaborating systems. New UML profiles and *Integration flow diagram* were presented. Examples of application of the model in SOA architecture were described. The configuration of the view for Domain-Driven Design approach and Micro-services architecture is considered. Moreover, refinement of Integration Platform Development Process is planned. Finally, description of security of an integration platform needs attention.

As far as blockchain technology is concerned, we have to deal with designing completely distributed solutions. Here, the 1+5 model fits perfectly, because we have collaborating parties acting according to rules defined in a smart contract. So, in course of further works, the model will be adjusted to architecturally describe smart contracts in *Contracts View* and blockchain nodes in *Deployment view*. In that way, the model will be verified in both architectures: centralized with ESB and distributed with point-to-point connections in blockchain.

References

1. Górski, T.: Architectural view model for an integration platform. J. Theor. Appl. Comput. Sci. **6**(1), 25–34 (2012)
2. Górski, T.: Integration platforms. Selected issues. Wydawnictwo Naukowe PWN (State Scientific Publishing House) (2012). (in Polish), ISBN 978-83-01-17071-4
3. Górski, T.: Architecture of integration platform for electronic flow of prescriptions. Ann. Collegium Econ. Anal. SGH Warsaw Sch. Econ. **25**, 67–83 (2012). (in Polish)
4. Górski, T.: Designing of integration platforms in service-oriented architecture. Wiadomości Górnicze **7–8**, 407–417 (2012)
5. Górski, T.: UML profiles for architecture description of an integration platform. Bull. Mil. Univ. Technol. **LXII**(2), 43–56 (2013)
6. Górski, T.: Model-to-model transformations of architecture descriptions of an integration platform. J. Theor. Appl. Comput. Sci. **8**(2), 48–62 (2014)
7. Górski, T.: The use of Enterprise Service Bus to transfer large volumes of data. J. Theor. Appl. Comput. Sci. **8**(4), 72–81 (2014)
8. Górski, T.: Model-driven development in implementing integration flows. J. Theor. Appl. Comput. Sci. **9**(2), 66–82 (2015)
9. Górski, T., Ziemski, G.: UML activity diagram transformation into BPEL integration flow. Bull. Mil. Univ. Technol. **LXVII**(3), 15–45 (2018)
10. Jain, R.: The Art of Computer Systems Performance Analysis: Techniques for Experimental Design, Measurement, Simulation, and Modeling. Wiley-Interscience, New York (1991). ISBN 0471503361
11. Rotem-Gal-Oz, A.: SOA Patterns, 1st edn. Manning Publications, Shelter Island (2012). ISBN 978-1933988269
12. ISO/IEC/IEEE 42010:2011, Systems and software engineering - Architecture description (2011). https://www.iso.org/standard/50508.html
13. github.com/drGorski/UMLProfileForIntegrationFlows/blob/master/EIP.epx
14. github.com/drGorski/M2MTransformations
15. github.com/drGorski/ePrescription
16. github.com/drGorski/ePharmacy

Modeling of Smart Contracts in Blockchain Solution for Renewable Energy Grid

Tomasz Górski$^{(\boxtimes)}$ (ORCID) and Jakub Bednarski (ORCID)

Institute of Naval Weapons and Computer Science,
Polish Naval Academy, Gdynia, Poland
{t.gorski,j.bednarski}@amw.gdynia.pl
https://amw.gdynia.pl/index.php/en/

Abstract. The paper presents the manner of smart contracts' modeling in blockchain solution. The authors illustrate the modeling approach using the example of a renewable energy system. The paper proposes extension of Unified Modeling Language (UML) in form of *UML Profile for Smart Contracts*, appropriate for smart contract design. Moreover, the standard smart contract design and implementation method, in Corda environment, has been made more flexible. The authors present static aspect of newly designed *Smart Contract Design Pattern*.

Keywords: Software architecture · Blockchain · Smart contract · Distributed ledger · Architectural views model 1+5

1 Introduction

A blockchain is a type of distributed ledger, composed of data in packages called blocks that are stored in append-only chain. Each block is linked to a previous one in a chain and cryptographically hashed so data remains unchanged. Blockchain solution is a network of distributed nodes and each of them has the same replica of a ledger. Efanov [2] noticed that blockchain has all-pervasive nature and occurs in many industries and applications. For example, Xia et al. [9] propose a blockchain-based system that addresses the problem of medical data sharing in a trust-less environment. Kaijun et al. [7] propose a public blockchain of agricultural supply chain system. Turkanović et al. [8] propose a globally trusted, decentralized higher education credit and grading platform named EduCTX which is based on the concept of the European Credit Transfer and Accumulation System (ECTS). On the other hand, Clack et al. [1] describe application of blockchain in financial sector but also consider aspect of definition of smart contract as an automatable and enforceable agreement.

We would like to apply blockchain technology for maximizing prosumer's benefits from renewable energy, [5]. The idea behind the Electricity Consumption and Supply Management System (ECSM) is to exploit the potential of renewable

R. Moreno-Díaz et al. (Eds.): EUROCAST 2019, LNCS 12013, pp. 507–514, 2020.
https://doi.org/10.1007/978-3-030-45093-9_61

energy generation sources so as to provide energy self-sufficiency and participation in a competitive energy market, [6]. Our aim is to provide flexible and easy to extend manner of modeling and implementation of smart contracts in such solutions. The article is focused on this very aim.

The paper is structured as follows. The second section introduces the assumptions of ECSM. The third section contains description of *Use cases*, *Contracts* and *Logical* views of architectural views model 1+5, [3], for smart contract design. The next section summarizes UML extensions proposed for smart contract modeling gathered in *UML Profile for Smart Contracts*. The fifth section reveals approach for smart contract implementation in Java programming language. The last one summarizes the paper and outlines directions for further work.

2   Electricity Consumption and Supply Management System

Electricity Consumption and Supply Management system provide functionality to monitor and record continuously information about inbound and outbound energy to/from prosumer's node in renewable energy grid, [6]. We have three types of nodes in such renewable energy grid (see Fig. 1): prosumer, energy stock exchange, power grid.

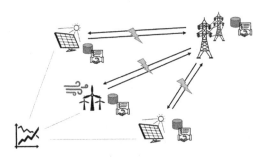

Fig. 1. Renewable energy grid.

Prosumer's node can sell energy to other nodes or to the power grid when it is economically justified. The energy stock exchange node has two main roles: confirms energy price for each transaction, provides real time energy price.

The authors have applied blockchain technology, in particular distributed ledger, to manage actions among renewable energy nodes and store transactions of selling energy. Each element in the system is actually blockchain node. It was designed and built with application of Corda Distributed Ledger Technology (DLT), [11]. Corda DLT is a specialized distributed ledger platform for financial market. Distributed Ledger Network has been designed where all inbound and outbound energy transactions are recorded. Information about transferred energy is a part of smart contract which is confirmed and stored in participating nodes.

3   Smart Contracts in Architectural Views Model 1+5

The Architectural views model 1+5 for integration solutions, [3], was designed to model collaborating systems in context of business processes (see Fig. 2).

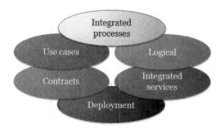

Fig. 2. Architectural views model 1+5 for integration solutions.

As far as blockchain technology is concerned the 1+5 model fits perfectly. In that kind of solution we have collaborating parties (e.g.: seller and buyer) that act on the basis of rules defined in a smart contract. So, we have proposed modeling elements for representing collaboration of parties through smart contracts. We have placed those elements within the 1+5 model mainly in *Contracts View*. As far as smart contracts are concerned we have proposed dedicated Unified Modeling Language stereotypes that describe the needed additional semantic structures and have included them in *UML Profile for Smart Contracts*.

3.1   Use Cases View

In the paper, we consider *Sell energy* use case of ECSM functionality, [6]. All diagrams have been prepared in Visual Paradigm, [14]. The design of ECSM is available at GitHub repository, [15]. The UML Use case diagram shows the use case (see Fig. 3). In the figure we can see an UML stereotype ≪IntegratedSystem≫ from *UML Profile for Integration Platform*, [4].

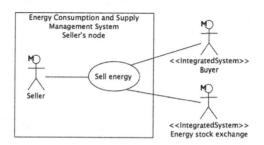

Fig. 3. The *Sell energy* use case.

The UML activity diagram shows flow of events of *Sell energy* use case (see Fig. 4).

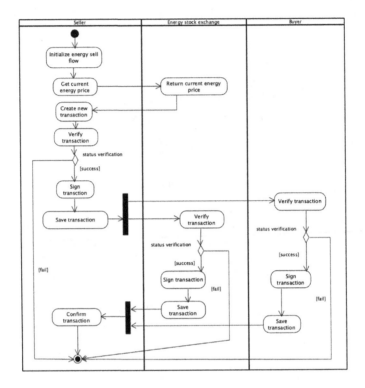

Fig. 4. The UML activity diagram for *Sell energy* use case.

3.2   Contracts View

The view presents contracts imposed on collaborating parties. New modeling means for smart contract were presented in the context of *Sell energy* use case.

Basic element in smart contract is *State* that represents object which is recorded in blockchain distributed ledger. We have identified the following attributes of state:

- value - represents amount of energy which is produced (by e.g., windmill) and send to renewable energy grid/buyer node,
- producer - node which sells energy,
- buyer - node responsible to manage energy grid and responsible for receiving produced energy/node which buys energy.

So, we have proposed new UML stereotype ≪State≫. Moreover, the design of a blockchain state requires using of *Party* class from Corda framework.

The UML class diagram presents both classes (i.e.: Party, State) connected by aggregation relationships (see Fig. 5).

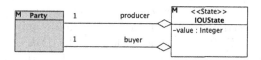

Fig. 5. Smart contract's state.

From Corda platform perspective contract is a concrete class that implements *Contract* interface. The class implements *verify()* method. Verification rules are implemented in the method body. But, we can not see verification rules explicitly. Moreover, in case of change we have to compile and build the contract again.

We have proposed more flexible approach *Smart Contract Design Pattern*. We have added two new stereotypes: for contract itself ≪Contract≫ and for verification rule ≪VerificationRule≫. The abstract class *VRContract* implements Corda *Contract* interface and defines list of rules. The *VRContract* uses definition of *VRule* interface which represent verification rule. The *VRule* defines *runRule()* method which must be implemented by concrete verification rule class. The concrete *IOUContract* class actually implements *verify()* method which iterates over list of verification rules. The UML class diagram presents classes and interfaces that constitute *Smart Contract Design Pattern* (see Fig. 6).

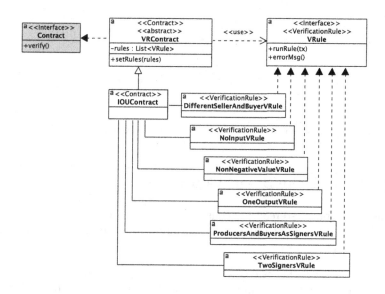

Fig. 6. Smart contract with verification rules.

3.3   Logical View

The view presents realizations of the use cases identified in the *Use case* view. To show the whole structure for the *Sell energy* use case realization we need one more element, Flow. It represents the process of agreeing ledger updates and defines communication between nodes which can be accomplish only in its context. So, we have proposed new UML stereotype ≪Flow≫. The UML Class diagram shows all modeling elements needed to design and implement smart contract (see Fig. 7). Classes and interfaces, which require implementation, have white background.

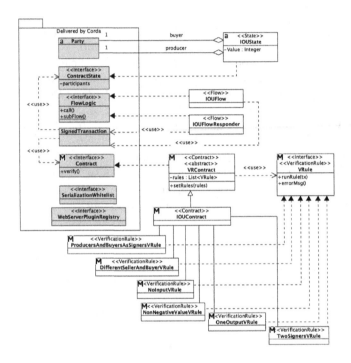

Fig. 7. Classes and interfaces for smart contract design.

4   UML Profile for Smart Contracts

Those newly proposed stereotypes constitute the new *UML Profile for Smart Contracts*. To sum up, we have identified the following UML stereotypes to model blockchain Smart Contract:

- ≪State≫ - object that is recorded in blockchain nodes/distributed ledger,
- ≪Contract≫ - agreement imposed on collaboration of blockchain nodes,
- ≪VerificationRule≫ - condition that must be fulfilled for the contract,
- ≪Flow≫ - process of agreeing updates and communication among nodes.

5   Smart Contract Implementation in Java

Proof-of-Concept (PoC) of ECSM was implemented in Java language with use of IntelliJ IDEA framework, [12]. The source code of PoC is available at GitHub repository, [13]. First of all, we have declared *VRule* interface (see Fig. 6). It declares *runRule(LedgerTransaction tx)* method. Within smart contract, for *Sell energy* use case, we have identified four business verification rules. One of business verification rules is that amount of sent energy must have a positive value. The Java *NonNegativeValueVRule* class for that rule implements *runRule(LedgerTransaction tx)* which is declared in *VRule* interface (see Fig. 8).

```
public class NonNegativeValueVRule implements VRule {

    private static final String ERROR_MSG = "The IOU's value must be non-negative.";

    @Override
    public boolean runRule(LedgerTransaction tx) {
        // IOU-specific constraints.
        final IOUState out = tx.outputsOfType(IOUState.class).get(0);
        final Party producer = out.getProducer();
        final Party buyer = out.getBuyer();
        return out.getValue() > 0;
    }

    @Override
    public String errorMsg() { return ERROR_MSG; }
}
```

Fig. 8. Source code of Java NonNegativeValueRule class.

Due to Corda requirements two technical verification rules have been also designed: transaction must have only one State and does not depend on any previous State. Very important role in implementation plays *VRContract* abstract class. The class defines attribute *rules* that holds collection of verification rules and implements *verify()* method from *Contract* interface. The *verify()* uses lambda expression to check whether all verification rules are met (see Fig. 9).

```
abstract public class VRContract implements Contract {

    protected List<VRule> rules = new ArrayList<>();

    public VRContract() { setRules(); }

    @Override
    public void verify(@NotNull LedgerTransaction tx) throws IllegalArgumentException {
        requireThat(check -> {
            rules.forEach(rule -> check.using(rule.errorMsg(), rule.runRule(tx)));

            return null;
        });
    }

    abstract void setRules();
```

Fig. 9. Source code of Java VRContract abstract class (fragment).

The last class in smart contract implementation is *IOUContract*. The class instantiates concrete verification rules classes and populates the *rules* attribute with immutable list of those verification rules objects.

6   Conclusions and Further Work

The paper shows modeling of smart contracts in context of Architectural views model 1+5. The authors present new UML stereotypes for smart contract modeling and place them in new *UML Profile for Smart Contracts*. Smart Contract Design Pattern was proposed and its static aspect has been presented in the paper. The pattern offers more flexible approach to adding new or modifying and removing existing verification rules within a smart contract.

As far as further work is concerned, we plan to show dynamic aspect of Smart Contract Design Pattern. Moreover, it is considered to implement transformation of model-to-code type to generate source code of classes and interfaces that implement smart contract. We plan to design such a transformation to generate source code in both Java and Kotlin languages. Furthermore, we will propose new stereotypes and tagged values to model blockchain network and blockchain node deployment configuration. We plan to design another transformation of model-to-code type to generate Gradle script for blockchain node deployment.

References

1. Clack, C.D., Bakshi, V.A., Braine, L.: Smart contract templates: foundations, design landscape and research directions, Barclays Bank PLC 2016–2017 (2017). http://arxiv.org/abs/1608.00771
2. Efanov, D., Pavel Roschin, R.: The all-pervasiveness of the blockchain technology. In: 8th Annual International Conference on Biologically Inspired Cognitive Architectures, BICA 2017, Procedia Computer Science, pp. 123 116–121 (2018)
3. Górski, T.: Architectural view model for an integration platform. J. Theor. Appl. Comput. Sci. **6**(1), 25–34 (2012)
4. Górski, T.: UML profiles for architecture description of an integration platform. Bull. Mil. Univ. Technol. **LXII**(2), 43–56 (2013)
5. Górski, T., Prus, P.: System architecture for maximazing prosumer's benefits from distributed generation. Bull. Mil. Univ. Technol. **LXII**(1), 101–113 (2013). (in Polish)
6. Górski, T., Bednarski, J., Chaczko, Z.: Blockchain-based renewable energy exchange management system. In: 26th International Conference on Systems Engineering, ICSEng 2018 Proceedings, Sydney, Australia (2018)
7. Kaijun, L., Ya, B., Linbo, J., Han-Chi, F., Van Nieuwenhuyse, I.: Research on agricultural supply chain system with double chain architecture based on blockchain technology. Future Gener. Comput. Syst. **86**, 641–649 (2018)
8. Turkanović, M., Hölbl, M., Košič, K., Heričko, M., Kamišalić, A.: EduCTX: a blockchain-based higher education credit platform. IEEE **6**, 5112–5127 (2018)
9. Xia, Q., Sifah, E.B., Asamoah, K.O., Gao, J., Du, X., Guizani, M.: MeDShare: trustless medical data sharing among cloud service providers via blockchain. IEEE **5**, 14757–14767 (2017)
10. www.omg.org/spec/UML/2.5.1/. Accessed 25 May 2019
11. www.corda.net/. Accessed 25 May 2019
12. www.jetbrains.com/idea/. Accessed 25 May 2019
13. github.com/drGorski/renewableEnergyBlockchain. Accessed 25 May 2019
14. www.visual-paradigm.com. Accessed 25 May 2019
15. github.com/drGorski/designECSM. Accessed 27 May 2019

Author Index

Printed in the United States
By Bookmasters